Springer Tracts in Civil Engineering

Springer Tracts in Civil Engineering (STCE) publishes the latest developments in Civil Engineering - quickly, informally and in top quality. The series scope includes monographs, professional books, graduate textbooks and edited volumes, as well as outstanding PhD theses. Its goal is to cover all the main branches of civil engineering, both theoretical and applied, including:

- Construction and Structural Mechanics
- Building Materials
- Concrete, Steel and Timber Structures
- Geotechnical Engineering
- Earthquake Engineering
- Coastal Engineering; Ocean and Offshore Engineering
- Hydraulics, Hydrology and Water Resources Engineering
- Environmental Engineering and Sustainability
- Structural Health and Monitoring
- Surveying and Geographical Information Systems
- Heating, Ventilation and Air Conditioning (HVAC)
- Transportation and Traffic
- Risk Analysis
- Safety and Security

Indexed by Scopus

To submit a proposal or request further information, please contact:
Pierpaolo Riva at Pierpaolo.Riva@springer.com (Europe and Americas) Wayne Hu at wayne.hu@springer.com (China)

More information about this series at https://link.springer.com/bookseries/15088

Sreevalsa Kolathayar · Indrajit Pal ·
Siau Chen Chian · Arpita Mondal
Editors

Civil Engineering for Disaster Risk Reduction

Springer

Editors
Sreevalsa Kolathayar
Department of Civil Engineering
National Institute of Technology Karnataka
Mangalore, Karnataka, India

Indrajit Pal
Disaster Preparedness, Mitigation and Management
Asian Institute of Technology
Klong Luang, Thailand

Siau Chen Chian
Department of Civil and Environmental Engineering
National University of Singapore
Singapore, Singapore

Arpita Mondal
Department of Civil Engineering
Indian Institute of Technology Bombay
Mumbai, Maharashtra, India

ISSN 2366-259X ISSN 2366-2603 (electronic)
Springer Tracts in Civil Engineering
ISBN 978-981-16-5311-7 ISBN 978-981-16-5312-4 (eBook)
https://doi.org/10.1007/978-981-16-5312-4

This Springer imprint is published by the registered company Springer Nature Singapore Pte Ltd.
The registered company address is: 152 Beach Road, #21-01/04 Gateway East, Singapore 189721, Singapore

Contents

About the Editors

Dr. Sreevalsa Kolathayar pursued M.Tech. from IIT Kanpur; Ph.D. from Indian Institute of Science (IISc) and served as International Research Staff at UPC BarcelonaTech Spain. He is presently Assistant Professor in the Department of Civil Engineering, National Institute of Technology, Surathkal, Karnataka, India. Dr. Sreevalsa has authored six books and over 75 research papers. He is Editorial Board member of two International Journals and Executive Guest Editor for one. His research interests include Seismic Hazard Assessment, Geohazards, Earthquake Preparedness, Landslides, and Water geotechnics. He is currently the Secretary Indian chapter of International Association for Coastal Reservoir Research (IACRR), and Executive Committee Member of Indian Society of Earthquake Technology. In 2017, The New Indian Express honoured Dr. Sreevalsa with South India's Most Inspiring Young Teachers Award. He is the recipient of ISET DK Paul Research Award from Indian Society of Earthquake Technology, IIT Roorkee in 2018. He has received Marie Curie Action Fellowship from European Union and Professional Recognition Award from Coastal Reservoir Center University of Wollongong Australia. He is in the roster of two technical committees of ASCE Geo-Institute namely, (i) Earthquake Engineering and Soil Dynamics (ii) Engineering Geology and Site Characterization. He received "IEI Young Engineers Award" by The Institution of Engineers (India), in recognition of his contributions in the field of Civil Engineering.

Dr. Indrajit Pal presently working as Assistant Professor and Chair at Disaster Preparedness, Mitigation, and Management program at the Asian Institute of Technology, Thailand. Prior to joining at AIT, Dr. Pal served as a faculty member at Centre for Disaster Management at Lal Bahadur Shastri National Academy of Administration, Mussoorie, India (Premier National Institute for training Indian Administrative Services Officers). Dr. Pal has done extensive work on capacity development of decision-makers, risk assessment and disaster risk governance. Dr. Pal has written six books and more than 55 articles in international and national peer-reviewed journals apart from supervising masters and doctoral research. Dr. Pal having about 18 years of experience in research and capacity development on Disaster Risk Management and Governance and Disaster Risk Science and Education. Research Expertise:

Disaster Risk Governance, Disaster Risk Reduction (DRR) and Development, Incident Command System (ICS), GIS/RS in disaster risk management, Climate Change Adaptation, Community Based Disaster Risk Management, Disaster Risk Science and Education, Sustainability, Public Health Risk assessment.

Dr. Siau Chen Chian is an Associate Professor at the Department of Civil and Environmental Engineering, National University of Singapore (NUS). He obtained his Ph.D. and B.Eng. (1st Class with Gold Medal) from Cambridge University and Nanyang Technological University respectively. Dr. Chian's contribution in earthquake engineering lies in the field of catastrophe modelling, landslides and damage vulnerability of underground structures. He was funded by the UK Engineering and Physical Sciences Research Council (EPSRC) as the geotechnical specialist to carry out reconnaissance missions at the Mw7.6 Padang Earthquake in 2009, Mw9.0 Great East Japan Earthquake in 2011 and Mw7.8 Muisne Earthquake in 2016. Dr. Chian was named Top 10 Innovators Under 35 in Asia by the MIT Technology Review for his novel contribution in catastrophe modelling in 2016. Dr. Chian is a nominated member of the Technical Committee on Earthquake Geotechnical Engineering under the International Society for Soil Mechanics and Geotechnical Engineering (ISSMGE). He is also an Integrated Research on Disaster Risk (IRDR) Young Scientist supported by The International Science Council (ISC) and the United Nations Office for Disaster Risk Reduction (UNDRR).

Dr. Arpita Mondal works as an Assistant Professor in the Department of Civil Engineering at the Indian Institute of Technology (IIT) Bombay. She is also an associate faculty member in the Interdisciplinary Program in Climate Studies at IIT Bombay. Arpita's research focuses on hydroclimatic extremes—how they can be characterized, what causes them, and how they are likely to evolve with climate change. She uses a combination of statistical analysis, physical understanding and computer model simulations. Arpita serves as an Associate Editor of the Springer journal Regional Environmental Change. She received the Early Career Research Award from the Science and Education Research Board (SERB), and the INSPIRE Faculty Award, both funded by the Department of Science and Technology, Government of India. She also received the Asian University Alliance (AUA) Scholars' Award. Arpita completed her Ph.D. at the Indian Institute of Science (IISc), Bangalore, M.Tech. at IIT Bombay and Bachelors in Civil Engineering at Jadavpur University, Kolkata. As a Ph.D. student, Arpita had received the Endeavour Research Fellowship (Government of Australia), Fulbright-Nehru Research Fellowship (USIEF), and the Berkner Travel Grant of the American Geophysical Union (AGU). Arpita scored the first rank in the university in both her Bachelors and Masters.

Chapter 1
Disaster Risk Reduction and Civil Engineering—An Introduction

Sreevalsa Kolathayar, Indrajit Pal, and Satya Venkata Sai Aditya Bharadwaz Ganni

1.1 Background

Disasters have resulted in large-scale destruction and loss of lives worldwide. The scenario is worse in many of the developing countries due to poor construction practices, lack of appropriate policies, and poor land use planning. Most of the casualties due to natural hazards happen because of the poorly built structures which are vulnerable to hazards. A significant portion of damage can be attributed to the lack of awareness, lack of preparedness, and lack of dissemination of knowledge on resilient construction techniques.

The steps toward building a disaster resilient society can be complementary measures, alongside other activities that focus on the progress of the community; with different stakeholders working in unison. Disaster resilience must not only be integrated into construction practices; but also community preparedness, emergency response, recovery, and long-term risk reduction efforts as well.

The UN Chengdu Declaration of Action, 2011 states that "There is no such thing as 'natural disasters.' Natural hazards—floods, earthquakes, landslides and storms—become disasters as a result of human and societal vulnerability and exposure, which can be addressed by decisive policies, actions and active participation of local stakeholders. Disaster risk reduction is a no-regret investment that protects lives, property, livelihoods, schools, businesses, and employment."

The disaster risk can be reduced by reducing the exposure and vulnerability. Disaster management cycle includes five phases.

S. Kolathayar (✉)
National Institute of Technology Karnataka, Surathkal, Mangalore, India
e-mail: sreevalsa@nitk.edu.in

I. Pal · S. V. S. A. B. Ganni
Asian Institute of Technology, Khlong Nueng, Thailand

S. Kolathayar et al. (eds.), *Civil Engineering for Disaster Risk Reduction*, Springer Tracts in Civil Engineering, https://doi.org/10.1007/978-981-16-5312-4_1

Disaster phase—This is the time when a disaster takes place. There is intense damage and loss to society during this phase.

Response phase—This phase includes activities as immediate response to the disaster. Along with rescue operations, immediate needs of the victims are addressed during this phase.

Recovery phase—In this phase, the impact of disaster is quantified, and psychological support is ensured for recovery.

Mitigation phase—Required steps needed to reduce the impact of damage during the next similar event are taken during this phase.

Preparedness phase—This phase involves education and awareness on warning signs, safe evacuation, first aid measures as well as Dos and Don'ts before, during, and after a disaster.

The preparedness phase is generally listed at the last in the cycle. There should be efforts to put preparedness phase at the beginning well before the disaster phase, so that community is prepared to face a disaster and recuperate fast [37, 39].

1.2 Terms and Definitions

This section presents the important terminologies used in disaster risk reduction and their definitions [43].

Natural hazard: Natural process or phenomenon that may cause loss of life, injury or other health impacts, property damage, loss of livelihoods and services, social and economic disruption, or environmental damage.

Disaster: The United Nations Office for Disaster Risk Reduction (UNDRR) defines a disaster as "a serious disruption of the functioning of a community or society involving widespread human, material, economic or environmental losses and impacts, which exceeds the ability of the affected community or society to cope with using its own resources."

Risk: The combination of the probability of an event and its negative consequences.

Risk assessment: A methodology to determine the nature and extent of risk by analyzing potential hazards and evaluating existing conditions of vulnerability that together could potentially harm exposed people, property, services, livelihoods, and the environment on which they depend.

Risk management: The systematic approach and practice of managing uncertainty to minimize potential harm and loss.

Disaster risk: The potential disaster losses, in lives, health status, livelihoods, assets, and services, which 10 could occur to a particular community or a society over some specified future time period.

Disaster risk reduction: The concept and practice of reducing disaster risks through systematic efforts to analyze and manage the causal factors of disasters, including through reduced exposure to 11 hazards, lessened vulnerability of people and property, wise management of land and the environment, and improved preparedness for adverse events.

Mitigation: The lessening or limitation of the adverse impacts of hazards and related disasters.

Preparedness: The knowledge and capacities developed by governments, professional response and recovery organizations, communities and individuals to effectively anticipate, respond to, and recover from, the impacts of likely, imminent or current hazard events or conditions.

Prevention: The outright avoidance of adverse impacts of hazards and related disasters.

Recovery: The restoration, and improvement where appropriate, of facilities, livelihoods and living conditions of disaster-affected communities, including efforts to reduce disaster risk factors.

Residual risk: The risk that remains in unmanaged form, even when effective disaster risk reduction measures are in place, and for which emergency response and recovery capacities must be maintained.

Resilience: The ability of a system, community, or society exposed to hazards to resist, absorb, accommodate to and recover from the effects of a hazard in a timely and efficient manner, including through the preservation and restoration of its essential basic structures and functions.

Response: The provision of emergency services and public assistance during or immediately after a disaster in order to save lives, reduce health impacts, ensure public safety, and meet the basic subsistence needs of the people affected.

Retrofitting: Reinforcement or upgrading of existing structures to become more resistant and resilient to the damaging effects of hazards.

Structural measures: Any physical construction to reduce or avoid possible impacts of hazards, or application of engineering techniques to achieve hazard resistance and resilience in structures or systems; non-structural measures: Any measure not involving physical construction that uses knowledge, practice, or agreement to reduce risks and impacts, in particular through policies and laws, public awareness raising, training, and education.

Early warning system: The set of capacities needed to generate and disseminate timely and meaningful warning information to enable individuals, communities, and organizations threatened by a hazard to prepare and to act appropriately and in sufficient time to reduce the possibility of harm or loss.

Exposure: People, property, systems, or other elements present in hazard zones that are thereby subject to potential losses.

Vulnerability: The characteristics and circumstances of a community, system, or asset that makes it susceptible to the damaging effects of a hazard.

Forecast: Definite statement or statistical estimate of the likely occurrence of a future event or conditions for a specific area.

Geological hazard: Geological process or phenomenon that may cause loss of life, injury or other health impacts, property damage, loss of livelihoods and services, social and economic disruption, or environmental damage.

Climate change: The United Nations Framework Convention on Climate Change (UNFCCC) defines climate change as "a change of climate which is attributed directly or indirectly to human activity that alters the composition of the global atmosphere and which is in addition to natural climate variability observed over comparable time periods."

Sustainable development: Development that meets the needs of the present without compromising the ability of future generations to meet their own needs.

1.3 Civil Engineering and DRR

It is important to have the infrastructure resilient to disaster. In addition, civil engineering has the potential to offer solutions that can reduce the risk during a disaster.

1.3.1 Risks to Infrastructures

Due to fast-changing evolution of both natural and man-made threats, adequate governance arrangements that are flexible and adaptive to the needs of people and communities is vital [25]. Risk governance frameworks are usually segregated into regional, national, and subnational levels in order to ensure a holistic engagement of stakeholders and comprehensive addressing of risks. The concept of critical infrastructure is closely linked with the risk governance process in that it utilizes risk assessment and risk management process to identify infrastructure that is vital for the continued functioning of society in both normal and emergency times.

Critical infrastructure has been described as the lifeblood of a society and as the world and society relies increasingly more on a digital and interconnected communication and utility system their importance and need for round the clock operation is becoming increasingly clear [25]. Critical infrastructures: Critical infrastructures (CIs) are those structures which are essential to make functioning of a community in normal circumstances and in the life-threatening circumstances of an emergency. The critical infrastructures may be technological, materials, assets, or facilities which are necessary for daily societal activities as well as economic and community functioning after a disaster event [8].

The damage or destruction of critical infrastructure may have far-reaching consequences for the community or society. There are two major reasons for this: 1. Infrastructures are becoming more complex as its interdependence, and people are becoming more reliant on infrastructure services. Different types of infrastructures are increasingly dependent on each other. It varies in country to country or community to community. Critical infrastructures are contextual. Electricity and water supply, waste (water) management, transport or information, and telecommunication technologies are basic infrastructures for the society which are inter-dependable. Critical infrastructure and its interdependence are the factors determining the resilience and vulnerability to single or multiple hazards. Aftermath of disaster the disruption of critical infrastructures may causes cascading impacts and hazards also. Uninterrupted and required supplies of basic humanitarian services are critical in disaster management, particularly in the very early chaotic phases when demand and available resources are highly uncertain, information systems are disrupted, and communication between communities, food suppliers, retail, and emergency authorities are difficult [34]. Assessing adaptative capacity for the city resilience is important to understand the structural and non-structural mitigation measures [41].

1.3.2 Civil Engineering Applications in DRR

The civil engineering professionals have an important role in disaster risk reduction in before, during, and after a disaster. The risk reduction schemes can be integrated in all phases of civil engineering projects, including planning, designing, construction, and maintenance in addition to urban planning and policy making [28]. The civil engineers can reduce the disaster risk in several ways such as proper land use planning, integrating efficient codal provisions, appropriate design reducing the vulnerability, quality improvement in construction, provision of sea walls, flood protection structures, drainage systems, and earth retention schemes [13, 28, 36]. UN Sustainable Development Goal 9 is to build resilient infrastructure, promote sustainable industrialization, and foster innovation. The UN explains: Investments in infrastructure—transport, irrigation, energy, and information and communication technology—are crucial to achieving sustainable development and empowering communities in many countries. It has long been recognized that growth in productivity and incomes, and improvements in health and education outcomes require investment in infrastructure. The UN has defined 8 targets and 12 indicators for SDG 9. Targets specify the goals, and indicators represent the metrics by which the world aims to track whether these targets are achieved. Target 1 is to develop sustainable, resilient, and inclusive infrastructures. UN definition: Develop quality, reliable, sustainable, and resilient infrastructure, including regional and transborder infrastructure, to support economic development and human well-being, with a focus on affordable and equitable access for all.

The role of civil engineers in the development of engineering codes and standards is of utmost important. The development of resources, tools, and approaches toward mitigating the impacts of natural and man-made hazards on the infrastructure is an important responsibility of the civil engineers.

1.4 Organization of the Book

This book discusses current civil engineering design and practice in context of disaster resilience in an environment-friendly manner. The reforms in civil engineering practice in order to achieve disaster resilience is highlighted in the book. The book covers the topics such as multi-hazard risk resilience, disaster mitigation, natural hazards, earthquakes and tsunami, floods and droughts, water security, landslide prevention, cloudbursts, hazard mapping, resilient infrastructures, sustainable cities, retrofitting and rehabilitation, early warning systems, critical infrastructure, emergency preparedness, etc.

This book is divided into four sections discussing diverse topics related to the role of civil engineering in disaster risk reduction and governance.

1.4.1 Water Security and Flood Risk

The first section is on water security and floods. This section includes nine chapters that discuss topics related to flood protection, drought monitoring, ground water storage, coastal reservoirs, coastal disasters, tsunami risk, and cloudbursts.

Chinnasamy et al. in their study *Satellite-based analysis of groundwater storage and depletion trends implicating climate change in South Asia: Need for groundwater security* had discussed about the limited groundwater in South Asia. The authors had used the remote sensing and global hydrological models to logically comprehend GW status and climatic patterns [22]. The explorative study on *Use of multi-sensor satellite remote sensing data for flood and drought monitoring and mapping in India* by Atasi et al. had aimed to formulate effective strategies for determining the losses due to flood and drought for decision support in regional-level assessment and management. The study has reviewed the use of satellite remote sensing in flood and drought monitoring and their risk assessments over various parts of India using near real-time data [24]. Fascetti in his study *Multiscale Data- and Physics-Based Modeling of Backward Erosion Piping in Flood Protection Infrastructure* had explained against the progressive evolution of internal damage caused by erosion mechanisms, with emphasis on backward erosion piping (BEP)-induced failures. Results of two virtual scenarios show that the model is inherently capable of accepting live data streaming from embedded sensors, and predict the response of the system in a computation ally efficient fashion [18]. In the study, *Kerala floods 2018: Causative factors that transformed single event to multi-hazard disaster* by Ramesh et al. had discussed about chronological events that have been provided indicating the amount of rain and the consequence of the multi-hazard event that have unfolded subsequently. Related to storm water maps of the city, this clearly exemplifies the scenario of water's natural flow being disrupted and proves beyond doubt that urbanization also had a role to play in the creation of floods [21].

Durin et al. in their study *River flow analysis—it's so easy, but it ain't?* had analyzed about the real case study of two hydrological measuring stations, average daily flow time series for the available duration of 20 years, from the year 1999 till the year 2018. Reliable analysis should involve progressive time series modeling, like decomposition of time series, ANOVA method, downscaling, etc. [11]. In the study, *Coastal Disasters and Mitigation Measures* by Neelamani had explained about the need, the stakeholders should select the suitable mitigation measure out of "Defend," "Adopt," and "Retrieve". This will help in saving life and property in the coastal area during short-term as well as long-term impacts from natural forces [31]. The chapter *Tsunami Flood Risk Management* by Barnett had discussed about the recognition of the regularity of tides, and corresponding need for living floor levels to be raised above high tide as well as river flood level. Solitary wave theory is proposed as the basis for standardization of a link between wave energy and wave height [9]. The authors Singh and Kansal in the study on *Cloudbursts—A Major Disaster in the Indian Himalayan States* had aimed to analyze the cloudburst events from 2001 to 2019 in the state of Uttarakhand. The scatter plot of aerosols optical depth versus elevation follows negative correlation, and most of the aerosols are concentrated at 100–1500 m elevations that facilitate the clouds forming mechanism [42].

1.4.2 Geohazards

The second section covers geohazards with seven chapters. This section discusses topics like landslide prediction and monitoring, interaction mechanism of landslides, perturbation of earth surface process, landslide early warning, local site effects, and post-earthquake reconnaissance.

In the study, *Review on landslide early warning system: A brief history, evolution, and controlling parameters* by Menon and Kolathayar had reviewed the previous research studies conducted in the soil slopes using manual instrumentations and the improved data acquisition systems (DAQ) that are used recently. The methods are involved in finding this threshold and using it for the development of the LEWS [12]. In the study, *Forecasting landslides for disaster risk reduction: Process based approaches and real time field monitoring measures* by Satyam and Treesa Abraham had explained about the occurrence of landslides for the development of a LEWS. From the results, it was observed that both the 2D Bayesian and the algorithm-based approaches have improved the performance of conventional thresholds considerably

[17]. In the study, *Interaction mechanisms of flow-like landslides impacting structures* by Cuomo had described about the examples of concrete buildings or protection barriers are explored by means of novel numerical approaches such as SPH and MPM. It concludes that one could observe that a key factor is the evolution of the pore water pressures inside the landslides, from the slope instability to the propagation stage [7]. Karjana et al. in their study, *Real-time Monitoring System Based on Wireless Sensor Networks and Remote Sensing Techniques for Landslide-prone Areas in the Northern Region of Thailand* had aimed to investigate potential landslides' environmental parameters and conditions using real-time monitoring based on wireless sensors networks (WSN) and remote sensing techniques. The integration of these approaches is expected to provide useful information for landslide-related disaster management [1]. In the study, *Perturbation of earth surface process by geophysical and meteorological process in the Nepal Himalaya* by Adhikari had discussed about the occurrences of different scale of earthquakes in Nepal. Study has also recommended to conduct multi-hazard risk assessment along this mountain chain for the sustainable risk reduction [20]. In the study, *Post Earthquake Reconnaissance: Theories versus Observations* by Chian presents the findings from the UK Earthquake Engineering Field Investigation Team's post-earthquake reconnaissance missions. It concludes that in comparison with structural failure, consequences of soil liquefaction such as foundation and ground settlements, lateral spreading, and uplifting of underground utilities have shown to be less lethal in the urban environment [10]. In the study, *Estimation of Local Site Effects: Lessons from Past Earthquakes, Current Practices and Future Trends* by Shreyasvi and Venkataramana had discussed about state-of-the-art practices in characterizing site/soil response to tectonically induced ground shaking. Study also states that empirical amplification functions can be developed for any given site if the ground motion recordings are available [5].

1.4.3 Resilient Infrastructures

The third section has eleven chapters focusing on resilient infrastructures. The topics discussed in this section include disaster resilience in the built environment, seismic assessment of soil arching, railway tracks exposed to floods, vibration isolation, climate resilient infrastructure, response reduction, restoration, retrofitting, and structural rehabilitation.

Shavindree et al. in the study *Disaster Resilient Properties: Built Environment Discourse* had a descriptive or mapping review which is conducted to identify disaster resilient properties in different contexts, and finally, a qualitative content analysis was conducted to derive a set of disaster resilient properties [27]. Krivtsov et al. in the study, *Multiple Benefits of Blue-Green Infrastructure and The Reduction of Environmental Risks: Case Study of Ecosystem Services Provided by a SUDS Pond* had highlighted the flood risk reduction and the value of multiple benefits and ecosystem services provided by BGI, which are relevant for a wide range of environmental, economic, and social risks. These aspects are vital for planning future developments and nature-based solutions, as well as for the management of the existing SuDS assets [33]. In the study, *Static and Seismic Assessment of Soil Arching in Piled Embankments* by Nimbalkar and Meena had discussed on addressing the effects of static loading and seismic excitation using finite element analysis (FEA) in two-dimensional (2D) state of stress. It concludes that the comparison with different design approaches reveals that current design approaches need to be improved including the seismic considerations in the piled embankment design [4]. Kaewunruen et al. in their study *Vulnerability of interspersed railway tracks exposed to flood and washaway conditions* had demonstrated the influences of ballast wash away on the vulnerability assessment of interspersed sleeper railway using nonlinear finite element simulations, STRAND7. This study considers the loss of support conditions as the consequence of flood and wash away conditions stemmed from extreme weather and climatic events [23]. In the study, *Modeling to Support Acceleration of Restoration of a Residential Building System in Southeastern B.C. due to Riverine Flooding* by Bristow and Ivy had explained about various components involved in functional failures of flood-damaged buildings to understand their recovery. The results demonstrate how restoration can be delayed and points to potential solutions to improve resilience through informed recovery planning of flooded buildings [35]. Hegde and Venkateswarlu in their study *Vibration Isolation of Foundation Systems using Geosynthetics Barriers* had attempted to quantify the performance of the barriers created using geosynthetics in mitigating the vibrations. In the geocell barrier condition, more than the degree of efficiency was observed as compared to the unreinforced barrier [29]. In the study, *Response Reductions in Base-Isolated Liquid Storage Tank under Far and Near Field Seismic Excitations* by Vern et al. had presented about the seismic control of a base-isolated liquid storage tank (LST). The maximum reduction of the order can be seen for the shear force and overturning moment at the optimum time of the isolator system [38]. In the study, *Estimating Service-Life Deterioration of RC Bridges Due To Multi-Hazards in Barak Valley Region, Assam, India* by Das and Sil had discussed the amalgamation of the deterioration hazard along with earthquake hazards directly affecting in the bridges of Barak valley. The prediction comprises two main damaging effects physical deterioration (aging effect) and sudden shocking effects (Earthquake) which are inter related by using the probabilistic method [15]. In the study, *Seismic Strengthening Solutions for Existing Buildings* by Hussaini et al. had discussed about various techniques that may be readily adopted for the seismic strengthening of existing URM and RCF buildings. Advanced methods of seismic retrofitting using base-isolation or energy dissipation devices have been addressed. In developed countries, these devices have found many applications in buildings [2]. Reddy in the study, *On Structural Rehabilitation and Retrofitting for Risk Reduction* had discussed about structures, which resist the normal as well as accidental loads due to natural hazards, or man-made hazards have major role on risk. To reduce the intense of disaster/risk all the normal, industrial and lifelines have to be continuously maintained and if required repair, rehabilitation, and retrofitting need to be performed on continuous mode [40].

1.4.4 Disaster Risk and Resilience

The fourth section includes six chapters which discuss disaster risk and resilience in the current scenario. This section focuses on topics like risk management, climate change, and sustainable development, ecosystem-based disaster risk reduction, tsunamigenic sources, impacts of cyclones, resilient urbanism, etc.

Integrated cost and risk management Enhancing supply chain Resilience by Lam and Yin had aimed to examine the existing strategies for enhancing resilience to transportation disruptions in supply chains. The approach is able to help decision makers to choose the optimal strategies under a given scenario [14]. Ramya et al. in their study on *Modeling climate change and glacier melt for sustainable development of a Himalayan region* had attempted to present the review of research done on context of evolving development plans under changing climate scenarios. The findings in this research are to be interpreted in light of assumptions and limitations inherent in the process of modeling [16]. Rodríguez et al. in their study on *Lessons from a century-tradition on Ecosystem-based Disaster Risk Reduction (Eco-DRR) in mountains: the case of the torrential system Los Arañones* (Canfranc, Pyrenees) had reviewed a long-term experience from implementing the Eco-DRR strategies, so-called Los Arañones, to protect the International Railway Station of Canfranc against snow avalanches and torrential floods. Under climate change conditions, and given the ongoing urbanization of the bottom valley, studies focused on the deterioration of the protective elements and their reliability will be required [3]. Higuera et al. in their study on *Filling in the gaps of the tsunamigenic sources* had presented about the landslide-generated wave model with the COMCOT model. Final conclusion would be that coastal landslides can explain most of the measurements and phenomena observed, and thus, they should be considered the main source of tsunami [6]. Kumar et al. in their study on *Demystifying Impacts of Cyclone Amphan 2019 amid COVID-19 Pandemic in West Bengal, India* had presented about the impact of Amphan on West Bengal and exemplify the steps taken by response agencies in managing this mega-emergency situation. Better coordination and communication among the departments of government could have eased out the suffering of people [19]. Machado et al. in their study on *Resilient urbanism from the perspective of climate change in Spain—the case of floods* had explained about resilient urbanism as the main tool for strengthening the urban environment, more specifically against the consequences of flooding. The results obtained in that research emphasize the certainty that decision makers must do much more to understand the challenge and improve the solutions adopted in a more integrated and effective way in the face of current and future risks [32].

1.5 Summary

The past disaster in the globe have made it evident that along with structural improvement, there is a need to build resilience as well, so that society is able to take up the impact of a disaster and also recover fast. This involves capacity building and building a community that is aware of natural hazards and preparedness measures. The Central and State Governments, City Corporations, and residents should maintain high disaster awareness, and the inhabitants of vulnerable regions should be able to help one another, forming a community where suitable action can be taken. A disaster management system needs to be established where associated institutions can coordinate for preparedness well before the disasters and protect lives during an event. Stable progress of process to build cities that is resilient against disasters should be undertaken, such as making structures earthquake resistant, flood protection schemes, evacuation plans, securing lifelines and other infrastructure, and road networks.

The disaster risk reduction and management plan should be comprehensive in nature; involving geologists, hydrologists, geotechnical engineers, and structural engineers to identify hazards, agencies to retrofit certain vulnerable infrastructure, and emergency response agencies to improve and exercise response plans. It is advised to be cautious of landslides in hilly areas and floods in cities which need special consideration for stabilization of slopes, proper drainage, etc., and relocation of important structures. A comprehensive plan will provide a broad picture, so that various organizations can act both independently and in coordination with each other before and during the hazard. Being prepared can lessen panic, worry, and losses that come with disasters. Communities, families, and individuals must know what to do before, during, and after a disaster. Involvement of local communities is fundamental and key to establish a sustainable and functional early warning system as well as preparedness [30]. Assessing social resilience for vulnerable communities is paramount for sustainable disaster risk reduction and management [26].

This book articulates various factors related to disaster preparedness and role of civil engineering in reducing the disaster risk out of different hazards. The book has been divided into mainly four sections (i) water security and flood risks, (ii) geohazards, (iii) resilient infrastructures, and (iv) disaster risk resilience. This book envisages knowledge dissemination on disaster risk reduction primarily focusing on civil engineering perspectives and cross-cutting issues. Research and innovations in civil engineering have the potential to offer solutions toward disaster resilient infrastructures. The vision of this book is in line with the priorities set by UN-SFDRR and UN-SDGs to promote a global culture of risk-awareness and disaster reduction.

References

1. Abellán A, Calvet J, Vilaplana JM, Blanchard J (2010) Detection and spatial prediction of rockfalls by means of terrestrial laser scanner monitoring. Geomorphology 119(3–4):162–171. https://doi.org/10.1016/j.geomorph.2010.03.016
2. Al-Hussaini TM, Hossain MNA (2008) Post-disaster management issues related to building collapse in Bangladesh. In: International disaster and risk conference
3. Allen SK, Ballesteros-Canovas J, Randhawa SS, Singha AK, Huggel C, Stoffel M (2018) Translating the concept of climate risk into an assessment framework to inform adaptation planning—insights from a pilot study of flood risk in Himachal Pradesh, Northern India. Environ Sci Policy 87:1–10. https://doi.org/10.1016/j.envsci.2018.05.013
4. Almeida MSS, Fagundes DF, Thorel L, Blanc M (2020) Geosynthetic-reinforced pile-embankments: numerical, analytical and centrifuge modelling. Geosynth Int 27(3):301–314. https://doi.org/10.1680/jgein.19.00011
5. Anbazhagan P, Kumar A, Sitharam TG (2013) Seismic site classification and correlation between standard penetration test N value and shear wave velocity for Lucknow city in Indo-Gangetic Basin. Pure Appl Geophys 170(3):299–318. https://doi.org/10.1007/s00024-012-0525-1
6. Arikawa T, Muhari A, Okumura Y, Dohi Y, Afriyanto B, Sujatmiko KA, Imamura F (n.d.) Coastal subsidence induced several tsunamis during the 2018 Sulawesi earthquake
7. Armanini A, Larcher M, Odorizzi M (2011) Dynamic impact of a debris flow front against a vertical wall. In: International conference on debris-flow hazards mitigation: mechanics, prediction, and assessment, proceedings, pp 1041–1049. https://doi.org/10.4408/IJEGE.2011-03.B-113
8. Bach C, Gupta, AK, Nair SS, Birkmann J (2013) Critical infrastructures and disaster risk reduction. In: National Institute of Disaster Management and Deutsche Gesellschaft für internationale Zusammenarbeit GmbH (GIZ), New Delhi, p 72
9. Ball JE, Babister MK, Nathan R, Weinmann PE, Weeks W, Retallick M, Testoni I (2016) Australian rainfall and runoff—a guide to flood estimation
10. Bendimerad F, Laurie Johnson P, Coburn A, Rahnama M, Morrow G (n.d.) RMS reconnaissance team architectural planner structural engineer
11. Brkić Ž, Kuhta M, Larva O, Gottstein S (2019) Groundwater and connected ecosystems: an overview of groundwater body status assessment in Croatia. Environ Sci Eur 31(1). https://doi.org/10.1186/s12302-019-0261-6
12. Casagli N, Morelli S, Frodella W, Intrieri E, Tofani V (2017) TXT-tool 2.039-3.2 ground-based remote sensing techniques for landslides mapping, monitoring and early warning. In: Landslide dynamics: ISDR-ICL landslide interactive teaching tools: volume 1: fundamentals, mapping and monitoring. Springer International Publishing, pp 255–274. https://doi.org/10.1007/978-3-319-57774-6_18
13. Chitrachedu RK, Kolathayar S (2020) Performance evaluation of coir geocells as soil retention system under dry and wet conditions. Geotech Geol Eng 38(6):6393–6406
14. Chopra S, Sodhi M (2004) Managing risk to avoid supply-chain breakdown. MIT Sloan Manage Rev
15. Das J, Sil A (2020) Condition assessment of superstructure component of reinforced concrete bridges through visual inspection in the Assam, India. Bridge Struct 16(1):39–57. https://doi.org/10.3233/BRS-200171
16. Devadas V (2001) Planning for rural energy system: part I. Renew Sustain Energy Rev. https://doi.org/10.1016/S1364-0321(00)00014-9
17. Dick GJ, Eberhardt E, Cabrejo-Liévano AG, Stead D, Rose ND (2015) Development of an early-warning time-of-failure analysis methodology for open-pit mine slopes utilizing ground-based slope stability radar monitoring data. Can Geotech J 52(4):515–529. https://doi.org/10.1139/cgj-2014-0028

18. Fascetti A, Oskay C (2019) Multiscale modeling of backward erosion piping in flood protection system infrastructure. Comput-Aided Civ Infrastruct Eng 34(12):1071–1086. https://doi.org/10.1111/mice.12489
19. Frontier articles on Society & Politics (n.d.) Retrieved 6 May 2021 from https://www.frontierweekly.com/views/may-20/25-5-20-Amphan-Maiden-Super-Cyclone-of-the-Century.html
20. Gnyawali KR, Adhikari BR (2017) Spatial relations of earthquake induced landslides triggered by 2015 Gorkha earthquake Mw = 7.8. In: Advancing culture of living with landslides. Springer International Publishing, pp 85–93. https://doi.org/10.1007/978-3-319-53485-5_10
21. Guntha R, Rao SN, Shivdas A (2020) Lessons learned from deploying crowdsourced technology for disaster relief during Kerala floods. Procedia Comput Sci 171:2410–2419. https://doi.org/10.1016/j.procs.2020.04.313
22. Hanasaki N, Fujimori S, Yamamoto T, Yoshikawa S, Masaki Y, Hijioka Y et al (2013) A global water scarcity assessment under shared socio-economic pathways—part 2: water availability and scarcity. Hydrol Earth Syst Sci 17(7):2393–2413. https://doi.org/10.5194/hess-17-2393-2013
23. Kohoutek R (1992) Of concrete and timber sleepers
24. Kulkarni SS, Wardlow BD, Bayissa YA, Tadesse T, Svoboda MD, Gedam SS (2020) Developing a remote sensing-based combined drought indicator approach for agricultural drought monitoring over Marathwada, India. Remote Sens 12(13):2091. https://doi.org/10.3390/rs12132091
25. Lidskog R, Lidskog R (2017) Risk governance. In: Oxford research encyclopedia of communication. https://doi.org/10.1093/acrefore/9780190228613.013.246
26. Lwin KK, Pal I, Shrestha S, Warnitchai P (2020) Assessing social resilience for flood-vulnerable communities in Ayeyarwady Delta, Myanmar. Int J Disaster Risk Reduct 51:101745. https://doi.org/10.1016/j.ijdrr.2020.101745
27. Malalgoda C, Amaratunga D, Haigh R (2014) Challenges in creating a disaster resilient built environment. Procedia Econ Financ 18:736–744. https://doi.org/10.1016/s2212-5671(14)00997-6
28. Malalgoda C, Amaratunga D, Pathirage C (2011) Exploring disaster risk reduction in the built environment. Building Economist, June 2011
29. Mandal A, Baidya DK, Roy D (2012) Dynamic response of the foundations resting on a two-layered soil underlain by a rigid layer. Geotech Geol Eng 30(4):775–786. https://doi.org/10.1007/s10706-012-9497-2
30. Mohanty A, Hussain M, Mishra M, Kattel DB, Pal I (2019) Exploring community resilience and early warning solution for flash floods, debris flow and landslides in conflict prone villages of Badakhshan, Afghanistan. Int J Disaster Risk Reduct 33:5–15. https://doi.org/10.1016/j.ijdrr.2018.07.01
31. Neelamani S, Al-Salem K, Taqi A (2017) Experimental investigation on wave reflection characteristics of slotted vertical barriers with an impermeable back wall in random wave fields. J Waterw Port Coast Ocean Eng 143(4):06017002. https://doi.org/10.1061/(asce)ww.1943-5460.0000395
32. Olcina J (2004) Riesgo de inundaciones y ordenación del territorio en la escala local: el papel del planeamiento urbano municipal
33. O'Donnell E, Thorne C, Ahilan S, Arthur S, Birkinshaw S, Butler D et al (2020) The blue-green path to urban flood resilience. Blue-Green Syst 2(1):28–45. https://doi.org/10.2166/bgs.2019.199
34. Pal I, Bhatia S (2017) Disaster risk governance and city resilience in Asia-Pacific region. In: Science and technology in disaster risk reduction in Asia: potentials and challenges, pp 137–159. https://doi.org/10.1016/B978-0-12-812711-7.00009-2
35. Pistrika A, Tsakiris G, Nalbantis I (2014) Flood depth-damage functions for built environment. Environ Process 1(4):553–572. https://doi.org/10.1007/s40710-014-0038-2
36. Ramkrishnan R, Kolathayar S, Sitharam TG (2019) Seismic hazard assessment and land use analysis of Mangalore City, Karnataka, India. J Earthq Eng 1–22

37. Rohith VR, Kolathayar S, Priyatham K, Kumar VK, Nikil S (2018) Disaster preparedness index: a valid and reliable tool to comprehend disaster preparedness in India. In: Urbanization challenges in emerging economies: resilience and sustainability of infrastructure. American Society of Civil Engineers, Reston, VA, pp 156–163
38. Sakai F, Nishimura M, Ogawa H (1984) Sloshing behavior of floating-roof oil storage tanks. Comput Struct 19(1–2):183–192. https://doi.org/10.1016/0045-7949(84)90217-7
39. Sitharam TG, Kolathayar S (2018) Preparing for earthquakes: lessons for India. Springer International Publishing
40. Textbook of Seismic Design (2019) Textbook of seismic design. Springer, Singapore. https://doi.org/10.1007/978-981-13-3176-3
41. Thanvisitthpon N, Shrestha S, Pal I, Ninsawat S, Chaowiwat W (2020) Assessment of flood adaptive capacity of urban areas in Thailand. Environ Impact Assess Rev 81. https://doi.org/10.1016/j.eiar.2019.106363
42. Thayyen RJ, Dimri AP, Kumar P, Agnihotri G (2013) Study of cloudburst and flash floods around Leh, India, during August 4–6, 2010. Nat Hazards 65(3):2175–2204. https://doi.org/10.1007/s11069-012-0464-2
43. UNISDR (2009) UNISDR terminology on disaster risk reduction. UNISDR

Part I
Water Security and Flood Risk

Chapter 2
Satellite-Based Analysis of Groundwater Storage and Depletion Trends Implicating Climate Change in South Asia: Need for Groundwater Security

Pennan Chinnasamy, Minna J. Hsu, and Agoramoorthy Govindasamy

2.1 Introduction

Groundwater (GW) is one of the most exploited natural resources in the world [1]. Without this precious commodity, the expansion of agriculture productivity to feed the global population would not have been a reality [2]. Nevertheless, the overuse of GW in irrigation and industrial sectors is putting more pressure to the already limited natural resource [3]. Studies have shown that the overuse worldwide had resulted in the depletion of GW below sustainable levels [4–6]. While incomes from agriculture-associated outputs have increased, the local and regional GW shortages have significantly gone up [7]. To make matters worse, the surface water across regions are drying up or polluted in urban areas [8].

The depletion of GW is dominated by countries such as the USA, China, Pakistan, Iran, India, Mexico, and Saudi Arabia [9]. Among the world's aquifers, the Upper Ganges aquifer including India and Pakistan has the highest area (0.48 million km^2) with the maximum water footprint showing the enormity of GW over-extraction [10]. Besides, 1.7 billion people inhabit South Asia that has a long history of population growth, high GW footprints, and depleting aquifers. This in turn impacts not only GW consumption rates at present, but also the future availability. Therefore, the looming

P. Chinnasamy (✉)
Center for Technology Alternatives for Rural Areas (CTARA) and Inter Disciplinary Program for Climate Sciences (IDPCS), Indian Institute of Technology Bombay, Mumbai, India
e-mail: p.chinnasamy@iitb.ac.in

M. J. Hsu
Department of Biological Sciences, National Sun Yat-Sen University, Kaohsiung 804, Taiwan
e-mail: hsumin@mail.nsysu.edu.tw

A. Govindasamy
College of Pharmacy and Health Care, Tajen University, Yanpu, Pingtung 907, Taiwan
e-mail: agoram@tajen.edu.tw

S. Kolathayar et al. (eds.), *Civil Engineering for Disaster Risk Reduction*, Springer Tracts in Civil Engineering, https://doi.org/10.1007/978-981-16-5312-4_2

water stressors will eventually aggravate GW depletion, and the global warming may even promote unsustainable GW usage scenarios [5].

Scientists have predicted that climate change will impact GW recharge through variable rainfall events and longer droughts catalyzing larger irrigation-water demands [11]. The increases in food production may further add stress to the already limited surface and GW resources. Therefore, the GW investigations will play a key role to increase awareness on water stress indicators other than climate change, which will aid in better formulation of future mitigation plans [12]. Such an investigation, if carried out at a mega-scale covering large geographic region by including more nations, can capture large-scale changes in climate patterns. It would lead to thorough understanding of GW depletion influenced by anthropogenic activities and unsustainable development. With projected increase in population, living standards, urbanization, and irrigated area, the water per capita will certainly dwindle [13]. To combat the above externalities and to sustain development, policy-makers are constantly altering management plans to increase efficiency. However, effective long-term planning is fundamental for a diverse region like South Asia, which relies on trans-boundary cooperation. In order to increase cooperation, data on GW usage and net availability are of utmost importance.

As a region, South Asia is a densely populated one, and the access to irrigation was responsible for the agrarian boom ever since green revolution off during the 1960s. It eventually contributed to the decline in GW quality and quantity in South Asia [14–16], besides environmental stress leading to severe land degradation and serious loss of soil quality [17]. Although the region has 2800 km^3 of renewable freshwater resources, high human density and inefficient water use have caused scarcity. For example, the water per capita has decreased from 3000 to 1000 L per day over the last 50 years [18]. In addition, the net irrigated area under surface irrigation has decreased by 24% during 1993–2001. But, the net GW irrigated area increased by 26%, which indicates a radical shift from the traditional canal/surface irrigation to intensive GW-pumped irrigation [16]. In 1940, the annual GW extraction of India, Pakistan, and Bangladesh was 15, 8, and 7 km^3, respectively, and it increased to 260, 70, and 75 km^3 in 2010. The GW withdrawal enormously increased over 20 times across India resulting in severe environmental stress [19].

The water storage estimates at regional and continental scales are limited for South Asia due to fewer monitoring stations. Besides, acquiring hydrological parameters for large trans-boundary areas can be complex issue due to lack of reliable field observation data in South Asia [20–23]. Therefore, remote sensing estimates of hydrologic variables have become a reliable source [4]. For example, [24] showed that anthropogenic stressors were responsible for GW loss at the rate of 54 km^3 per year in north India. Likewise, [25] concluded that the current GW withdrawal rates are high in South India. The GW extraction rate in Tamil Nadu, south India went up to 8%, which is more than the natural recharge rate, according to [20]. All these studies used remote sensing tools to estimate GW extraction trends and provided recommendations for better monitoring plans.

The ongoing water stressors are often neglected or downplayed in South Asia when discussing availability in terms of future climate change projections. As a result, most studies were focused on water stress due to climate change without understanding the ongoing water stressors and future trends. To answer this overlooked aspect, we have designed this study to analyze the decadal trends covering climatic (rainfall and temperature) and hydrologic (GW storage, runoff, and evapotranspiration) variables. The results shown in this paper will stimulate a paradigm shift in thinking on indicators linking future climate change and water stressors. Accordingly, large-scale GW adaptation plans will weigh current use trends while considering in the context of future climate change projections. This study for the first time has analyzed climate patterns and water storage trends for the South Asian region at large.

2.2 Methods

The South Asian region has an area of 5 million km^2, and it includes Afghanistan, Bangladesh, Bhutan, India, Nepal, and Pakistan (Fig. 2.1). Small islands/nations such as Andaman, Nicobar, Maldives, and Sri Lanka were excluded due to high leakage errors of satellite data. India is the largest in terms of land area and Bhutan,

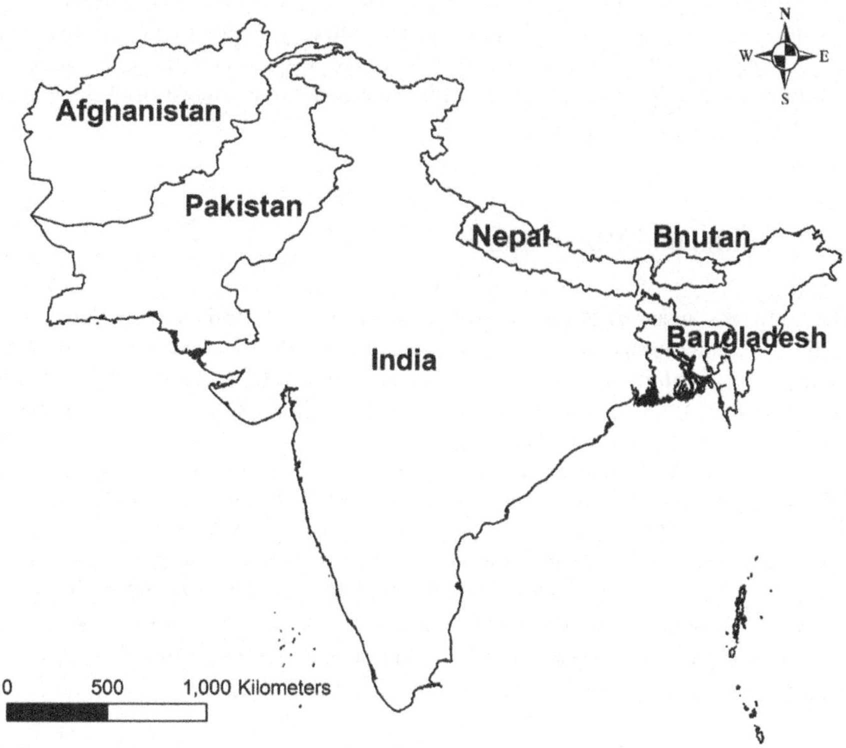

Fig. 2.1 Countries in the South Asian study site

the smallest. Major river basins include the Indus, Ganges, and Brahmaputra. Most of the water resources for South Asia originate in the Himalayas, flowing through Indus, Ganges, and Brahmaputra rivers, finally draining into the sea [26].

The surface runoff, evapotranspiration (ET), air temperature, total soil moisture (SM), and rainfall parameters were collected from the Noah Land Surface Model (LSM) driven by the Global Land Data Assimilation System (GLDAS) [4]. Data were downloaded at monthly (spatial) and $1.0 \times 1.0°$ (temporal) resolutions from NASA's Goddard Earth Sciences Center for the period from 2002 to 2014 (http://disc.sci.gsfc.nasa.gov/datacollection/GLDAS_NOAH10_M_V020.shtml). Data on terrestrial water storage (TWS) were collected from the Gravity Recovery And Climate Experiment (GRACE) satellite.

GRACE estimates TWS anomalies as equivalent water thickness at monthly intervals, and the accuracy increases with area [4, 22, 23, 27, 28], and therefore, it is more suitable for South Asia. The monthly land mass grids data (level V-RL05) for South Asia were downloaded from the NASA Jet Propulsion Laboratory (JPL) distributed data archives (http://gracetellus.jpl.nasa.gov/data) [29, 30]. The GW storage anomalies are isolated from TWS anomaly data using total soil moisture variations (SM) information from GLDAS-NoaH. The GW is computed as the difference between TWS and SM. May (peak summer month) and August (peak monsoon month) GW values were used to net compute recharge and discharge for the basin, respectively. The net GW discharge of the Xth year was estimated by subtracting August GW of the previous year (X-1th year) from the May of the Xth year. Similarly, the net GW recharge of the Xth year was estimated by subtracting the August of the Xth year from the May of the Xth year.

2.3 Results and Discussion

The rainy season started at the end of May, peaked in July and ended in September (Table 2.1). The average annual precipitation was 731 mm (equivalent to 3656 BCM for the basin), with the lowest of 544 mm recorded in 2002 and highest of 892 mm recorded in 2013. July was the wettest month with 175 mm, while December was the

Table 2.1 Hydro-climate parameters such as temperature, rainfall, runoff, ET, and GW for South Asia from 2002 to 2014

Parameter	Temperature (°C)	Rainfall (mm/mth)	Runoff (mm/mth)	ET (mm/mth)	GW (mm/mth)	Recharge (BCM/yr)	Discharge (BCM/yr)
Average	21.36	60.93	3.89	44.43	−0.60	289.66	357.85
Std. dev	5.82	60.98	5.06	17.59	43.08	302.80	243.97
Minimum	10.68	2.51	0.00	23.33	−114.45	61.45	93.22
Maximum	30.38	211.59	21.43	88.39	130.55	1070.65	834.77

yr, mth—refer to year and month, respectively

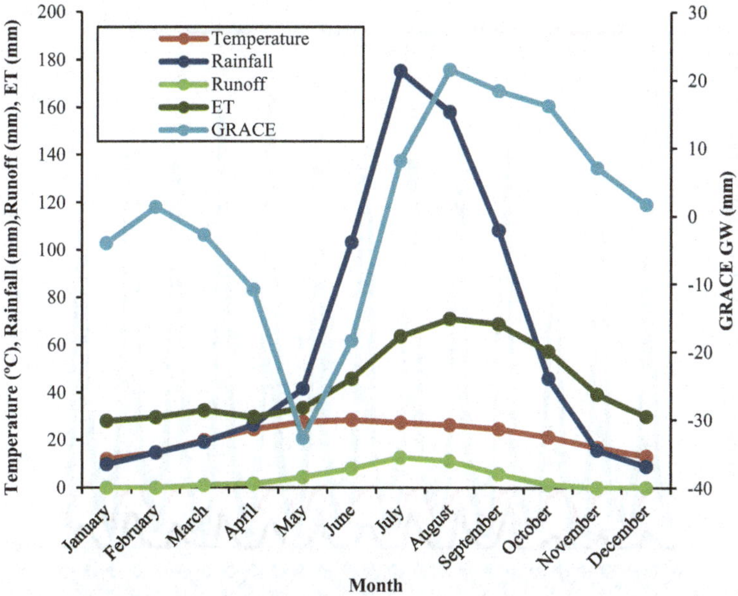

Fig. 2.2 Seasonal cycle of hydro-climate parameters (temperature, rainfall, runoff, ET, and GW) for South Asia from 2002 to 2014

driest with an average of 8.93 mm (Fig. 2.2). The average temperature was 21 °C, with January and June being the coldest and hottest with 12 and 28 °C, respectively. The highest rainfall was 211.59 mm recorded in July 2010, while the lowest of 2.51 mm was recorded in February 2004. The hottest temperature of 30.38 °C was observed in June 2014 and the coldest in January 2005 (10.68 °C). The peak summer extended from May to July.

The climate-guided hydrologic parameters for South Asian on average are depicted in Fig. 2.3. The highest runoff (21.43 mm), ET (88.39 mm), and GW (130.55) were estimated in July 2005, August 2010, and June 2002, respectively. The least ET (23.33 mm) and GW (−114.45 mm) were noted in April 2010 and May 2009, respectively. The average monthly runoff, ET, and GW were 3.89, 44.43, and 0.60 mm, respectively (Table 2.1). The winter period of November to February had negligible runoff, while July had the most (12.72 mm). The highest and lowest ET months were January (27.9 mm) and August (70.98 mm). Due to hot/dry conditions, May recorded the lowest GW (−32.72 mm), while the monsoon peak in August (21.54 mm) recorded the highest GW (Fig. 2.2). The GW and ET both peaked in August, after intense rains in July that showed the lag period of one month. However, the runoff peaked with rainfall in July (Fig. 2.2), which showed on average a flashy system for South Asia. The rainfall versus runoff resulted in a clockwise recession curve indicating the region to have flashier rains on average. It, therefore, has the potential to increase water storage in South Asia, and it can be done by constructing large number of water harvesting structures.

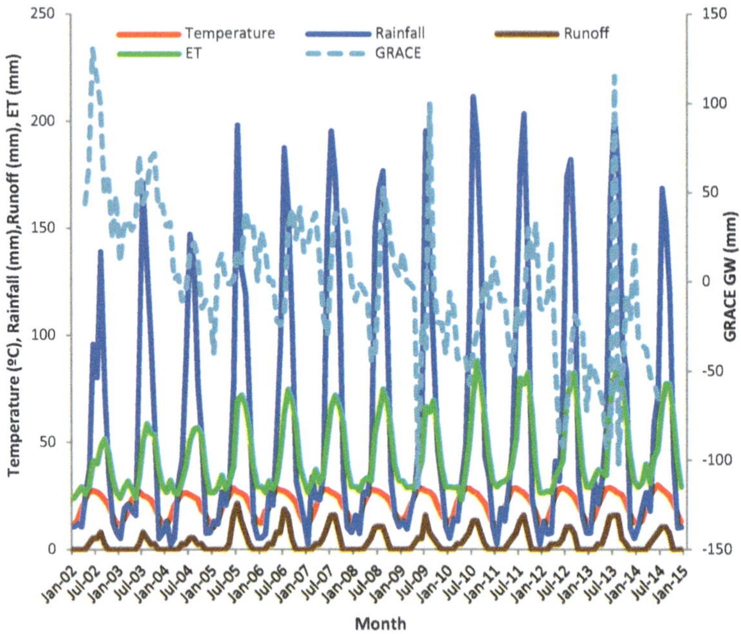

Fig. 2.3 Monthly time series of hydro-climate parameters (temperature, rainfall, runoff, ET, and GW) for South Asia from 2002 to 2014

South Asia's average recharge and discharge were 290 and 358 Billion Cubic Meter (BCM), respectively. Half of the time, the discharge was more than recharge, with the exception of 2010 that recorded high discharge. It could be due to prolonged increase of discharge in 2009 and 2010. The least recharge (62 BCM) and discharge (93 BCM) were noted in 2005 and 2011, correspondingly. As shown in Fig. 2.3, the annual rainfall has not changed much. But, there was considerable variation in annual GW use and recharge indicating high variability in water availability and use in South Asia. The long-term trend indicated an increase in the maximum annual temperature by 0.2 °C, average air temperature by 0.09 °C and rainfall by 2 mm, runoff 0.2 mm, and ET by 2 mm (Fig. 2.3). The ET increase might have been due to global warming combined with expansion in irrigated area. Although the rainfall has not declined, the GW storage showed a massive yearly decline of 15 mm indicating the reality on the absence of GW recharge while water usage continues to go upwardly. Therefore, land and water resource managers, policy-makers, and politicians need to think about building large number of rainwater harvesting structures to capture excess rainfall during monsoons, and then only the GW can be recharged effectively [7, 31–34].

The dry and cool seasons reported above coincided with the El Niño (2002–03) and La Niña (2010–11) for the region [35, 36]. A report reviewed the rain/temperature trend and indicated that there was variability in rains across India, with some regions showing an increase while others showing a decrease [35]. Similarly, positive annual rainfall was recorded in central/west India during 1901–1982, but a decreasing trend in the east was noted in other studies [37, 38]. Moreover, some showed negligible or no trend in the annual rains for India [36]. Analysis of 135 years (1871–2005) of rainfall data showed only a slight decrease of 0.8 mm per year [36]. The study concluded that no rainfall trend over long term for India, but small pockets of variations were noted at smaller scales. Moreover, Sheikh et al. [39] used data from 265 rainfall stations between 1961 and 2000 and showed the annual rainfall increasing by 3.7 mm per year in South Asia. Hence, the results of previous and current study further indicate that there may not be much decrease in total rainfall for South Asia with the exception of high variability and shifts in monsoon. Therefore, future management plans should incorporate methods to capture this variability in rainfall by building many water harvesting structures so that GW can be recharged [7, 32].

The temperature trends in South Asia between 1971 and 2000 indicated an increase in the maximum daily air temperature by 0.01 °C per year from 40.02 to 40.19 °C [39]. This is a troublesome trend since the climate can heat up more in the Indo-Gangetic Plains leading to stressful conditions across India, Pakistan, Nepal, and Bangladesh. Another study analyzed India's air temperature from 1881 to 1997, which showed 0.57 °C increase during the last century [40]. Similarly, [41] reported an increase of 0.35 °C over the last century for India. Our results compliment these findings and portray the fact that the South Asian region is certainly warming up faster by 0.09 °C yearly. The current decade has encountered the highest maximum and average air temperatures when compared to previous decades. So, future strategies should incorporate appropriate sustainable development activities, and then only water and land resources can be managed wisely [6, 42, 43].

The GW has drastically decreased in South Asia over the past decade, mostly due to an increase in agriculture and development activities influenced by globalization [6, 9, 17]. Increases in GW irrigation have supported dynamic crop yield per area in South Asia, especially in India, Pakistan, and Bangladesh [16]. In all countries, irrigation represents one-third of agricultural land, with disparities across national boundaries that range from 45% in Bangladesh and Pakistan to 92% in Afghanistan [15]. In terms of water volume, on average, irrigation consumes about 90% of the total water consumed in the region, with variations that range from 87% in Sri Lanka to 98% in Afghanistan and Nepal [4]. The growth in GW use across the agrarian nations, such as in India, Pakistan, and Bangladesh is unsustainable as the extraction rates are higher than natural GW recharge rates, as evidenced by widespread drop in GW levels.

Although the acceleration in agricultural activities led to food security and economic growth in South Asia, the GW is severely depleted in recent decades. For example, 21.4 km^3 per year is depleting in the Tamil Nadu state of India influenced by high agricultural activity [20]. The GW abstraction rate of 0.14, 0.24, 2.44, and 21.70 km^3 per year was reported for Afghanistan, Bangladesh, Pakistan, and India, respectively, from 1960 to 2000 [44]. From 2000 to 2009, the rates have enormously increased to 0.28, 1.92, 3.61, and 43.31 km^3 per year for Afghanistan (100% increase), Bangladesh (700% increase), Pakistan (48% increase), and India (100% increase), respectively. Therefore, within nine years, there has been terrific stress on

GW since the water usage has skyrocketed in the region. The skyrocketing trend in fact satisfies nearly 60% of India's and 40% of Pakistan's today's agricultural demands.

2.4 Conclusion

It is important to realize the fact that political and aquifer boundaries are different; therefore, each country's water policy will greatly influence the neighbors' GW resources. Hence, future conflicts can be greatly averted if all nations in South Asia develop an integrated long-term science-based environmental policy to protect the integrity of surface and GW resources at large. As a matter of fact, the GW contribution to South Asian agricultural economy, including China, has been estimated at USD 25–30 billion annually. But, desertification and land degradation are increasing in South Asia and, thus, compromising agriculture productivity and environmental integrity. For example, over 30–40% of India's land area has been degraded. If the environmental integrity, including the land and freshwater resources, is not strategically managed by all countries using the latest science and technology tools, the water-dependent development in South Asia will certainly backfire in near future.

Acknowledgements The authors thank the GLDAS data team for encouraging discussions on GRACE and GLDAS data. The authors gratefully acknowledges the Sir Dorabji Tata Trust, Mumbai for awarding the Tata Visiting Chair status to carry out research at NM Sadguru Water and Development Foundation in Gujarat, India. This research was supported by the Programmatic Cooperation between the Directorate-General for International Cooperation (DGIS) of the Dutch Ministry of Foreign Affairs and IHE Delft in the period 2016–2023, also called DUPC2., through the 2019/089/108483/EWH (GRACERS project). This research was supported by the Programmatic Cooperation between the Directorate-General for International Cooperation (DGIS) of the Dutch Ministry of Foreign Affairs and IHE Delft in the period 2016–2023, also called DUPC2., through the 2019/089/108483/EWH (GRACERS project).

References

1. Russo TA, Lall U (2017) Depletion and response of deep groundwater to climate-induced pumping variability. Nat Geosci 10:105–110
2. World Bank (2010) Deep wells and prudence: towards pragmatic action for addressing groundwater overexploitation in India. World Bank, Washington, DC
3. ADB (2007) Asian development water outlook 2007: achieving water security for Asia. http://ncrpb.nic.in/NCRBP%20ADB-TA%207055/Toolkit-Resources/Annexure%209_Asian%20Water%20Development%20Outlook%202007.pdf
4. Rodell M, Velicogna I, Famiglietti JS (2009) Satellite-based estimates of groundwater depletion in India. Nature 460:999–1002
5. Famiglietti JS (2014) The global groundwater crisis. Nat Clim Change 4:945–948
6. McNutt M (2014) The drought you can't see. Science 345:1543–1543

7. Agoramoorthy G (2016) Sadguru model of rural development: elevates food security and ease poverty. Astral International, Delhi
8. Qiu J (2010) China drought highlights future climate threats. Nature 465:142–143
9. Gleeson T, Wada Y, Bierkens MF, van Beek LP (2012) Water balance of global aquifers revealed by groundwater footprint. Nature 488:197–200
10. MacDonald AM, Bonsor HC, Ahmed KM, Burgess WG, Basharat M et al (2016) Groundwater quality and depletion in the Indo-Gangetic Basin mapped from in situ observations. Nat Geosci 9:762–776
11. Taylor RG et al (2013) Ground water and climate change. Nat Clim Change 3:322–329
12. Mirumachi N (2015) Transboundary water politics in the developing world. Routledge, London
13. Hanasaki N, Fujimori S, Yamamoto T, Yoshikawa S, Masaki Y, Hijioka Y, Kainuma M, Kanamori Y, Masui T, Takahashi K, Kanae S (2013) A global water scarcity assessment under shared socio-economic pathways—part 2: water availability and scarcity. Hydrol Earth Syst Sci 17:2393–2413
14. Shah T, Roy AD, Qureshi AS, Wang J (2003) Sustaining Asia's groundwater boom: an overview of issues and evidence. Nat Resour Forum 27:130–141
15. Shah T, Singh OP, Mukherji A (2006) Some aspects of South Asia's groundwater irrigation economy: analyses from a survey in India, Pakistan, Nepal Terai and Bangladesh. Hydrogeol J 14:286–309
16. Shah T (2010) Taming the anarchy: groundwater governance in South Asia. Routledge-Taylor and Francis Group
17. Shiva V (2016) The violence of the green revolution: third world agriculture, ecology, and politics. University Press, Kentucky
18. Subramanian V (2004) Water quality in south Asia. Asian J Water Environ Pollut 1:41–54
19. CGWB (2015) Ground water year book India 2014–15. http://www.indiaenvironmentportal.org.in/files/file/Ground%20Water%20Year%20Book%202014-15.pdf
20. Chinnasamy P, Agoramoorthy G (2015) Groundwater storage and depletion trends in Tamil Nadu state, India. Water Resour Manage 1–14. https://doi.org/10.1007/s11269-015-0932-z
21. Chinnasamy P (2017) Depleting groundwater—an opportunity for flood storage? A case study from part of the Ganges River basin, India. Hydrol Res 48:431–441
22. Chinnasamy P, Ganapathy R (2018) Long-term variations in water storage in Peninsular Malaysia. J Hydroinf 20(5):1180–1190
23. Chinnasamy P, Maheshwari B, Prathapar SA (2018) Adaptation of standardised precipitation index for understanding watertable fluctuations and groundwater resilience in hard-rock areas of India. Environ Earth Sci 77(15):1–16
24. Tiwari VM, Wahr J, Swenson S (2009) Dwindling groundwater resources in northern India, from satellite gravity observations. Geophys Res Lett 36. https://doi.org/10.1029/2009GL039401
25. Ferrant S, Caballero Y, Perrin J, Gascoin S, Dewandel B, Aulong S, Dazin F, Ahmed S, Maréchal JC (2014) Projected impacts of climate change on farmers' extraction of groundwater from crystalline aquifers in South India. Sci Rep 4:43697. https://doi.org/10.1038/srep03697
26. CWC (2014) Ganga Basin. Central Water Commission, Ministry of Water Resources, Delhi
27. Chinnasamy P, Hubbart JA, Agoramoorthy G (2013) Using remote sensing data to improve groundwater supply estimations in gujarat India. Earth Interact 17(1):1–17. https://doi.org/10.1175/2012EI000456.1
28. Chinnasamy P, Hsu MJ, Agoramoorthy G (2019) Groundwater storage trends and their link to farmer suicides in Maharashtra State India. Front Public Health 7. https://doi.org/10.3389/fpubh.2019.00246
29. Swenson S, Wahr J (2006) Post-processing removal of correlated errors in GRACE data. Geophys Res Lett 33:L08402
30. Landerer F, Swenson S (2012) Accuracy of scaled GRACE terrestrial water storage estimates. Water Resour Res 48(W04531):2012. https://doi.org/10.1029/2011WR011453
31. Agoramoorthy G (2012) India's dam dilemma: an eco-friendly remedy. Environ Sci Technol 6:1–592

32. Agoramoorthy G, Hsu MJ (2008) Small size, big potential: check dams for sustainable development. Environ Sci Policy Sustain Dev 50:22–35
33. Agoramoorthy G, Hsu MJ (2015) Irrigation-based social work relieves poverty in India's drylands. Int Social Work 58:23–31
34. Agoramoorthy G, Hsu MJ (2016) Small dams revive dry rivers and mitigate local climate change in India's drylands. Int J Clim Change Strateg Manag 8:271–285
35. Kumar V, Jain SK (2004) Trends in rainfall amount and number of rainy days in river basins of India, 1951–2004. Hydrol Res 42:290–306
36. Kumar V, Jain SK, Singh Y (2010) Analysis of long-term rainfall trends in India. Hydrol Sci J 55:484–496
37. Parthasarathy B, Dhar ON (1974) Secular variations of regional rainfall over India. Q J Res Meteorol Soc 100:245–257
38. Pant GB, Hingane LS (1998) Climatic changes in and around the Rajasthan desert during the 20th century. Int J Climatol 8:391–401
39. Sheikh MM, Manzoor N, Ashraf J, Adnan M, Collins D, Hameed S, Manton MJ et al (2014) Trends in extreme daily rainfall and temperature indices over South Asia. Int J Climatol 35:1625–1637. https://doi.org/10.1002/joc.4081
40. Pant GB, Kumar KR (1997) Climates of South Asia. Wiley, Chichester
41. Kothyari UC, Singh VP (1996) Rainfall and temperature trends in India. Hydrol Process 10:357–372
42. Reyer CPO, Rigaud KK, Fernandes E, Hare W, Serdeczny O, Schellnhuber HJ (2017) Turn down the heat: regional climate change impacts on development. Reg Environ Change 17:1563–1568
43. Chinnasamy P, Prathapar SA (2016) Methods to investigate the hydrology of the Himalayan springs: a review. Colombo, Sri Lanka: International Water Management Institute (IWMI). p 28. (IWMI Working Paper 169). https://doi.org/10.5337/2016.205
44. Doll P, Schmied HM, Schuh C, Portmann FT, Eicker A (2014) Global-scale assessment of groundwater depletion and related groundwater abstractions: combining hydrological modeling with information from well observations and GRACE satellites. Water Resour Res 50:5698–5720

Chapter 3
Use of Multi-sensor Satellite Remote Sensing Data for Flood and Drought Monitoring and Mapping in India

Atasi De, Deepti B. Upadhyaya, S. Thiyaku, and Sat Kumar Tomer

3.1 Introduction

Natural calamities are caused by earth processes like floods, droughts, cyclones, tsunamis, earthquakes, and epidemics [1]. There are two types of natural disasters: natural and man-made. Natural disasters occur on a regular interval due to earth's natural process, whereas man-made disasters are caused by human activities. As an example, when two earth plates approach toward a common fault line, it gives rise to earthquakes and tsunamis. On the other hand, drastic climate changes give rise to cyclones, floods, and droughts. Natural calamities often impact human settlements in a very impactful way, leading to widespread loss of lives, property, building, agriculture, as well as economy. Such a natural calamity is then considered a disaster. At the same time, a natural calamity such as an earthquake or tsunami which happens in the middle of the sea or ocean, and people have not been affected by it, cannot be called a disaster [2]. Based on National Policy on Disaster Management, India is prone to floods, earthquakes, droughts, cyclones, tsunamis, landslips, and avalanches. It is reported that nearly 59% of India's land is prone to earthquakes, about 12% of land is flood prone, approximately 76% of coastline is prone to cyclones, and 68% of agricultural areas are prone to droughts [3].

Floods can be considered as one of the most common natural disasters during the last few decades resulting in loss of lives, property, and economic loss. Every year, monsoon comes in India during the month of June–September, and subsequently, most of the rivers overflow. The rains are important in the agricultural sector, but excessive rainfall turns into flood, leading to environmental and socioeconomic losses. Nearly 40 million ha of farmland in India is flood prone, and about 8 million

A. De (✉) · D. B. Upadhyaya · S. Thiyaku · S. K. Tomer
Satyukt Analytics Private Limited, Bangalore, Karnataka 560094, India
e-mail: atasi@satyukt.com

D. B. Upadhyaya
Indian Institute of Science, Bangalore, Karnataka, India

S. Kolathayar et al. (eds.), *Civil Engineering for Disaster Risk Reduction*, Springer Tracts in Civil Engineering, https://doi.org/10.1007/978-981-16-5312-4_3

ha of land surface area is affected annually every year. For the last few decades, many parts of India have been affected by severe flood situations. Based on the data available in Rajya Sabha in 2018, India lost 1808 lives, and about 957 billion of Indian economy has been destroyed across India [3]. There has been an increase in the frequency and intensity of flood due to the warming climate. Even though the number of death tolls has decreased over the past decade, the number of affected people and our country's economic loss have increased drastically [4].

Flood monitoring and assessment are very difficult to handle using traditional approaches for larger scales of area with high accuracy. Remote sensing is steadily becoming a useful tool in disaster management, providing temporal and spatial information for a wide range of applications. This technology can easily map hazardous areas, drainage networks, and land cover. It is reported that water information extraction is one of the most common uses of remote sensing for quick access of flood hazard assessment, as satellite imageries offer a large area of coverage and periodic repeatability [5, 6]. For the last few decades, researchers have been using the optical remote sensing data for flood inundation monitoring in order to determine the water information such as normalized difference water index (NDWI) and automated water extraction index (AWEI). Optical sensors can only detect the earth's signal naturally detected or emitted by the surface of earth in visible and infrared spectral bands. However, they are not capable of providing information in the presence of cloud cover or during nighttime. In order to overcome such hurdles, active microwave sensors such as synthetic aperture radar (SAR) are introduced which consist of their own illumination and record the amount of incident signal, operating under all-weather conditions. They are even able to penetrate clouds, emergent aquatic plants, and forest canopies, and can detect standing water [7–10].

Like floods, drought is another natural calamity which can occur frequently, lead to massive damages in agriculture, environment, economy, and property, and affect thousands of lives. Nowadays, millions of farmers in India are struggling with crop failure due to drought mainly in western parts of India. More than 80% parts of Karnataka and 70% parts of Maharashtra have been observed as drought-affected zones. The main reason for drought is insufficient precipitation which reduces soil moisture and destroys the plant growth when prolonged over a long period of time. As drought is a recurring calamity and common for all climate zones, it is not an easy task to monitor and predict the drought based on the conventional techniques over the large areas. In this chapter, several proxy indicators such as rainfall amount, soil moisture (SM), land surface temperature (LST), NDVI, and evapotranspiration (ET) have been used for monitoring the drought events [11–13].

The entire article is divided into four sections: (a) flood monitoring using remote sensing where it is focused on monitoring and mapping the flood events based on optical, microwave as well as a combined approach, (b) drought monitoring and assessment, (c) flood and drought risk assessment, and (d) conclusions.

3.2 Flood Monitoring Using Remote Sensing

3.2.1 Applications of Optical Remote Sensing

In the era of modern technology, the integration of remote sensing and GIS has become the most important tool in flood inundation monitoring and assessment [5]. Several multiple satellites can be used to offer timely and valuable information, based on their water recognition capability. Mainly, optical and radar systems can be utilized where the radar systems opt larger wavelengths compared to optical sensors. In most of the studies, optical sensors can only monitor the post-flood scenario but are not able to analyze during flood events because of cloud cover. Further, the revisit cycle of many sensors is so long; that is, it is very difficult to assess flood monitoring frequently. Hence, optical sensors are mainly utilized in order to understand the extent of the flood [14].

Researchers have been using the Landsat including Multi Spectral Scanner (MSS), Thematic Mapper (TM), SPOT, LISS-IV, IRS, MODIS, etc., to compute the flood extent based on different algorithms [14–18]. However, a few numbers of optical sensors are significant in flood assessment due to their fine resolutions [15]. Landsat is one of the most widely used remote sensing data since it offers continuous free data at a global scale of coverage. Landsat MSS data with 80 m resolution is important to deal with the flood-prone areas, and they are found most suitable for extracting water or moist soil from dry surface because their band 7 has a strong capability of water absorption in the near-infrared range of the spectrum. In the early 1980s, Landsat TM, having 30 m spatial resolution, was the primary source of data for inundation monitoring and delineating the flood boundary. Landsat TM band 4 is the most suitable in discriminating water from dry land surface as it is almost equivalent to MSS band 7, whereas its band 4 cannot be utilized optimally in land use areas like industrial, downtown commercial areas, etc. It has been observed that normalized difference water index (NDWI) is the best method to map the flood areas among all other methods. Interestingly, IRS-D provides finer spatial resolution as compared to other optical satellite sensors. Patel and Srivastava further explained in identifying the safe and unsafe zones for Tapi river basin, as well as the entire Surat district by integrating the IRS-D and DEM data as thousands of lives have already been lost in this area for the last five decades [18].

Nowadays, coarse resolution data such as Advanced Very High Resolution Radiometer (AVHRR) proves to be the most useful for flood mapping on a regional scale because they have high frequency of global coverage, wide swath, and high repetivity [15–18]. This helps to determine the flooding in near-real time.

In order to detect water more efficiently, normalized difference vegetation index (NDVI) can be opted to assess the river inundation from AVHRR image sets, based on the potentiality of near-infrared band. Several studies have already been done in flood mapping and monitoring using NDVI as we know that water has a unique spectral signature in NIR band that is completely different from other surface features.

3.2.2 Application of Microwave Remote Sensing

The use of optical remote sensing data is restricted in severe weather conditions, especially in the presence of cloud cover. Recently, many scientists are adopting the radar remote sensing, where the radar signals can penetrate through cloud cover for collecting earth information, for constant monitoring and mapping in various parts of India [19–21]. So, in case of a radar system, the long wavelength helps to penetrate through not only clouds but also some ground objects such as tree canopy, below the grassland, under the soil surface, etc. Currently, the most reliable approach to flood monitoring and mapping is to adopt synthetic aperture radar (SAR) due to its all-weather capability and unique sensitivity to geometrical as well as structural orientation, providing reliable and accurate information in managing flood events [7, 16, 22]. This happens when a smooth water surface provides no backscatter signal to the antenna and appears black in SAR imagery [20]. At the same time, a wind-ruffled surface can give larger backscatter signals compared to the surrounding land. This complicates the detection of water surfaces on SAR images for flood applications. This chapter will talk about the radar system for monitoring and mapping the flood phenomenon.

A concept on thresholding has become the most popular technique in SAR remote sensing to differentiate flood-prone zones from non-flooded zones using radar image datasets [4, 23–28]. Usually, radar backscattered signal is computed in terms of a function of the incidence angle of the sensor and digital number (DN) [23]. The threshold values are determined by a set of processes based on the study areas and the overall spectral signature of the imagery [4]. Sentinel-1 has VV and VH polarizations in which VV backscatter signal is found to be better delineation water in comparison with VH backscattered signal [4, 25, 26].

A few researchers have started the near-real-time flood inundation mapping during the flood disaster using SAR technology in order to assess the property loss and damage. Kundu et al. successfully monitored the flood extent, and its depth from 2003 to 2008 in Kendrapara districts of Odisha using RADARSAT-1 satellite data [26]. Nakmuenwai et al. monitored multi-temporal dual-polarized RADARSAT-2 data for mapping the flood zones using clustering-based threshold techniques [27]. Cian et al. performed normalized difference flood index (NDFI) using Sentinel-1 data for rapid flood mapping [9]. Capolongo et al. (2019) investigated the real-time, long-term flood disaster monitoring from multi-temporal and high-resolution SAR datasets for post inundation recovery. Agnihotri et al. mentioned the flood mapping and monitoring before, during, and after the flood in Ganga and Ramganga rivers [25].

3.2.3 Application of a Combined Approach

There are three combinations of satellite remote sensing data used for flood phenomenon detection which is actually applied before and after the inundation as follows: (a) optical imagery/optical imagery, (b) optical imagery/SAR imagery, and (c) SAR imagery/SAR imagery. Although we have three types of combinations available, the second combination (b) is found to be the most suitable in identifying the flood inundation. A few studies have been proposed for the analysis of flood progression which rely on multi-temporal and multi-source data [29–32]. However, these methods are particularly used in special cases and these combinations of approaches need to be required for further study of flood analysis. A list of advantages and disadvantages of various flood indices has been briefly described in Table 3.1.

3.3 Drought Monitoring and Assessment

Drought is a complex natural disaster as it is caused by a multitude of factors. Drought can be explained as a lack of presence of water at a particular location and time compared to average value at that climatic zone. Drought can be characterized based on severity, intensity, duration, and areal extent [35]. Thus, when monitoring and assessing drought multiple aspects should be considered in an integrated manner to give mitigative measures. Remote sensing forms a promising platform which provides real-time spatial observations of several atmospheric and land surface variables that can be used to estimate precipitation, evapotranspiration, soil moisture, and vegetation conditions. When combined with modeling and forecasting of the water cycle, information on future drought can also be generated for drought preparedness.

Drought is measured using indices which are used in assessment and monitoring procedures, and many authors have shown the usefulness of the drought indices reviewed in the literature (e.g., [36, 37] among others). There are several indices: standardized precipitation index (SPI) [38] and palmer drought severity index (PDSI) [39] which are measured based on precipitation to monitor meteorological drought. The index indicates the severity due to changing rainfall pattern in any location. To estimate SPI, the precipitation record is first fitted to a gamma distribution, which is then transformed into a normal distribution using an equal probability transformation. After standardization,

$$\text{i.e., SPI} = \frac{(X_i - m[X])}{\text{std}[X]} \tag{3.1}$$

where X_i is the precipitation variable, $m[X]$ the average (typically over a period of 30 years), and std$[X]$ the standard deviation, SPI values above zero indicate wet periods and values below zero indicate dry periods. PDSI is calculated by estimating a constant or coefficient which are dependent on climate variations of the

Table 3.1 Advantages and disadvantages of various flood indices

S. No.	Indices	Parameter	Advantages	Disadvantages	References
1.	Normalized difference vegetation index (NDVI)	Vegetation	This is easily determined without assumptions regarding soil varieties, land cover classes, or climate conditions	Inherent nonlinearity due to having ratio-based index; additive noise effects, extremely sensible to canopy background brightness, asymptotic signals over high biomass conditions	[33]
2.	Normalized difference water index (NDWI)	Water	High resolution and good coverage at all terrains	Stress to plant canopies can be caused by impacts other than drought, and it is difficult to discern them using only NDWI. The period of record for satellite data is short, with climatic studies being difficult	[34]
3.	Normalized difference flood index (NDFI)	Flood, water	Proposed to highlight the flooded areas	Mixed pixel problem hamper in classification	[9]
4.	Normalized difference sigma-naught index (NDSI)	Open water	Useful for open water flooding detection based on automatic thresholding and Bayesian probability	Only applicable on SAR datasets. Not eligible for optical imageries	[10]

area under investigation. This method requires estimation of hydrological components like runoff, evapotranspiration, recharge, and loss which can be accomplished by using models with proper forcing and soil hydraulic properties.

Standardized runoff index (SRI) [40] which uses runoff or streamflow is useful in drought monitoring in regions having a majority of snow cover. The runoff and streamflow data required for this index can be obtained from observations or from hydrological models applied over drought-affected areas. These are useful in areas having high variation in terrain. The SRI can be estimated in a similar way to SPI.

Standardized soil moisture index (SSI) [41] and soil moisture percentile [42] are mostly used indices for agricultural drought monitoring. Xu et al. (2018) computed SSI using SMAP and NLDAS datasets as

$$\mathrm{SSI} = \frac{X_{\mathrm{SMAP}} - \mu_{\mathrm{NLDAS}}}{\sigma_{\mathrm{NLDAS}}} \tag{3.2}$$

where X_{SMAP} is daily soil moisture content from SMAP L3 product, μ_{NLDAS} means soil moisture for the day from North American Land Data Assimilation System (NLDAS), and σ_{NLDAS} is standard deviation.

Many other indices based on impact are caused by drought, and these indices indicate the surface condition rather than showing quantitative measurement of drought. These indices include normalized difference vegetative index (NDVI) [43], vegetation condition index (VCI) [16], drought severity index (DSI) [44], vegetation temperature condition index (VTCI) [45], standardized precipitation evapotranspiration index (SPEI) [43] which are commonly used for agricultural drought monitoring. A list of advantages and disadvantages of commonly used drought indices is provided in Table 3.2.

3.4 Flood and Drought Risk Assessment

Flood information and experience during early flood events give valuable insights into future events. The most important factors such as flood hazard, flood vulnerability, and integrated risk are required to assess for flood risk management. For such applications, remote sensing plays a significant role in understanding the situation and providing cost-effective and reliable solutions as demonstrated in Fig. 3.1.

Drought risk identification is done based on studying the historical data about hydro-climatic variables like precipitation, evaporation, and irrigation. The remote sensing approaches developed show the impact of drought and scale of impact. For example, using NDVI values the crop-affected area can be identified very effectively. To prevent the loss, the drought preparedness measures need to be practiced which include activities like modeling and forecasting the situations beforehand. The future predictions of climatic forcing variables can be used as forcing in hydrological modeling knowing the soil hydraulic properties at the investigating area. Thus, the possible risk can be assessed using such methods. One of the programs include the 'National Agricultural Drought Assessment and Monitoring System (NADAMS)' project which was conceptualized and developed by the National Remote Sensing Centre (NRSC), ISRO, Department of Space. Drought assessment is done by evaluating drought indices (Fig. 3.2) on a fortnightly and monthly basis [51].

Table 3.2 Advantages and disadvantages of different drought indices

S. No.	Indices	Parameter	Advantages	Disadvantages	References
1	Standard precipitation index	Precipitation	Quantifying the impacts of different precipitation shortages to different water resources on different timescales	Spatial distribution features cannot be identified	[46, 47]
2	Moisture adequacy index	Evapotranspiration	Considering the water balance, soil characteristics, and crop growths	More data needed and complex computation	[48]
3	Drought severity index	Precipitation evapotranspiration runoff	Considering rainfall, latent evaporation, antecedent soil moisture, and runoff; quickly reflecting the change of soil moisture	Complex computation	[42]
4	Hydrological drought severity index	Runoff	Analyzing the time integral flow of concrete section of the river	Low resolution	[12, 13]
5	Temperature and vegetation index	Surface temperature normalized differential vegetation index	Directly obtaining the parameters from the satellite image, simple and convenient calculation	Only representing the relative values of the same image moisture state but not comparable in time	[49]

3.5 Case Studies

3.5.1 Case Studies for Flood Assessment

Flood assessment models have been built by estimating the function of hazards, vulnerability, exposure, and adaptive and resilience capacity. Several numerical

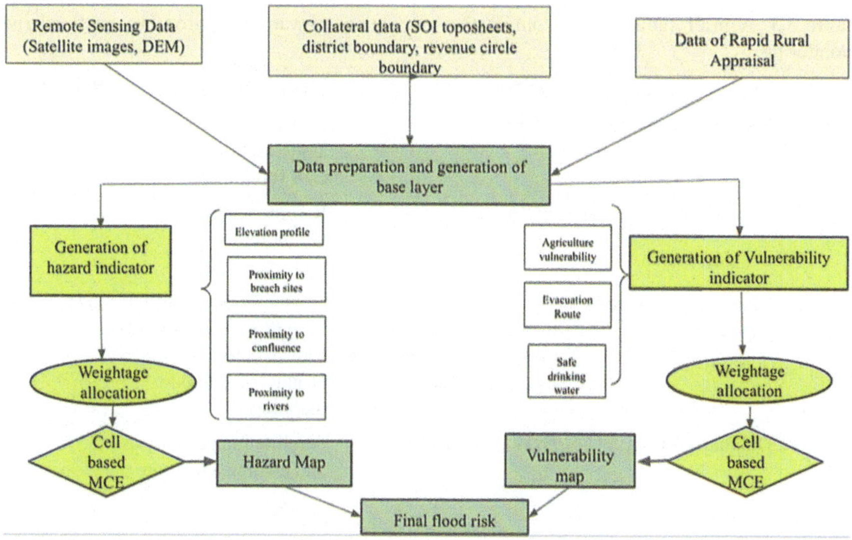

Fig. 3.1 Approaches for flood risk assessment [50]

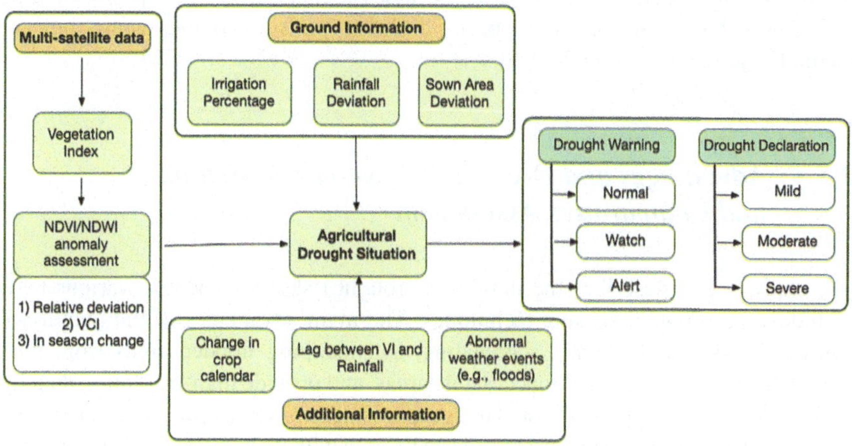

Fig. 3.2 Approaches for drought assessment under NADAMS program [51]

models integrated with remote sensing and GIS have been implemented as listed in Table 3.3, for different Indian cities, using different indices.

Table 3.3 A brief summary on comparative assessment of various approaches on flood risk monitoring

S. No.	City	Indices	Methodology	References
1	Mumbai, Maharashtra	NDVI	Integrated flood assessment model and GIS framework	[52]
2	Kolkata, West Bengal	NDFI	Remote analysis of vulnerability for non-linearization	[53]
3	Chennai, Tamil Nadu	Tropical rainfall measuring mission (TRMM), global enable forecast system (GEFS)	Combination of global rainfall data and global forecast precipitation products for flood mapping	[54]
4	Dhemaji, Assam	NDVI and NDWI	Satellite image analysis and change detection	[55]

3.5.2 Case Studies for Drought Assessment

Similarly, case studies where drought assessments have been performed by researchers have been carried out in different parts of India, using multiple methodologies, as listed in Table 3.4.

3.5.3 Advantages and Drawback of Research Methods and Comparative Assessment

The study briefly describes the flood and drought risk assessment in various parts of India using remote sensing techniques. The main advantages of the mentioned methodologies are detecting, segregating, and mapping flooded areas from non-flooded areas, cover vast and inaccessible areas, and provide high-resolution data for flood and drought monitoring and damage assessment. Optical remote sensing data is easily available at reasonable price and can be used with well-defined data processing techniques. Optical data cannot penetrate through cloud or vegetation covers where radar sensors can provide timely and effective flood warning information through clouds. However, they are costly and tough to analyze the data.

3.6 Conclusions

Floods and droughts are the most devastating natural hazards in the world specially in India. With a huge population like that of India, where water shortage is an acute

Table 3.4 A brief summary on comparative assessment of various approaches on drought monitoring

S. No.	City	Index	Methodology	References
1	Bihar	Soil moisture index and NDVI	Drought assessment in Bihar during kharif season 2013	[8]
2	Rajasthan	Standardized precipitation index (SPI), NDVI, water supply vegetation index (WSVI), and vegetation condition index (VCI)	The interlink of NDVI-SPI and WSVI-SPI at multiple timescales correlation analysis were conducted for the NDVI/WSVI and SPI at 1-, 2-, 3-, 6-, 9-, and 12-month timescale using these three parameters for the year 2002–2004	[56]
3	Marathwada (Maharashtra)	SPI, land surface temperature (LST), soil moisture (SM), and NDVI	Analysis was done on a monthly scale from 2002 to 2018. Two approaches were used to combine indices, an expert judgment-based weight of each parameter (Method-I) and principal component analysis (PCA)-based weighting approach (Method-II)	[57]
4	Madurai (Tamil Nadu)	SPI, reconnaissance drought index (RDI), streamflow drought index (SDI)	Drought index analysis was done for 100 years to assess the nature of drought in the region	[58]
5	Gujrat	Vegetation condition index and temperature condition index (TCI)	Drought analysis was done from 1981 to 2010 GIS-based study has used heat wave and temperature data	[11]

problem, it definitely needs strong initiatives at all levels to monitor drought and flood conditions. Better water management calls for early and constant managing of flood and drought conditions, via advanced monitoring with latest technological tools. One such readily available technological tool is remote sensing, an early warning, early response, and an intelligent monitoring system. The study has reviewed the use of satellite remote sensing in flood and drought monitoring and their risk assessments over various parts of India using near-real-time data. Remote sensing approach is cost-effective and efficient for identifying the flood hazard zones, differentiating flooded crops from non-flooded crops, recognizing the damaged built-up areas, and pre-positioning of necessary foodstuff and their delivery. Remote sensing techniques have been in both optical and microwave wavelengths, but a combined approach has been found to be more suitable for both identifying and risk assessing flood inundation conditions. Similarly, remote sensing has proved to be very beneficial for drought monitoring, especially because of its ability in identifying local-level variations. This is a marked improvement from previous methods such as climate-based drought indices, which were of limited use as they provided at best a generalized spatial knowledge of drought variations. We have also mentioned how remote sensing has been extensively used for risk assessment for flood and drought monitoring, which greatly boosts individual reliance than reliance on government and policy makers.

References

1. Faisal A, Kafy A, Roy S (2018) Integration of remote sensing and GIS techniques for flood monitoring and damage assessment: a case study of Naogaon district, Bangladesh. J Remote Sens GIS 7(236):2
2. Foudi S, Osés-Eraso N (2014) Flood risk management: assessment for prevention with hydro-economic approaches. In: Routledge handbook of the economics of climate change adaptation. Routledge, pp 331–348
3. Dhiman R, VishnuRadhan R, Eldho TI, Inamdar A (2019) Flood risk and adaptation in Indian coastal cities: recent scenarios. Appl Water Sci 9(1):5
4. Sanyal J, Lu XX (2004) Application of remote sensing in flood management with special reference to monsoon Asia: a review. Nat Hazards 33(2):283–301
5. Haq M, Akhtar M, Muhammad S, Paras S, Rahmatullah J (2012) Techniques of remote sensing and GIS for flood monitoring and damage assessment: a case study of Sindh province, Pakistan. Egypt J Remote Sens Space Sci 15(2):135–141
6. Wang Y, Li Z, Tang Z, Zeng G (2011) A GIS-based spatial multi-criteria approach for flood risk assessment in the Dongting Lake Region, Hunan, Central China. Water Resour Manage 25(13):3465–3484
7. Chen P, Liew SC, Lim H (1999) Flood detection using multitemporal Radarsat and ERS SAR data. In: Proceedings of 20th Asian conference of remote sensing, Hong Kong, 22–25 Nov 1999
8. Choudhary K, Goel I, Bisen PK, Mamatha S, Ray SS, Chandrasekar K, Murthy CS, Sesha Sai MVR (2015) Use of remote sensing data for drought assessment: a case study for Bihar state of India during kharif, 2013. In: High-impact weather events over the SAARC region. Springer, Cham
9. Cian F, Marconcini M, Ceccato P (2018) Normalized difference flood index for rapid flood mapping: taking advantage of EO big data. Remote Sens Environ 209:712–730
10. Ulloa NI, Chiang S-H, Yun S-H (2020) Flood proxy mapping with normalized difference sigma-naught index and Shannon's entropy. Remote Sens 12(9):1384
11. Bhuiyan C, Saha AK, Bandyopadhyay N, Kogan FN (2017) Analyzing the impact of thermal stress on vegetation health and agricultural drought—a case study from Gujarat, India. GISci Remote Sens
12. Dracup JA, Lee KS, Paulson EG(1980a) On the definition of droughts. Water Resour Res 16:297–302
13. Dracup JA, Lee KS, Paulson EG (1980b) On the statistical characteristics of drought event. Water Resour Res 16:289–296
14. Klemas V (2015) Remote sensing of floods and flood-prone areas: an overview. J Coast Res 31(4):1005–1013

15. Lin L, Di L, Yu EG, Kang L, Shrestha R, Rahman MS et al (2016) A review of remote sensing in flood assessment. In: 2016 fifth international conference on agro-geoinformatics, July 2016. IEEE, pp 1–4
16. Liu R, Liu N (2002) Flood area and damage estimation in Zhejiang, China. J Environ Manage 66:1–8
17. Islam MM, Sadu K (2000) Flood hazard assessment in Bangladesh using NOAA-AVHRR data with geographical information system. Hydrol Process 14(3):605–620
18. Patel DP, Srivastava PK (2013) Flood hazards mitigation analysis using remote sensing and GIS: correspondence with town planning scheme. Water Resour Manage 27(7):2353–2368
19. Shiva Prasad Sharma SV, Roy PS, Chakravarthi V (2018) Assessment of social vulnerability to the impact of flood hazard: a case study of Kopili River Basin, Assam, India. Int Arch Photogramm Remote Sens Spat Inf Sci
20. Rees WG (2001) Physical principles of remote sensing. Cambridge University Press, Cambridge
21. Tiwari V, Kumar V, Matin MA, Thapa A, Ellenburg WL, Gupta N, Thapa S (2020) Flood inundation mapping-Kerala 2018; harnessing the power of SAR, automatic threshold detection method and Google Earth Engine. PLos ONE 15(8):e0237324
22. Honda KC, Francis XJ, Sah VP (1997) Flood monitoring in central plain of Thailand using JERS-1 SAR data. In: Proceedings of 18th Asian conference of remote sensing, Malaysia, 20–24 Oct 1997
23. Kussul N, Shelestov A, Skakun S (2011) Flood monitoring from SAR data. In: Use of satellite and in-situ data to improve sustainability. Springer, Dordrecht, pp 19–29
24. Clement MA, Kilsby CG, Moore P (2018) Multi-temporal synthetic aperture radar flood mapping using change detection. J Flood Risk Manag 11(2):152–168
25. Agnihotri AK, Ohri A, Gaur S, Das N, Mishra S (2019) Flood inundation mapping and monitoring using SAR data and its impact on Ramganga River in Ganga basin. Environ Monit Assess 191(12):760
26. Kundu S, Aggarwal SP, Kingma N, Mondal A, Khare D (2015) Flood monitoring using microwave remote sensing in a part of Nuna river basin, Odisha, India. Nat Hazards 76(1):123–138
27. Nakmuenwai P, Yamazaki F, Liu W (2017) Automated extraction of inundated areas from multi-temporal dual-polarization RADARSAT-2 images of the 2011 central Thailand flood. Remote Sens 9(1):78
28. Tong X, Luo X, Liu S, Xie H, Chao W, Liu S et al (2018) An approach for flood monitoring by the combined use of Landsat 8 optical imagery and COSMO-SkyMed radar imagery. ISPRS J Photogramm Remote Sens 136:144–153
29. Psomiadis E, Diakakis M, Soulis KX (2020) Combining SAR and optical earth observation with hydraulic simulation for flood mapping and impact assessment. Remote Sens 12(23):3980
30. Martinis S, Twele A, Strobl C, Kersten J, Stein E (2013) A multi-scale flood monitoring system based on fully automatic MODIS and TerraSAR-X processing chains. Remote Sens 5(11):5598–5619
31. Zhang Y, Zhang H, Lin H (2014) Improving the impervious surface estimation with combined use of optical and SAR remote sensing images. Remote Sens Environ 141:155–167
32. Huete A, Didan K, Miura T, Rodriguez EP, Gao X, Ferreira LG (2002) Overview of the radiometric and biophysical performance of the MODIS vegetation indices. Remote Sens Environ 83(1–2):195–213
33. Chandrasekar K, Sesha Sai MVR, Roy PS, Dwevedi RS (2010) Land surface water index (LSWI) response to rainfall and NDVI using the MODIS vegetation index product. Int J Remote Sens 31:3987–4005
34. Shah RD, Mishra V (2015) Development of an experimental near-real-time drought monitor for India. J Hydrometeorol 327–345
35. Zargar Z, Sadiq R, Naser B, Khan FI (2011) A review of drought indices. Environ Rev 19:333–349

36. van Hoek M (2016) Drought monitoring from space: a focus on indicators, early detection and development of a web-based integrated portal. Ph.D. thesis, Chinese Academy of Sciences, 168 pp
37. McKee TB, Doesken NJ, Kleist J (1993) The relationship of drought frequency and duration to time scales. In: Eighth conference on applied climatology, Anaheim, CA, 17–22 Jan 1993, pp 179–184
38. Palmer WC (1965) Meteorological drought. Weather bureau research paper no. 45. Weather Bureau, Washington, DC
39. Shukla S, Wood AW (2008) Use of a standardized runoff index for characterizing hydrologic drought. Geophys Res Lett 35:L02405
40. Hao Z, AghaKouchak A (2013) Multivariate standardized drought index: a parametric multi-index model. Adv Water Resour 57:12–18
41. Wang A, Bohn TJ, Mahanama SP, Koster RD, Lettenmaier DP (2009) Multimodel ensemble reconstruction of drought over the continental United States. J Clim 22:2694–2712
42. Ji L, Peters AJ (2003) Assessing vegetation response to drought in the northern Great Plains using vegetation and drought indices. Remote Sens Environ 87:85–98
43. Beguería S, Vicente-Serrano SM, Reig F, Latorre B (2013) Standardized precipitation evapotranspiration index (SPEI) revisited: parameter fitting, evapotranspiration models, tools, datasets and drought monitoring. Int J Climatol 34(10):3001–3023
44. Mu Q, Zhao M, Kimball JS, McDowell NG, Running SW (2013) A remotely sensed global terrestrial drought severity index. US Department of Energy Publications, p 361
45. Wang P-X, Li X-W, Gong J-Y, Song C (2001) Vegetation temperature condition index and its application for drought monitoring. In: IGARSS 2001. Scanning the present and resolving the future. Proceedings. IEEE 2001 international geoscience and remote sensing symposium (Cat. No. 01CH37217)
46. Munger TT (1916) Graphic method of representing and comparing drought intensities. Mon Weather Rev 44:642–643
47. Kincer JB (1919) The seasonal distribution of precipitation and its frequency and intensity in the United States. Mon Weather Rev 47:624–631
48. McGuire JK, Palmer WC (1975) The 1957 drought in the eastern United States. Mon Weather Rev 85:305–314
49. Sandholt I, Rasmussen K, Andersen J (2002) A simple interpretation of the surface temperature/vegetation index space for assessment of surface moisture status. Remote Sens Environ 79:213–224
50. Hazarika N, Barman D, Das AK, Sarma AK, Borah SB (2018) Assessing and mapping flood hazard, vulnerability and risk in the Upper Brahmaputra River valley using stakeholders' knowledge and multicriteria evaluation (MCE). J Flood Risk Manag 11:S700–S716
51. Ray SS, Sesha Sai MVR, Chattopadhyay N (2015) Agricultural drought assessment: operational approaches in India with special emphasis on 2012. In: Ray K, Mohapatra M, Bandyopadhyay B, Rathore L (eds) High-impact weather events over the SAARC region. Springer, Cham
52. Kulkarni AT, Mohanty J, Eldho TI, Rao EP, Mohan BK (2014) A web GIS based integrated flood assessment modeling tool for coastal urban watersheds. Comput Geosci 64:7–14
53. Chakraborty T (2010) Flood inundation study of Kolkata Metro City with remote analysis of vulnerability by non-linearization algorithm (RAVANA). Doctoral dissertation, Jadavpur University
54. Sharma VK, Mishra N, Shukla AK, Yadav A, Rao GS, Bhanumurthy V (2017) Satellite data planning for flood mapping activities based on high rainfall events generated using TRMM, GEFS and disaster news. Ann GIS 23(2):131–140
55. Das K (2017) NDVI and NDWI based change detection analysis of Bordoibam Beelmukh wetlandscape, Assam using IRS LISS III data. ADBU J Eng Technol 6(2)
56. Jain SK, Keshri R, Goswami A, Sarkar A (2010) Application of meteorological and vegetation indices for evaluation of drought impact: a case study for Rajasthan, India. Nat Hazards 54(3):643–656

57. Kulkarni SS, Wardlow BD, Bayissa YA, Tadesse T, Svoboda MD, Gedam SS (2020) Developing a remote sensing-based combined drought indicator approach for agricultural drought monitoring over Marathwada, India. Remote Sens 12:2091
58. Surendran U, Kumar V, Ramasubramoniam S, Raja P (2017) Development of drought indices for semi-arid region using drought indices calculator (DrinC)—a case study from Madurai District, a semi-arid region in India. Water Resour Manage 31(11):3593–3605
59. Blumenstock G (1942) Drought in the United States analyzed by means of the theory of probability. Technical bulletin no. 819. United States Department of Agriculture, Washington, DC

Chapter 4
Data- and Physics-Based Modeling of Backward Erosion Piping

Alessandro Fascetti

4.1 Introduction

Quantification of resilience in Flood Protection Infrastructure (FPI) is of vital importance in the operational management of these widespread civil engineering systems, in order to provide safety to communities and guarantee the transport of goods and people on the transportation systems that are commonly interdependent with the FPI [1]. Moreover, FPI are notoriously difficult to maintain, as a result of different layers of uncertainties associated with inspection operations [2]: (1) presence of vegetation and fauna in river beds and water streams, (2) difficulty in characterizing material and geometrical parameters to characterize the system, (3) uncertainties related to the spatial and temporal distribution of such parameters that change significantly across the infrastructure, and (4) general lack of detailed information on the original design and construction documents.

In this framework, the present manuscript presents a review of recently developed numerical tools to enable science-based approaches to formalize the resilience assessment of FPI at the system scale (i.e., 10^5 m), providing a coherent tool for the evaluation of the state of the infrastructure. This manuscript will focus on the so-called Backward Erosion Piping (BEP) phenomenon, which is defined as the progressive loss of the finer fraction of soil from the foundation of a FPI. This mechanism is triggered on the protected side of the infrastructure (i.e., downstream with respect to the river level) by local damage to the foundation of the FPI that induces significant local hydraulic gradients and then progresses toward the upstream side of the infrastructure (hence, the term *backward*). If hydraulic equilibrium is not attained in the system, the erosion develops for for the entire length of the levee and undermines the stability of the whole structure. BEP is one of the most common reasons for FPI

A. Fascetti (✉)
Department of Civil and Environmental Engineering,
University of Pittsburgh, 3700 O'Hara Street,Pittsburgh, PA15261, USA
e-mail: fascetti@pitt.edu

S. Kolathayar et al. (eds.), *Civil Engineering for Disaster Risk Reduction*, Springer Tracts in Civil Engineering, https://doi.org/10.1007/978-981-16-5312-4_4

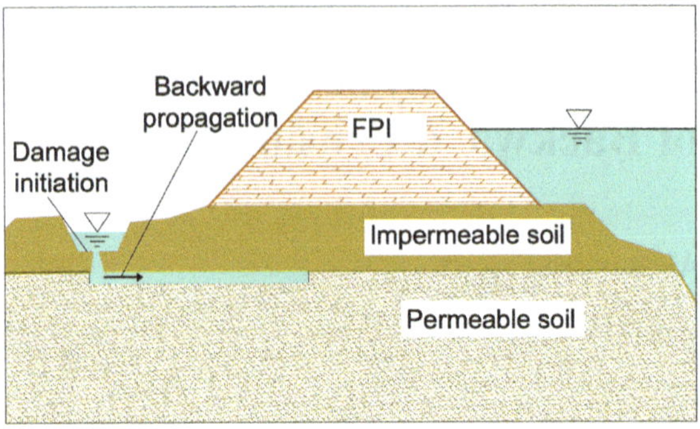

Fig. 4.1 Schematic description of the backward erosion piping phenomenon

failures [3], as it accounts for approximately one third of all the recorded failures over the last century [4]. As illustrated in Fig. 4.1, the phenomenon generally shows very little visual precursory signs, which makes it hard to contrast in a timely fashion. Moreover, the complexity associated with the prediction of its evolution lies in the fact that a correct description of BEP requires multi-physics mathematical models to account for the soil mechanics, sediment transport, and water flow.

The traditional approaches for the design of FPI against BEP generally rely on the quantification of the expected vertical hydraulic gradient in the downstream portion of the structure, providing geometrical conditions to limit the likelihood of BEP initiation and progression. Among these methods, the most widely used are Bligh's [5, 6] and Terzaghi's methods [7], as well as the well-known blanket theory developed by the USACE [8]. More recently, Sellmeijer proposed an analytical method to evaluate the critical conditions for BEP progression to occur in the presence of the so-called sand boils (i.e., local damage at the downstream side of the FPI) [9], which was complemented by further experimental tests in subsequent investigations [10]. Other available methods also include empirical considerations, based on available experimental and field data, which are used together with the geometry parameters of the structure in exam to assess the likelihood of BEP evolution in the system [11, 12].

Several experimental works are available in literature, which aimed at investigating the influence of different geometrical and mechanical parameters on the evolution of the phenomenon. Predicting the evolution of BEP is, in fact, a complex task. A first set of studies investigated the effect of the soil characteristics and stress state in laboratory settings to quantify the likelihood of initiation of erosion [13–15], or its evolution under different flow conditions [16, 17]. A comprehensive experimental campaign at different scales was carried out at the Ijkdijk facilities in the Netherlands, where the research team extensively studied BEP and internal erosion mechanisms at different length-scales [18, 19], with the ultimate goal of defining optimal sensor placement and enable early warning systems to be deployed.

Such experimental studies are often complemented by computational approaches, in view of the complex geometrical and mechanical conditions to be quantitatively defined, as well as the need for effective parametrization of the problem.

The numerical efforts available in literature can generally be subdivided in three macro-categories:

1. Permeability-based models, in which the progression of erosion is modeled by adjusting the apparent hydraulic conductivity of the soil [20–23];
2. Coupled DEM-Flow models, in which the fluid phase and the soil skeleton are simulated separately, while their interactions are modeled by means of ad hoc coupling schemes [24, 25];
3. Multiphase models, in which the evolution over time of the different phases (i.e., solid particles, interstitial water and the fluidized fraction of the soil) is modeled by gradient-based constitutive relationships, and the coupling is imposed by means of mass balance equations [26–31].

This manuscript presents the results of a recent numerical investigation that leverages both deterministic physics-based and data-based approaches to define a multi-scale framework for the resilience assessment of FPI. The local-scale physics simulations are based on a multiphase description of the computational domain. In view of the computational cost associated with these local-scale simulations, a large dataset of numerical results is used to train a multilayer machine learning model, to allow for quasi-real-time predictions. This capability is leveraged to explore the parameters hyperspace and perform a full sensitivity analysis of the numerical model, which is a prohibitive task for the numerical approach. Moreover, the multiscale approach was designed to be capable of accepting live data obtained from in situ and remote sensors, and such capability is demonstrated through a projected real-life case, where a large portion of a selected FPI is analyzed over the span of one year.

4.2 Physics-Based Model

The deterministic physics-based model employed at the local scale describes erosion as a mass transfer between three different constituents (i.e., the soil skeleton, the pore water and the mobilized- or *fluidized*-particles thought of as suspended in the pore water). Firstly, a nonlinear diffusion equation is employed to model the flow of the groundwater [20, 22, 23, 32]:

$$\frac{\partial h(\mathbf{x},t)}{\partial t} = \nabla \cdot (D(h(\mathbf{x},t))\nabla h(\mathbf{x},t)) \qquad \mathbf{x} \in \Omega\,, t \in (0,T) \tag{4.1}$$

with Dirichlet and Neumann boundary conditions:

$$\begin{aligned} h &= h_b(t) \qquad \text{on } \Gamma_b \subset \partial\Omega \\ \mathbf{q} \equiv -D\frac{\partial h}{\partial \mathbf{n}} &= \mathbf{q}_b \qquad \text{on } \Gamma_q \subset \partial\Omega \end{aligned} \tag{4.2}$$

In the previous equations, Ω represents the computational domain, h is the hydraulic head field, T is the total time, and $D = k/S_S$ is the diffusivity coefficient, which is a measure of the transmissivity of the soil (k is the soil permeability, and S_S its specific storage). Based on previous works available in literature who investigated the mobilization of soil particles as a result of local hydraulic gradients [33, 34], the proposed mass balance equation for the fluidized particles reads:

$$\frac{\partial(\gamma(\mathbf{x},t)\phi(\mathbf{x},t))}{\partial t} + \frac{\partial(\gamma(\mathbf{x},t)\mathbf{q}(\mathbf{x},t))}{\partial \mathbf{x}} = \frac{\partial \phi(\mathbf{x},t)}{\partial t} \tag{4.3}$$

in which γ is the concentration of fluidized particles (i.e., the ratio between the volume of the particles and the total pore space), and ϕ is the porosity. In Eq. 4.3, the right hand side represents the change in porosity over time, which is equal to the ratio of the mass generation term $\dot{m}$ and the particle density ρ_s. Based on works available in literature which investigated different erosion problems [33–36] as well as common traits identified in the description of BEP in literature, the following constitutive relation is proposed:

$$\frac{\dot{m}}{\rho_s} = \frac{\partial \phi(\mathbf{x},t)}{\partial t} = \begin{cases} 0 & \text{if} \quad |\mathbf{q}| \leq q_{crit} \\ \lambda\gamma\sqrt{1 - \frac{\phi^2}{\phi_{max}^2}}(q_{crit} - |\mathbf{q}|) & \text{if} \quad |\mathbf{q}| > q_{crit} \end{cases} \tag{4.4}$$

In the previous equation, the maximum allowable porosity value ϕ_{max} is considered to be a material property related to the amount of fine fraction in the soil, $\mathbf{i}$ represents the hydraulic gradient (i.e., the gradient of the hydraulic head field), i_{crit} is the value of such gradient at onset of erosion, and λ is a parameter that can be linked to the soil microstructure [34]. The value chosen for i_{crit} has a strong influence on the numerical results and needs to be validated by means of appropriate experimental calibration. Lastly, the model is fully defined by introducing the Darcy and Kozeny–Carman equations:

$$\mathbf{q} = -k\mathbf{i} \tag{4.5}$$

$$k = c_{\text{kc}} \frac{\phi^3}{(1-\phi)^2} \tag{4.6}$$

in which c_{kc} represents the Kozeny-Carman parameter.

To solve the described system of equations, the response field is discretized along the one-dimensional domain represented by the erosion path (see Fig. 4.1). A mid-point finite difference integration is used for space discretization, and a forward difference algorithm is employed to discretize time. Accuracy of the proposed approach was validated numerically by convergence analysis with mesh refinement for the case of highest admissible global gradient.

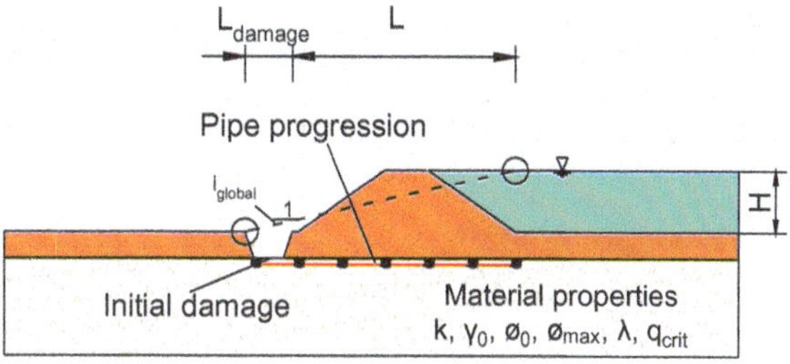

Fig. 4.2 Graphical description of the parameters of the physics-based model

4.2.1 *Model Parameters and Solution Strategy*

The response of the proposed physics-based model depends on the initial and boundary conditions for the simulations (length of the erosion path L, extent of the local downstream damage L_{damage}, global hydraulic gradient i_{global} – derived from the total hydraulic head), as well as the mechanical parameters governing the response of the cohesionless soil (initial and maximum porosity ϕ_0 and $\phi_{\max}$, permeability k, initial concentration γ_0, critical flux at onset of erosion q_{crit}, microstructure parameter λ), for a total of nine parameters to be quantified in the simulations. The relevance of such parameters and their physical description will be discussed together with the adopted ML-based multiscale strategy. The physical description of these quantities is given in Fig. 4.2. It is worth mentioning that it is possible to take into account the bi-dimensional groundwater flow by evaluating the distribution of the hydraulic head according to provisions given in the blanket theory [37] and assign such values at the computational nodes of the 1-dimensional erosion path considered in the model. Each physics-based simulation is considered to start at the onset of erosion evolution, which is initiated by a local damage at the downstream side of the FPI that causes an immediate drop in the hydraulic head [38].

The coupled system of equations presented in the previous is solved by means of the the computational algorithm reported in Fig. 4.3.

4.3 Data-Based Homogenization Scheme

Machine learning (ML) techniques for the solution of various engineering problems have steadily gained increasing attention over the last decades [39–44], as a result of increased efficiency in constructing and manipulating large datasets comprised of both real (measured) and synthetic (simulated) data. Properly defined ML models can represent an alternative to traditional deterministic methods for the solution of computationally expensive problems, particularly when the computational time of a given model does not allow to retrieve numerical solutions in a timely fashion to act on the problem under investigation. In this framework, the previously presented physics-based model was used to train an ad-hoc designed multilayer ML algorithm

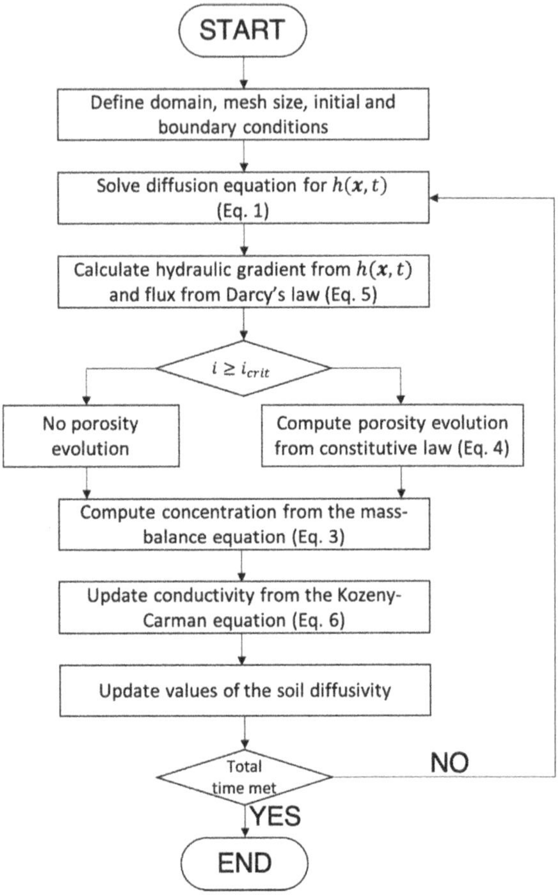

Fig. 4.3 Graphical description of the solution strategy used for the solution of the physics-based model

to perform both classification and regression, in order to bridge the local-scale information to the system scale of the FPI. The two layers of the ML model are constructed as follows: (1) the first layer is tasked with the classification of the current state of the system, based on a given set of values for the input parameters, (2) according to the results of such classification, a regression layer is activated to perform predictions on the response of the system and quantify the evolution of BEP. A graphical interpretation of the algorithm is given in Fig. 4.4.

Based on the available literature [45–47], and on considerations related to the expected composition of the synthetic dataset that will be used for the training of the different ML techniques in the multilayer model, in this research a support vector machine (SVM) classifier and a Gaussian process regression (GPR) model were used. Discussing the details of these two methods is beyond the scope of this manuscript, but the reader is referred to the full derivation of the model presented in [27].

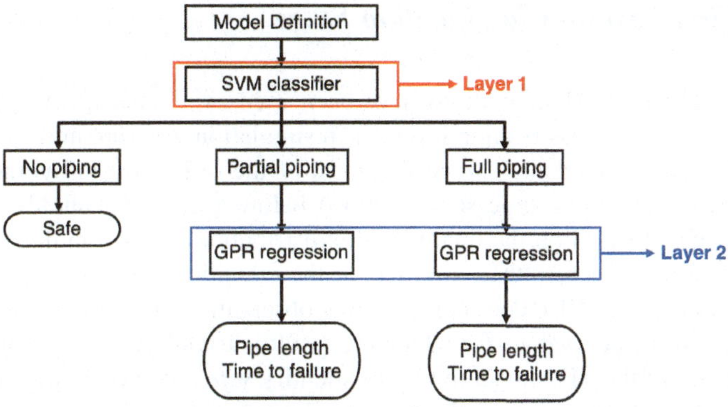

Fig. 4.4 Multilayer machine learning approach

Table 4.1 Definition of the model parameters and observation intervals

Parameter	Lower bound	Upper bound
Domain length [m]	10	20
Critical flux [m/s]	$2.5e^{-7}$	$8e^{-5}$
Microstructural parameter [m^{-1}]	$1e^{-5}$	$1e^{-4}$
Initial damage [m]	0	0.5
Global gradient [-]	0.01	0.225
Initial concentration [-]	0.001	0.01
Initial porosity [-]	0.35	0.45
Hydraulic conductivity [m/s]	$1e^{-6}$	$2e^{-4}$
Maximum porosity [-]	0.75	0.9

4.3.1 Data and Training

The ML model was trained by means of 7400 simulations performed by sweeping the entire parameters hyperspace using Monte-Carlo sampling with constant probability density functions for all the variables. The 9 parameters that uniquely define each local-scale simulation, together with the considered observation intervals, are reported in Table 4.1. The strategy employed for the definition of the ML-based multiscale model is the following: the proposed local-scale multiphase model was first validated by means of experimental data available in literature (both at the material and cross-sectional scales). Each simulation was run until steady-state conditions were attained. The numerical results were then used to train both the SVM and the GPR algorithms, using a 5-fold cross-validation in both cases to test the accuracy of the model.

4.3.2 First Layer—Classification Stage

The results of the 7400 simulations have been used as a dataset to train the support vector machine classification layer. Each simulation was run until steady-state conditions were attained (i.e., when the norm of the difference in hydraulic head between two consecutive time steps dropped below a given threshold) and automatically labeled based on the final porosity distribution attained in the mesh. The author also wants to stress that, in order for the classification model to be able to predict cases in which BEP does not occur, the observation intervals for the different variables must include values for which the result is trivial, resulting in no piping. The capability of the ML model to rapidly identify *safe* (i.e., no piping) conditions is, in fact, of vital importance in the correct solution of the problem, and therefore, relatively low values of the upstream hydraulic head and damage length must be included in the training set (see Table 4.1). Based on the possible outcomes of any given simulation, three classes were defined and assigned to each parameters vector: (1) *no piping*, when BEP does not initiate as a result of lower hydraulic gradients and/or contained (or null) damage length; (2) *partial piping*, when BEP initiates and progresses until an equilibrium condition is met and the pipe does not fully develop; (3) *full piping*, when the given hydraulic and geometrical conditions allow for a full development of BEP in the domain.

4.3.3 Second Layer—Regression Stage

In the second stage of the proposed multilayer model, all the data points labeled as *partial* and *full* piping were used as a training set for the Gaussian process regression. For each simulation in this dataset, the following two parameters were evaluated: (1) the total length of the erosion path (which is equal to the total domain length for the case of full piping, or the furthest point undergoing erosion in the case of partial piping) normalized by the total length of the domain; (2) the time at which steady-state conditions were attained. The ML prediction is then used in the data-based homogenization technique discussed in the following.

4.3.4 Data-Based Homogenization

The results obtained from the previously described multilayer model are then utilized to quantify the system-scale response of FPI systems in a computationally efficient manner. In this context, the infrastructure is discretized along its centerline, and each computational point is thought of as a cross-section that can undergo BEP (such cross-section need not be constant throughout the length of the system). At each computational time step, the data-based predictions are obtained from the multilayer model, and the current eroded length is computed at each cross-section:

$$L_{\text{bep}}(t = n_t \text{d}t) = \sum_{i=1}^{n_t} v_{\text{bep}}(i) \cdot \text{d}t \tag{4.7}$$

where i is the loop variable for the time steps, n_t the current time step counter, dt the given computational time interval, and v_{bep} is a measure of the averaged speed of BEP evolution, computed as follows:

$$v_{\text{bep}}(i) = \begin{cases} 0 & \text{if case} = \text{'no piping'} \\ \frac{L_\text{p}}{t_\text{p}} & \text{if case} = \text{'partial' } \& \, L_{\text{bep}} < L_\text{p} \\ 0 & \text{if case} = \text{'partial' } \& \, L_{\text{bep}} \geq L_\text{p} \\ \frac{L}{t_\text{f}} & \text{if case} = \text{'full'} \end{cases} \tag{4.8}$$

where L_p and t_p are the regression estimates for the length and time to failure for *partial* piping and t_f is the regression estimate time to failure obtained from the full piping cases. Based on the results of the full-scale model, a synthetic measure for the safety factor against BEP is proposed as follows:

$$\text{SF}_{\text{bep}}(\mathbf{x}, t) = 1 - \frac{L_{\text{pipe}}(\mathbf{x}, t)}{L(\mathbf{x})} \tag{4.9}$$

Such safety factor, defined as a function of time and position along the system domain, is equal to 1 when the structure is undamaged, and 0 when BEP has fully developed.

4.4 Numerical Results

The capabilities of the proposed multiscale computational framework need to be evaluated both at the *local* scale (i.e., that of the multiphase model) and the *global* scale (i.e., the system level). For this reason, the multiphase model was first validated by means of different experimental results available in literature. After the accuracy of the local-scale model has been assessed, the system-scale model was validated by comparing results of the multiphase model with the data-driven predictions. Once the model was validated at both scales, a real-life scenario was simulated, which focused on the assessment of a portion of the Nashville Metro Levee System against two fictitious scenarios, to prove capabilities and scalability of the proposed model.

4.4.1 Physics-Based Model Validation

The multiphase local-scale model was validated by means of available data obtained from the experimental campaign performed at the IJkdijk facilities [18, 48, 49] on four full-scale levees composed of soils with increasing hydraulic conductivity. The experimental setup for the specific tests used in such validation is reported in Fig. 4.5, while Table 4.2 presents the reported soil properties used in the simulations.

The experimental tests were conducted by increasing the upstream hydraulic head at a constant rate until BEP was observed. The reported water level at which piping started developing was defined as the critical hydraulic head for the specific soil in exam. The computational approach for the simulation of the tests was the following: all the parameters reported in [18] were used *as is*, while the remaining parameters required for the simulations were obtained by best fit with the first test (i.e., the one with the lowest soil permeability), as reported in Table 4.3. Such parameters were then kept constant in the simulations of all the remaining tests. The numerical results reported in Fig. 4.6 exhibit a substantial agreement with the experimental observations, proving that the multiphase model is capable of predicting BEP progression at the cross-sectional scale of the FPI.

4.4.2 Data-Based Model Validation

In this section, the accuracy metrics for the proposed ML model are presented first, in terms of confusion matrix (for the classifier layer) and scatter diagram (for the regression layer). Regarding the classification stage, the composition of the dataset was as

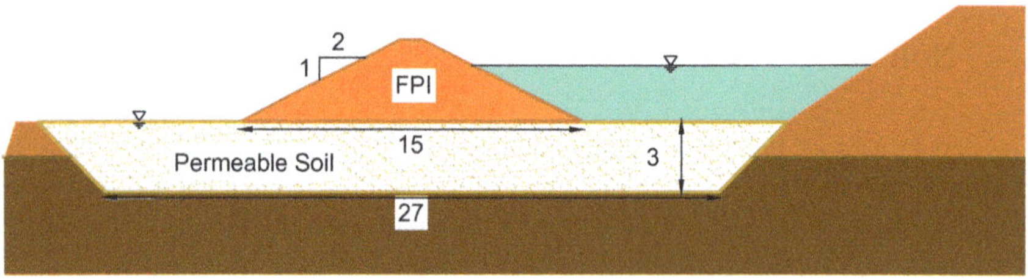

Fig. 4.5 Geometrical arrangement of the full-scale Ijkdijk tests (all dimensions in m)

Table 4.2 Reported soil parameters for the IJkdijk experimental campaign [19]

Type [-]	d_{70} [m]	d_{60}/d_{10} [-]	ϕ [-]	k [m/s]
Fine sand	$1.8e^{-4}$	1.6	0.35–0.45	$8e^{-5}$
Coarse sand	$2.6e^{-4}$	1.8	0.34–0.45	$1.2–1.4e^{-4}$

Table 4.3 Model parameters used in the local model validation

Experiment [-]	L [m]	q_{crit} [m/s]	λ [m^{-1}]	L_{damage} [m]	γ_0 [-]	k [m/s]	ϕ_0 [-]	ϕ_{max} [-]
1	15	$2.72e^{-5}$	$8e^{-5}$	$1e^{-2}$	$9e^{-3}$	$8e^{-5}$	0.4	0.8
2	15	$3.06e^{-5}$	$8e^{-5}$	$1e^{-2}$	$9e^{-3}$	$8e^{-5}$	0.4	0.8
3	15	$4.08e^{-5}$	$8e^{-5}$	$2e^{-2}$	$9e^{-3}$	$8e^{-5}$	0.39	0.8
4	15	$4.76e^{-5}$	$8e^{-5}$	$2e^{-2}$	$9e^{-3}$	$8e^{-5}$	0.39	0.8

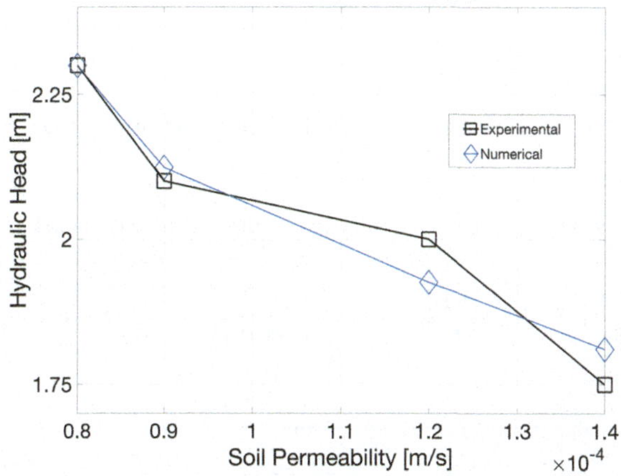

Fig. 4.6 Comparison between experimental and numerical results for the local-scale model validation (adapted from [27])

follows: 4881 data points pertained *no piping* cases, 441 to *partial piping*, and 2078 to *full piping* cases, respectively. The imbalance in the size of the different classes is a direct consequence of the fact that the results of the local-scale simulations are not known a-priori, and the different cases are the results of the sampling performed on the input variables. The results reported in Fig. 4.7 show that the proposed quadratic support vector machine exhibits an overall accuracy greater than 90%. The author also wants to stress the point that the most important feature of such classifier is the capability of discerning *no piping* cases from the other two groups, in which the SVM performs with accuracy greater than 90%.

Figure 4.7 also reports the scatter plot of the ground truth (i.e., the multiphase numerical results) against the ML predictions for the speed of BEP progression of all the numerical simulations in the testing data (10% of total dataset size).

After the accuracy of the multilayer ML model has been established, validation of the global scale model was performed by comparing the multiphase simulation results and the ML predictions for a typical case of BEP progression. Particularly, 200 simulations were performed to evaluate the influence of the porosity of the soil on the time required for BEP to fully develop. By using the parameters reported in Table 4.4, the two models (i.e., the physics-based multiphase model and the multilayer ML

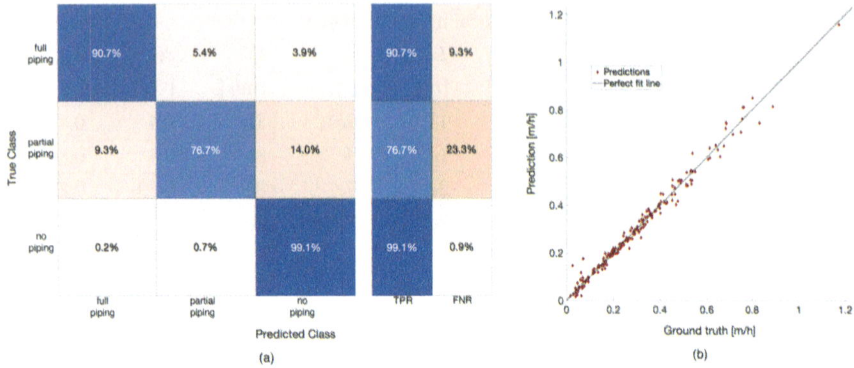

Fig. 4.7 (**a**) Support vector machine test confusion matrix and (**b**) Gaussian process regression performance plot

Table 4.4 Model parameters used in the validation of the system scale model

q_{crit} [m/s]	λ [m^{-1}]	γ_0 [-]	ϕ_0 [-]	ϕ_{max} [-]	i_{global} [-]
$2.72e^{-5}$	$8e^{-5}$	$9e^{-3}$	0.35–0.45	0.85	0.2043

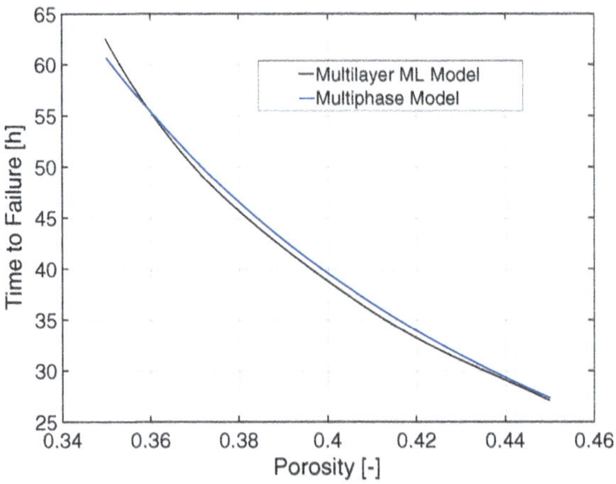

Fig. 4.8 Influence of the initial porosity on the time to failure: numerical model versus ML model predictions (adapted from [27])

model) were run using an Intel i7 desktop computer. Figure 4.8 shows the excellent agreement between the local-scale model and the ML predictions, with the machine learning model only requiring $\sim$ 3.2 s of runtime versus the $\sim$ 58 min required for the multiphase simulations, for a three orders of magnitude speedup obtained by employing the data-based model.

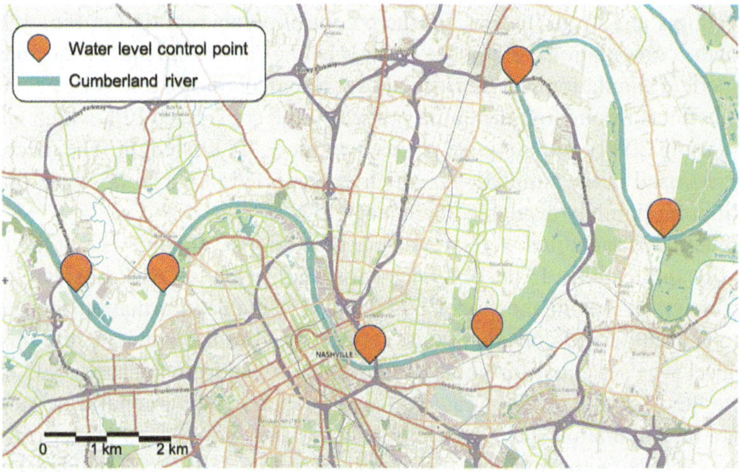

Fig. 4.9 Planar view of the simulated portion of the Nashville Metro Levee System on the Cumberland river

4.5 Full-Scale Model and Computational Scenarios

Having validated both the multiphase local model and the multilayer ML model at different scales, the capabilities of the approach proposed herein are demonstrated by performing numerical simulation of the Nasvhille Metro Levee System (see Fig. 4.9). This portion of the FPI has suffered extensive damage as a result of the severe flooding event that struck the city of Nashville in 2010 [50].

Simulating FPI systems at the structural scale of the infrastructure (i.e., 10^5 m scale) is a task associated with multiple uncertainties and difficulties. While the planar arrangement of the infrastructure follows the riverbed closely and can therefore be retrieved from widely available satellite data, the three-dimensional geometry of the structure (i.e., the variation of the cross-sectional dimensions along the centerline of the FPI) is a complex task that requires accurate quantification of different measures (e.g., the crown height and base width of the levee, upstream, and downstream slopes). The geometry of the specific FPI section in exam was retrieved by means of a remote sensing campaign performed by means of both laser-based (i.e., LiDAR) and RGB (i.e., photogrammetry) information. Details on such effort can be found in previous work by the author [27].

One of the most important parameters in the assessment of the FPI against BEP is the upstream water level time history. This measure, in fact, has direct influence on the global hydraulic gradient that drives the erosion initiation and progression. The capability of the proposed multiscale model of accepting live data received from embedded and remote sensors is demonstrated by utilizing the measured water level tables from the Cumberland river that are publicly available through the USGS National Water Information System [51]. This web-based interface provided water level readings at 15 min interval for the 6 control points reported in Fig. 4.9. As a result of the computational efficiency granted by the data-based multiscale approach, the numerical simulation was performed over a span of one year to fully account for seasonal effects throughout the entire portion of the considered FPI, which was discretized using 2000 computational points over the $\sim$ 47 km length. The mechanical parameters used in the simulations, which are reported in Table 4.5, were obtained from different publicly available documents [52–54].

Two fictitious scenarios were simulated, pertaining to conditions that are typical of real-life applications. In the first one, a normal distribution was considered for all the parameters governing the model, based on the available information on the infrastructure and numerical considerations. As described in the previous, the upstream water level was set equal to the measured quantity, and this measure was used to compute the global hydraulic gradient i_{global} at each computational time step. The results of the simulation show that the spatial variability of the parameters results in a highly variable factor of safety distribution (see Fig. 4.10), with the development of different localization areas.

In the second scenario, all the parameters were kept constant and equal to their mean value (see Table 4.5), except for one portion of the FPI which was idealized as experiencing high levels of downstream damage. This fictitious scenario could represent the real-life case in which an inspection reveals the presence of extensive damage in localized area of the infrastructure. The results show that the proposed multiscale model is able to correctly enforce the evolution of BEP in the areas interested by the initial local damage, with values of $\mathrm{SF_{bep}}$ approaching the 0 (i.e., fully developed BEP) at the end of the simulation, as reported in Fig. 4.11.

Table 4.5 Values of the parameters used in the system scale simulations

Parameter	Mean Value (Both Scenarios)	Distribution (Scenario 1)	Parameters (Scenario 1)	Variability (Scenario 2)
L [m]	16.1	Gaussian	$\sigma = 0.2$	–
q_{crit} [m/s]	$2.75e^{-5}$	Gaussian	$\sigma = 2.5\mathrm{e}^{-2}$	–
λ [m^{-1}]	$3e^{-4}$	Gaussian	$\sigma = 1\mathrm{e}^{-5}$	–
L_{damage} [m]	0.4	Weibull	$a = 5e^{-2}$, $b = 0.25$	0.75 (localized)
γ_0 [-]	0.005	Gaussian	$\sigma = 1\mathrm{e}^{-3}$	–
k [m/s]	$1e^{-4}$	Gaussian	$\sigma = 1\mathrm{e}^{-4}$	–
ϕ_0 [-]	0.35	Gaussian	$\sigma = 1\mathrm{e}^{-2}$	–
ϕ_{max} [-]	0.85	Gaussian	$\sigma = 1\mathrm{e}^{-1}$	–

Fig. 4.10 Resulting safety factor distribution for Scenario 1

Fig. 4.11 Resulting safety factor distribution for Scenario 2

4.6 Conclusions

In this manuscript, a review of recently developed data- and physics-based approaches has been presented, in the context of simulating backward erosion piping in geotechnical infrastructure. First, a novel multiphase numerical model was discussed for the solution of the local-scale problem. Such model was employed to construct a rich dataset of numerical results for the training of a machine learning model capable of predicting the response of the infrastructure at the system scale. The ML approach is based on a multilayer strategy, in which the first stage is used to classify the current state of the system and the possibility of BEP to progress by means a support vector machine, while the second uses Gaussian process regression to quantify to extent of the damage. The efficiency of the approach was demonstrated by analyzing a large portion of the Nashville Metro Levee System over one year. Results of two virtual scenarios show that the model is inherently capable of accepting live data streaming from embedded sensors and predict the response of the system in a computationally efficient fashion. The multiscale framework can, therefore, be embedded in resilience and risk analysis platforms, providing an efficient tool for the quasi-real-time assessment of FPI systems. The author wishes to emphasize that the scenarios used for the case study on the Nashville Metro Levee System are entirely fictional and do not indicate areas of increased flood risk.

References

1. Ward PJ, Jongman B, Aerts JC, Bates PD, Botzen WJ, Loaiza AD, Hallegatte S, Kind JM, Kwadijk J, Scussolini P et al (2017) A global framework for future costs and benefits of river-flood protection in urban areas. Nat Clim Change 7(9):642–646
2. Jordan P, Manojlovic N, Fröhle P (2019) Maintenance of flood protection infrastructure in the North Sea region—an analysis of existing maintenance strategies. Coast Struct 2019:654–662
3. Foster M, Fell R, Spannagle M (2000b) The statistics of embankment dam failures and accidents. Can Geotech J 37(5):1000–1024
4. Richards KS, Reddy KR (2007) Critical appraisal of piping phenomena in earth dams. Bull Eng Geol Environ 66(4):381–402
5. Bligh W (1910) Dams, barrages and weirs on porous foundations. Eng News 64(26):708–710
6. Lane EW (1934) Security from under-seepage masonry dams on earth foundations. Proc Am Soc Civ Eng ASCE 60:929–966
7. Terzaghi K, Peck RB, Mesri G (1996) Soil mechanics in engineering practice. Wiley
8. USACE (1956) Investigation of underseepage and its control. Lower Mississippi Rivers, Vicksburg, MS
9. Sellmeijer JB (1988) On the mechanism of piping under impervious structures. Technical University of Delft
10. Sellmeijer H, de la Cruz JL, van Beek VM, Knoeff H (2011) Fine-tuning of the backward erosion piping model through small-scale, medium-scale and ijkdijk experiments. Eur J Environ Civ Eng 15(8):1139–1154
11. Foster M, Fell R, Spannagle M (2000a) A method for assessing the relative likelihood of failure of embankment dams by piping. Can Geotech J 37(5):1025–1061
12. Schmertmann JH (2000) The no-filter factor of safety against piping through sands. In: Judgment and innovation: the heritage and future of the geotechnical engineering profession. American Society of Civil Engineers, pp 65–132
13. Bendahmane F, Marot D, Alexis A (2008) Experimental parametric study of suffusion and backward erosion. J Geotech Geoenviron Eng 134(1):57–67
14. Richards K, Reddy KR (2012) Experimental investigation of initiation of backward erosion piping in soils. Géotechnique 62(10):933
15. Sharif YA, Elkholy M, Hanif Chaudhry M, Imran J (2015) Experimental study on the piping erosion process in earthen embankments. J Hydraul Eng 141(7):04015012
16. Fleshman MS, Rice JD (2014) Laboratory modeling of the mechanisms of piping erosion initiation. J Geotech Geoenviron Eng 140(6):04014017
17. Ke L, Takahashi A (2014) Triaxial erosion test for evaluation of mechanical consequences of internal erosion. Geotech Test J 37(2):347–364

18. van Beek VM, Knoeff H, Sellmeijer H (2011) Observations on the process of backward erosion piping in small-, medium- and full-scale experiments. Eur J Environ Civ Eng 15(8):1115–1137
19. De Vries G, Koelewijn A, Hopman V (2010) Ijkdijk full scale underseepage erosion (piping) test: evaluation of innovative sensor technology. In: Scour and erosion, pp 649–657
20. Fascetti A, Oskay C (2019a) Dual random lattice modeling of backward erosion piping. Comput Geotech 105:265–276. https://doi.org/10.1016/j.compgeo.2018.08.018
21. Rahimi M, Shafieezadeh A (2020) Coupled backward erosion piping and slope instability performance model for levees. Transp Geotech 24:100394
22. Robbins B (2016) Numerical modeling of backward erosion piping. In: Applied numerical modeling in geomechanics, pp 551–558
23. Vandenboer K, van Beek V, Bezuijen A (2014) 3D finite element method (FEM) simulation of groundwater flow during backward erosion piping. Front Struct Civ Eng 8(2):160–166. https://doi.org/10.1007/s11709-014-0257-7
24. El Shamy U, Aydin F (2008) Multiscale modeling of flood-induced piping in river levees. J Geotech Geoenviron Eng 134(9):1385–1398
25. Wang Y, Ni X (2013) Hydro-mechanical analysis of piping erosion based on similarity criterion at micro-level by PFC3D. Eur J Environ Civ Eng 17(sup1):s187–s204
26. Bi J, Zhang H, Luo X, Shen H, Li Z (2021) Modeling of internal erosion using particle size as an extra dimension. Comput Geotech 133:104021
27. Fascetti A, Oskay C (2019b) Multiscale modeling of backward erosion piping in flood protection system infrastructure. Comput-Aided Civ Infrastruct Eng 34(12):1071–1086
28. Fujisawa K, Murakami A, Nishimura S (2010) Numerical analysis of the erosion and the transport of fine particles within soils leading to the piping phenomenon. Soils Found 50(4):471–482
29. Luo Y (2013) A continuum fluid-particle coupled piping model based on solute transport. Int J Civ Eng 11(1B):38–44
30. Wang DY, Fu XD, Jie YX, Dong WJ, Hu D (2014) Simulation of pipe progression in a levee foundation with coupled seepage and pipe flow domains. Soils Found 54(5):974–984
31. Zhang X, Wong H, Leo CJ, Bui TA, Wang J, Sun W, Huang Z (2013) A thermodynamics-based model on the internal erosion of earth structures. Geotech Geol Eng 31(2):479–492
32. Hagerty D, Curini A (2004) Impoundment failure seepage analyses. Environ Eng Geosci 10(1):57–68
33. Stavropoulou M, Papanastasiou P, Vardoulakis I (1998) Coupled wellbore erosion and stability analysis. Int J Numer Anal Methods Geomech 22(9):749–769
34. Vardoulakis I, Stavropoulou M, Papanastasiou P (1996) Hydro-mechanical aspects of the sand production problem. Transp Porous Media 22(2):225–244
35. Sakthivadivel R, Irmay S (1966) A review of filtration theories. University of California, College of Engineering, Hydraulic Engineering Laboratory
36. Vardoulakis I, Papanastasiou P, Stavropoulou M (2001) Sand erosion in axial flow conditions. Transp Porous Media 45(2):267–280
37. US Army Corps of Engineers (2000) Design and construction of levees
38. De Wit J, Sellmeijer J, Penning A (1981) Laboratory testing on piping. In: Proceedings of the 10th international conference on soil mechanics and foundation engineering, Stockholm, pp 15–19
39. Adeli H (2001) Neural networks in civil engineering: 1989–2000. Comput-Aided Civ Infrastruct Eng 16(2):126–142
40. Bogdanor MJ, Oskay C, Clay SB (2015) Multiscale modeling of failure in composites under model parameter uncertainty. Comput Mech 56(3):389–404. https://doi.org/10.1007/s00466-015-1177-7
41. Nabian MA, Meidani H (2018) Deep learning for accelerated seismic reliability analysis of transportation networks. Comput-Aided Civ Infrastruct Eng 33(6):443–458
42. Rafiei MH, Adeli H (2018) A novel unsupervised deep learning model for global and local health condition assessment of structures. Eng Struct 156:598–607
43. Reich Y (1997) Machine learning techniques for civil engineering problems. Comput-Aided Civ Infrastruct Eng 12(4):295–310

44. Zhang X, Oskay C (2017) Material and morphology parameter sensitivity analysis in particulate composite materials. Comput Mech 62(3):543–561. https://doi.org/10.1007/s00466-017-1512-2
45. Dong L, Wesseloo J, Potvin Y, Li X (2016) Discrimination of mine seismic events and blasts using the Fisher classifier, Naive Bayesian classifier and logistic regression. Rock Mech Rock Eng 49(1):183–211
46. Lasisi A, Attoh-Okine N (2019) Machine learning ensembles and rail defects prediction: multilayer stacking methodology. ASCE-ASME J Risk Uncertainty Eng Syst Part A Civ Eng 5(4):04019016
47. Loh WY (2011) Classification and regression trees. Wiley Interdiscip Rev Data Min Knowl Discov 1(1):14–23
48. van Beek V, Bezuijen A, Sellmeijer H (2013) Backward erosion piping. In: Erosion in geomechanics applied to dams and levees, pp 193–269
49. Koelewijn A, De Vries G, Van Lottum H, Förster U, Van Beek V, Bezuijen A (2014) Full-scale testing of piping prevention measures: three tests at the ijkdijk. G & White. In: Physical modeling in geotechnics. Taylor & Francis Group, London
50. Moore BJ, Neiman PJ, Ralph FM, Barthold FE (2012) Physical processes associated with heavy flooding rainfall in Nashville, Tennessee, and vicinity during 1–2 May 2010: the role of an atmospheric river and mesoscale convective systems. Mon Weather Rev 140(2):358–378
51. Goodall JL, Horsburgh JS, Whiteaker TL, Maidment DR, Zaslavsky I (2008) A first approach to web services for the national water information system. Environ Model Softw 23(4):404–411
52. Metropolitan Nashville-Davidson County (2015) Metropolitan Nashville-Davidson County Multi-hazard mitigation plan
53. Nashville Area Metropolitan Planning Organization (2015) Building resilience: a climate adaptation plan
54. Nashville Metro Water Services (2013) Unified flood preparedness plan

Chapter 5
Kerala Floods 2018: Causative Factors that Transformed Single Event to Multi-hazard Disaster

Maneesha Vinodini Ramesh, V. C. Sudarshan, Geethu Thottungal Harilal, Balmukund Singh, Amritanand Sudheer, and Hari Chandana Ekkirala

5.1 Introduction

Kerala, a state in south-western India, also known as God's own country, situated between the Arabian Sea to the west and the Western Ghats to the east having a coastline of 580 km in length varies between 35 and 120 km in width [1]. Bestowed with natural beauty, lined with golden pristine beaches, mapped with criss crossing emerald blue backwaters, beautiful green mountain ranges which have sustained human life and habitat for centuries, home to some of the elephants, langur monkeys and tigers. A state where the monsoon is an annual event and people are accustomed to experiencing heavy rainfall for days at a stretch was surprised when it received more than its usual share of rainfall during the monsoon of 2018. The rain which started off as a single event initiated a multi-hazard event affecting lives of the people. The criss-crossing blue water channels turned brown and black, flooded the entire region and swept homes, lives and livestock, submerging large tracts of habitable land along its path [1–4].

The festive season Onam, which is usually filled with joy and happiness, for celebrating harvest season, turned one's feeling from joy into deep distress caused

M. V. Ramesh (✉) · V. C. Sudarshan · G. T. Harilal · B. Singh
Amrita Center for Wireless Networks and Applications (AmritaWNA), Amrita School of Engineering, Amritapuri, Amrita Vishwa Vidyapeetham, Kollam, India
e-mail: maneesha@am.amrita.edu; maneesha@amrita.edu

A. Sudheer
Department of Computer Science and Engineering, Amrita School of Engineering, Amritapuri, Amrita Vishwa Vidyapeetham, Kollam, India

H. C. Ekkirala
Amrita School for Sustainable Development, Amritapuri, Amrita Vishwa Vidyapeetham, Kollam, India

S. Kolathayar et al. (eds.), *Civil Engineering for Disaster Risk Reduction*, Springer Tracts in Civil Engineering, https://doi.org/10.1007/978-981-16-5312-4_5

by (i) loss, (ii) disappointment and (iii) non-measurable misfortune due to this large-scale destruction. This paper is a reflection on multi-hazard events due to a single rainfall event which in turn caused the Kerala floods of 2018.

5.2 Chronology of Events During Kerala Floods 2018

Kerala has two monsoon periods: (i) south-west monsoon and (ii) north-east monsoon [5]. During the south-west monsoon of 2018, the state was experiencing heavy rainfall from July 2018 to August 2018 [6, 7]. The populace was unaware of the expected multi-hazard event that may be initiated due to this continuous heavy rainfall. It was only on 8 August 2018 that people started experiencing flood like situation [8]. Water levels in the villages, towns and cities increased. Assumed to be a regular feature, communities did not respond even when the alerts were provided. People noticed water level increase due to the increase in rainfall and assumed that the water level would reduce soon. But the intensity of rain increased day by day from 1905 to 2400 mm/day [6, 7] from 8 August to 19 August which had an effect on lives of the populace of the state of Kerala. The chronology of events is shown in Fig. 5.2, affecting all the 14 district names of which are italicized in Fig. 5.1.

The analysis of the events described in Fig. 5.2 clearly indicates that the people were not aware of the gravity of the situation even though regular rainfall alerts were released from 29 July 2018 and flood alert due to dam opening from 9 August 2018 day. Additionally, the population was completely ignorant towards responses for the different phases of alerts as provided by Indian Meteorological Department (IMD) and the state government agencies [9]. This has made families remain in their homes even when the alerts were issued leading to increase in risk due to impending multi-events which has led to massive impact, evacuation and loss of life affecting all 14 districts of Kerala. This paper aims to understand the triggers and pathways to multi-hazard events which have impacted more than 10 million citizens with a death toll of more than 500 lives.

5.3 Multi-hazard Impact Due to Spatial and Temporal Overlap

During the last ten years, the state has received an average 2039.7 mm, cumulative rainfall during the south-west monsoon period, i.e. June to September [7, 10]. The month of August 2018 has experienced more than 200% departure from the normal rainfall. Districts of Idukki, Palakkad, Malappuram, Kottayam and Kollam observed more than 50% of rainfall in August 2018. However, the districts which have experienced the largest impact due to Kerala floods 2018 are Idukki, Ernakulam, Alappuzha, Pathanamthitta, Malappuram, Wayanad and Thrissur (Fig. 5.3). This indicates

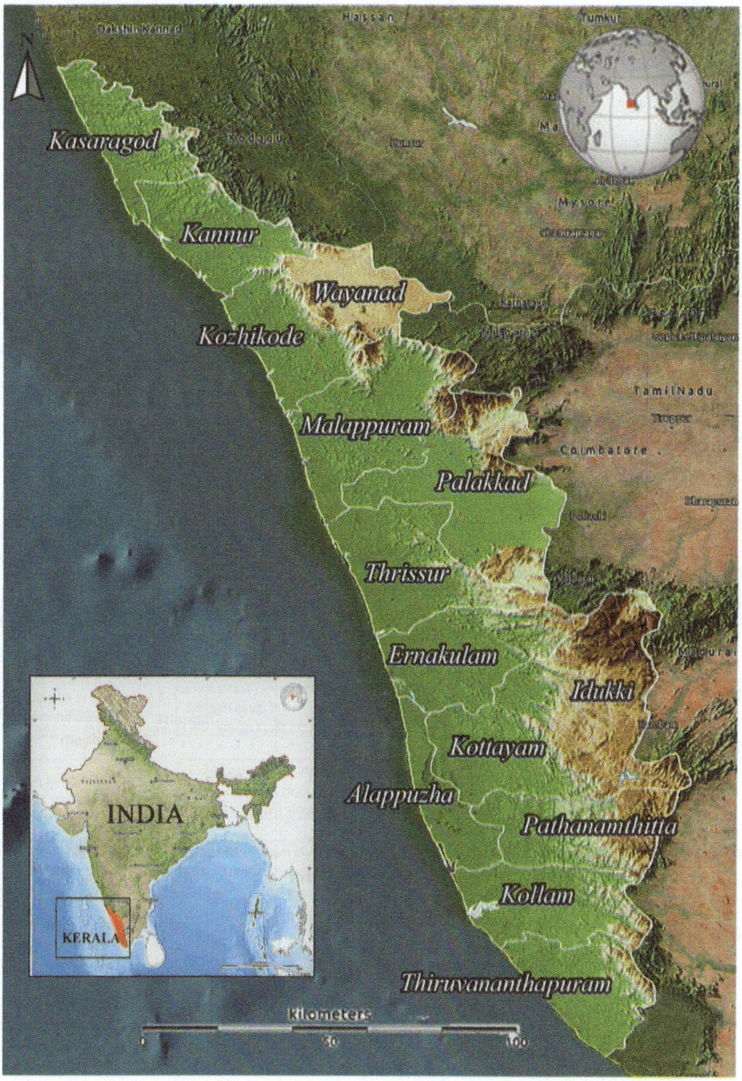

Fig. 5.1 Location map of Kerala

non-existence of a linear relationship between spatially distributed regions of rainfall departure and the impact of disaster, which reveals an existence of multi-hazard scenario.

Our analysis has shown that the continuous rainfall in the month of August 2018 has contributed to the initiation of landslides, flood, collapse of infrastructure, houses and ruination of agricultural lands. This occurrence has contributed to creation of combined multi-hazard events predominantly in Kannur, Kozhikode, Malappuram,

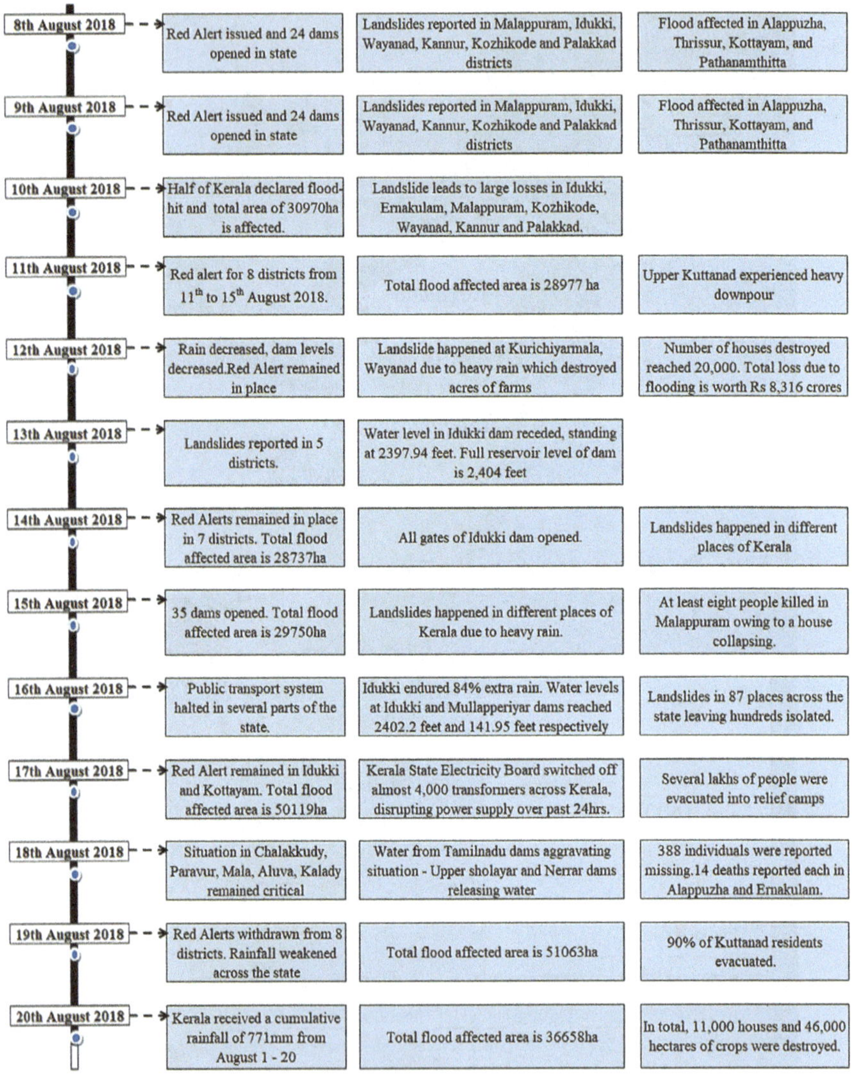

Fig. 5.2 Detailed events during Kerala flood 2018

Wayanad, Palakkad, Thrissur, Ernakulam, Idukki, Alappuzha and Pathanamthitta districts of Kerala. This has prompted us to study and evaluate spatio-temporal correlation of events such as (i) rainfall, (ii) floods, (iii) landslides and (iv) dam opening, leading to multi-hazard scenario.

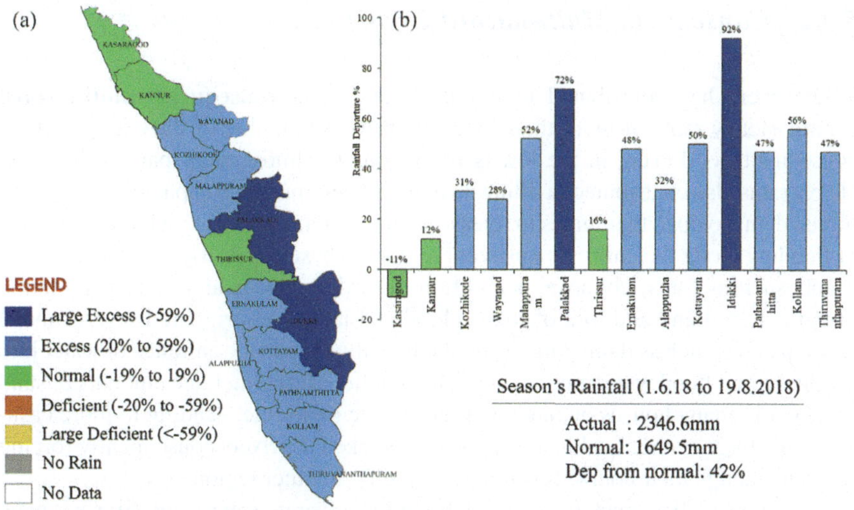

Fig. 5.3 Rainfall performance of Kerala during August 2018. **a** Rainfall departure percentage of each district in Kerala during August 2018. **b** District-wise rainfall departure percentage during 1.6.2018–19.8.2018

5.3.1 Single Event

During Kerala floods 2018, the state has experienced multiple phenomena like single event, combined multi-hazard event, consequent multi-hazard event, subsequent multi-hazard event and simultaneous multi-hazard event. The major **single event** experienced by the state is the heavy rainfall with unexpected rainfall departure. Table 5.1 indicates the south-west monsoon percentage departure of rainfall from normal at various time periods during 2018.

Table 5.1 Percentage of departure from normal for different time periods during south-west monsoon 2018 [6]

Period	Actual rainfall in mm	Normal rainfall in mm	% Departure from normal in mm
1 June 2018–30 September 2018	2515.7	2039.7	23
1 June 2018–19 August 2018	2346.6	1649.5	42
1 August 2018–19 August 2018	758.6	287.6	164
9 August 2018–19 August 2018	349.7	98.4	255

5.3.2 Consequent Multi-hazard Events

Single event had contributed to initiation of several **consequent multi-hazard events**. Heavy rains of more than 2515.7 mm for 80 number of days triggered the consequent flood event in the low-lying regions in almost every part of the state. Heavy rains also contributed in filling up dams to their threshold capacity [7, 10]. This forced dams to open their gates to release excess water, leading to an increase in area of flooding along the entire downstream regions. These floods triggered consequent events such as forest collapse, demolition of bridges, roads and houses, ruination of agricultural land and loss of lives. Flood and landslides triggered several social consequences such as damaging the medical facility due to the entrance of water into hospitals, and flood water entered up to the first floor bringing huge amounts of sand and mud in many houses damaging all electronic items, etc., leading to more casualties and huge economical loss. Landslides blocked the rivers consequently forcing rivers to change their course resulting in flooding in other regions.

In regions of Wayanad, heavy rainfall led to (i) rise in water level, (ii) river bank erosion and (iii) weak weathered geology of terrain that had previous human interventions. These led to a number of geohazards such as (i) landslide, (ii) land sinking and (iii) flood. Most of the high-range areas in the districts of Idukki, Wayanad, Malappuram and Kannur experienced the combined effect of landslide and flood worsening the conditions of this area. A large number of shallow and massive landslides were identified in all the districts, and a total of 474 new landslides were identified in Wayanad district alone as per DSC/NDEM Map no: 2018/03, NRSC/ISRO, Hyderabad, dated 27.09.2018.

Maximum cumulative rainfall experienced in the last week of July 2018 is less than 100 mm per day. Continuous rainfall from last week of July to the first week of August led to accumulation of water in dams, consequently leading to opening of dam gates on 9 August 2018. In the same day, rainfall has triggered consequent landslide events. From 6 August to 16 August, heavy increase in rainfall has led to all dams reaching their maximum threshold levels, causing the gates of dams to be opened, eventually leading to floods and landslides. The flood impact in areas around the dam is much less due to the fact that Idukki dam is an upstream dam (Fig. 5.4).

Similar chronology of event has taken place in the Idamalayar dam region; since this dam is located on the downstream, combined water release from Idukki and Idamalayar dam has caused huge impact starting from 9 August till 28 August 2018, with heavy flooding between 17 August and 20 August due to simultaneous opening of multiple gates of several dams as shown in Fig. 5.5.

5.3.3 Subsequent Multi-hazard Events

During the month of August 2018, several **subsequent multi-hazard** events were witnessed. Anthropogenic activities increased the incidences of multi-hazards.

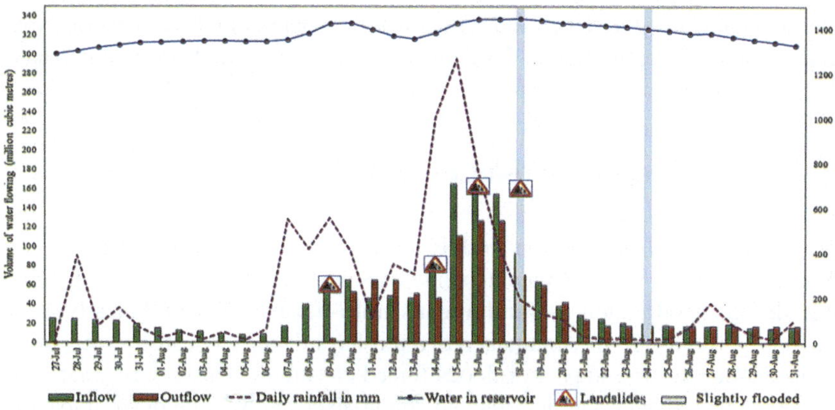

Fig. 5.4 Idukki dam opening timeline including daily rainfall, landslide events and flood conditions

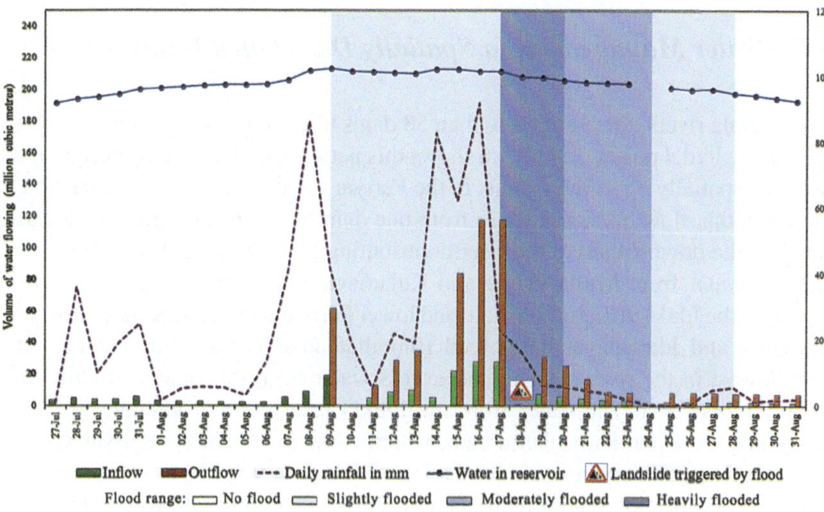

Fig. 5.5 Idamalayar dam opening timeline including daily rainfall, landslide events and flood conditions

Events such as (i) removal of toes of the hill or mountain altered to fit in human needs, along with subsequent rain event causes a consequent landslide event, which otherwise would not have occurred, (ii) sand mining in the river basin leads to change in ecology and river morphology [11] during subsequent heavy rainfall scenarios, the water overflows the shrunk river channel creating consequent flood scenario, (iii) quarrying of mountains for getting natural resources for human consumptions, weakens the hill slope stability increasing the vulnerability of landslides during subsequent heavy rainfall and (iv) urbanization of catchment areas reduces the spatial area for groundwater infiltration and water retention, increasing the risk of excess water flowing leading to floods and infrastructure collapse, all these scenarios were clearly visible during the Kerala floods 2018, increasing the multi-hazard impact.

5.3.4 *Simultaneous Multi-hazard Events*

As days progressed from last week of July to 8 of August, the disaster event has transitioned from single to consequent to simultaneous events of rainfall, dam opening, landslides and flood leading to **simultaneous multi-hazard**. Figure 5.5 clearly shows that from 9 August 2018, simultaneous multi-hazard impacts can be seen in the downstream areas. The impact of floods forced closure of transportation networks such as roads, railways and airports and simultaneous power shutdown impacting communication networks as well.

5.3.5 *Water Management in Spatially Distributed Dams*

Across Kerala rivers, there are more than 58 dams to feed the population with water and generate hydel power. Figure 5.6 shows this network of dams in the Periyar river basin. The spatially distributed dams in the Periyar river basin are interconnected in such a way that flow of excess water from one dam in the upper region feeds those located on the downstream of the river contributing to multi-hazard scenario.

Excess water from Mullaperiyar and Kulamavu dams fills up Idukki reservoir. Water from the Idukki/Cheruthoni dam and lower Periyar dam feeds Kallarkutty dam. Kallarkutty and Idamalayar dams feed Bhoothathankettu dam. Bhoothathankettu dam is lowest in the river system, and excess water from this dam will ultimately flow to Arabian Sea. The water from the Idukki reservoir and Idamalayar drains directly into the Bhoothathankettu dam, which is the lowest in the Periyar system, just 15 km from Kothamangalam town.

When the gates of Cheruthoni dam in Idukki are opened, water will flow to the Cheruthoni River reaching Aluva within six hours. Water eventually merges with Arabian Sea after flowing through dedicated channels in Ernakulam, Idamalayar, Malayattoor and Kalady. As a result, the Cheruthoni water channel on Thodupuzha–Puliyanmala state highway overflows. Water then flows through Thambiyad, Karimban 'chappaths' to enter Ernakulam district at Periyar Pamba dam, which in turn flows to Neriamangalam, Bhoothathankettu, Idamalayar, Malayattoor, Idamalayar, Malayattoor and Kalady. The flow of water from there will be via Aluva, Thuruth in Chengamanad panchayat, Thottumugham in Keezhmad panchayat, Mulavukad panchayat, Panambukad, Vallarpadam, Mulavukad and Ponnarimangalam.

The torrential rainfall caused the dams to rise their level leading to simultaneous opening of multiple gates of these dams due to which the spatial impact of flood and landslide has widened covering all the areas as shown in Fig. 5.7.

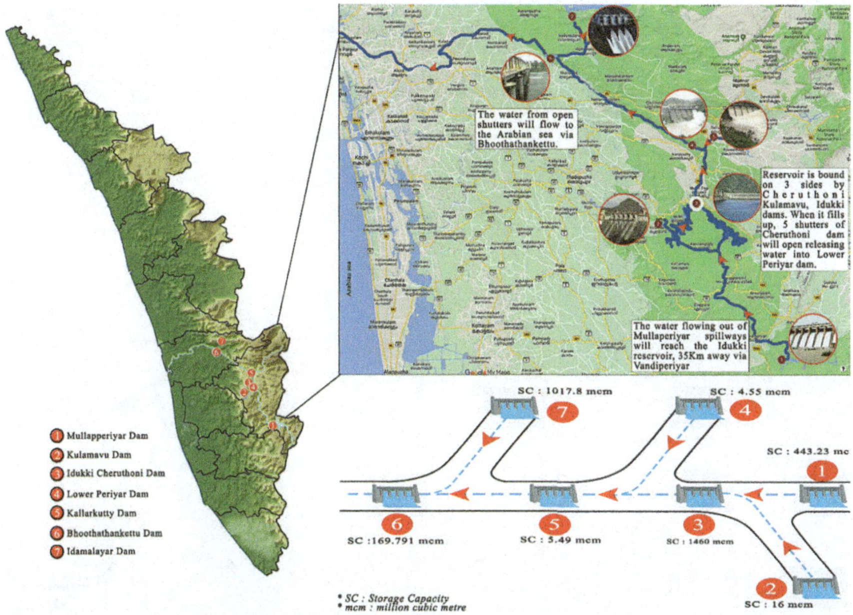

Fig. 5.6 Example of dam network: **a** location of 7 dams situated in Periyar river basin in Kerala map, **b** water flow direction of Periyar river and location of 7 dams in Idukki and Ernakulam districts, **c** dam interconnection of major dams in Idukki and Ernakulam

Considering Ernakulam, the impact due to flood was only covering the areas such as Aluva, Muvattupuzha and North Paravur. As the Idukki dam was opened on 9 August 2018, releasing 50 m^3 of water every second [12], within a few hours the flood risk increased covering additional areas such as Kalady, Perumbavoor, Nedumbassery, Angamali and Varappuzha increasing the impact by on people. This increased the coverage of risk by 20-folds. This impact has magnified since Idukki dam is networked with Kulamavu, Lower Periyar, Kallarkutty, Mullaperiyar, Bhoothathankettu and Idamalayar dams, leading to the opening of them from 9 to 14 August 2018. The analysis of this event alone shows that the current Kerala floods in 2018 are a multi-hazard event, consisting of single, combined, consequent, subsequent and simultaneous multi-hazard events. Even though it started as a single event triggered by rainfall, it transformed itself into a simultaneous event leading to multi-hazards.

Indeed, the event analysis for the period between 8 August 2019 and 20 August 2018 clearly depicts the interrelationship, where the single event of rainfall transformed into consequent, subsequent and simultaneous events which led to Kerala floods 2018 as shown in Fig. 5.8.

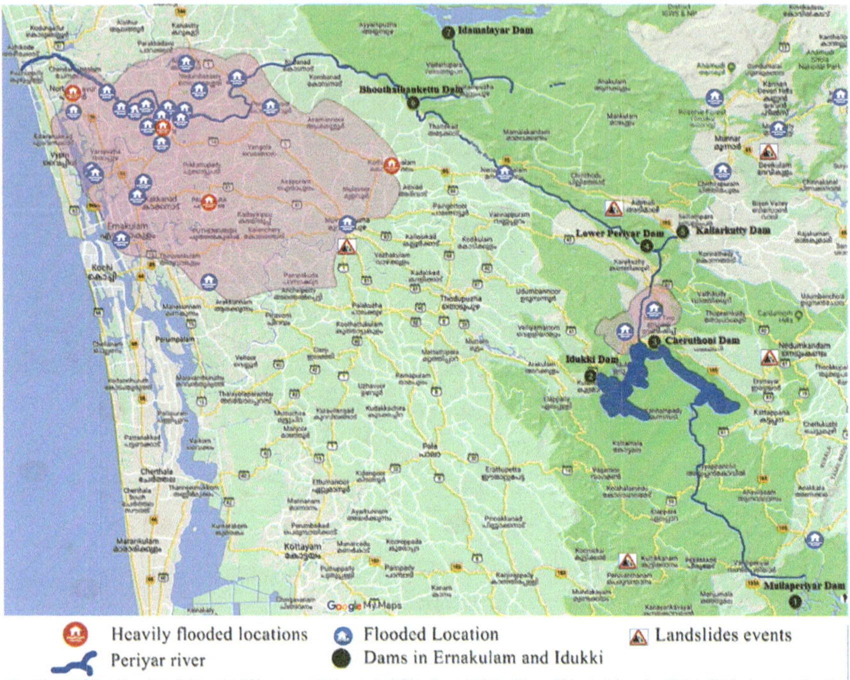

Fig. 5.7 Flood and landslide affected areas in Idukki and Ernakulam

5.3.6 *Multi-hazard Interactions in Kerala Foods*

Analysis of single hazards, where each hazard is analysed as a separate entity, does not provide sufficient understanding of the nature of hazards in a spatial extent [13, 14]. A robust multi-hazard approach includes the potential interactions of all the hazards in the study region [13, 14]. The two main interactions that took place during this disaster are triggering and increased probability interactions. Triggering relationships occur when a primary hazard results in triggering one or more secondary hazards. Increased probability relationships occur when a primary hazard increases the probability of occurrence of a secondary hazard [13, 14]. There are cases where both interactions have the potential of occurring; i.e. a primary hazard can both trigger and increase the probability of a secondary hazard.

Gill and Malamud [13] have developed a tool to illustrate hazard interactions in a matrix form, where the same set of hazards are considered as primary and secondary hazards. A colour code is assigned to each category of interactions, i.e. triggering, increased probability and both triggering and increased probability. This type of visualization is important in understanding the interconnected nature of interactions between hazards in a multi-hazard event such as the 2018 Kerala floods. The natural hazards considered in this study are heavy rainfall, floods and landslides. The

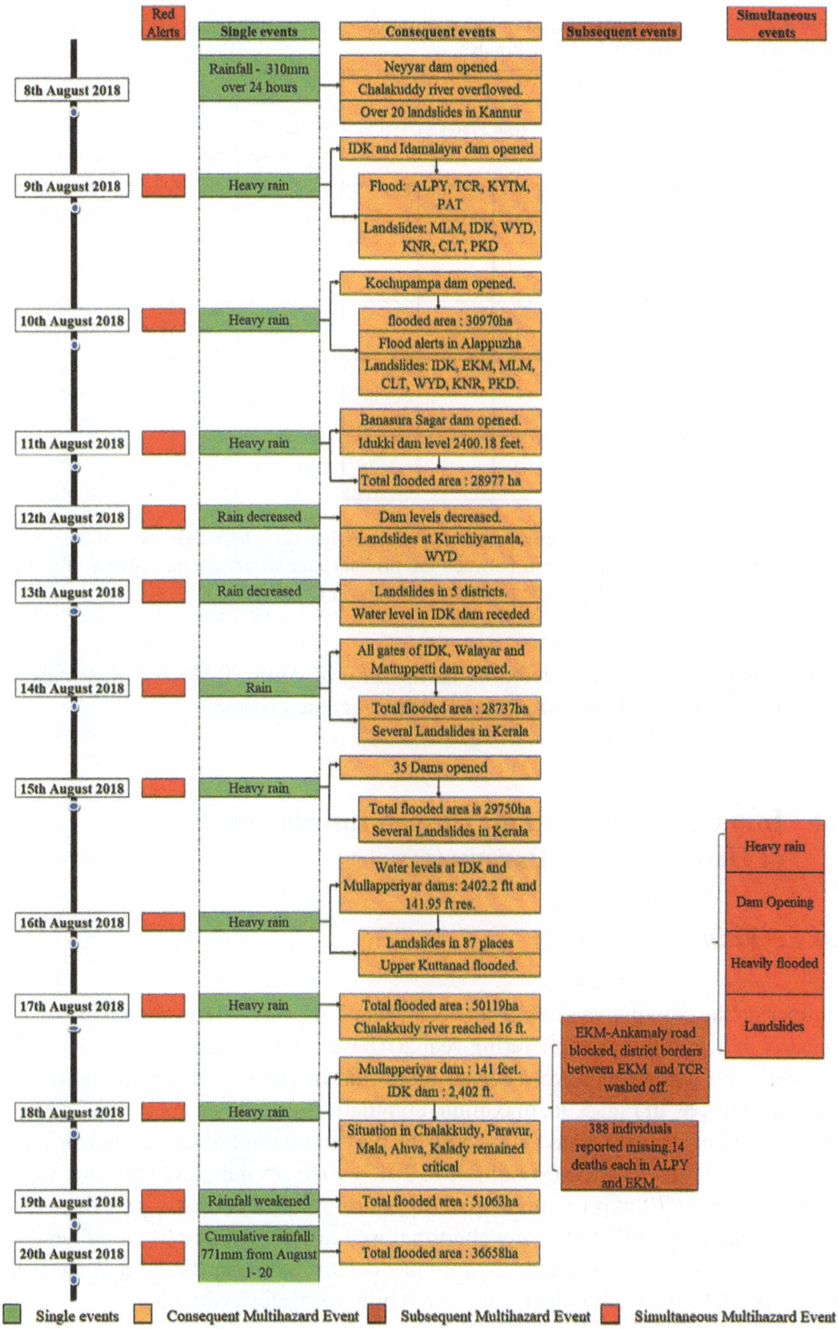

Fig. 5.8 Detailed chronology of Kerala flood. KNR—Kannur, CLT—Kozhikode, WYD—Wayanad, MLM—Malappuram, PKD—Palakkad, TCR—Thrissur, EKM—Ernakulam, ALPY—Alappuzha, IDK—Idukki, KYTM—Kottayam, PAT—Pathanamthitta

Table 5.2 A matrix of natural hazards and anthropogenic processes, along with the interactions that occurred during the 2018 Kerala floods

Secondary Hazards and Anthropogenic Processes					
Primary Hazards and Anthropogenic Processes		HR	DO	FD	LS
	HR				
	DO				
	FD				
	LS				

Hazard/ Anthropogenic Process Name	Key Code	Type	
Heavy Rainfall	HR	Natural Hazard	
Dam Opening	DO	Anthropogenic Process	
Flood	FD	Natural Hazard	
Landslide	LS	Natural Hazard	
Nature of Interaction	Triggering	Increased Probability	Both Triggering and Increased Probability
Colour Code			

The natural hazards and anthropogenic process types are assigned a colour code and a corresponding abbreviation. The type of interaction between the primary and secondary hazards has also been assigned a corresponding colour code as described in the key

anthropogenic process considered is the opening of dams. Table 5.2 illustrates the interactions between these hazards and anthropogenic processes.

5.4 Impact Intensification: Unseen Interlacing of Heterogeneous Parameters

5.4.1 Rain

Comparing the cumulative rainfall between years 2017 and 2018 as shown in Fig. 5.9, it clearly indicates that rainfall during year 2018 was higher when compared to 2017. Thirteen out 14 districts have received more rain in 2018 when compared to year 2017. In 2017, as in Fig. 5.9a, maximum cumulative rainfall was between 2500 and 3000 mm, whereas in 2018, as in Fig. 5.9b the maximum rainfall has been in the range of 3500–4000 mm. The lowest rainfall in 2017 is 763.75 mm, and the same region has received 967.70 mm of cumulative rain during the 2018 monsoon. The excess rainfall cannot be attributed to any abnormal weather pattern like (i) cloudburst, (ii) cyclone and (iii) storm, because the rain has occurred during the normal monsoon period of 2018. What is evident is the availability of water everywhere.

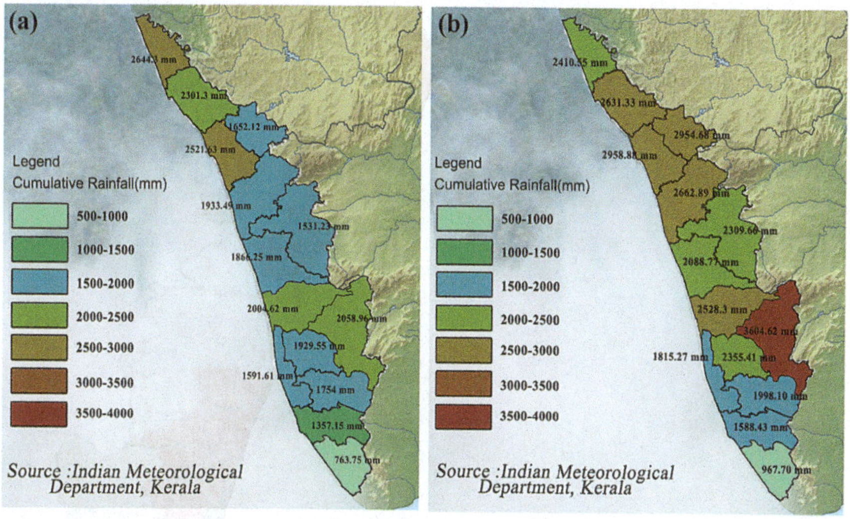

Fig. 5.9 Performance of south-west monsoon during **a** 2017 and **b** 2018

5.4.2 Urbanization

Rapid urbanization has had an effect on the number of people being affected by the floods. From Fig. 5.10, regions with very high urban population concentration are those where the flood impact on the population has been maximum. People in Thrissur, Ernakulam, Kozhikode and Kannur regions were the areas greatly affected. Kannur and Kozhikode were affected by rain and landslide, whereas in Thrissur and Ernakulam it was the floods which had an impact on lives of people. This clearly indicates that rapid urbanization or unscientific growth of cities has also contributed to the magnification of the disaster event. Water from highlands flows to midland and finally towards lowland before flowing into the sea. Urbanization in midland and lowland has prevented the natural flow of water both over the surface and prevented water seepage into the surface, thereby creating artificial tanks which acted as water storage tanks and creating flood scenarios across these regions. Looking at Fig. 5.9, related to storm water maps of the city of Ernakulam, this clearly exemplifies the scenario of water's natural flow being disrupted and proves beyond doubt that urbanization also had a role to play in the creation of Kerala floods 2018.

5.4.3 Terrain

Terrain plays a vital role in shaping the drainage network of a river system. It is also a key contributor to the stream power which is generated by flowing of water in the surface drainage. Kerala landscape has (i) highlands, (ii) midlands and (iii) lowlands.

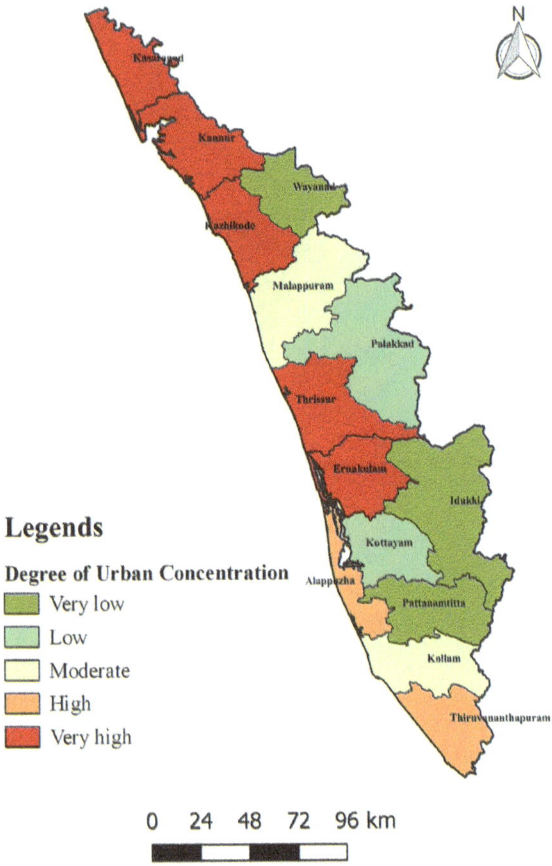

Fig. 5.10 Degree of urban concentration in Kerala

Potentially, water from catchments of the higher region flows to the lower regions and the scenario is similar to what has happened during the intervening periods of 9–20 August 2018 in addition to water being released from the dams at the same time. Figure 5.11 shows the topography and drainage map of Kerala showing the rivers which comes from elevations above 1500 m from the districts of Idukki, Wayanad and Palakkad and flows through the multiple structurally controlled drainage networks coming to the low-lying areas such as Ernakulam and Alappuzha. As we clearly see from Fig. 5.11, most of the area has elevation below 100 m in these districts, during the above-mentioned period the huge discharge in rivers led to accumulation and lateral spreading of water creating the flood situation.

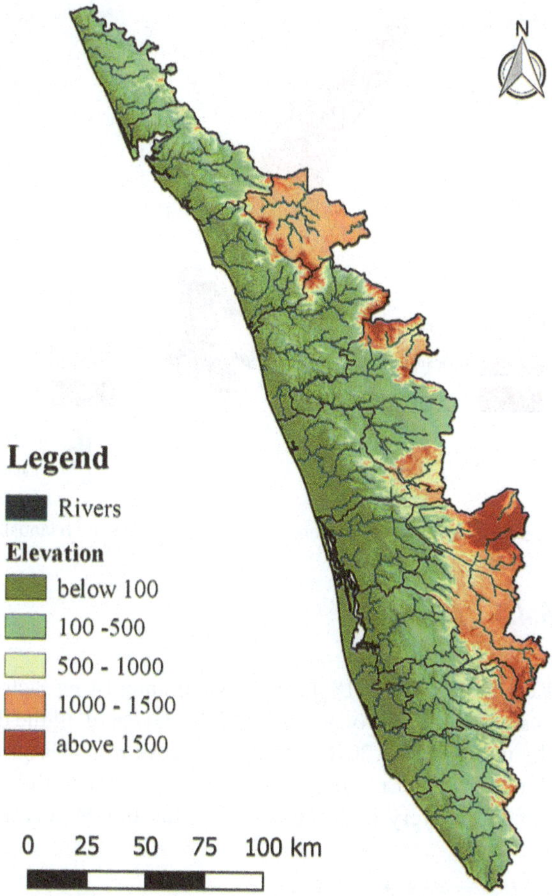

Fig. 5.11 Topography and drainage map of Kerala

5.4.4 Landslide

Inferences from the susceptibility map of landslide, as indicated in Fig. 5.12a, state of Kerala is susceptible to landslide. Susceptibility is decided based on various underlying factors like (i) rain, (ii) topology, (iii) slope stability, (iv) water retention in soil and so on. In 2018, it was the rain that was the main factor in landslide occurrences across the regions. Due to topology of the region, slope stability has been affected which in turn had an impact on the water retention capability. Even though Fig. 5.12b indicates major landslide occurrences, there have been many landslides of minor nature that have occurred, which is not indicated here. Indication of these minor landslide events in the map would have made the floods look minor. This clearly indicates that the region being susceptible was vulnerable and waiting for an opportunity for landslide occurrences and rain was detrimental in landslide occurrences across the region.

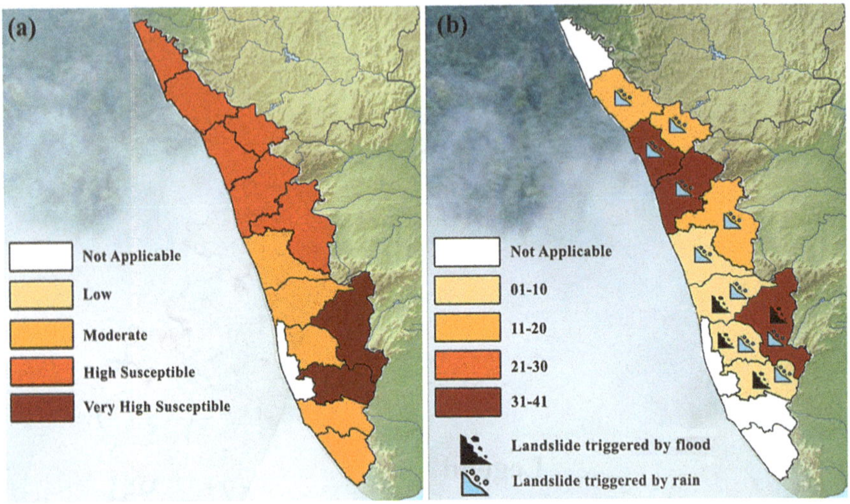

Fig. 5.12 **a** Landslide susceptibility map of Kerala and **b** landslide inventory of Kerala during 2018

5.4.5 *Flood*

Kerala floods of 2018 have shown an interesting factor when comparing the impact versus susceptibility as per Fig. 5.13. Gates of all the dams have been opened during the monsoons of 2018 to release excess water. Regions which were susceptible to floods between moderate and very high susceptibility as shown in Fig. 5.13a had a high and very high impact when the floods occurred as indicated in Fig. 5.13b.

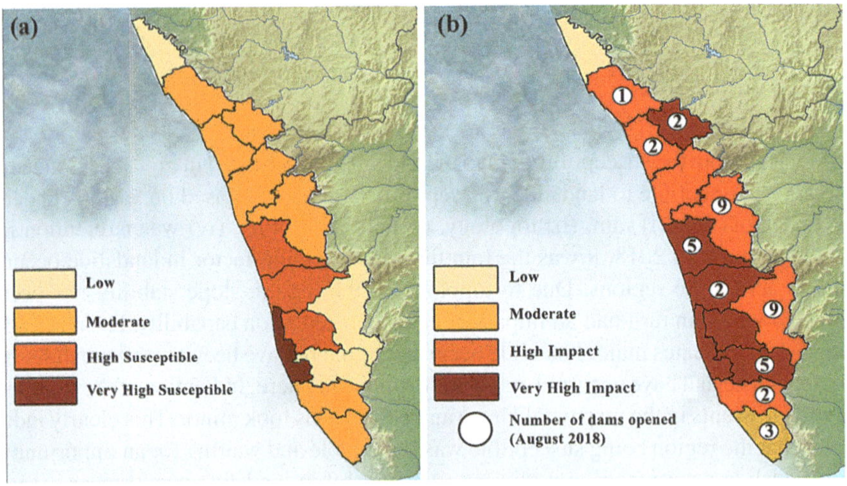

Fig. 5.13 **a** Flood susceptibility map of Kerala and **b** flood impact map of Kerala during 2018

This indicates either (i) excess water from rain or release of excess water from dams and (ii) obstructions in water flow pathways, which may have been due to natural or man-made obstructions causing the floods. In Fig. 5.13a, one district in the north of the state showed low susceptibility to floods and the same was not impacted by flood as shown in Fig. 5.13b. This region has higher levels of urbanization with no major dams. Two districts in the state are indicated as low susceptible to floods and had a high impact due to floods when gates of 9 dams were opened to release water. The flow of water had a very high impact in regions along the pathway of water from these dams.

5.5 Results

5.5.1 Multi-hazard Scenario

Taking into account not just rain but combination of (i) rain, (ii) terrain, (iii) river interlacing the regions, (iv) flood plain area, (v) opening of dams, (vi) landslides and (vii) urbanization, which has led to Kerala floods 2018, a multi-hazard scenario map of Kerala has been proposed in Fig. 5.14. The events and the causes need to be looked at from the zoning at each parameter that has contributed to the floods. These are the unseen heterogeneous parameters that have interlaced to create a situation of disaster. The zoning map shows in layers how the different layers when joined together create the combined hazardous effect. It is easier to comprehend the situation at individual level, and only when all factors are joined together, the interlacing effect comes into focus. By interlacing these factors, the result obtained shows that Kozhikode, Thrissur, Ernakulam, Kottayam and Pathanamthitta have very high multi-hazard index whereas Kannur, Malappuram, Wayanad, Palakkad, Idukki and Alappuzha have high multi-hazard index. On the other hand, districts like Kasaragod, Kollam and Trivandrum have low multi-hazard index. It is clear from Fig. 5.14 that the multi-hazard index shows high in regions where the impact of flood and landslides has also been same as indicated in Figs. 5.13b and 5.14b; hence, this can be considered as a representative of Kerala 2018 multi-hazard scenario.

5.5.2 Chronology of Multi-hazard Events

Summarizing the entire chronology of events, the period between 8 August 2018 and 20 August 2018 is of importance to us, illustrated in Fig. 5.15. It is during this period that single event transformed into other events which led to loss of more than 400 human lives with over a million being displaced.

From Fig. 5.10 (Sect. 5.4.2), regions with very high urban population concentration are those where the flood impact on the population has been maximum. People in

Fig. 5.14 Multi-hazard map of Kerala

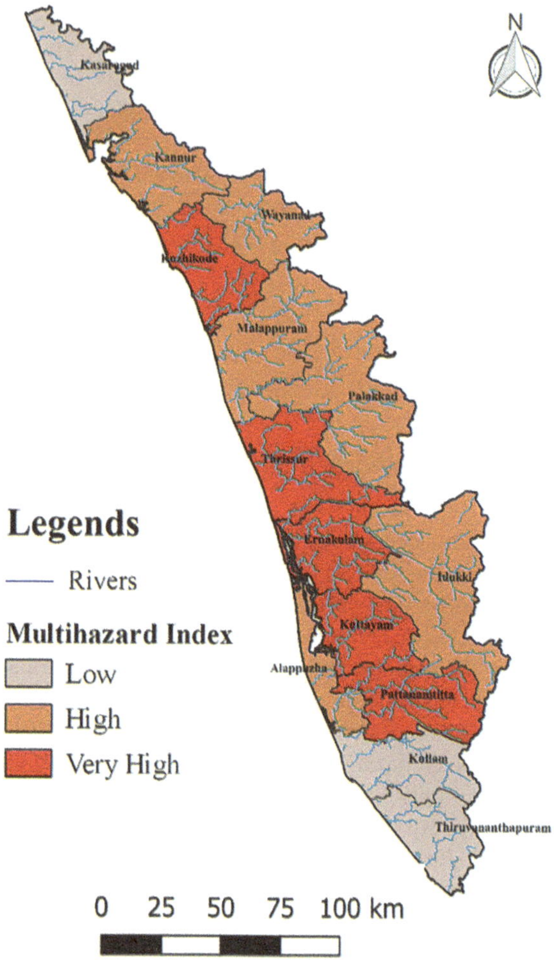

Thrissur, Ernakulam, Kozhikode and Kannur regions were the areas greatly affected. Kannur and Kozhikode were affected by rain and landslide, whereas in Thrissur and Ernakulam it was the floods which had an impact on lives of people. This clearly indicates that rapid urbanization or unscientific growth of cities has also contributed to the magnification of the disaster event. Water from highlands flows to midland and finally towards lowland before flowing into the sea. Urbanization in midland and lowland has prevented the natural flow of water both over the surface and prevented water seepage into the surface, thereby creating artificial tanks which acted as water storage tanks and creating flood scenarios across these regions. Looking at Fig. 5.11 (Sect. 5.4.3), related to storm water maps of the city of Ernakulam, this clearly exemplifies the scenario of water's natural flow being disrupted and proves beyond doubt that urbanization also had a role to play in the creation of Kerala floods 2018.

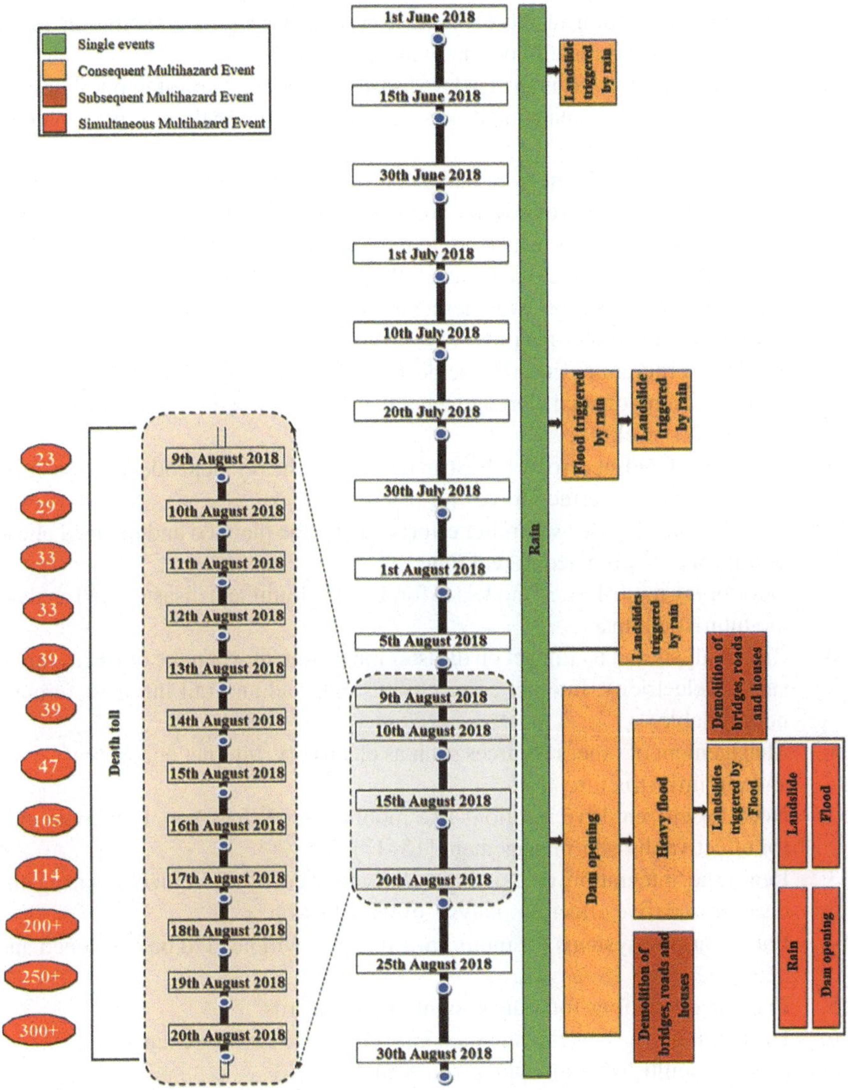

Fig. 5.15 Chronology of multi-hazard events

5.6 Discussion—Reflections and Learning

Looking back at the events that have unfolded during the south-west monsoon of 2018, Kerala experienced more than usual rainfall, leading to flooding. Rainfall also caused landslides in areas where human intervention had made the ecology fragile. Release of water from dams led to floods affecting millions of people.

Points that can reflect upon the event of Kerala floods 2018 are:

1. Human intervention in the ecological system has given ammunition to the heavy rainfall to create floods and landslides.
2. Protocols and processes to manage disasters of wider spatio-temporal impact.
3. Effective water management systems for operationalizing the spatially distributed dams.
4. Impact of water release from dams could have been simulated and the effect analysed before release of water and the resultant could have been communicated to the population which could have helped in better planning for rescue and relief much earlier, than wait till the event to occur.
5. Enhancing community awareness on impacts of water release from dams during such extreme scenarios.
6. Effective programmes need to be developed for making the population aware of the meaning of each alert and what the steps they need to implement based on the alert level.
7. Collapse of rail and road infrastructure led to heavy dependencies on local fishermen and the armed forces for rescue.
8. Location-specific disaster relief efforts need to be planned and mapped ahead of time and coordinated in real time.
9. Developing multi-hazard models to forecast the imminent disasters and analyse the futuristic impacts.
10. Risk analysis and its impact on disaster management when the other resources such as electricity, Internet connectivity, and road and rail infrastructure are not available.
11. Management of other resources such as electricity, Internet connectivity, road and rail infrastructure.
12. Need for citizen drive solutions like mobile apps WhatsApp, Facebook, etc., for effective disaster management [15–17].
13. Real-time integration of existing solutions used by the citizens during the disaster to derive effective analysis of the situation.
14. Early warning systems for multi-hazard events will need to be developed and deployed.
15. Integrated solutions for managing multiple hazards.
16. Developing solutions to anticipate impacts due to climate change.
17. Detailed multi-hazard mapping and zoning.
18. Developing integrated models which needs rapid application across all hazard spectrum.
19. Empowering community to become disaster resilient.

5.7 Conclusion

Floods during monsoon 2018 impacted the lives of common people. What caused the flood is debatable and cannot be pointed towards a particular event. It is accepted that it rained 200% above the normal in the month of August 2018 during the monsoon of 2018 over the state of Kerala. This rain filled up dams and in parallel caused landslides in several regions. Rain water filled dams and when the gates of the dams were opened to release excess water also created landslides and floods. Landslides created debris which blocked flow of water creating floods. Landslides also changed the direction of the water pathway of a river. Intense mining of sand from river bed has affected the river morphology and increased threat to floods. The territorial map drawn for this specific paper clearly indicates that areas affected by floods and landslides have high impact for multi-hazard events. This makes it difficult to identify the precise reason for flooding; was it landslide and its debris or just release of water from dams or the rapid urbanization which blocked the flow of water.

Instead of focusing our efforts towards identifying the root cause for the floods, it is best for a country like India to focus its effort and energy on building a disaster resilience community model that can be used across any type of hazardous events. Integrated simulations to assess the exponential multiple events that trigger disaster needs to be developed and implemented at different scales. Early warning systems for multi-hazard events will need to be developed and deployed. Models developed can be simulated and can be used to indicate the extent of damage an event can create. Need of the hour is for communities to be resilient to multi-hazard events which will require enhanced community engagement model development and implementation. Lot has been learnt from Kerala floods 2018, and applying this learning onto the resilient model can improve chances of the state overcoming the next disaster in a way that only a minimum number of people get affected, thereby increasing the resilience of the population to any natural or man-made disaster(s).

Acknowledgements The authors would like to express gratitude for the immense amount of motivation and guidance provided by Dr. Sri. Mata Amritanandamayi Devi, Chancellor, Amrita Vishwa Vidyapeetham. The authors are thankful to Mr. Ratheesh Kumar and other staff and students of Amrita Vishwa Vidyapeetham for his valuable support and hard work.

References

1. Amritanand S, Anand S, Amritesh AR (2020) Dynamic and time critical emergency management for level three disaster: a case study analysis of Kerala floods 2018. In: Proceedings of the 21st international conference on distributed computing and networking, Jan 2020, pp 1–6
2. Chondol T, Bhardwaj S, Panda AK, Gupta AK (2020) Multi-hazard risk management during pandemic. In: Integrated risk of pandemic: covid-19 impacts, resilience and recommendations. Springer, Singapore, pp 445–461
3. Shaji E, Gopinath G (2020) 'Kerala deluge of 18': a geo-climatological caveat. Social
4. Sarun S, Kumar RA, Jayalakshmi VK (2020) A geographic analysis of natural hazard profile of Kerala, India
5. Wikipedia (2006) Geography of Kerala. Wikipedia.org, 16 Jan 2006. Retrieved from https://en.wikipedia.org/wiki/Geography_of_Kerala
6. Performance of south west monsoon 2018 over Kerala (2018) Indian Meteorological Department (IMD). Retrieved from https://www.imdtvm.gov.in

7. IMD (2018) 19 Aug 2018. Retrieved from https://www.imdtvm.gov.in/images/rainfall%20over%20kerala%20during%20monsoon%20season-2018%20and%20forecast%20for%20next%205%20days.pdf
8. Wikipedia (2018) 2018 Kerala floods. Wikipedia.org, 10 Aug 2018. Retrieved from https://en.wikipedia.org/wiki/2018_Kerala_flood
9. India Today Web Desk (2018) Kerala floods: red alert in 12 districts, death toll touches 174. India Today, 18 Aug 2018. Retrieved from https://www.indiatoday.in/india/kerala/story/kerala-floods-red-alert-in-12-districts-death-toll-touches-174-1317406-2018-08-18
10. Performance of Monsoon (2018) IMD, 31 Aug 2018. Retrieved from https://www.imdtvm.gov.in/index.php?option=com_content&task=view&id=63&Itemid=89
11. Kamboj V, Kamboj N, Sharma S (2017) Environmental impact of river bed mining—a review. Int J Sci Res Rev
12. India Today Web Desk (2018) Kerala's Idukki dam shutter opened after 26 years. India Today, 9 Aug 2018. Retrieved from https://www.indiatoday.in/india/story/kerala-idukki-dam-1309651-2018-08-09
13. Gill JC, Malamud BD (2014) Reviewing and visualizing the interactions of natural hazards. Rev Geophys 52(4):680–722
14. Gill JC, Malamud BD (2016) Hazard interactions and interaction networks (cascades) within multi-hazard methodologies. Earth Syst Dyn 7(3):659–679
15. Guntha R, Rao SN, Ramesh MV (2020) Deployment experiences with Amrita Kripa: a user-friendly feature rich crowdsourced humanitarian application. Procedia Comput Sci 171:302–311
16. Guntha R, Rao SN, Ramesh MV (2020) Architectural considerations for building a robust crowdsourced disaster relief application. In: 2020 international conference on COMmunication Systems & NETworkS (COMSNETS), Jan 2020. IEEE, pp 638–641
17. Guntha R, Rao SN, Shivdas A (2020) Lessons learned from deploying crowdsourced technology for disaster relief during Kerala floods. Procedia Comput Sci 171:2410–2419
18. Kerala population 2011–2018 census. Retrieved from https://www.census2011.co.in/census/state/kerala.html
19. Rivers. Kerala State Irrigation Department. Retrieved from http://www.irrigation.kerala.gov.in/index.php/resources/water-bodies/rivers
20. Preparedness of disaster management and geo-scientific strategies, 23
21. Wikipedia (2001) Kerala. Wikipedia.org, 31 Oct 2001. Retrieved from https://en.wikipedia.org/wiki/Kerala

Chapter 6
River Flow Analysis—It Is So Easy, but It Isn't?

Bojan Đurin, Lucija Plantak, Sara Dadar, and Atena Pezeshki

6.1 Introduction

Due to the stochastic nature of the river flow, as well as the climate changes which nowadays drastically affects their amount, as well as intensity of changing, it is very difficult to predict the amount and intensity of the flow in the future. Such will be seen at the example of local River Bednja in Croatia, Fig. 6.1, on the most upstream measuring station Lepoglava, and Ludbreg, which is the most downstream measuring station.

River Bednja is located in northwestern Croatia and its basin is 596 km^2 in size, while it belongs to the Drava and Danube basins and the Black Sea basin [2]. The Bednja basin area stretches from Trakošćan in the west to Ludbreg in the east. It is predominantly a lowland area with elements of mountains in the western and southern parts. Due to the fan-shaped upper part of the river basin, it has large and fast concentrations of inflows, and intense precipitation contributes to this. In addition to precipitation, the amount of water in Bednja also depends on the melting of snow in the source part of Bednja [3].

B. Đurin (✉)
Department of Civil Engineering, University North, Varaždin, Croatia
e-mail: bojan.durin@unin.hr

L. Plantak
Faculty of Geotechnical Engineering, University of Zagreb, Varaždin, Croatia
e-mail: lucija.plantak@gfv.unizg.hr

S. Dadar · A. Pezeshki
Faculty of Agriculture, Department of Water Science and Engineering, Ferdowsi University of Mashhad, Mashhad, Iran
e-mail: sara.dadar@mail.um.ac.ir

A. Pezeshki
e-mail: atena.pezeshki@mail.um.ac.ir

S. Kolathayar et al. (eds.), *Civil Engineering for Disaster Risk Reduction*, Springer Tracts in Civil Engineering, https://doi.org/10.1007/978-981-16-5312-4_6

Fig. 6.1 Location of Bednja River in Croatia [1]

According to [4], the recorded water waves on the River Bednja last on average four days (Lepoglava), and five days on other hydrological measuring stations). The time of raising the water wave at the hydrological station Lepoglava is 24 h, while at the other measuring stations is about 30 h.

6.2 Case Study

Figure 6.2 shows the location of the River Bednja in Croatia with measurement stations Lepoglava (Fig. 6.3) in Ludbreg (Fig. 6.4).

6.3 Discussion

Figure 6.5 shows the hydrogram of the River Bednja for the most upstream measuring station Lepoglava, while Fig. 6.6 shows the hydrogram of the River Bednja for the most downstream station Ludbreg, all for the years 1999 till 2018.

Table 6.1 shows main statistical parameters of River Bednja for two measuring stations, Lepoglava and Bednja.

It could be observed, not only for this case but also in most of the analyzed situations, that the definition of the linear trend, as a most simple procedure for prediction of the value of the flows in the future, does not have a firm foothold

Fig. 6.2 Location of the River Bednja in Croatia [5]

Fig. 6.3 Measuring station Lepoglava

for obtaining the dependable value. Values of the coefficient of determination, R^2, indicate treasurable correlations. In accordance with that, other procedures must be accomplished for the purpose of better prediction of flow values.

Next, the most common step is calculation of the correlation that shows the connection between the values of the flow for measuring stations, in this case Lepoglava and Ludbreg, Fig. 6.7.

Fig. 6.4 Measuring station Ludbreg

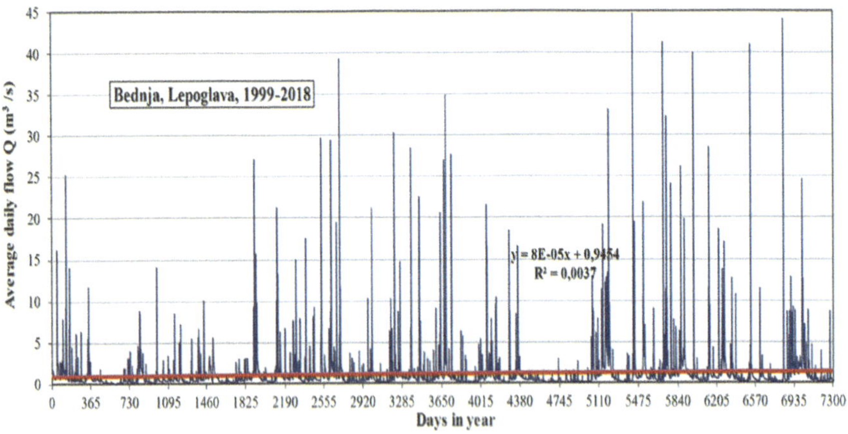

Fig. 6.5 Hydrogram of Bednja River on Lepoglava measuring station [6]

In this case, correlation between average daily flows of the two measuring stations is a (theoretically) moderate correlation, with respect to the value of R^2. Due to the hydrological experience and knowledge, such correlation should be taken with the wariness because it could be seen from Fig. 6.5 that large input values of the flow

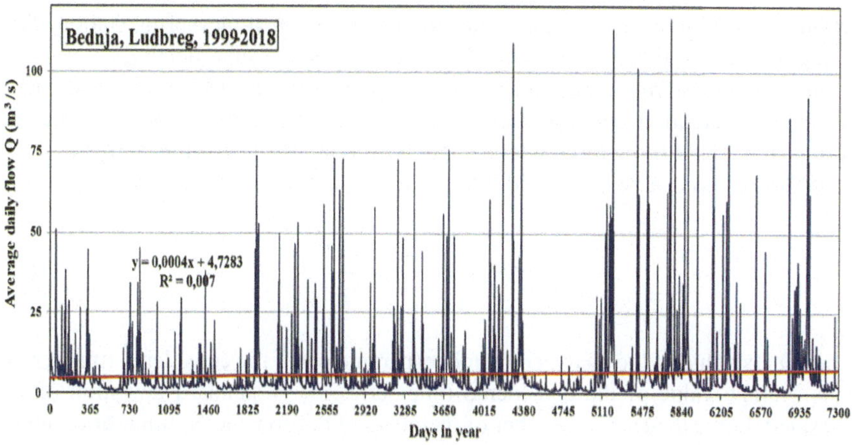

Fig. 6.6 Hydrogram of Bednja River on Ludbreg measuring station [6]

Table 6.1 Main statistical parameters of River Bednja for two measuring stations, Lepoglava and Bednja

Measuring station	Lepoglava	Ludbreg
Average (m^3/s)	1.23	6.17
Standard deviation (m^3/s)	2.68	9.98
Minimum (m^3/s)	0.03	0.43
Maximum (m^3/s)	44.70	116.00

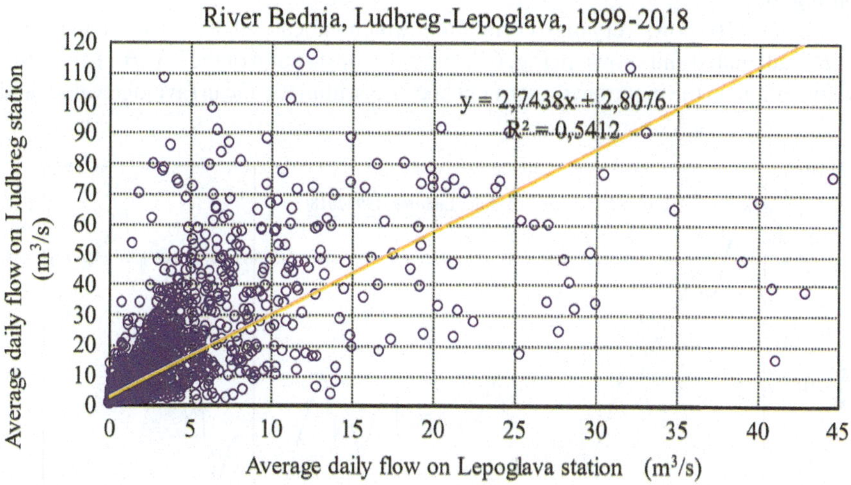

Fig. 6.7 Correlation of average daily flow between data from Ludbreg and Lepoglava measuring stations

within one station lead to the large output value of the next measuring station. This is also an obvious example if correlation analysis is taken superficially.

In order to determine the given values by visual occurrence of the trends, data grouping, fluctuation and appearances during the time appear, rescaled adjusted partial sums (RAPS) method will be applied for both the measuring stations.

RAPS values can be calculated through Eq. (6.1) [7],

$$\mathrm{RAPS}_k = \sum_{t-1}^{k} \frac{Y_t - \overline{Y}}{S_y} \tag{6.1}$$

where Y_t is value of the analyzed member (parameter) of the considered time series, $\overline{Y}$ is average value of the considered time series, S_y is standard deviation of the considered time series, n is number of members of the considered time series, and $k = 1, 2, \ldots, n$ is counter during summation.

Graphical presentation of the obtained RAPS values, obtained from Eq. (6.1), are shown in Fig. 6.8 and 6.9.

It can be seen that there are four subperiods on data from both measuring stations which give insight in periods of low and high flows, which cannot be determined from hydrograms and linear trends. First begins in the first part of 1999 year, second from beginning of 2000 year and last till the end of 2012 year, while third comes after and last till the end of 2017 year. Fourth subseries has its start at the end of 2017 year and extends further. Correlation of RAPS values of both measuring stations is shown on Fig. 6.10.

Figure 6.10 shows very high correlation, which indicates that mistrust in the value of R^2, calculated and shown in Fig. 6.7, is valid. Constitution of the Eq. (6.1) gives full justification on the mentioned averment because arithmetic mean (average value) and

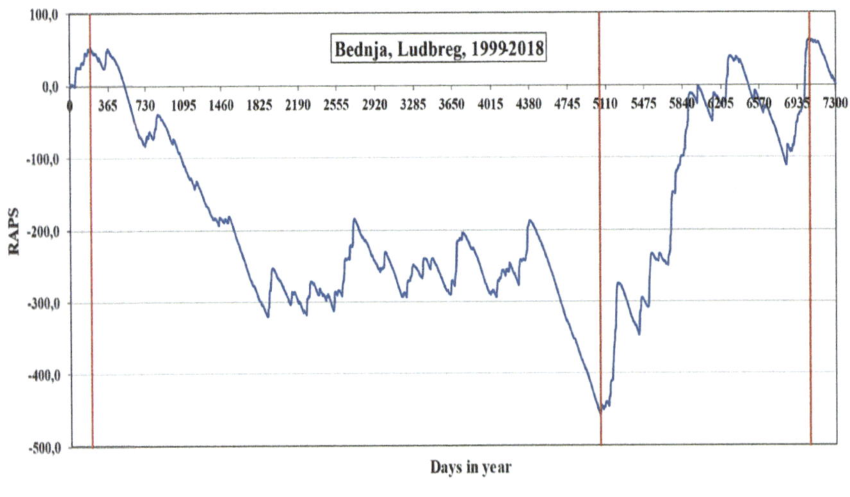

Fig. 6.8 RAPS on values obtained on Ludbreg measuring station

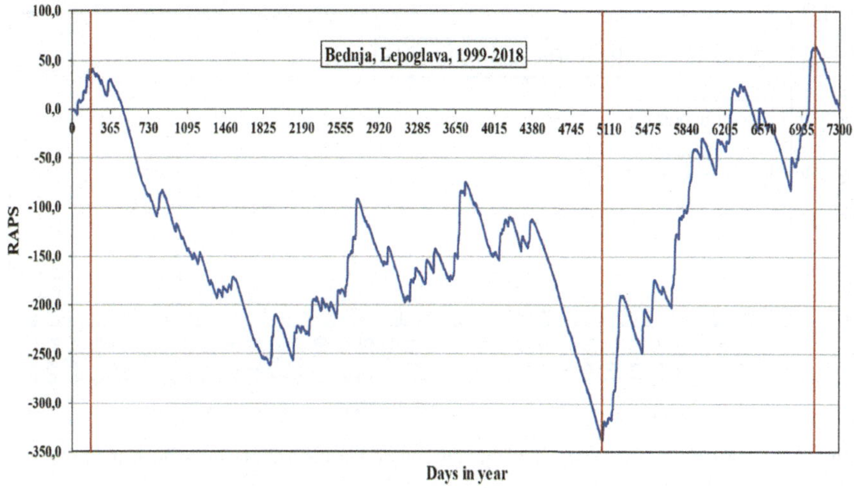

Fig. 6.9 RAPS on values obtained on Lepoglava measuring station

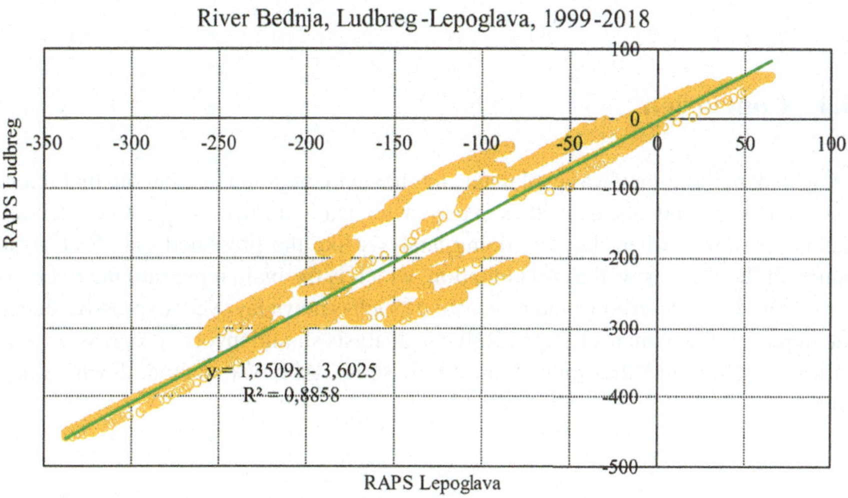

Fig. 6.10 Correlation of RAPS values from Ludbreg and Lepoglava measuring stations

standard deviation provide insight into the aberration and errors caused by natural or anthropogenic reasons.

Between analyzed hydrological measuring stations Lepoglava and Ludbreg, there are three other measuring stations. It is expected that presented analysis could be more precise by taking into the account values on mentioned stations. River Bednja has torrential characteristics, as well as relatively small values of the flows, so between all measuring stations, there are drastic changes of the flows, caused by many tributaries,

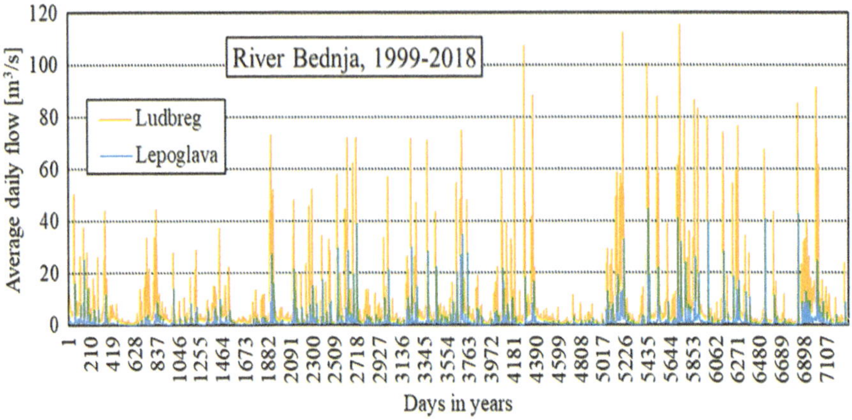

Fig. 6.11 Comparison of the average daily flows on measuring stations Lepoglava and Lubreg

as well as inflow from the catchment. Such drastic differences could be obtained by comparison of the flows for both analyzed measuring stations, Fig. 6.11.

6.4 Conclusion

Presented methodology was explained and shown in intention to point out that scientist, engineers, students, as well as experts who deal with hydrology, very often are making mistakes when they are doing analysis like the presented one. Stochastic nature of the flow, as well as precipitation, which is the main input into the rivers, is not so simple for describing and forecasting. Such is nowadays very expressive due to the impact of the climate changes. Reliable analysis should involve progressive time series modeling, like decomposition of time series, ANOVA method, downscaling, etc.

References

1. Geographic guide, Available: http://www.geographicguide.com/europe-maps/croatia.htm. Accessed 12 Feb 2021
2. Marković T, Larva O, Brkić Ž, Dolić M, Kuhta M (2015) State of groundwater in the Danube River Basin with regard to the natural metal content and their anthropogenic impact. HGI-CGS, Zagreb
3. Zaninović K, Gajić-Čapka M, Perčec Tadić M et al (2008) Climate Atlas of Croatia 1961–1990, 1971–2000. State Hydrometeorological Institute, Zagreb

4. Paladin M, Vidaković Šutić R, Vrcelj B, Ričković V (2015) Preparation of danger maps and calculation of damage to the basin river of poverty. Institute of Electric Power and Energy d.d. Zagreb, Hrvatske vode. Available: https://bib.irb.hr/datoteka/764772.Izrada_karata_opasnosti_i_izracun_steta_na_slivu_rijeke_Bednje.pdf
5. Wikiwand map, Available: https://www.wikiwand.com/en/Bednja_(river). Accessed 4 Feb 2021
6. Croatian Meteorological and Hydrological Service (CMHS) (2020) Daily flow for Lepoglava and Ludbreg from 1999 till 2018. Croatian Meteorological and Hydrological Service (CMHS): Zagreb, Croatia
7. Bonacci O, Andrić I (2008) Sinking karst rivers hydrology—case of the Lika and Gacka (Croatia). Acta Carsologica 37(2–3):185–196

Chapter 7
Coastal Disasters and Mitigation Measures

S. Neelamani

7.1 Introduction

Global warming as well as the melting of ice from North and South Pole and other high altitude hills and mountains are the main reasons for the global sea level rise (SLR). SLR is causing permanent inundation of coastal area, seawater intrusion, accelerated erosion and coastal area degradation, which are considered as long-term coastal disaster. This is considered as a slow event but will lead to disaster when permanent inundation happens with loss of high investment infrastructures such as ports, harbors, marinas, beach houses, coastal industries, etc. Scientifically acceptable and economically viable mitigation measures are required to reduce the socioeconomic impact. The other major vulnerability is short-term coastal disaster, which is due to cyclones, storm surge, heavy rain in the coastal area, tsunami, high wind speed, etc. However, they are short-term events, but the impact in terms of loss of life and property is significant. It is necessary to understand and apply suitable mitigation measures to reduce the losses. This article is prepared to understand some of the scientific aspects of the short-term and long-term coastal vulnerability and the suitable mitigation measures to be followed to reduce the loss of life and property during risks due to the natural forces as described above.

First, some important steps to be followed for clear understanding of coastal vulnerability are discussed. Then, a detailed description on long-term as well as short-term coastal vulnerability and adaptation principles are described.

S. Neelamani (✉)
Coastal Management Program, Kuwait Institute for Scientific Research, Kuwait, Kuwait
e-mail: nsubram@kisr.edu.kw

S. Kolathayar et al. (eds.), *Civil Engineering for Disaster Risk Reduction*, Springer Tracts in Civil Engineering, https://doi.org/10.1007/978-981-16-5312-4_7

7.2 Coastal Vulnerability Assessment—Important Steps

For selection of proper mitigation measures against coastal vulnerability due to future sea level rise and associated action of natural forces on the coast, it is necessary to understand the following points:

- Define what type of coastal vulnerability you wish to investigate (flooding, erosion, socioeconomic, seawater intrusion, etc.)
- The interest is on short-term vulnerability (cyclones, storm surge, tsunami, high wind speed, etc.,) or long-term vulnerability due to sea level rise
- Whether the interest is in assessing vulnerability alone or on the assessment of the risk also
- If you want to assess the coastal risk, then it is necessary to assess the degree of exposure, which is a strong function of space
- List the parameters that are important for vulnerability assessment [E.g., some coastal countries have very high waves (UK) and some countries with very little waves (Gulf countries)]
- Identify the source of data (or) collect the data based on field works
- Assure the quality of data
- Divide the total coast length into many segments (If spatial variation of the parameter is high within one segment of the coast, then select smaller segments. For example, if the spatial variation of geomorphology is very high, say, rocky coast for some distance and sandy coast for some distance, then select smaller coastal segments).
- Select correct weightage for each parameter.
- Use analytic hierarchy process [6, 7], if needed.
- Take the help of subject experts to assign values from 1 to 9 [6, 7] based on relative importance between parameters.
- Engineers and scientists understand physical vulnerability better.
- For decision makers, it is required to reveal the economic vulnerability of each coastal cell.
- Assess the economic loss of each coastal cell for different sea level rise scenario and for the intercepting natural forces (wave, tide, wind, tsunami, storm surge, current speed etc.,); coastal characteristics (geomorphology, land elevation, seabed slope, sea bed soil character and strength, etc.,) and socioeconomic parameters (population, land use, land cover, cultural heritage, recreation area, tourist area, road network, etc.)
- An area with less flood/inundation risk at present may become more flood/inundation risk area, since the government/coastal authority may build a city for future population.

The above steps will help for assessing the degree of short-term and long-term coastal vulnerability and risks. Once the hot spots are identified (the coastal cells which would encounter high risk of flooding/inundation, coastal erosion, sea water intrusion, socioeconomic loss etc.,), then it is possible to look into different mitigation options and select the appropriate one based on science, art and economics.

7.3 Long-Term and Short-Term Coastal Vulnerabilities and Its Impacts

7.3.1 Long-Term Coastal Vulnerability

The long-term coastal vulnerabilities are mainly due to sea level rise (SLR), ocean acidification, coastal land subsidence, etc. According to IPCC 2014 report, the sea level rise can be in the range of 0.52 to 0.98 by 2100 for RCP 8.5 pathway. Some of the questions in the minds of stakeholders, who use the coastal area for different purposes (commercial, residential, etc.,) are:

What will be the level of coastal flooding/inundation and its spatial variation for different sea level rise scenarios?

What will happen to all the existing coastal infrastructures, which are already closer to high water level?

Which coastal location is suitable for future coastal infrastructure development by considering the climate change and future global warming scenarios?

Is it possible to clearly indicate the hotspot of erosion/inundation/socioeconomically vulnerable coastal areas for different sea level rise scenarios so that proper action can be taken now, which will improve adaptation and resilience of all classes of stakeholders?

For selection and application of suitable mitigation plans, a clear understanding of the type of vulnerabilities are important. The following are the major long-term coastal vulnerabilities due to sea level rise:—coastal inundation; coastal erosion and accretion; socio-economic loss; seawater intrusion and biological vulnerability. In general, the loss of life is not expected to happen due to sea level rise since natural adaptation to survive happens by nature for the vulnerability induced by sea level rise.

It is essential to understand and educate people on different vulnerabilities. For example, if we know the future level of inundation at different cells of the coast, then it is possible to plan suitable inundation preventing structures. It is also possible to plan now to avoid building coastal infrastructures in areas prone for high level of inundation. Hence, the basic requirement is to prepare inundation map for different sea level rise scenarios like 0.25, 0.5, 1.0, 2.0 m etc., since it is difficult to predict accurate sea level rise in future, say by 2100.

Regarding coastal erosion, if we know the future level of erosion at different cells of the coast, then it is possible to plan suitable erosion protection measures and budget allocation by governments for such activities. Knowing spatial variation of erosion vulnerability will help to avoid building coastal infrastructures such as beach houses, hotels etc., in areas prone for high level of coastal erosion. Since coastal erosion as well as accretion are considered as disasters in many coastal engineering practices, it is required to scientifically understand the reason for erosion and accretion and plan suitably for its effective management [5].

A clear estimate of the socioeconomic loss for different level of physical vulnerability needs many inputs such as cost of coastal land per unit area, the coast of existing commercial and noncommercial infrastructures, etc. It will help to take a decision, whether existing coastal infrastructures need to be moved landward and the investment for such operation is worth or not.

Assessment of long-term vulnerability based on seawater intrusion helps to understand the severity of seawater intrusion issues at different cells along the coast. Some areas with impervious substrata may not experience any seawater intrusion problem even with sea level rise of 1.0 m. If the vulnerability of seawater intrusion for a particular coastal cell is known, then it is possible to take proper action to prevent significant seawater intrusion into the ground water aquifer.

7.3.2 Short-Term Coastal Vulnerability

The short-term coastal vulnerabilities are due to cyclone, storm surge, tsunami, flash rain-induced flooding during cyclone events, earthquake in the coastal area, etc. The intensive scientific developments during the past three decades has resulted in better atmospheric models to predict the formation area of cyclones, path and region of coastal crossing, intensity of wind speed and rainfall intensity with respect to space and time. However, further accurate prediction is needed for storm surge intensities and storm induced intense rainfall and the associated flooding and damages. Tsunami prediction has improved significantly but the prediction of level of damage due to tsunami-induced flow needs further accurate prediction. The main problem with short-term coastal vulnerability due to cyclone, tsunami and earthquake is that the response time is less for transmission of the message to the population, convincing the people to take actions to protect their life and property. Simple mitigation measures is required and is discussed in the following sections.

7.4 Mitigation Measures for Long-Term and Short-Term Coastal Vulnerabilities

There are two important management approaches available for mitigations of coastal disasters due to long-term and short-term vulnerabilities. The first one is defend, adopt and retrieve, which is one important framework to be adopted to prevent socioeconomic losses. The other possible mitigation measurement strategy is reduce, reuse and recycle. They are briefly discussed in the following section.

7.4.1 Mitigation Measures Based on the Management Principle: Do Nothing/Defend/Adopt and Retrieve

Mitigation Measures-Do Nothing

Do nothing is the simplest measure but may or may not help for mitigating the problems of coastal vulnerability in terms of erosion, inundation or flooding, sea water intrusion, biological degradation etc. due to climate change, sea level rise and increased intensity of cyclones etc. Many economically backward countries cannot afford to spend money for mitigations and hence they will go for "Do Nothing" framework. In such situation, at least they need to learn the type of coastal vulnerabilities, the basic science behind the nature's forces and response, the mechanisms of escaping and protecting their properties in case of future emergencies.

Mitigation Measures-Defend

Many coastal countries around the work has allowed the citizens to invest and develop coastal infrastructures near the high water line. Kuwait and many Gulf countries fall under this category. Only recently, the governments are trying to enforce construction of coastal infrastructures after leaving certain distance from high water line called buffer zone so that any possible increase in sea level rise will not affect the planned infrastructure. Hence, for the infrastructures, which were already built near the high water line, the adaptation policy is "Defend" only. Based on the environmental conditions of waves, currents, wind and seabed soil conditions, it is scientifically possible to select a suitable defending scheme such as Groin fields, sea walls, offshore breakwater or beach nourishment activities. The economics and degree of tolerance of the owner for partial failures of the existing infrastructure decides the solution at the end. A typical example for "Defend" for an existing coastal structure is as shown in Fig. 7.1.

The stakeholders need to analyze the cost of the mitigation solution against the return from the costal infrastructure. If it is profitable or acceptable, then it is worth going for "Defend" solution. If it is economically not viable, then the mitigation is to demolish the infrastructure and try to reuse whatever materials are worth for such a purpose.

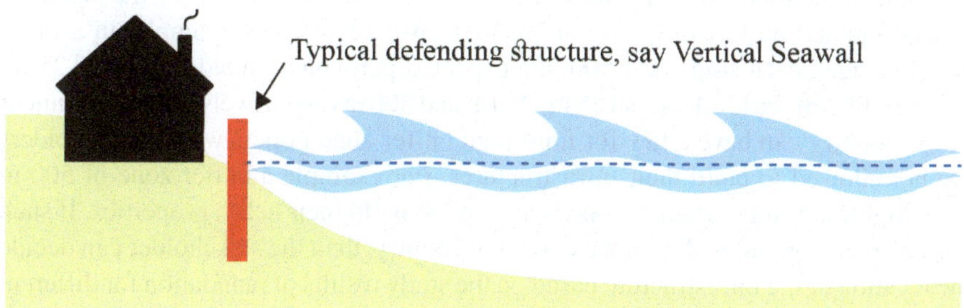

Fig. 7.1 Mitigation for a structure, which is built closer to high water line

Fig. 7.2 Mitigation measure for a structure, which needs to be built near high water line

Mitigation Measures-Adopt

The mitigation mechanism "Adopt" is needed, if the stakeholder is compelled or required to create the coastal infrastructure near the high water line (it is known to the stakeholder that in future, the sea level will rise and will damage the infrastructure). Such compulsions are inevitable for projects such as new ports, harbors, marina or beach hotel or coastal tourism parks, where it is required to build the infrastructure near the beach and near the high water line. In such situation, it is necessary to assess the engineering impacts expected on the planned coastal structure for future sea level rise and make sure the structure is stable even for increased sea level rise and more frequent and harsh action from natural wave induced forces. A typical example is construction of a beach hotel, where the complete structure can be designed to build on strong and thick foundations and the loads can go down to ground by suitable pile groups as shown in Fig. 7.2. The expected future problem is erosion and advancement of shoreline near or beyond the structure. The initial investment will be higher to make sure the structure would not fail due to erosion, sub soil liquefaction and mining effort. The bed level of the structure need to be raised based on the future expected sea level rise. For example, if the life of the structure is 50 years and the expected sea level rise if 0.5 m during its lifetime then, the foundation level of the structure should be raised by at least 0.5 m compared to the base level without any impact due to sea level rise induced inundation.

Mitigation Measures-Retrieve

The safe mitigation mechanism is "Retrieve". Here, the estimate of the width of coastal inundation for different sea level rise is assessed. For example, in a place with average beach slope of 1:500, the expected permanent inundation for 0.25 m, 0.5 m and 1.0 m sea level rise is 125 m, 250 m and 500 m respectively. The government of the country can have a law for minimum buffer zone policy, where stakeholders are not allowed to build their infrastructures. For example a buffer zone of 500 m from high water line, where no one is allowed to build their beach properties. If such coastal zone regulation does not exist in the country, then the stakeholder can decide the location of the infrastructure based on the study results of inundation for different sea level rise expected in future (Fig. 7.3).

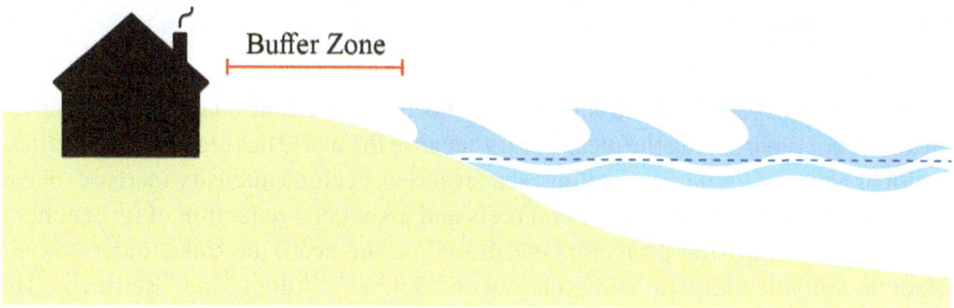

Fig. 7.3 Mitigation for a structure, which is to be built on the coastal area based on future expected sea level rise and width of inundation

7.4.2 Mitigation Strategy Based on "Reduce, Reuse and Recycle"

The other management strategy for mitigation of coastal disaster is "Reduce, Reuse and Recycle." The basics of this strategy is as follows: If we "Reduce" the usage of all resources for coastal infrastructure developments, then it will help in reduction of power consumption, which in turn will reduce the emission of greenhouse gas which in turn help for reducing the sea level rise and frequency of cyclones with intense power. Hence, it is possible to reduce the order of sea level rise as well as the increased level of coastal erosion. Reuse and recycle also will help in this direction of management strategy.

For example, it is good to use floating breakwaters for coastal erosion protection or creating relatively calm pool of water for open sea construction or for reducing wave climate around open sea loading and unloading terminals or for protecting open sea aquacultural cages. Floating breakwaters are suitable for moderate wave energy level and for short-period wave conditions with wave periods less than eight s. It is proved that floating breakwater with skirt walls are efficient wave dampers and the volume of materials required for its construction will be much less when compared to say concrete caissons or rubble mound breakwaters [4]. It is also proved that floating breakwater made out of scrap tires are also effective and cost competitive [1, 2].

Another example from "Reduce" point of view is replace rubble mound and caisson type breakwaters by using series of porous walls [3, 8]. Further development on new type of wave barriers are available at [9].

In coastal engineering projects, "Reuse" policy is applicable only for materials like scrap tires as floating breakwaters. Reusing of civil engineering materials from demolished structures may not serve the purpose since they easily pollute the coastal waters and exposure of steel reinforcement from the demolished material also create accelerated corrosion and its environmental effect on the coastal waters. Hence, as on now, "Reuse" type management policy did not gain momentum. Similarly, "Recycle" policy has not gained much importance for coastal infrastructure projects. Hence, it is recommended to follow the "Reduce" policy, which will help for reducing the greenhouse gas emission and will reduce the sequence of consequences.

7.5 Conclusions

The disasters of coastal infrastructures and selection of suitable adaptation measure is one of the hot topic during the recent years because the world has recognized the effect of global warming, polar ice melting, sea level rise, cyclone intensity increase, ocean acidification and degradation of coral reefs and associated reduction of its beneficial effect as natural erosion protectors etc. Based on the need, the stakeholders should select the suitable mitigation measure out of "Defend," "Adopt" and "Retrieve." The stakeholders need to be educated in the management strategy based on "Reduce," "Reuse" and "Recycle" policy. Suitable selection of these management strategies are based on economics, environmental impacts, aesthetics, level of acceptance, degree of risk taking, etc. An effective effort is needed to educate the stakeholders the merits and demerits of the mitigation strategy. This will help in saving life and property in the coastal area during short-term as well as long-term impacts from natural forces.

References

1. Al-Yousif A, Neelamani S, Levinson AV (2019) Wave transmission, reflection, and dissipation of scrap tire floating breakwater in random waves. ISH J Hydraul Eng. https://doi.org/10.1080/09715010.2019.1664342
2. AlYousif A, Neelamani S, Levinson AV (2020) Hydrodynamic performance of Tire-based Floating Breakwater. Marine Geores Geotechnol. Taylor and Francis. https://doi.org/10.1080/1064119X.2020.1795761
3. Neelamani S, Al-Salem K, Altaf Taqi K (2017) Experimental investigation on wave reflection characteristics of slotted vertical barriers with an impermeable back wall in random wave fields. J Waterway Port Coastal Ocean Eng ASCE 143(4). https://doi.org/10.1061/(ASCE)WW.1943-5460.0000395
4. Neelamani S, Josko L (2017) Experimental study on the hydrodynamic performance of floating pontoon type breakwater with skirt walls. J Offshore Mech Arctic Eng 140(2): 021303–9. Paper No. OMAE-17–1029, 9. https://doi.org/10.1115/1.4038343
5. Neelamani S (2017) Coastal erosion and accretion in Kuwait—problems and management strategies. In: Nelson R-B, Williams AT, Pranzini E, Anfuso G (eds) Ocean and coastal management. Management strategies for coastal erosion processes. vol. 156, pp 76–91
6. Saaty TL (1990) How to make a decision: the analytic hierarchy process. Eur J Oper Res 48:9–26
7. Saaty TL (2008) Decision making with the analytic hierarchy process. Int J Serv Sci 1(1):83–98
8. Vijay KG, Neelamani S, Sahoo T (2019) Wave interaction with multiple slotted barriers inside harbour. Phys Numerical Modell Ocean Eng 193. https://doi.org/10.1016/j.oceaneng.2019.106623
9. Vijay KG, Nishad CS, Neelamani S, Sahoo T (2020) Gravity wave interaction with a wave attenuating system. Appl Ocean Re 101:102206

Chapter 8
Tsunami Flood Risk Management

Alastair Barnett

8.1 Introduction

8.1.1 The Development of Flood Risk Management

Flood risk management has evolved with human habitation over thousands of years, based on a need to avoid frequent flooding of floors designated as living areas in a built environment. Some cultures have sought naturally drained land for settlement, progressively reducing the frequency of remaining inundation events by enhancing natural drainage systems, or by relocating to areas observed to have fewer problems. Others have adapted to occupy areas subject to frequent ground inundation by raising their houses on poles.

All these ancient design strategies have in common a need to understand the local *variation of water surface levels* associated with a flood event, relative to the fixed floor level of a living space. The basic disciplines of flood hydraulics, therefore, date back at least to the prehistoric times of legends such as Hercules and the Hydra.

Inland settlements need only to understand drainage pathways as discharging to a lower downstream level, but coastal people have learned that the sea is subject to gravity waves which may periodically raise that downstream level enough to threaten their floor levels. In an area called the *fluvial-tidal zone of a drainage system* [1], there has long been recognition of the regularity of tides and the corresponding need for living floor levels to be raised above high tide as well as river flood level.

In extreme cases such as in Hangzhou on the Qiantang River, the rising tide forms a sudden bore wave. The total level increase may approach 9 m, a flood wave with an incoming hazard of the same order as the largest urban tsunami. However, because of the regularity of tidal events and the late Holocene sea-level stationarity [2], close

A. Barnett (✉)
HYDRA Software Ltd, PO Box 9463, Hamilton 3240, New Zealand

S. Kolathayar et al. (eds.), *Civil Engineering for Disaster Risk Reduction*, Springer Tracts in Civil Engineering, https://doi.org/10.1007/978-981-16-5312-4_8

observation over many years has enabled the risk at Hangzhou to be contained within an empirically designed channel where no human habitation is attempted.

8.1.2 Zones of Increasing Vulnerability

However, we are now experiencing the end of the Holocene stable mean sea levels, as accelerating rises comparable with observations elsewhere are being recorded worldwide in coastal communities. Further, increases in storm intensities are also forecast in response to global warming, so we need to plan for increased storm surge levels as well as greater flood discharges. Again both high tides and river flood levels need to be considered, but now in fluvial-tidal zones of unprecedented expansion.

The Hong Kong Drainage Services Department has responded [1] by proposing the design of drainage systems that should be based on a return period of T-years. A range of return periods is tabulated, depending on the scale of the assets put at risk by design failure.

Concurrent analysis is recommended of two separate cases:

Case I: a T-year sea level in conjunction with an X-year rainfall,

Case II: an X-year sea level in conjunction with a T-year rainfall

The T-year flood level is taken as the higher of the flood levels from Cases I and II.

In these cases, pragmatic experience with many designs has suggested that $X = 10$ when $T = 50$ years or more (corresponding with design for densely populated urban areas and for rural drainage design where failure hazards would be significant) while $X = 2$ for $T = 10$ years or less (for rural villages and areas of intensive farming).

8.2 Extension to Tsunami Events

8.2.1 Hong Kong Practice

Hong Kong practice does not extend to tsunami flooding, probably because there is little historical evidence there of significant tsunami hazards. The joint probability is also small of a major tsunami coinciding with a major rainfall event, so that design of surface drainage capacity need not consider high tsunami levels downstream.

However, the tsunami threat is significant in many cities, and the drainage system is a potential entry point for a major flood event originating downstream. This especially applies to open channels, including natural watercourses. If coastal flood barriers are under consideration, the question of a design coastal flood must be resolved. In these respects, the tsunami hazard has too many similarities to the storm surge hazard (and indeed tidal hazards of the Hangzhou type) to be treated in isolation from drainage system design.

A key assumption in the Hong Kong analysis is that a design return period can be assigned to a sea-level record on the same basis as to a rainfall record. Statistical methods have been well established in hydrology, but the probabilities of rainfall and runoff are not the same, and the assumptions required to establish local relationships between them often turn out not to be transferable to other sites.

A similar situation prevails in geophysics, in that the magnitudes of earthquake events correspond only weakly with the magnitudes of the resulting tsunami events, leading to significant differences in the ranking of the scale of the events and the resulting derived recurrence intervals.

8.2.2 Flood Frequency Analysis

This is a relatively new procedure for flood probability estimation based on river channel discharge records rather than rainfall probability distributions. An excellent discussion of the advantages and disadvantages of this alternative was presented in 1987 [3] by one of the leaders of the development.

The key advantage is the removal of a requirement for any assumptions about the relationship between the probabilities of rainfall and runoff. Indeed, rainfall records are no longer required. The corresponding disadvantage is a complete reliance on an assumption that peak discharge is the variable that "can be considered to be drawn randomly from a well-behaved statistical distribution, which is, thus amenable to frequency analysis". This assumption becomes less dependable when available peak discharge records are too short to include major events or are derived by rainfall-runoff methods rather than direct discharge measurement.

Indeed, the reliable measurement of upstream discharge peaks has been identified as a "weak point" in flood modelling by the flood risk management technical committee of the IAHR [4] and recommendations for improvement have been made.

8.2.3 Tsunami Magnitude Ranking

Tsunami frequency analysis becomes possible by analogy with flood frequency analysis if a single scalar can be found to represent incoming tsunami magnitude in the way peak discharge at some adjacent upstream point represents incoming flood magnitude. Then tsunami events can be ranked by the procedures discussed in [3]. However, "annual series analysis" considers only the highest event in each year of record, and this implied link with an annual meteorological cycle has no justification for tsunami generation by earthquakes. Instead, "partial series analysis" considers all events greater than an arbitrary base value, consistent with usual practice with records of tsunami events. A "recurrence interval" or "return period" is then an appropriate term for tsunamis in this case, leading to an understanding of the Hong Kong "T- year flood level" as referring to a tsunami generated flood level which (strictly speaking) would occur with an Annual Recurrence Interval (ARI) of T-years.

Tsunami "height" is a popular description of magnitude, but this term needs careful qualification as tsunami heights were measured for the *same* tsunami at between 0.6 and 30 m in reports about the 2004 Indian Ocean tsunami [5]. This is partly because tsunami wave heights are dependent on the ocean depth beneath the wave, with the lowest of the above values being measured by satellite in mid-ocean depths of the order of 3 km. Partly also, wave heights are increased by interaction with the shore, and the highest of the above values resulted from impact against cliffs. Analysis is needed to reach a standardized objective definition of "wave height".

8.3 Numerical Model Analysis

8.3.1 Numerical Model Analysis Validation: The Pukaki Canal

The largest canal in the Upper Waitaki hydro project, New Zealand, is the Pukaki canal, with a width of 80 m, a depth of 10 m, a length of 12 km and a flow capacity of 560 m^3/s. This is large enough to be visible from space, as in the satellite photograph in Fig. 8.1.

Commissioning tests of the canal flow controls included dispatching a metre-high artificial tsunami down the canal, offering a chance to validate the wave modelling accuracy of the *AULOS* package [6] used for canal design. A 1 m tsunami in 10 m water depth is a moderate-sized event in a coastal or estuarine channel, so propagation of such a wave could be replicated and studied at full scale over a distance of 12 km. Further, the Y-shaped canal configuration with the smaller Ohau canal joining near the downstream end gave the opportunity to study reflections of gravity waves at channel junctions, an important aspect of wave modelling.

An example of the resulting match between the output from a water level recorder and the *AULOS* numerical model is shown in Fig. 8.2.

The model was started with the water level slightly higher than the observed level, a condition maintained throughout. Otherwise, the match is excellent, particularly with reference to the arrival time of the various abrupt level changes. Wave amplitudes are slightly overestimated in the peaks, and a sudden brief dip at 2 h is not replicated, possibly because the computational time step used was a little coarse to resolve that detail.

Fig. 8.1 Pukaki canal as seen from space. The scale of the photograph can be estimated from the 15 km length of Lake Pukaki, the large pale coloured lake in the centre of the picture

8.3.2 Case Study: The 1947 Tatapouri Tsunami

Two closely spaced significant local tsunami events [7] occurred on the East Coast of New Zealand in 1947. The resulting peak wave levels recorded along the coast are shown in Fig. 8.3. The maximum recorded height was > 10 m during the first (March 26) event at Tatapouri, a small settlement located in the centre of the Figure.

It is clear that wave heights vary considerably along the coast. Increasing distance from the relevant epicentre accounts for part of this variation, but not for the doubling at a few points (especially Tatapouri) of generally recorded adjacent wave heights.

Using *AULOS*, numerical modelling of the tsunami propagation was used to study possible reasons for this discrepancy. The model reconstruction of initial tsunami development on March 26 is presented in Fig. 8.4.

The results in Fig. 8.4 are shown for every third 12 s time step. The initial displacement shape is unsustainable in stationary water, so the peak immediately accelerates water away in both directions as shown by the profile at 0.6 min.

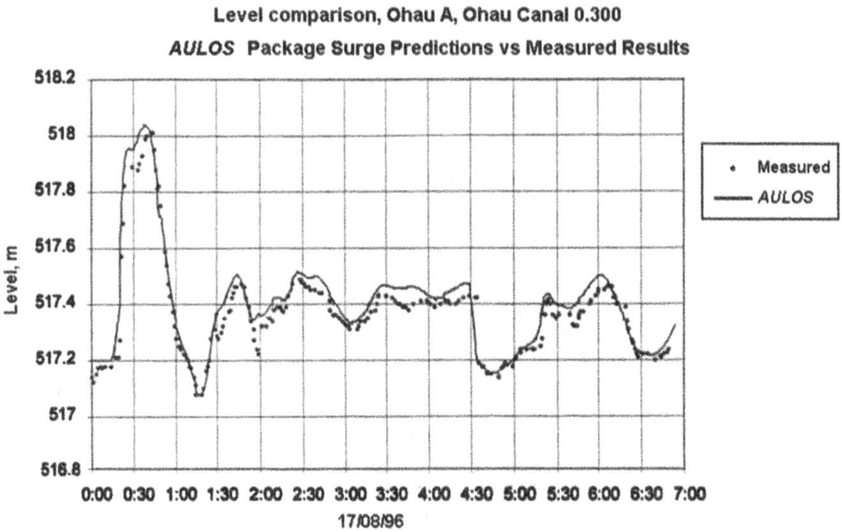

Fig. 8.2 Validation of *AULOS* surge predictions against levels measured at a recorder station on the Pukaki canal

Once accelerated, the two waves will keep moving by inertia, so a trough starts to develop between them at 1.2 min. This is asymmetrical, partly because the initial wave is asymmetrical, but mainly because the seawards wave (towards the left of the diagram) meets a submerged local seamount, the Tolaga Knoll, while the landwards wave initially expands freely over the deep flat basin. The inclusion of the Tolaga Knoll in this model is not consistent with its exclusion from the laterally averaged approach to setting up the effect of horizontal displacement, but the local influence of the Knoll was considered to be of more significance for initial wave development, and it was therefore included. The higher peak of the seaward wave is, therefore, the result of partial reflection from the Tolaga Knoll.

At 1.8 min, the seaward wave is starting to dissipate rapidly into the deep water over the Knoll, while the rear of the landward wave is almost back to normal (zero) sea level. At 2.4 min, the seaward wave has almost totally dispersed into the boundary zone, which was designed to damp out incoming waves. Meanwhile, the landward wave peak has risen slightly in height as it reaches shallowing water at around 40 km.

Finally, at 3.0 min, the seaward wave is largely a returning flow over Tolaga Knoll to fill the trough between the diverging waves, while the landward wave has reached the shape of a stable translation wave, although the peak has again risen as the water shallows further. Note, however, that the peak is still only approximately half the height of the initial displacement.

The next stage of the tsunami development is translation to the coast, as shown in Fig. 8.5. This is simply a continuation of the results presented in Fig. 8.4.

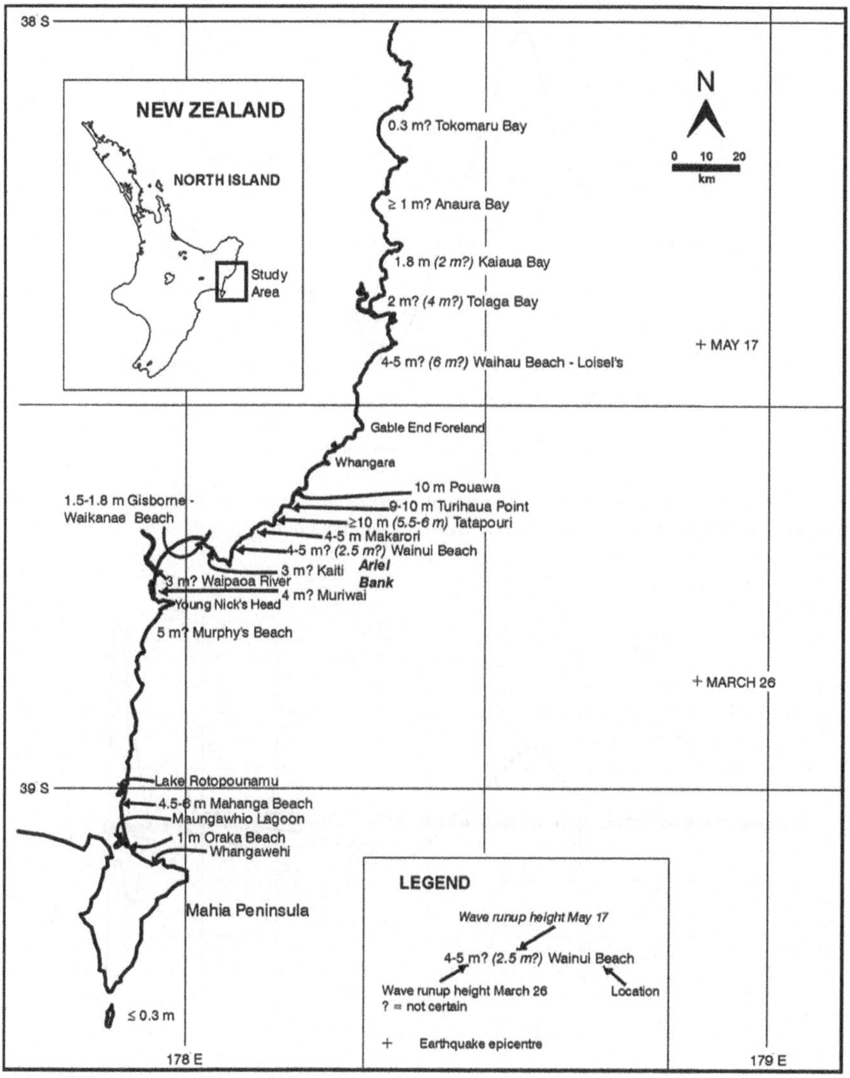

Fig. 8.3 Estimated wave runup heights for the 26 March 1947 and 17 May 1947 tsunami

Now changes are happening much more slowly, as the translation wave will propagate unchanged indefinitely in deep water. As the water shallows, however, two opposite effects come into play.

First, the shallowing reduces the wave speed. Since to a first approximation, all parts of a translation wave travel at the same speed over the same depth, and successive crests cannot overtake each other. Therefore, the time for the waveform to pass any fixed point must stay constant, which means the length of the wave reduces. If the

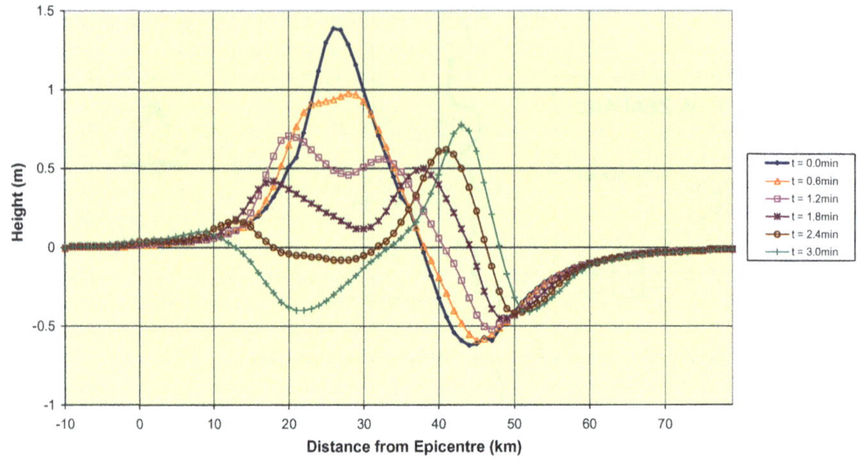

Fig. 8.4 Initial tsunami development. The initial profile ($t = 0$ min) is derived from the March 26 seabed displacement profile reconstructed by earthquake modellers

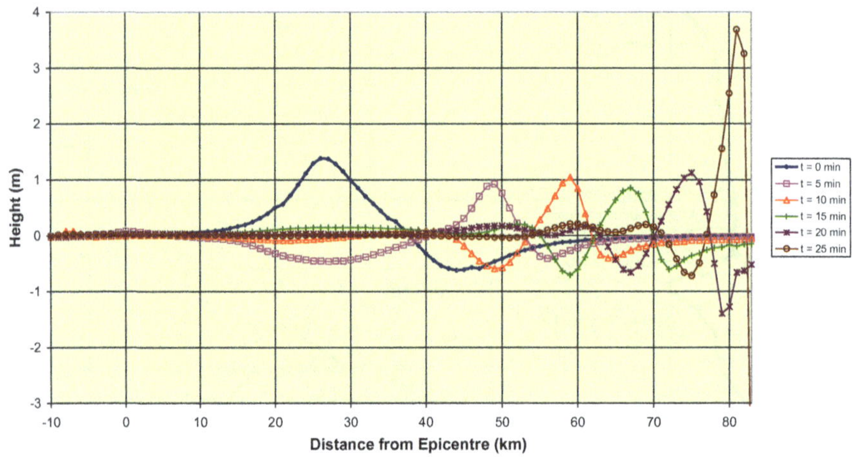

Fig. 8.5 Tsunami translation to the coast

wave volume is unchanged, the wave has to increase in height. However, the model shows little change in height, so considerable volume is being reflected back offshore.

Second, the effect of frictional resistance starts to become significant, so that the above first approximations become less accurate. This means the wave speed does not reduce as much as expected from simple theory, so the wave height increases less. Also, resistance has a dissipative effect, which lengthens the tail of the wave. This effect will reduce the wave height, and in some cases, more than counteracting the effect is expected from any reduction in wave speed.

The first result at 0 min is again the initial displacement profile from Fig. 8.4. At 5 min, the wave has the same essential shape as at 3 min (Fig. 8.4) but has now grown to almost 1 m in height as the water shallows to 150–200 m depth. At 10 min, the wave has grown slightly higher, but it then enters a long stretch of almost constant depth of 70 m from 58 to 70 km. The effect of frictional resistance then operates alone, reducing the wave height at 15 min below that at 5 min.

At this point, this wave is clearly not going to approach matching the observed 10 m height at Tatapouri, even allowing for a doubling in height on reflection from the coast. Resonance is a mechanism which is known to increase wave heights greatly, but this appears to be ruled out for the initial wave at least, as a resonance effect has to build up over several cycles.

Another mechanism known to increase wave heights is a funnelling effect in a channel of tapering width such as in Hangzhou Bay. There is some evidence of this from the 10 m depth contour suggested by limited soundings at Tatapouri Point and again in a constriction offshore between Ariel Bank and the Monowai Rocks. Channel tapering can be reproduced in a 3D scalar model [4], but insufficient information was found to relate the nearshore area around Tatapouri to some kind of channel.

Therefore, a different approach has been adopted. This is to work backwards from the observed wave height to determine what order of channel taper is required for a model match. If this is not unreasonable in the light of the dimensions of the widths of flow available to an approaching wave, it is arguable that initial seabed displacements, as reconstructed by the earthquake modellers, are not inconsistent with the observed tsunami response. This in turn would justify further investigation with refined models, in particular running on higher resolution shallow bathymetric soundings.

As shown in Fig. 8.6, a tsunami height of some 10 m at Tatapouri can indeed be reproduced by tapering the modelled approach channel. No tapering was used

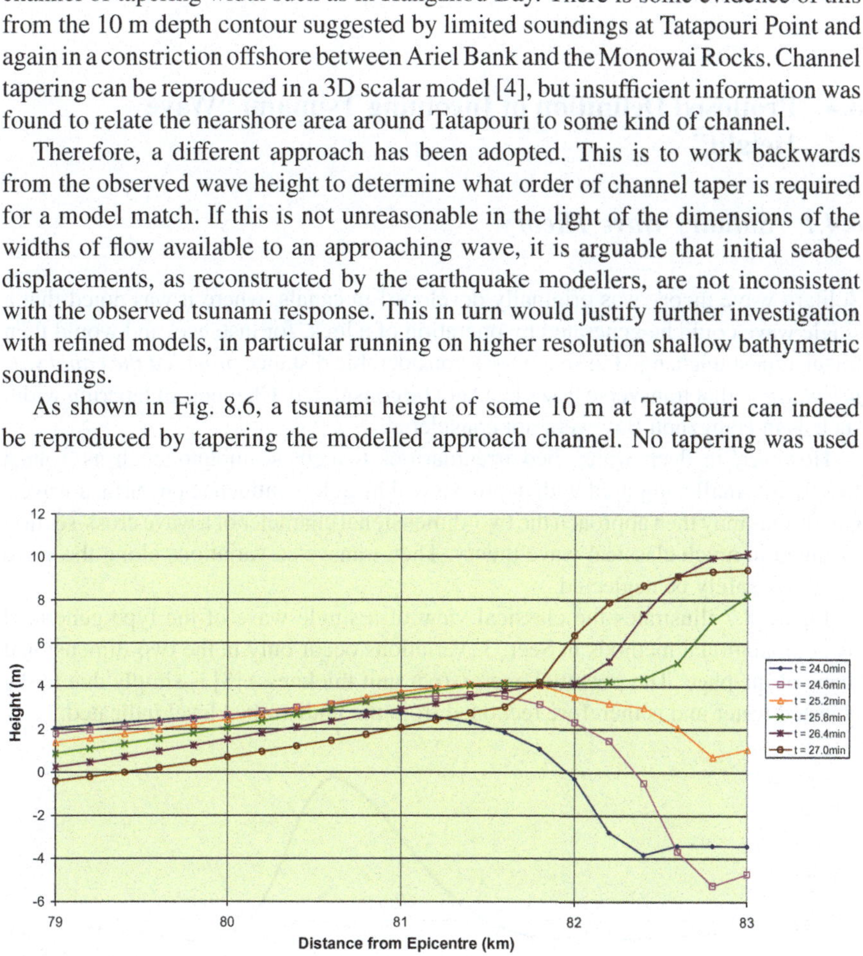

Fig. 8.6 Tsunami impact on the coast, showing the effect of an assumed channel taper

to 70 km, so the wave until 15 min is unaffected (except for the leading trough). From 70 to 79 km, the figures used were a five times reduction in width to match the narrows between Ariel Bank and the Monowai Rocks, followed by a further eight times reduction in width between 79 and 82 km to simulate the tapering inlet sketched by the indicated 10 m contour off Tatapouri.

The continuation of the results from Fig. 8.5 is shown for this "fine grid" area in Fig. 8.6. As in Fig. 8.4, the waveform is changing rapidly in this area, so the results are again shown at intervals of every three 12 s time steps.

From 24.0 min, there is a short-term withdrawal from the foreshore originating from the initial depression inshore of the peak. Between 24.6 and 25.2 min, the wave starts to surge upwards at the coast (83 km). The surge then rises rapidly to 25.8 min, peaking at 26.4 min at just over 10 m as observed. Finally, a reflected wave starts to develop at 27.0 min.

8.4 Proposed Definition of Incoming Tsunami "Wave Height"

8.4.1 Solitary Wave Theory

Solitary wave theory was originally developed in canals, where it was noted that a single wave could be generated by operation of a lock, for instance, and would then travel almost unchanged in shape for a considerable distance *provided the canal was prismatic* so that transverse flows need not be considered. Channels of tapering width such as in Hangzhou Bay were not considered.

However, in deep water, bed irregularities (except seamounts such as Tolaga Knoll) are small compared with depth, so will have less influence on surface waves. Conditions may then approach the two-dimensional character of a wave cross-section assumed in much classical wave theory. Then transverse variations along the wave crest may safely be neglected.

Figure 8.7 illustrates the classical view of a single wave of the type generated by computational methods in Sect. 3. Variations occur only in the two-dimensional plane of the page. The potential energy (per unit thickness) [8] is simply due to the elevated water and is therefore reckoned from the undisturbed level indicated.

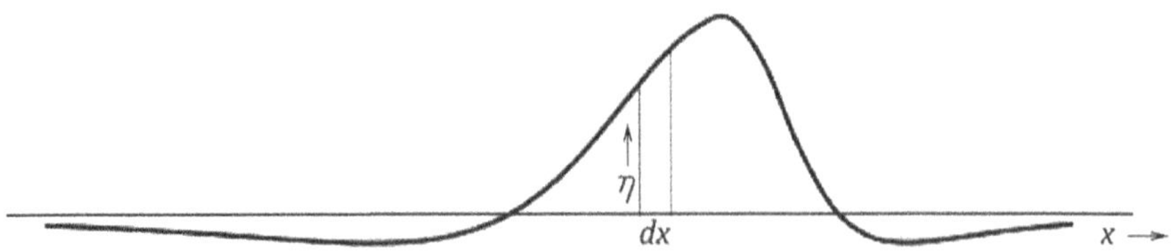

Fig. 8.7 Classical two-dimensional view of a single wave

$$V = \frac{1}{2}\rho g \int_{-\infty}^{\infty} \eta^2 \mathrm{d}x \tag{8.1}$$

Here V is the potential energy (as above), ρ is the mass density, g is the acceleration of gravity, η is the vertical distance of the wave surface from the undisturbed level, and x is the horizontal distance. The origin of x makes no difference to the calculation but is normally placed near the peak level for convenience.

Field measurement is almost certain to record a profile at least as irregular as in Fig. 8.7, but numerical integration may be used to evaluate V in this case. Judgement must be used to limit the integration to distinguish selected waves from the profiles of preceding or following waves, as infinite integration limits apply only to analytical functions which give finite values of V with these limits. Such a function is the sech^2 function used in the first solitary wave analysis developed by Boussinesq in 1872.

8.4.2 Deriving the Wave Height H

It is convenient to recognize this early analytical advance by deriving a relationship between maximum wave height and potential energy based on the Boussinesq profile:

$$H = \frac{3}{h}\left(\frac{V}{4\rho g}\right)^{\frac{2}{3}} \tag{8.2}$$

Here H is the height of the wave peak above the undisturbed level, and h is the depth of the seabed below the undisturbed level.

8.4.3 Choice of Standard Depth

A standard depth h_0 is needed to allow for computation from Eq. (8.2) of the standard wave height H_0. This depth should be.

- Close enough inshore to surround any significant land mass with a contour reasonably comparable in shape with the coastline
- Far enough offshore for frictional processes to be minimal to simplify the transition to onshore propagation models
- Much larger than any reasonable value of H, so local wave speed can be treated as approximately dependent on h alone
- A round number for simplicity

A standard depth of 1000 m meets all these requirements (assuming the metric system is accepted as defining the international length standard). This also has the practical advantages of a round number for wave celerity (about 100 m/s = 360 km/hr) and for typical wave heights (about 1 m for a major tsunami at 1000 m, for example, the 1947 Tatapouri tsunami—see Fig. 8.5).

A standard depth $h_0 = 1000$ m is therefore proposed for the standard single wave. Design can then proceed from a single wave height for each return period, as wave shape variations should not greatly alter the potential energy, and hence, the wave energy transmitted towards the coast.

Note this methodology is limited to remotely generated tsunami, classified [9] as Type III. For earthquake shocks which actually displace the seabed under the area of interest, two other types of tsunami will occur (typically in a harbour) before a Type III event. First, a Type I tsunami, described as an "immediate wave", will be generated by horizontal displacement of submerged vertical walls during an earthquake shock, and then a Type II tsunami, a "seismic seiche" will be generated in response to initial variations in vertical displacement across the local water body. Both of these tsunami types require special adaptations of the above procedures.

8.5 Conclusions

1. Tsunami waves are not inherently different from flood waves closely and successfully studied for design purposes in many engineering schemes, with one exception.
2. This is the availability of a hazard magnitude index to provide a ranking scale for comparing different events for design purposes.
3. Wave heights are used now; but in great recent tsunami disasters, these often differ by almost two orders of magnitude during the progress of a single wave.
4. This makes it almost impossible to index hazard magnitude to wave height unless a definitive procedure is agreed to standardize reference conditions.
5. Solitary wave theory is proposed as the basis for standardization of a link between wave energy and wave height.
6. The standard depth contour for the conversion of incoming wave energy to a standard wave height is proposed to be 1000 m, as this is a round figure close enough inshore to surround all significant land masses, but far enough offshore to allow simplification of the transition from oceanic to inshore models.
7. It is proposed that differences of incoming wave shape alone will not greatly alter the reflection and frictional processes between the 1000 m contour and the coast.
8. Therefore, a single standard incoming wave height for each return period should be sufficient basis to model the tsunami hazard magnitude at each location.

References

1. Drainage Services Department (2018) Stormwater drainage manual, 5th edn. DSD, Government of the Hong Kong Special Administrative Region, Hong Kong
2. Earle S (2015) Physical geology, 2nd edn. BC Campus Open Education, British Columbia
3. Pilgrim DH, Doran DG (1987) Flood frequency analysis. In: Pilgrim DH (ed) Australian rainfall and runoff, Revised Edition. The Institution of Engineers, Australia, Barton, ACT
4. Barnett AG, Muste M, Molinari D (2020) Weak points in the flood risk modelling chain. In: Haun S (ed) IAHR 85th anniversary summit webinars session 14, IAHR. https://www.iahr.org/en/lives/details?live_id=32&video_id=244
5. Barnett AG (2006) Required steps for evaluating tsunami protection schemes. In: Proceedings water conference 2006, Water New Zealand. Christchurch
6. HYDRA Software Ltd (2021) The AULOS package: www.auloshydraulics.com
7. Downes G, Webb T, McSaveney M, Darby D, Doser D, Chagué-Goff C, Barnett AG (2001) The 26 March and 17 May Gisborne earthquakes and tsunami: implications for tsunami hazard for the East Coast, North Island, New Zealand. In: Proceedings ITS tsunami risk assessment beyond 2000: theory, practice and plans. Moscow, Russia (2001)
8. Milne-Thomson LM (1968) Theoretical hydrodynamics, 5th edn. MacMillan, London
9. Barnett AG, Beanland S, Taylor RG (1991) Tsunami and Seiche computation for Wellington harbour. In: Proceedings pacific conference on earthquake engineering. Auckland

Chapter 9
Cloudburst—A Major Disaster in The Indian Himalayan States

Sachchidanand Singh and Mitthan Lal Kansal

9.1 Introduction

Cloudbursts and related flash flooding are one of the Himalayan regions' utmost significant natural calamities. The cloudbursts result from monsoon clouds originating from the Bay of Bengal, bringing an immense amount of rainfall across the Gangetic Plain to the Himalayas, and under the topographic influence "bursting" into a torrential downpour [1]. In India, cloudbursts usually occur in steep orographic regions during the monsoon season, such as the Himalayan region, the north-eastern states, and the Western Ghats. Cloudburst usually occurs where a steep topography experiences a deep and fast uplift of the saturated thermodynamic unstable atmosphere [2]. Convective clouds formed may lie up to 15 km height in cloudburst-like systems and because of its extremely localized occurrence, and ground observation stations are barely able to capture the features of the storm [1]. Generally, cloudbursts occur in inaccessible hilly areas and are not reported until a significant socio-economic disturbance is caused by the subsequent flash flooding and landslides [3]. A cloudburst, also known as rain gush or rain gust, is an intense form of rainfall where a high rain intensity falls over a localized region of about 20–30 km^2 [2]. The India Meteorological Department (IMD) defines cloudburst as a 100 mm/h intensity rainfall event accompanied by lightning and strong winds. This excessive amount of rain falling within a short period further leads to an intense runoff in the form of flash-floods [4]. The conditions and physical processes leading to flooding in the Himalayas are extremely complex, involving multiple genes, and diverse atmospheric and terrestrial surface processes. Geological instability of the region, together with high atmospheric precipitation, is responsible for widespread landslide occurrences. Despite their devastating effects on human life and the Himalayan socio-economy,

S. Singh · M. L. Kansal (✉)
Water Resources Development & Management, Indian Institute of Technology Roorkee, Roorkee, India
e-mail: mlk@wr.iitr.ac.in

S. Kolathayar et al. (eds.), *Civil Engineering for Disaster Risk Reduction*, Springer Tracts in Civil Engineering, https://doi.org/10.1007/978-981-16-5312-4_9

the hydro-meteorological analysis of the "cloudbursts" is given little consideration. [5] have illustrated the usage of remote sensing knowledge to link the space–time variability of the cloudbursts with precipitation, topography, and large-scale patterns of circulation.

In this paper, an attempt has been made to utilize the satellite-derived hydro-meteorological datasets for understanding the characteristics and spatiotemporal variation of cloudbursts over the Himalayan region, especially Uttarakhand. This would not only help in developing a more comprehensive understanding of the mechanism associated with these cloudbursts events but would also shed some light on the identification of these events in future.

9.2 Study Area

The Uttarakhand state is located from latitude 28.44–31.28 N, longitude 77.35 to longitude 81.01 E (Fig. 9.1). The area of this hilly state is 53,483 km^2, which makes up 1.63% of India's landmass. There are 13 districts and 95 development blocks in the State and the whole state is divided into two regions the eastern part, known as Kumaun and, the western part as Garhwal. Dehradun, in the district of Garhwal, is the state capital. Uttarakhand is the initial source of many perennial Indian rivers.

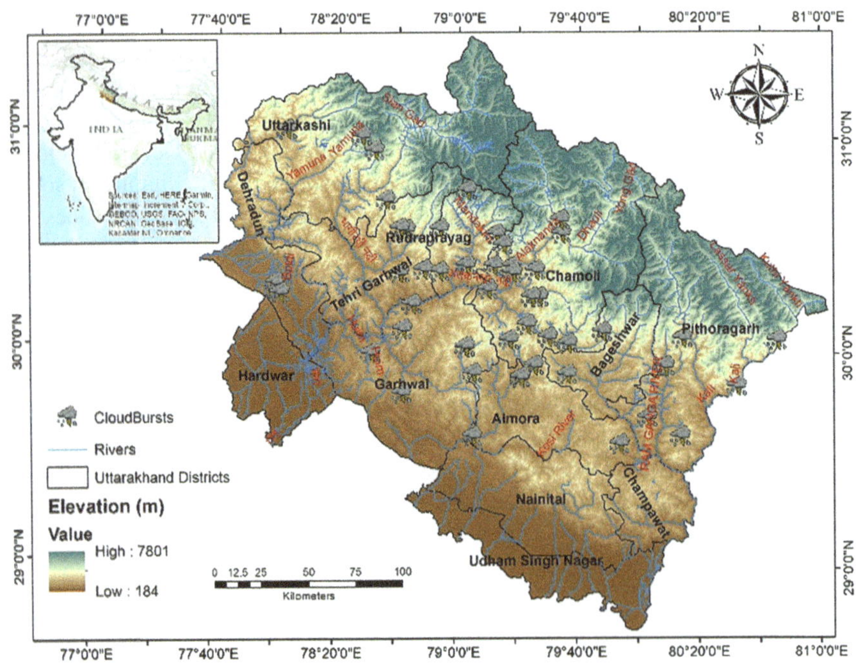

Fig. 9.1 The elevation map of Uttarakhand with the cloudbursts location

The region is dominated by three major river systems, namely the Ganga system, the Kali River network, and the Yamuna-Tons river system. In high altitudes as well as in the lower Uttarakhand basins, the climate conditions are very distinct. The climate and landscape differ considerably with the elevation. The state has a high estimated annual precipitation of 1229 mm. Cloudburst is now a common problem in the high-altitude area of the Himalayas during the moon and pre-monsoon. In the past 20 years during the monsoon, the state alone witnessed more than 40 incidents of cloudbursts.

9.3 Datasets and Methodology

9.3.1 Cloudbursts Occurrences

To estimate the hydro-meteorological characteristics of the recent cloudbursts in the region, firstly, the collection of occurrences of the cloudbursts, which were documented by various agencies, authors, and news reports, namely National Institute of Disaster Management (NIDM), Uttarakhand State Disaster Management Authority (USDMA), [6, 7] and India Meteorological Department (IMD). Around 44 locations were identified, in which 14 incidents were reported in Chamoli, 5 in Pauri Garhwal and Pithoragarh each, 4 in Almora, Tehri, and Uttarkashi each, 3 in Rudraprayag, 2 in Bageshwar and Dehradun each, and 1 incident in Yamunotri. The detail of some major incidents (Fig. 9.2) and their estimated impacts are as follows:

a. August 11, 2001—Around 27 humans, 64 animals, 22 houses were lost due to the cloudburst-induced flash flood near the Phata village of Rudraprayag.
b. August 31, 2001—28 houses, 7 animals, and 7 humans were reported missing in the Gona village, Tehri.
c. September 15, 2010—Around 2 villages were washed away due to the cloudbursts induced flash-floods in Almora.
d. August 3, 2012—A cloudburst incident was reported in Pandrasu ridge of Uttarkashi. The resulting flash-floods affected around 35 humans, 436 livestocks and damaged around 591 houses.
e. September 12–14, 2012- A series of cloudburst occurred in the Rudraprayag district that resulted in the loss of about 66 peoples.
f. June 16–17, 2013—Numerous incidents of cloudbursts and resulting flash-floods were reported in Kedarnath, Rambada, Badrinath region of Uttarakhand. This incident killed more than 10,000 people, around 400 houses were damaged and more than 1 lakhs people were affected.
g. May 28, 2016—Around 120 houses and100 animals were lost in the cloudburst incident in Kemra village of Tehri.
h. July 2–19, 2018—Multiple incidents of cloudbursts and flash-floods were reported in the Uttarakhand at Garhwal, Pithoragarh, Dehradun, Chamoli, and

Fig. 9.2 Major cloudburst incidents in Uttarakhand followed by flash flooding caused huge damages at **a** Kedarnath, Rudraprayag-2013 **b** Thipak village, Chamoli-2016 **c** Pithoragarh-2018 **d** Uttarkashi-2019; Image Source: https://www.hindustantimes.com/; https://www.firstpost.com/; https://primenewslive.com/; https://www.indiatoday.in/

Uttarkashi district. Near about 20 people, 15 houses, 15 shops, 4 roadway bridges, several vehicles, and cowsheds were damaged.

i. August 8–18, 2019—Several cloudburst events were reported in Chamoli, Rudraprayag, Tehri, Garhwal, Bageshwar, Uttarkashi, and Pouri district of Uttarakhand. More than 30 people were reported dead.

9.3.2 Dataset Collection

To evaluate the trends and correlation of various cloudbursts influencing factors, the satellite observations were incorporated. The datasets were retrieved using various algorithms in Google Earth Engine [8]. The datasets used in this study are as follows:

- The Digital Elevation Model has obtained from Shuttle Radar Topography Mission (SRTM) satellite datasets at 30 m resolution.
- The European Centre for Medium-Range Weather Forecasts (ECMFW) ReAnalysis (ERA5) from year 2001 to 2019 at 0.25° was used for the analysis of air temperature, dew point temperature and pressure.
- The Climate Hazards Group InfraRed Precipitation with Station data (CHIRPS) from year 2001 to 2019 at 0.05° resolution was used for the analysis of rainfall.

- The Land Aerosol Optical Depth Daily (MCD19A2.006: Terra and Aqua) at 1000 m resolution from the year 2001 to 2019 was used to study the Aerosol Optical Depth.

9.4 Methodology to Assess the Cloudbursts

The study of the characteristics of the cloudburst requires the identification of numerous influencing factors. These factors have been categorized into three division, each representing a vital cause for the occurrence of cloudbursts. They are, namely Climatic, Orographic, and Anthropogenic cause. The climatic factor being the primary cause, includes precipitation, air and dew point temperature, surface pressure over a region, and their interaction with the environment.

The orographic factor mainly includes the elevation of the region. In the Himalayan regions, the orography along with the hydro-meteorology generates large-scale forcing leading to synoptic-scale extreme events that further lead to flooding. The anthropogenic cause includes the changes in the land-cover patterns and aerosols and its impact on the natural and atmospheric environment. The monthly time series of these factors were evaluated and correlations were plotted to study their variability with the topography. The overall methodology to assess the cloudbursts for the region is shown in Fig. 9.3.

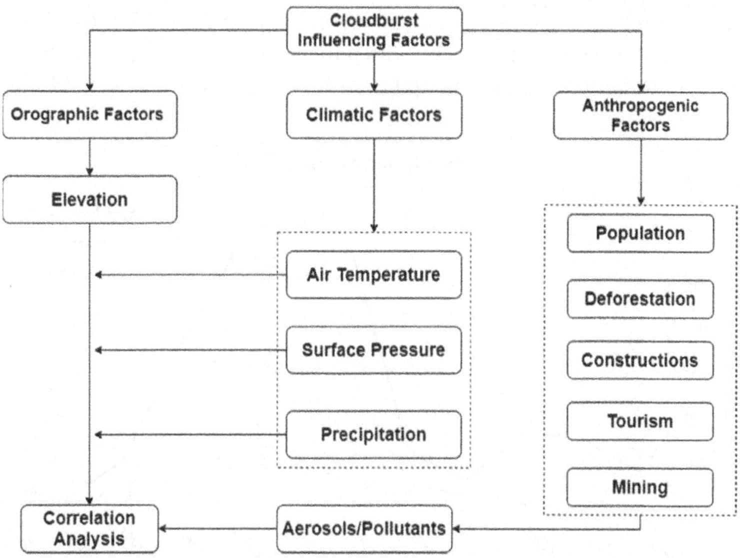

Fig. 9.3 The overall methodology adopted to assess the cloudburst impact over the region

9.5 Results and Discussion

Extreme rainfall associated with cloudbursts and resulting in flash-floods over the Himalayan region is a result of numerous complex and non-linear processes. The atmospheric flows together with orographic forcing interactions invigorate flash-floods. The GFI-Global Flood Inventory by [9], shows that most of the floods (64%) events occur due to heavy rainfall of shorter duration, followed by torrential rains (11%). These heavy rain occurrences are primarily attributed to regional factors such as lake outbursts, cloudbursts, orographic forcing, and landslides.

9.5.1 Climatic Factors

The climate of India is generally governed by the summer and winter monsoon. On account of varying elevation and rugged topography, the climate of the Himalayas differs across the regions [10, 11].

Precipitation Active monsoon is the most common extreme rainfall events in the Himalayas that occur every year. The winds from the southwest cause rainfall during the summers while during winters the region receives the rainfall and snowfall from the Mediterranean region.

The annual precipitation time series of Uttarakhand from 2000 to 2019 shows that the minimum and maximum annual rainfall ranged from 1535 to 2338 mm in the year 2001 and 2011, respectively, with a mean annual rainfall of 1865 mm. The maximum, minimum and mean precipitation curve was plotted (Fig. 9.4) and it was

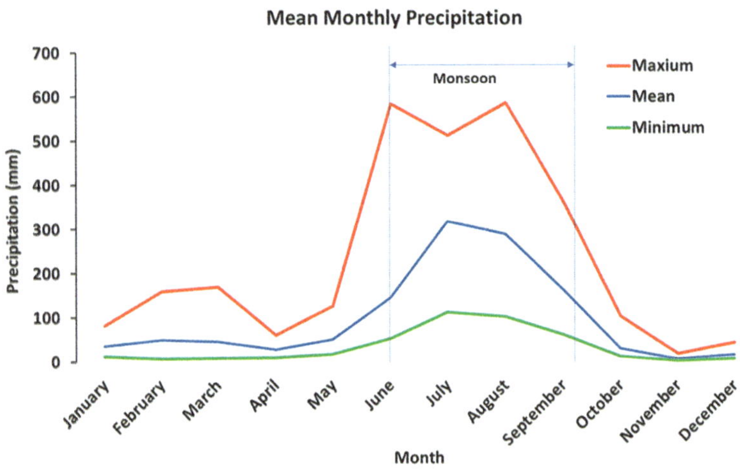

Fig. 9.4 The minimum, maximum, and mean monthly rainfall curve for Uttarakhand

found that in the monsoon season, the monthly precipitation ranges from 364 to 586 mm, which may cause the occurrences of frequent flashfloods.

Air and Dew Point Temperature Increase in temperature increases the atmospheric humidity and leads to extreme rainfall occurrences [12]. During a rainfall event, the temperatures can decrease due to the movement of cold air associated with the rainfall event and evaporative cooling. Also, the distribution of rainfall also varies throughout the months as shown in Fig. 9.5. In this study, the region was divided into six seasons based on the air temperature statistics, viz, pre-summer, summer, post-summer, pre-winter, winter, and winter as shown in Table 9.1.

The amount of moisture and humidity in the air is indicated by the dew point temperature. The increase in dew point temperature increases the moisture content of the air, such that when it becomes equal to the air temperature, precipitation occurs [13]. From the time series analysis of the air temperature from 2000 to 2019 for Uttarakhand, it was found that the mean air temperature ranges from −4.72 °C

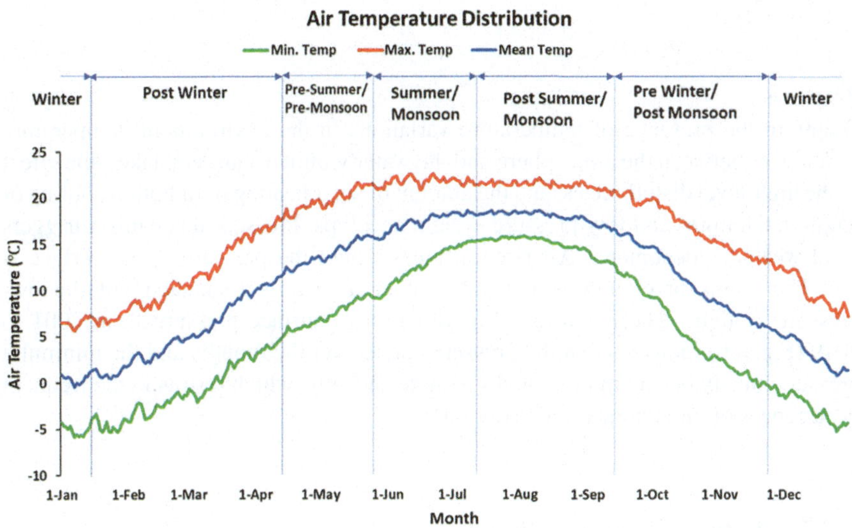

Fig. 9.5 The distribution of Air temperature w.r.t. the seasons for Uttarakhand

Table 9.1 Distribution of seasons based on Air Temperature

Pre-summer/pre-monsoon Mean 14.3 °C /Min 12.0 °C/Max 16.2 °C		Summer /monsoon Mean 17.8 °C/Min. 16.4 °C/Max 18.7 °C		Post-summer /monsoon Mean 17.9 °C/Min 16 °C/Max 18.6 °C	
16-Apr	31-May	1-Jun	15-Jul	16-Jul	15-Sep
Pre-winter /post monsoon Mean 10.2 °C/Min 5.0 °C/Max 16.2 °C		Winter Mean 2.1 °C/Min – 0.4 °C/Max 5.2 °C		Post-winter Mean 5.7 °C/Min 0.5 °C/Max 12.6 °C	
16-Sep	30-Nov	1-Dec	15-Jan	16-Jan	15-Apr

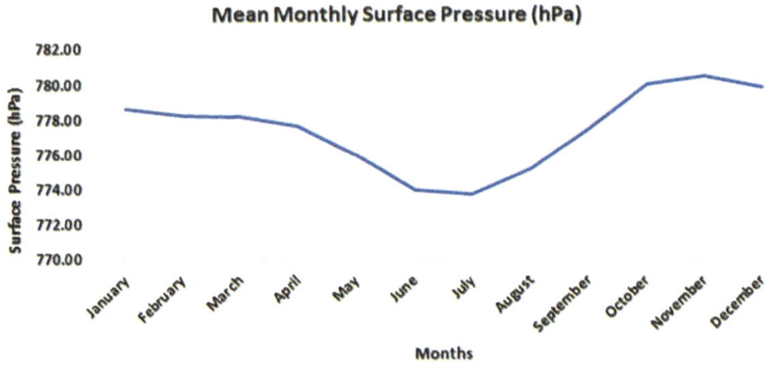

Fig. 9.6 The mean monthly surface pressure distribution for Uttarakhand

to 21.86 °C in the months of January and June, respectively, and the mean dew point temperature ranges from −5.42 °C to 15.94 °C in the months of January and July, respectively.

Pressure
Owing to the presence of temperature variations in the environment, temperature differences between the atmosphere and the water within oceans and lakes, the effect of the high-level disturbances, and the amount of solar heating or radiation cooling of a region the high- and low-pressure system develops. Pressure mechanism triggers local weather phenomena. Air rises in areas where the pressure at the surface is low. The clouds, precipitation, and bad weather are mainly associated with the low-pressure systems. The mean monthly variation of surface pressure from 2001 to 2019 (Fig. 9.6) indicates that the pressure varies over the months and the minimum pressure mostly occurs in the months of June and July, which results in the frequent occurrences of cloudbursts in this region.

9.5.2 Anthropogenic Activities

The region's vulnerability to disasters has increased due to unscientific development and patterns of land use, deforestation, increased pressure on humans and cattle populations, poor socio-economic conditions, increased tourism. According to the census, the population increased to 100.86 lakhs in 2011 from 84.89 lakhs in 2001 and the average population density in Uttarakhand rose to 189 in 2011 from 159 in 2001 (19%). The extensive population growth over a decade has adversely affected the already fragile hill state environment by increasing demand for development and forcing people to engage in more agriculture and other developmental activities in sensitive areas. As the tourist influx has increased, many multistoried hotels and other amenities were built even in the environmentally sensitive areas of the city.

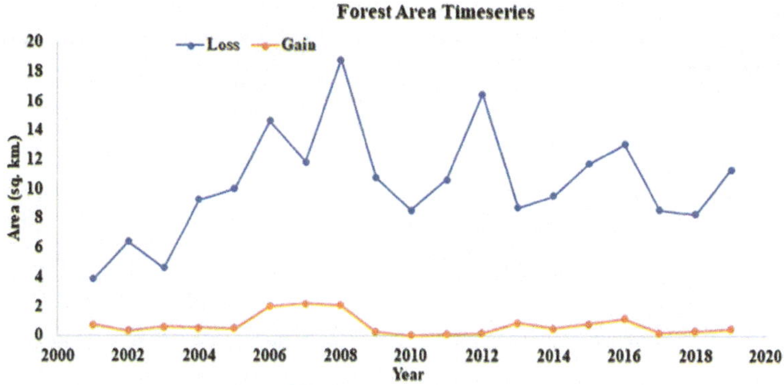

Fig. 9.7 The timeseries forest cover change for Uttarakhand from 2001–2019

Some of the hotels are designed for tourists/pilgrims on the banks of rivers and even in flora of the plain [14, 15]. The forest degradation due to these activities has led to increase in the events of landslides and soil erosion which further silts the reservoirs. The time series analysis of forest cover from 2001–2019 using Global Forest Change datasets [16] revealed that in the Uttarakhand region, every year around 9.63 km^2. of forest loss occurs on an average (Fig. 9.7). This loss in forest cover hinders the evapotranspiration and thus leads to disturbance in the water cycle and affects the precipitation pattern throughout the region.

All these anthropogenic activities lead to the development of aerosols. The connection between aerosol clouds and precipitation is a concern in the field of aerosol and has a direct impact on the atmosphere. The rise in anthropogenic aerosol emissions in recent years and the aerosol-cloud-precipitation interaction analysis has become increasingly relevant in the study of atmospheric pollution. The aerosols directly influence earth-atmosphere layer radiation by absorbing and scattering sun radiation; the indirect effect of aerosols relates to the presence of cloud condensation nuclei (CCN) or ice nuclei (IN) in aerosols. In addition, mesoscale weather patterns, such as cloudbursts and thunderstorms, create changes in the properties of atmospheric aerosols over short time scales, either by removing them from the environment or by spatially redistributing them.

9.5.3 Orographic Influence

The mountains act as a barrier to the winds carrying moisture and force them to ascend. The windward side of the mountains receives the maximum precipitation. The interaction of mountains with the large-scale atmospheric flow and localized events determines the rainfall amplitude. The Indian Summer Monsoon (ISM) generates strong rainfall on the southern slopes of the Himalayas, while ISM precipitation is deprived in the northern Himalayas. Due to the altitudinal variation, the rainfall

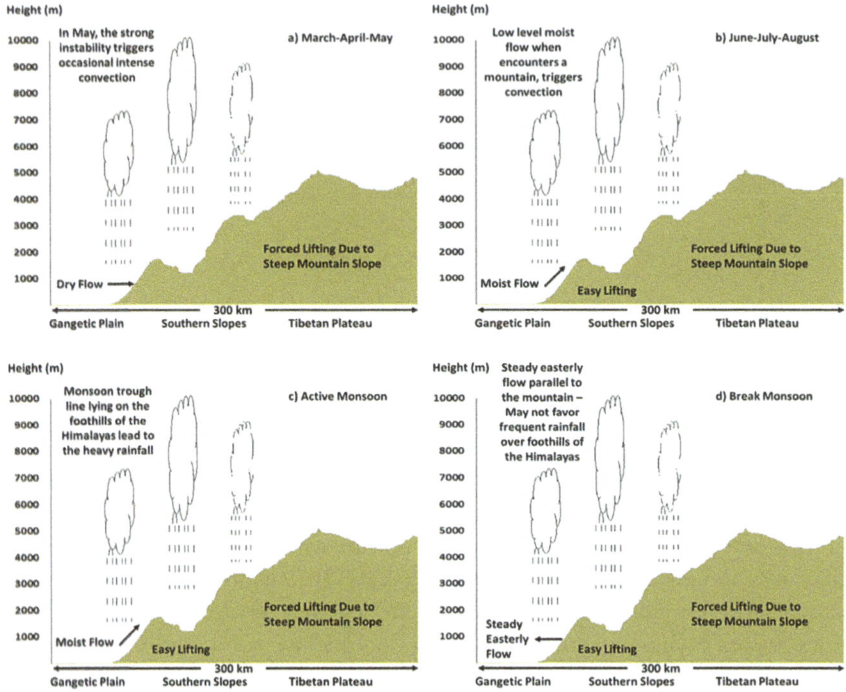

Fig. 9.8 Schematic illustrations of spatiotemporal variation in rainfall for **a** pre-monsoon (March–April-May), **b** summer monsoon (June-July–August), **c** active monsoon, and **d** break monsoon modified from [19]

distribution becomes complex. Some studies indicate that the trend of rainfall rises with altitude up to around 2500 msl and then starts to decrease. Studies the Himalayas relating to rainfall characteristics during the monsoon season suggest that rainfall decreases with altitudes between 2800 and 4500 m for Nepal [17, 18]. The seasonal spatiotemporal distribution of rainfall is shown in Fig. 9.8.

In the present study, around 500 random points were taken over Uttarakhand, and the scatter plots of Elevation vs Mean precipitation, Air temperature, Surface pressure, and Aerosol Optical depth were plotted for 2000–2019. The scatter plot of mean precipitation vs elevation (Fig. 9.9a) range shows that the mean precipitation showed negative correlation with the elevation and most of the precipitation was concentrated in the elevation range of 1000–2000 m. A scatterplot between air temperature and elevation (Fig. 9.9b) also follows strong negative correlation, which explains the presence of higher temperature at lower elevated areas. Also, the scatter plot of surface pressure vs elevation (Fig. 9.9c) also follows a negative correlation which indicates that at higher elevation the presence of low-pressure causes increases in evaporation which further leads to cloud formation. Furthermore, the scatter plot of aerosols optical depth vs elevation (Fig. 9.9d) follows negative correlation, and

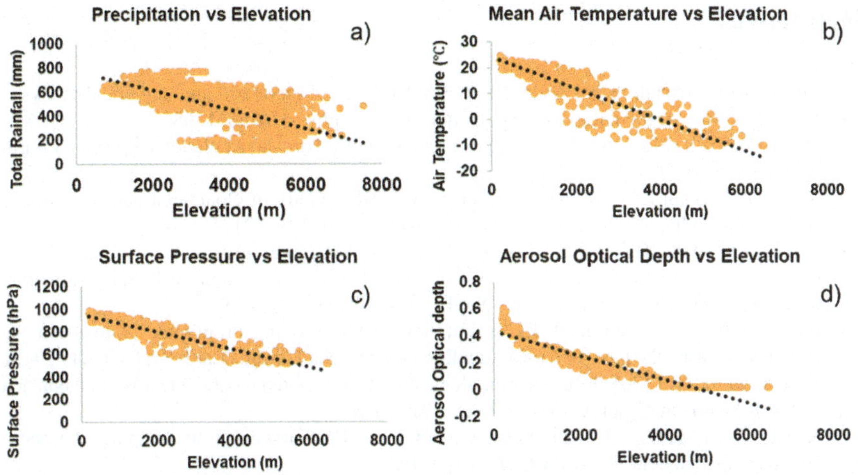

Fig. 9.9 The variation of **a** Precipitation with Elevation **b** Air temperature with Elevation **c** Surface pressure with Elevation **d** Aerosol Optical Depth with Elevation

most of the aerosols are concentrated at 100–1500 m elevations that facilitate the clouds forming mechanism.

9.6 Conclusion

Cloudburst and associated disaster are prominent in the Uttarakhand region of the Himalayas, affect thousands of people every year, and cause loss of life, property, livelihood, infrastructure, and environment. Most of the cloudburst events are concentrated during the monsoon months, and majority of these incidents occur during night times. The southwest and Mediterranean winds along with regional moisture are dominant during this period, and they descend over the low-pressure region and combine with the local humidity are further orographically lifted and cause cloudburst and consequently flash-floods. The region's vulnerability to disasters has also increased due to poor patterns of land use & socio-economic conditions, deforestation, increased pressure on humans and cattle populations, and increased tourism. These activities in these areas have further led to the development of aerosols, which influence the earth-atmosphere layer radiation by absorbing and scattering sun radiation and further impacts the cloud formations. While cloudburst events are difficult to predict, a dense network of rain gauges, especially in areas prone to cloudburst, is needed to better understand this phenomenon. The ground stations datasets coupled with satellite imageries can prove to be a powerful tool for the assessment of cloudbursts.

References

1. Chaudhuri C, Tripathi S, Srivastava R, Misra A (2015) Observation- and numerical-analysis-based dynamics of the Uttarkashi cloudburst. Ann Geophys 33(6):671–686
2. Das S, Ashrit R, Moncrieff MW (2006) Simulation of a Himalayan cloudburst event. J Earth Syst Sci 115(3):299–313
3. Thayyen RJ, Dimri AP, Kumar P, Agnihotri G (2013) Study of cloudburst and flash floods around Leh, India, during August 4–6, 2010. Nat Hazards 65(3):2175–2204
4. Singh S, Dhote PR, Thakur PK, Chouksey A, Aggarwal SP (2020) Identification of flash-floods-prone river reaches in Beas river basin using GIS-based multi-criteria technique: validation using field and satellite observations. Nat Hazards 105(3):2431–2453
5. Barros AP, Kim G, Williams E, Nesbitt SW (2004) Probing orographic controls in the Himalayas during the monsoon using satellite imagery. Nat Hazards Earth Syst Sci 4(1):29–51
6. Joshi V, Kumar K (2006) Extreme rainfall events and associated natural hazards in Alaknanda valley, Indian Himalayan region. J Mt Sci 3(3):228–236
7. Kansal ML, Thakur A (2017) Flood estimation for a cloudburst event in an ungauged western Himalayan catchment. Int J Hydrol 1(6):163–168
8. Gorelick N, Hancher M, Dixon M, Ilyushchenko S, Thau D, Moore R (2017) Google earth engine: planetary-scale geospatial analysis for everyone. Remote Sens Environ 202:18–27
9. Adhikari P, Hong Y, Douglas KR, Kirschbaum DB, Gourley J, Adler R, Robert BG (2010) A digitized global flood inventory (1998–2008): compilation and preliminary results. Nat Hazards 55(2):405–422
10. Mishra AK (2020) Remote-sensing monitoring of a cloudburst event in north India. Remote Sens Lett 11(4):383–387
11. Sati VP (2013) Extreme weather related disasters: a case study of two flashfloods hit areas of Badrinath and Kedarnath Valleys, Uttarakhand Himalaya, India. J Earth Sci Eng 3:562–568
12. Funk C, Peterson P, Landsfeld M, Pedreros D, Verdin J, Shukla S, Husak G, Rowland J, Harrison L, Hoell A, Michaelsen J (2015) The climate hazards infrared precipitation with stations - A new environmental record for monitoring extremes. Sci Data 2:1–21
13. Qasem SN, Saeed S, Hamed SN, Mosavi A, Shamshirband S, Chau K (2019) Estimating daily dew point temperature using machine learning algorithms. Water 11(3):582
14. Kansal ML, Shukla S, Tyagi A (2014) Probable role of anthropogenic activities in 2013 flood disaster in Uttarakhand, India. In: World Environ. Water resource congress 2014 water without borders—Proceedings 2014 World Environment Water Resource Congress, May, pp 924–937
15. Pandey VK, Mishra A (2018) Trends of hydro-meteorological disaster in Uttarakhand India. Int J of Cur Res in Sci and Tech 4(12):1–7
16. Hansen MC, Potapov P, Moore M, Tubanova SA, Tyukavina A, Thau D, Stehman SV, Goetz SJ, Loveland TR, Kommareddy A, Egorov A, Chini L, Justice CO, Townshend RG (2013) High-resolution global maps of 21st-century forest cover change. Science 342(6160):850–853
17. Bookhagen B, Burbank DW (2010) Toward a complete Himalayan hydrological budget: Spatiotemporal distribution of snowmelt and rainfall and their impact on river discharge. J Geophys Res Earth Surf 115(3):1–25
18. Nandargi S, Dhar ON (2011) Extreme rainfall events over the Himalayas between 1871 and 2007. Hydrol Sci J 56(6):930–945
19. Shrestha D, Singh P, Nakamura K (2012) Spatiotemporal variation of rainfall over the central Himalayan region revealed by TRMM precipitation radar. J Geophys Res Atmos 117(22):1–14

Part II
Geohazards

Chapter 10
Review on Landslide Early Warning System: A Brief History, Evolution, and Controlling Parameters

Varun Menon and Sreevalsa Kolathayar

10.1 Introduction

Extreme natural events are an inevitable on earth. The best way is to mitigate the disaster by reducing exposure and vulnerability. But the lack of information regarding the incursion of these disasters always causes chaos in terms of infrastructural damages and loss of life. Some researchers consider a landslide as a fall, topple, and spread of landmass, and the landmass flow failure as a flow slide [1]. It is an impending disaster triggered by either excess precipitation [2], seismic activities [3, 4], or artificial ground alteration processes. Out of these, 93% of landslides between 1972 and 2007 are caused by heavy rainfall. The major share of landslides is observed in the monsoon season in most parts of the world [5]. Nature has a special bond with our surroundings, and whenever there is a change in climate our surroundings tend to adapt to that conditions. The landscapes tend to change according to the climatic changes and that affects the meteorological conditions and results in a landslide [6]. Southern states of India like Kerala, Karnataka, and north-eastern states like Meghalaya, Tripura, etc., are blessed with the high amount of rainfall each year caused by south-western monsoon winds. This blessing sometimes becomes the reason for disasters like floods, soil erosion, and landslides. Because of the lack of relevant long-term data in these areas, a physical model-based study to evaluate the nature-human interactions is required to assess the landslide initiation process. The insight from these studies will help to develop a better Landslide early warning system (LEWS).

The major duty of a geotechnical engineer is to protect the landslide-prone areas or give necessary warning to avoid loss of life with minimum cost [7]. The controlling of the landslide is quite difficult, rather it is advised to avoid it by suitable LEWS. The birth of LEWS has been in a transition stage. A successful LEWS consists of a continuous monitoring system with the ability to withstand the weathering actions. Slope

V. Menon (✉) · S. Kolathayar
National Institute of Technology Karnataka, Surathkal, India

S. Kolathayar et al. (eds.), *Civil Engineering for Disaster Risk Reduction*, Springer Tracts in Civil Engineering, https://doi.org/10.1007/978-981-16-5312-4_10

monitoring is a key component for hazard assessment of a landslide for the protection against the aftermath of the disaster [8]. LEWS was suggested for the prediction and monitoring of landslide-prone areas since the study about landslide mechanics was started. There are various instrumentation techniques available. Instrumentation is the key to successful monitoring, control, and evaluating the effectiveness of remedial measures in any type of landslides. There can be different parameters used by the field measuring instrument, but most of the LEWS is based on rainfall threshold [9]. The important factor regarding this threshold is its quantity and quality. As the quantity and clarity of the data increase, the threshold value also changes. And there are different rainfall thresholds calculated for almost similar regions by past researchers [6, 10, 11]. The performance can always be increased as new sets of data is available. This is the main justification for the laboratory parametric study. In laboratory studies, one can control the rainfall parameters as well as the soil parameters including the tilt. It can provide a unified solution based on the results rather than a probabilistic approach. The major drawback of LEWS with the parameters based on rainfall threshold is the poor spatial resolution. The LEWS can give a general warning for the whole area, but cannot pinpoint the exact location of the failure [9]. The rainfall threshold can be established based on moisture content in the soil, or the increase in the pore water pressure. The moisture content is based on the weight of the soil, whereas the pore pressure variations affect its shear strength based on the intrinsic properties of the soil. Some researchers have also considered temperature variation studies to incorporate the effects of snowfall in landslides. The monitoring is done for different features of the soil as shown in Fig. 10.1 [7].

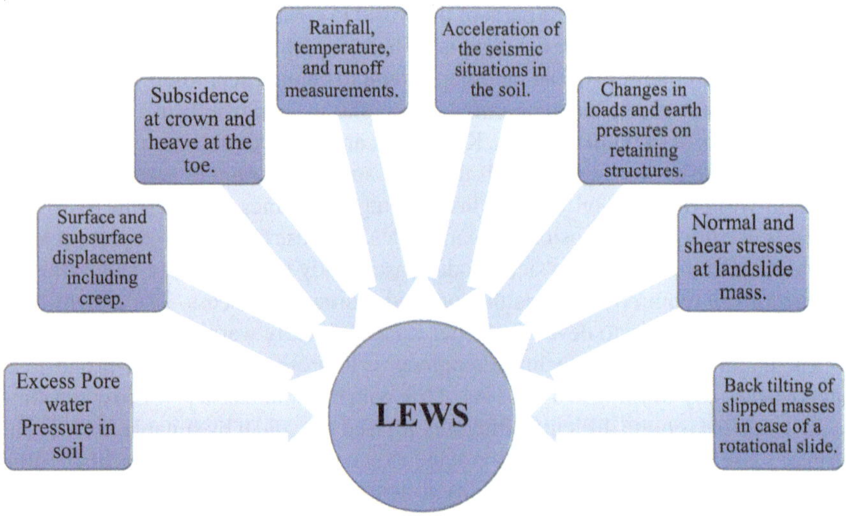

Fig. 10.1 Parameters that can be utilized for developing a LEWS

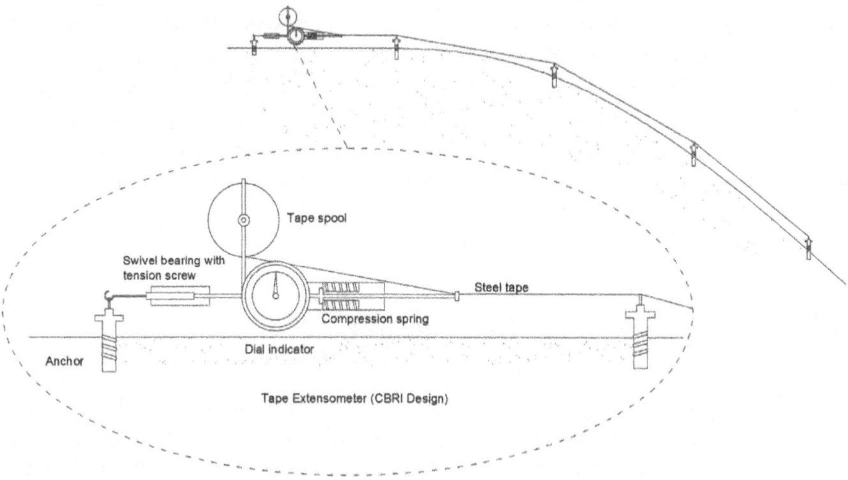

Fig. 10.2 Tape extensometer used for slope monitoring

The pioneer works have been done using slide fences incorporated with alarms [12]. Then, other mechanical instrumentation techniques like Extensometers, Tilt-meters, Deflectometer, Linear Variable Differential Transformer (LVDT), Strain meters came into practice which can be used as displacement measuring instruments [7]. Lately, photogrammetry techniques like Light Detection and Ranging images (LiDAR) have been used for finding surface deformations and displacement in unsafe slopes. Figure 10.2 shows the earliest model developed for slope monitoring by Central Building Research Institute, India [7]. The major limitations of this method were: the measurements are relative displacements, and the tape lengths are available up to 30 m long only. Most of the recent works available in the literature focus on electronic measurements like sensors incorporated with DAQ, as they do not have such constraints for measurement. Also, rain gauges are an important accessory in case the focus is on rainfall-induced landslides. Sometimes, just by simple observations an expert can detect the possibility of an impending landslide. The natural deformations in the locality like the behavior of the vegetative cover around the slope area are an important accessory for this. As the ground starts to move, the trees standing on the surface tend to back tilt, which can be observed further by the shape of the trunk, which will always grow vertically from the tilted portion of the tree making a bend on the trunk. In some places, the trees tend to forward tilting as well, allowing the upper portion to move toward the slide direction. The vegetation can be used as slope protection. An example of this was found in the site investigation of Dehradun-Mussorie landslide in India, where tall vertical trees near tension cracks made the slide inactive for more than 30 years [7].

Landslide studies can be further improved by creating a good database. A hazard database like this should include relevant data regarding the event in a unified pattern that is accepted by researchers around the world. The Global Fatal Landslide Database

(GFLD) is a perfect example of this. GFLD consists of a landslide database from 2004 to 2016. This has been continuously updated for almost a decade on a worldwide scale, and it is openly available in many Geographical Information System (GIS) platforms [13]. As new information becomes available through government aided agency reports, academic papers and personal communications, the database will be updated. But this database lacks the geotechnical properties of the location to learn about geotechnics in the landslide mechanism.

10.2 Landslide-Rainfall Relationship

The influence of water will create changes to the shear strength of the soil. The pore pressure rise in the already saturated soil will decrease its shear strength and evidence from the unconfined compression test shows that there is an 84% decrease in the shear strength with an increase in 30% in moisture content [14]. The major parameter that influences the landslide during torrential rain is the decrease in the shear strength with the increase in the pore pressure when the loose soil layer contracts [15–17]. Landslide flume studies were conducted in laboratories around the world to understand the landslide mechanics during rainfall with various intensities [18]. The duration and intensity of rainfall also seem to affect the landslide. A high-intensity short-duration rainfall after a low-intensity long-duration rainfall can cause significant landslides in the tropical regions [19]. The parametric studies can be conducted using previously established methods like flume study incorporated with a rainfall simulator, and the pore pressure variations are monitored using pore pressure sensors or hydraulic piezometers as shown in Fig. 10.3.

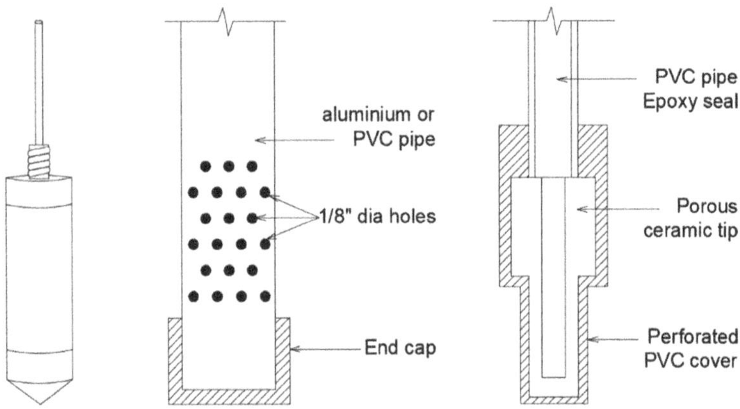

Fig. 10.3 Hydraulic Piezometers

Researchers tried to establish some relationship between the failure and moisture content for a long time. Approximate mobility index (AMI) is a factor corresponding to the soil layer contraction leading to flow failure. It is the ratio between the percentage of saturated moisture content and liquid limit [20]. If AMI > 1, the soil mass fluidizes and doesn't fluidizes when AMI < 0.45. Between 0.45 and 1, the soil mass fluidizes only when abundant water is supplied. AMI is established with the help of laboratory experimentation. One other method of finding out the relationship between rainfall and landslide is from the recorded data regarding rainfall-initiated landslides.

Rainfall threshold is a related term when it comes to the rainfall-induced landslide [6, 11, 21–25]. The word threshold means the point above which the probability of landslide is very high. The threshold may be the duration, intensity, or quantity of rainfall with reference to the soil properties. But the soil properties may be site-specific, hence finding a unified method is very difficult. The triggering of a landslide is mostly caused by accumulative rainfall rather than a single-day rainfall [22, 25, 26]. Many threshold empirical results have been developed, $I = 14.82\ D^{-0.39}$, $I = 15.58\ D^{-0.52}$, and $I = 13.35\ D^{-0.44}$ are few of the examples [10, 27]. An entire inventory of threshold values focused on Italy has been made from the data collected in the past [11, 28]. The threshold value and its corresponding equation are only site-specific because a case study in Wayanad district in the state of Kerala, India shows various thresholds for different locations from 0.75 to 1 mm/h for an entire day to cause landslide in those regions using frequentist method [29]. A threshold can be obtained only from three distinct ways as shown in Fig. 10.4. Field monitoring and finding the threshold are the most superior method, but any problem demands a solution rather than waiting for the impending disaster. An incomplete database will cause problems in the threshold calculation, and the current data available for the public in India does not contain the intensity and duration (in hours) of rainfall. Rather the available data consists of the quantity of precipitation per day. When the available data regarding the rainfall is on a coarser scale, the chances of misinterpreting the threshold value are higher [29] that is why, a better version of rainfall gauges is required for data collection of landslide-prone areas. Hence, the laboratory parametric study is also very important. That is where, the geotechnical engineers come into

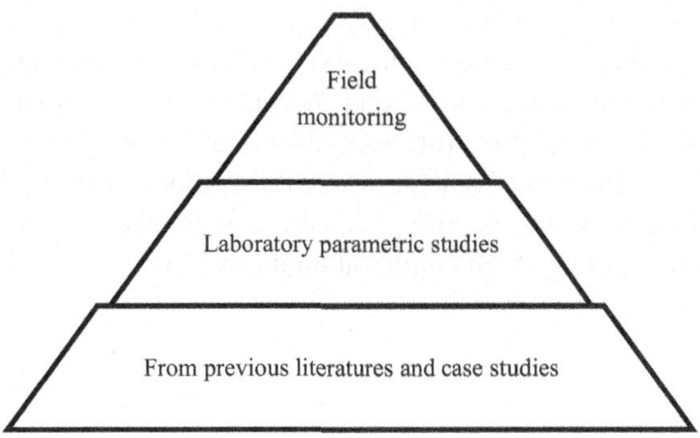

Fig. 10.4 Methods to find rainfall threshold in a hierarchical form

the play for the identification of the rainfall threshold using available resources and knowledge about the landslide mechanics. The data collected during an event is much more valuable because the chances are very rare to obtain such data. In a mudslide recorded previously, the water content at the slip surface seems to be higher than the water content below it and slightly higher than the one above it [30, 31]. And creep movements due to rainfall can come in many varieties of rates like from 1 mm per year to 100 mm per year [32], [33], hence precision instrumentation is necessary.

Here is an example of an early warning system developed using empirical relations. A shallow rainfall-induced landslide is a hydraulically driven first-time failure [23]. In-situ pore water pressure, water content, and rainfall measurement are incorporated in the early warning system and added an alarm on the thresholds of these field variables. The following Eqs. (10.1) and (10.2) are used to establish a threshold value in different slopes for developing an early warning system [23].

$$FoS = \frac{\tau_f}{\tau_m} = \frac{c' + \sigma_n' \tan\varphi'}{\tau_m} = \frac{c' + \left[\left(\gamma z \cos^2\beta - p_w\right)\right].\tan\varphi'}{\gamma z \cos\beta \sin\beta} \tag{10.1}$$

$$FoS = \frac{c' + \left[\left(\overline{\gamma} z \cos^2\beta - p_a\right) + S_r(p_a - p_w)\right].\tan\varphi'}{\overline{\gamma} z \cos\beta \sin\beta} = \frac{c' + \left[\left(\overline{\gamma} z \cos^2\beta\right) + S_r s\right].\tan\varphi'}{\overline{\gamma} z \cos\beta \sin\beta} \tag{10.2}$$

Here, τ_f is the available shear strength of the soil, τ_m represents the mobilized shear strength, c' is the cohesion intercept, σ'_n represents the effective stress normal to the failure plane, γ is the soil's total saturated unit weight, β is the slope angle, and z represents the considered depth of the Mohr–Coulomb failure plane. Where, $\overline{\gamma}$ is the average soil unit weight down to depth z of the failure plane. The effective stress acting normal to the failure plane has a total stress component that is dependent on the slope angle (β), average soil unit weight, and capillary stress component ($S_r s$).

Even though such methods are available, a database of the landslides with the intensity and duration of rainfall in that location is very important for the implementation of a LEWS on a global scale. As the data increases, the database becomes more reliable, and accurate threshold values can be calculated. A better understanding of the landslide causes and mechanics is very much necessary before assembling the monitoring instrumentations to get meaningful results [34]. The model study is considered an advantage where the availability of field data is very poor. In addition, the pore water pressure threshold can be taken much precisely. Physical models are more reliable than empirical methods [23].

10.3 Landslide Monitoring and Instrumentation Techniques

A successful LEWS start from good monitoring capabilities. LEWS are in a trance state, better technology is being introduced each day and those can be implemented for better monitoring capabilities. Resolution, ruggedness, reliability, and durability are important traits of good instrumentation [35]. The major objective of instrumentation is to provide a timely warning. Cost is also a very important factor when it comes to field deployment, the instrumentations provided should be as economical as possible. The instrumentation is helpful to measure the in-situ pore pressure, identifying slip surface, its depth, and shape, monitoring the existing structures, measurement of cracks, identify subsidence and heave in the soil surface, change in load and earth pressure, acceleration of soil during seismic situations, displacements in the soil surface, and to serve as evidence for a lawsuit. Each type of instrument comes in different models and resolutions, and hence, it is necessary to identify the best among the available tools for the specific type of monitoring [35].

The easiest way of finding relative displacements in the slope surface is to conduct the survey using triangulation or electronic distance measurement systems (EDM) toward the nearest landmarks like pillars, moving pins, trees, mortar patches, monuments, etc. [35]. Most of the movements in an unprotected slope happen in a matter of seconds and then there will be a quiet time. Hence, a continuous monitoring system is very important. Wire or tape extensometers (Fig. 10.2) have been used for the identification of slope stability for a long time using a fixed point or monument as reference. But wire extensometer cannot properly interpret the movement pattern and identify the slip surface and depth. By using a potentiometer and incorporating a global positioning system (GPS), the measuring sensitivity can be increased and the horizontal and vertical components can be found [36, 37]. The displacement can be measured by making a monument on the ground as a reference, using electronic distance measurement (EDM) equipment [36]. This method assumes that the displacement that occurred is at a constant pace. But this assumption leads to an inaccurate interpretation of the actual landslide behavior. In-situ hydraulic pore pressure sensors (Fig. 10.3) can also be used to monitor the pore pressure differences during the measurement of the surface deformations.

Measurement of the variations in the inclination of the slope is another way to detect the probability of landslide. An automatic inclinometric system (AIS) has been used in the Cabella Ligure, Italy for early warning [38]. AIS is an evolved form of conventional tiltmeters and inclinometers. AIS consists of accelerometer sensors, load cells, and electric motors to move the probe that measures the slope, up and down. The results obtained from this system have a good correlation with cumulated rainfall and cumulated displacement that occurred at a depth where the sliding surface has located. But there seems to be a time lag between the rainfall peak and the peak movement as reported by the researchers. To reduce the errors and for continuous monitoring and cross-checking, the target detection and tracking method can be used. TDT can be used to identify the surface displacements in the slope surface [39]. This method consists of using high-quality cameras with proper positioning to capture the images and using image processing tools available in mathematical software like MATLAB to prepare codes based on computer vision and image processing that give even smaller surface deformations and record them.

As time progresses, the technology follows, the interdisciplinary behavior of engineering subjects has come in handy for the sustainable development of the future. Basic knowledge in electronics and sensor technologies is necessary for any researcher in modern society. There are many examples for such studies, ultra-wideband radiofrequency can be used to measure the distance between a transmitter and receiver. This principle can be used for measuring slope surface deformations. The receiver is placed at an inert location, and the transmitter is placed at the site prone to landslide. If any movement occurs then the transmission radio wave's time of flight (ToF) will be changed and according to that the displacement can be measured [40, 41]. Previously, such systems have been developed with commercially available 2.4 GHz Radio [42]. The major disadvantage of this method is that if any disturbances happen to the transmission medium, which lies between a straight path between transmitter and receiver that will affect the time of flight of the radio waves. Examples of such problems would be rainfall and obstructions to the line of sight [43]. Also, the power source used for such a sensor-based monitoring system should be able to run the system for at least a full season without any disruption to the continuous monitoring, only then it will be useful in the landslide probable seasons like a monsoon.

Wireless sensor network (WSN) is the future of LEWS. In the last decade, WSN has been used extensively because of its simplicity and low cost of installation. Hence, they are widely used for environmental monitoring works [44–46]. These sensors are embedded deep inside the earth up to a depth of 20 m for monitoring purposes. They consist of piezometers, tiltmeters, dielectric moisture sensors, strain gauges, and geophones [47]. These sensors were deployed for a pilot study at Munnar, where the rainfall-induced landslides were a threat. The study focuses more on the electronics components and calibration rather than studying the landslide mechanics of the failure. Also, dielectric moisture sensors can be used to measure the change in volumetric water content in the soil during the rainfall event [47]. Another widely used terminology in the studies regarding LEWS is the Micro-Electro-Mechanical Systems (MEMS). MEMS capable of measuring the tilt of the slope incorporated with moisture sensors that can measure the water content indirectly from the change in the dielectric constant of the soil has been previously used [48].

The most recent techniques are using LiDAR and embedded sensors for the most accurate measurements. Digital image capture, Optical satellite sensors, Google earth platform, Radar satellite sensors, Terrestrial laser scanner, Airborne LiDAR, and Ground-based radar are a few of the recent advances in the sensor capabilities and Data processing methods that have been useful for landslide investigation these days [49]. Photogrammetry is the earliest approach by taking pictures of the ground using aircraft. And recently, the technology is so improved that unmanned aerial vehicles (UAV) are also being used for taking site details by various organizations. The LiDAR stands for light detection and ranging. The latest satellites have cameras with the capability of taking three-dimensional photographs of the surface of the earth by using this technology. Hence, this technique can be used for measuring small deformations in the ground. Ground-based interferometric synthetic aperture radar (GB-InSAR) is one of the modern ways to monitor slope stability and landslide. This method doesn't give any warnings prior but this is useful to identify the smallest of deformations in the surface of the soil, from 0.1 mm can be observed with this method. The principle is similar to the working of a radar, the signals will be backscattered from the soil surface continuously to form a surface image, and the deformations can be observed from these images [50]. An attempt has been made to use Sentinel mission-ESA satellites to collect data of the ground and act like a real-time monitoring system and establish an early warning system [51]. The displacements in the surface of the slope are being measured from the two consecutive interferometric observation images of the same area. These satellites have a very short revisiting time and because of that the possibility of early warning is possible. The Sentinel sensors are capable of measuring the velocity up to 42.6 cm/year [52–54]. Other than this Terrestrial laser scanners (TLS), Infrared thermography, Robotic total stations, Digital photogrammetry workstations (DPW) are also some image-based survey methods to get the displacement of the slope surface. Methods like these have disadvantages like high cost and the need for a clear line of sight between targets [43]. Also, vegetations and human activities provide a shield against this type of monitoring [55].

10.4 Laboratory Parametric Studies

The landslide fluidization occurs in three stages, volumetric compaction of the soil layer due to topsoil or shear, generation of the excess pore water pressure in the saturated zones of the soil, and induction of fast shearing near the slip surface [18]. There seems to be very high-pore pressure generation just after the sliding failure. The pore pressure variation with the change in the basic properties showed that the finer particle will have more pore pressure generation [56]. Further studies are recommended to study the parameters that affect the excess pore water pressure that causes slope instability by changing the grain size and particle distribution [57, 58]. Most of the previous works have been conducted on the sand to simulate the liquefaction conditions easily, whereas some researchers have also tried flume tests using silty sand and silty clay [59, 60]. The term 'liquefaction' has been used by many researchers to describe the solid–fluid transformation. But the loading is considered static only and the condition before the pore pressure generation causing the thin shear zone to lose its strength is termed as liquefaction [61]. Static liquefaction is not a cause, it is the consequence of the impending instability causing landslide [62]. Most of the understanding regarding the liquefaction phenomenon has been studied by simple element testing of smaller samples under idealized conditions, and hence, the uncertainties that come with the location-specific characteristics are not included [61]. The landslide triggered by static liquefaction has been studied in the past and it has been defined as the loss of effective stress in the shear zone due to undrained monotonic loading, and this will develop large plastic strain due to the inability of the soil to sustain given stress [1, 56, 57, 58].

A threshold value is very important for a LEWS. That is the reason for the parametric study using the laboratory flume test being used widely. Even the rainfall threshold can be occurred using an experimental approach using a laboratory flume test. In a previous study using a small tilting flume setup, a threshold has been calculated, which is $I = 80.065\ \mathrm{D}^{-0.596}$ [63], where I is the intensity of rainfall and D is the duration. Another factor is that the soil has a heterogeneous structure, and the properties of the soil are specific to a particular location. These properties can be controlled and changed according to our needs in a laboratory environment. A typical laboratory test setup consists of a flume with two pressurized nozzles to work as a rainfall simulator. The excess pore water pressure variation, according to the change in soil properties like grain size, initial sample thickness, void ratio, and other basic factors can be studied in the case of a simulated environment [18, 56–58, 63–67]. This method of study is also suitable for recommending a remedial measure for the impending disaster. Usually, the failure occurs after a period of 30 min to 2 h, even 24 h long experiments can be done with laboratory experiments with the help of a data acquisition system support to record the data. The major objective of laboratory parametric study is to replicate the soil layer contraction and generation of excess pore water pressure [18]. Most of the landslide flume tests will include instrumentation like pore pressure sensors to measure the excess pore water pressure, a rainfall simulator with flow control valves, and video cameras to record the surface deformations [18, 61]. LiDAR cameras being one of the easily available commodities in the market can be used as a means for measuring surface deformation in future studies. Flume test incorporated with two-nozzled Norton type rainfall simulator has been as a similar study in the past [64]. The calibration of this rainfall simulator was done using a laser disdrometer (OTT-Parsivel). Disdrometers are electronic rain gauges capable of measuring and recording precipitation, intensity, and duration. Several highly sensitive pore pressure sensor with 20 kPa capacity (KYOWA PGM-02 kg) has been used to measure the excess pore pressure. 64 conducted a landslide flume test with the help of Fiber Bragg Grating Sensors (FBG). Slope strain measurements can be conducted using FBG sensors laid along the length of the slope. The strain, as well as the temperature, can be measured with such sensors. FBG sensors along with TDR probes and pressure gauges were used to measure the pore pressure and soil volumetric content. Surface deformations are filmed using high-quality cameras, and this has been used for cross-referencing with the events that occurred during the test. Such facilities are required to conduct a successful laboratory parametric study.

10.5 Principles of Developing a LEWS

The major steps involving in developing an early warning system are given in Fig. 10.5 out of which the response of the people is one of the major concerns

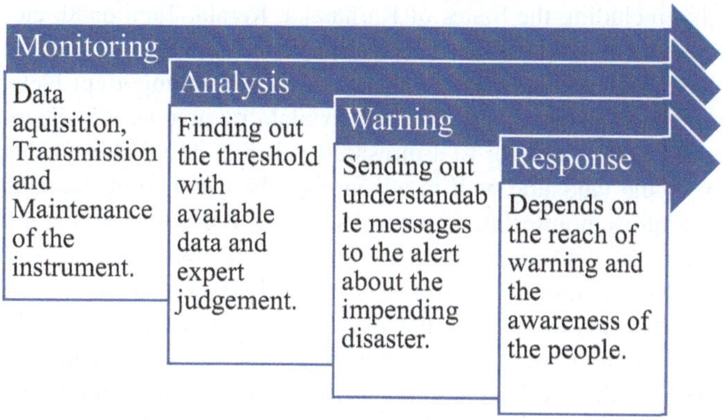

Fig. 10.5 LEWS work chart

in terms of developing an early warning system. How they will react to the warning given by the system is one of the major issues to be faced, while designing a LEWS. Hence, training the public must be considered. The important awareness to the people should be such that the warning should not be a single process, and it should consist of multiple levels. Previous researchers [44] proposed a three-level warning system namely, Ordinary level, Attention level, and Alert level. Depending on the reliability of the data available, expert judgment is also necessary to establish a threshold value for the LEWS. The threshold value depends on the type of climate, land use, geology of the location, and the landslide [68]. The monitoring systems should overcome some of the problems during further development. First one being, identifying the location where the soil is weak. Hence, the monitoring devices can be placed in correct positions where the slip surface likely to take place. The second one being the identification of the controlling parameter [69]. One of the major drawbacks of providing continuous monitoring is the lack of proper arrangement of real-time monitoring data [70]. Proper management of the data is also as important as proper instrumentation. In previous works, the flow of data was managed by a data collection and processing center (DCPC). The accuracy of the data available is very much important for the development of a LEWS because a minor variation can lead to a vast difference from the threshold value predicted. Also, the joint effort of the researchers and the government authorities is required for the data collection. In previous studies, a minor change in the landslide data caused a significant change in the threshold value [71]. Also, the threshold value for different sites may be different as well. Hence, an ideal LEWS should be able to incorporate that variable as an input so that the value can be found out by laboratory methods for the site that requires the LEWS and adjust the threshold value in the LEWS device directly before the deployment.

The analysis stage requires the understanding of a proper threshold that can be used for the development of the LEWS in that region. For example, in southern parts of India, including the States of Karnataka, Kerala, Tamilnadu, etc. Rainfall-induced landslides are in dominance, as well as Himalayan regions and north-eastern provinces have been studied earlier [22, 48, 72–75]. The important factor for such studies is that the need for a long-term data availability of at least 50 years to understand the trend before coming to a threshold value. It is possible to find the time of initiation of the landslide well in advance of the event from the data collected from the continuous monitoring. In previous studies, it is observed that the tilt of the slope surface is not dependent on the peak rainfall intensity rather the continuous precipitation that causes the soil to lose strength by increased pore pressure. And also, the lack of proper drainage will cause soil erosion as well [76].

The instrumentation of a device that is capable of identifying the required threshold, and predict the landslide is an important step to be done, while developing LEWS. Recommendations for any future study in this field should be to develop a system that can measure the displacements in the surface of the landslide and according to that come to the conclusion that the slope surface is going to be unstable according to the material properties of the soil. Such a system can be developed on the principle of an Inertial Navigation System (INS). INS is a self-contained navigation system that measures the displacement without any reference points with the help of sensors like accelerometers and gyroscopes [77]. INS can be used as a final warning before the slope failure. The initial warning can be given according to the change in rainfall parameters, which will be measured using a rain gauge that is capable to record both the intensity and duration of the rainfall. The site parametric will be also provided to DAQ to get site-specific results of the slope stability. Figure 10.6 shows an illustration of the recommended LEWS. In real-time tilt measuring operations, the voltage variation that occurred will be converted to acceleration and angle [78]. Also, there are different methods for calibration of the accelerometer for getting more precise results suggested by previous literature, out of which six face method is most widely used [79]. The INS devices are made with wireless data transmission capability with propriety technologies like LoRaWan or Zigbee. This will enable the possibility of measurement from the most remote locations to the adjacent data acquisition systems (DAQ). Also, an integrated rain gauge will help to identify the intensity and duration of the rainfall in the region to support the early warning system. The LEWS is designed such that it consists of various instruments together and functions as a single system, collecting data from each instrument and giving a single output from the interrelationship.

Figure 10.6 is only an illustration of the surface displacement-based LEWS. The measuring parameters as given in Fig. 10.1 may change according to the function. As there is a scarcity in the availability of field or laboratory data, this is the suitable method that can be adopted. And methods like wire extensometers and tiltmeters can also be suggested, but the era is changing and modern instrumentation methods have to evolve with it.

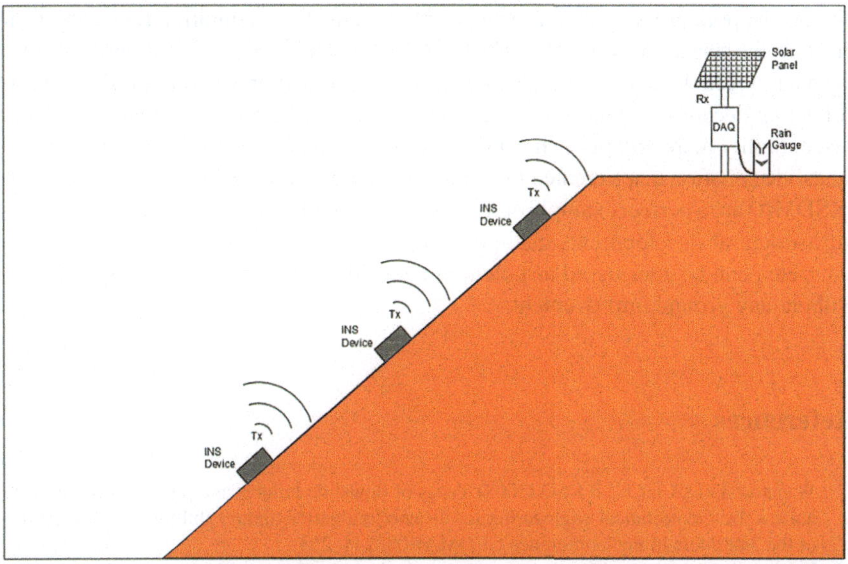

Fig. 10.6 Representative illustration of the recommended LEWS based on the surface displacement

10.6 Conclusion

The landslide early warning system is a necessity in the current century, as the infrastructural development is increasing and more and more land is being acquired for road construction and other developments. The valleys and landslide-prone locations are under threat. The implementation of a LEWS will radically reduce the danger in the region and help to save lives. The development of a successful LEWS will also consist of understanding the landslide parameters and establishing a threshold value for the early warning. This is possible only by laboratory parametric studies involving landslide flume or by collecting the historical data for a large duration and using mathematical techniques for establishing probabilistic thresholds. After obtaining the threshold values, the next step in developing a LEWS is to make a prototype which would be able to identify these values in the field. The prediction accuracy comes second, rather the system architecture is more important [9]. The performance can be calibrated in future with the abundance of data available from various locations and studies. The accuracy of the system is directly proportional to the size of available data or the accuracy of the laboratory tests conducted.

Previously, various techniques have been used as the instrumentation to work as LEWS. But none of them provide a unified method to be used to find the threshold values for the early warning. Currently, the landslide is a site-specific anomaly, the further studies should focus on developing a unified relationship between the rainfall and the landslide probability of soil slope surface. With the development of such a theory, any site can be used for a single type of LEWS. The further study should focus on how the pore pressure increase and rainfall intensity and duration relate and from that how the surface movement can be related using INS devices. The identification of the locations to deploy is the main question of concern after developing a LEWS. By collecting decades of data, some local government agencies have created landslide susceptibility maps and published them online in open access for anyone to use. The landslide zonation maps created by the Kerala State Disaster Management Authority (KSDMA) are a perfect example of that. Also, sites with unprotected slope where new infrastructural developments take place like road constructions, tunneling, mining sites, etc., can be considered as potential landslide-prone areas if the slopes are left without any ground improvement.

References

1. Wanatowski D, Chu J, Lo RSC (2008) Types of flowslide failures and possible failure mechanisms. In: Geotechnical engineering for disaster mitigation and rehabilitation—proceedings of the 2nd international conference GEDMAR08, 244–253.
2. Eckersley JD (1985) Flowslides in stockpiled coal. Elsevier Science Publishers B.V., Amsterdam—Printed in The Netherlands 22:13–22
3. Seed HB (1966) Landslides during earthquakes due to soil liquefaction. J Soil Mech Found Div ASCE 94(5):1055±1122
4. Ishihara K, Okusa S, Oyagi N, Ischuk A (1990) Liquefaction-induced own slide in the collapsible loess deposit in Soviet Tajik. Soils Found 30(4):73±89
5. Osanai N, Yoko T, Kazuya A, Tomoaki M (2009) Realty of cliff failure disaster. Tech Note Natl Instit Land Infrastruct Manag 530:204p
6. Guzzetti F, Peruccacci S, Rossi M, Stark CP (2008) The rainfall intensity-duration control of shallow landslides and debris flows: an update. Landslides 5(1):3–17
7. Bhandari RK (1970) Mudflows in stiff fissured clays. PhD Thesis, University of London (Unpublished)
8. Eberhardt E (2012) Landslide monitoring: the role of investigative monitoring to improve understanding and early warning of failure. Cambridge
9. Segoni S, Rosi A, Fanti R, Gallucci A, Monni A, Casagli N (2018) A regional-scale landslide warning system based on 20 years of operational experience. Water (Switzerland) 10(10):1–17
10. Caine N (1980) The rainfall intensity-duration control of shallow landslides and debris flows. Geografiska Annaler Series A 62(1–2):23–27
11. Peruccacci S, Brunetti MT, Gariano SL, Melillo M, Rossi M, Guzzetti F (2017) Rainfall thresholds for possible landslide occurrence in Italy. Geomorphology Elsevier 290(March):39–57
12. Watari M (1977) Mechanism of Neocene mudstone landslide-Sarukuyogi landslide. Proc 9th Int Conf Soil Mech Foundat Eng 1:963–1012
13. Froude MJ, Petley DN (2018) Global fatal landslide occurrence 2004 to 2016. Nat Hazards Earth Syst Sci Discuss 2012:1–44
14. Kuriakose SL, Sankar G, Muraleedharan C (2009) History of landslide susceptibility and a chorology of landslide-prone areas in the Western Ghats of Kerala, India. Environ Geol 57(7):1553–1568
15. Casagrande A (1971) On liquefaction phenomena, report of lecture. Geotechnique 21:197–202
16. Vuillermet E (1992) Caracteristiques geotechniques des argiles glacio-lacustres du Trieves. These de l'Universite Joseph Fourier de Grenoble, 191 pp
17. Nieuwenhuis JD (1991) Variations in stability and displacements of a shallow seasonal landslide in varved clays. University of Utrecht, Balkema, Rotterdam, p 144

18. Okura Y, Kitahara H, Ochiai H, Sammori T, Kawanami A (2002) Landslide fluidization process by flume experiments. Eng Geol 66(1–2):65–78
19. Senthilkumar V, Chandrasekaran SS, Maji VB (2018) Rainfall-Induced Landslides: Case Study of the Marappalam Landslide, Nilgiris District, Tamil Nadu, India. Int J Geomech 18(9):05018006
20. Ellen SD, Fleming RW (1987) Mobilization of debris flows from soil slips, San Francisco Bay region, California. Geol Soc Am Rev Eng Geol 7:31–40
21. Aleotti P (2004) A warning system for rainfall-induced shallow failures. Eng Geol 73(3–4):247–265
22. Dikshit A, Satyam DN (2018) Estimation of rainfall thresholds for landslide occurrences in Kalimpong, India. In: Innovative infrastructure solutions. Springer International Publishing 3(1):1–10
23. Eichenberger J, Ferrari A, Laloui L (2013) Early warning thresholds for partially saturated slopes in volcanic ashes. Comput Geotech 49:79–89
24. Garcia-Urquia E, Axelsson K (2015) Rainfall thresholds for the occurrence of urban landslides in Tegucigalpa, Honduras: an application of the critical rainfall intensity. Geogr Ann Ser B 97(1):61–83
25. Wang H, Zhang YC, Hu HY (2013) A study on relationship of landslide occurrence and rainfall. Appl Mech Mater 438–439:1200–1204
26. Mandal P, Sarkar S (2020) Estimation of rainfall threshold for the early warning of shallow landslides along National Highway-10 in Darjeeling Himalayas. Nat Hazards, Springer, Netherlands 105(3):2455–2480
27. Hong Y, Alder R, Huffman G (2006) Evaluation of the potential of NASA multi-satellite precipitation analysis in global landslide hazard assessment. Geophys Res Lett 33(22):1–5
28. Brunetti MT, Peruccacci S, Rossi M, Luciani S, Valigi D, Guzzetti F (2010) Rainfall thresholds for the possible occurrence of landslides in Italy. Nat Hazards Earth Syst Sci 10(3):447–458
29. Abraham MT, Satyam N, Rosi A, Pradhan B, Segoni S (2020c) The Selection of rain gauges and rainfall parameters in estimating intensity-duration thresholds for landslide occurrence: Case study from Wayanad (India). Water (Switzerland) 12(4)
30. Bhandari RK (1988) Special lecture: some practical lessons in the investigation and field monitoring of landslides. In: Landslides = glissements de terrain : proceedings of the fifth international symposium on Landslides, 10–15 July 1988, 1435–1457
31. Hutchinson JN (1969) A reconsiderations of the coastal landslides at Folkestone Warren Kent. Geotechnique 19:6–13
32. Palshin, Trzhtsinki (1963) Horizontal beds of 60m thick sandstones & Limestones lying over Cambrian Alenrolites & Argillites Costitute slope of Augara river in Siberia, Stiff, overlying blocks Creep downslope. In: Case record, IV international symposium on landslides, 1984, vol. 1. Toronto, p 257
33. Rapp (1967) On the measurement of Solifluction movements. Revue de Geomorphologie dynamique, pp 162–163
34. Angeli MG, Pasuto A, & Silvano S (February, 2000) A critical review of landslide monitoring experiences. Eng Geolog 55(3):133–147. https://doi.org/10.1016/S0013-7952(99)00122-2
35. Bhandari RK (1984) Simple and economical instrumentation and warning systems for landslides and other mass movements. In: IV international symposium on landslides = IV symposium international surles Glissements de Terrains : ISL, 1984, Toronto, 251–305
36. Corominas J, Moya J, Lloret A, Gili JA, Angeli MG, Pasuto A, Silvano S (2000) Measurement of landslide displacements using a wire extensometer. 55:149–166
37. Gili JA, Corominas J, Rius J (2000) Using Global Positioning System techniques in landslide monitoring. 55:167–192
38. Lollino G, Arattano M, Cuccureddu M (2002) The use of the automatic inclinometric system for landslide early warning : the case of Cabella Ligure (North-Western Italy) 27:1545–1550.
39. Gance J, Malet J, Dewez T, Travelletti J (2014) Target detection and tracking of moving objects for characterizing landslide displacements from time-lapse terrestrial optical images. Eng Geol Elsevier BV 172:26–40

40. Win MZ, Scholtz RA (1998) Impulse radio: how it works. IEEE Commun Lett 2(2):36–38
41. Win MZ, Dardari D, Molisch AF, Wiesbeck W, Zhang J (2009) History and applications of UWB [scanning the issue]. Proc IEEE 97(2):198–204
42. Lanzisera S, Zats D, Pister KSJ (2011) Radio frequency time-of-flight distance measurement for low-cost wireless sensor localization. IEEE Sensors J IEEE 11(3):837–845
43. Intrieri E, Gigli G, Gracchi T, Nocentini M, Lombardi L, Mugnai F, Frodella W, Bertolini G, Carnevale E. Favalli M, Fornaciai A, Marturià Alavedra J, Mucchi L, Nannipieri L, Rodriguez-Lloveras X, Pizziolo M, Schina R, Trippi F, Casagli N (2018) Application of an ultra-wide band sensor-free wireless network for ground monitoring. Eng Geol Elsevier BV 238:1–14
44. Intrieri E, Gigli G, Mugnai F, Fanti R, Casagli N (2012) Design and implementation of a landslide early warning system. Eng Geol Elsevier BV 147–148, 124–136
45. Liu H, Meng Z, Cui S (2007) A wireless sensor network prototype for environmental monitoring in greenhouses. Int C Wirel Comm Net
46. Yoo S, Kim J, Kim T, Ahn S, Sung J, Kim D (2007) A2S: automated agriculture system based on WSN, I Symp. Consum Electr
47. Ramesh MV, Vasudevan N (2012) The deployment of deep-earth sensor probes for landslide detection. Landslides 9(4):457–474
48. Abraham MT, Satyam N, Bulzinetti MA, Pradhan B, Pham BT, Segoni S (2020) Using field-based monitoring to enhance the performance of rainfall thresholds for landslide warning. Water 12(12):3453
49. Petley D (2012) Remote sensing techniiques and landslides. Cambridge
50. Casagli N, Morelli S, Frodella W, Intrieri E, Tofani V (2017) TXT-tool 2.039–3.2 ground-based remote sensing techniques for landslides mapping, monitoring and early warning. In: Landslide dynamics: ISDR-ICL landslide interactive teaching tools: volume 1: fundamentals, mapping and monitoring, 255–274
51. Intrieri E, Carla T, Farina P, Bardi F, Ketizmen H, Casagli N (2019) Satellite interferometry as a tool for early warning and aiding decision making in an open-pit mine. IEEE J Selected Topics Appl Earth Observat Remote Sensing Instit Electri Electron Eng 12(12):5248–5258
52. Barra O, Monserrat P, Mazzanti C, Esposito MC, Scarascia Mugnozza G (2016) Potentiality of SENTINEL-1 for landslide detection: first results in the Molise Region (Italy). In: Proceedings Europe geoscience union general assembly conference, 2016, vol. 18, p 2916
53. Novellino FC, Brahmi M, Sowter A, Bateson L, Marsh S (2017) Assessing the feasibility of a national. In: SAR ground deformation map of Great Britain with Sentinel-1. Geosciences 7(2):1–14
54. Raspini F et al (2018) Continuous, semi-automatic monitoring of ground deformation using Sentinel-1 satellites. Sci Rep 8, Art. no. 7253
55. Zhang Y, Tang H, Lu G, Wang Y, Li C, Zhang J, An P, Shen P (2020) Design and testing of inertial system for landslide displacement distribution measurement. Sensors (Switzerland) 20(24):1–16
56. Wang G, Sassa K (2001) Factors affecting rainfall-induced flowslides in laboratory flume tests. Géotechnique 51(7):587–599
57. Wang G, Sassa K (2003) Pore-pressure generation and movement of rainfall-induced landslides: effects of grain size and fine-particle content. Eng Geol 69(1–2):109–125
58. Wang G, Sassa K (2007) On the pore-pressure generation and movement of rainfall-induced landslides in laboratory flume tests
59. Bishop AW (1973) The stability of tips and spoil heaps. Q J Eng Geol 6(3–4):335–376
60. Kramer SL (1988) Triggering of liquefaction flow slidesin coastalsoil deposits. Eng Geol Elsevier Science Publishers BV Amsterdam 26:17–31
61. Eckersley JD (1991) Instrumented laboratory flowslides. Geotechnique 41(2):489–502
62. Kramer SL, Seed HB (1988) Initiation of soil liquefaction under static loading conditions. J Geotech Eng Div Am Soc Civ Eng 114(4):412–430
63. Cogan J, Gratchev I (2019) A study on the effect of rainfall and slope characteristics on landslide initiation by means of flume tests Landslides. Landslides 16(12):2369–2379

64. Acharya G, Cochrane TA, Davies T, Bowman E (2009) The influence of shallow landslides on sediment supply: a flume-based investigation using sandy soil. Eng Geol Elsevier BV 109(3–4):161–169
65. Li HJ, Zhu HH, Li YH, Hu W, Shi B (2021) Fiber Bragg grating-based flume test to study the initiation of landslide-debris flows induced by concentrated runoff. Geotech Testing J 44(4).
66. Neumann M, Zumr D, Laburda T, Kavka P, Johannsen LL, Balenovic N, Chladova Z, Fiser O, Strauss P, Dostal T, Klik A (2018) Comparison of the rainfall kinetic energy measured by different distrometers. Poster. EGU General Assembly 2018 20(1):EGU2018–13821
67. Sangjum I, Song E, Dongyeob K (2017) Stochastic investigation of the feasibility of using remotely sensed moisture data for rainfall induced landslide hazard assessment. Advancing Culture of Living with Landslides
68. Intrieri E, Gigli G, Casagli N, Nadim F (2013) Brief communication landslide early warning system: toolbox and general concepts. Nat Hazards Earth Syst Sci 13(1):85–90
69. Uchimura T, Towhata I, Wang L, Nishie S, Yamaguchi H (2015) Precaution and early warning of surface failure of slopes using tilt sensors. Soils Foundat Elsevier 55(5):1086–1099
70. Intrieri E, Bardi F, Fanti R, Gigli G, Fidolini F, Casagli N, Costanzo S, Raffo A, Di Massa G, Capparelli G, Versace P (2017) Big data managing in a landslide early warning system: Experience from a ground-based interferometric radar application. Nat Hazards Earth Syst Sci Copernicus GmbH 17(10):1713–1723
71. Gariano SL, Brunetti MT, Iovine G, Melillo M, Peruccacci S, Terranova O, Vennari C, Guzzetti F (2015) Calibration and validation of rainfall thresholds for shallow landslide forecasting in Sicily, southern Italy. Geomorphol Elsevier BV 228:653–665
72. Teja TS, Dikshit A, Satyam N (2019) Determination of rainfall thresholds for landslide prediction using an algorithm-based approach: case study in the Darjeeling Himalayas India. Geosciences 9:302
73. Soja R, Starkel L (2007) Extreme rainfalls in Eastern Himalaya and southern slope of Meghalaya plateau and their geomorphologic impacts. Geomorphology 84:170–180
74. Dikshit A, Satyam DN (2018) Estimation of rainfall thresholds for landslide occurrences in Kalimpong India. Innov Infrastruct Solut 3:24
75. Abraham MT, Satyam N, Pradhan B, Alamri AM (2020) Forecasting of landslides using rainfall severity and soil wetness: A probabilistic approach for Darjeeling Himalayas. Water 12:804
76. Abraham MT, Satyam N, Pradhan B (2020b). IoT-based geotechnical monitoring of unstable slopes for landslide early IoT-based geotechnical monitoring of unstable slopes for landslide early warning in the Darjeeling Himalayas
77. Woodman OJ (2007) An introduction to inertial navigation. University of Cambridge Computer Laboratory
78. Cina A, Manzino AM, Bendea IH (2019) Improving GNSS landslide monitoring with the use of low-cost MEMS accelerometers. Appl Sci (Switzerland) 9(23)
79. Titterton DH, Weston JL (1997) Strapdown Inertial Navigation Technology. Peter Peregrinus Ltd. on behalf of the Institution of Electrical Engineers, London, UK

Chapter 11
Forecasting Landslides for Disaster Risk Reduction: Process-Based Approaches and Real-Time Field Monitoring

Neelima Satyam and Minu Treesa Abraham

11.1 Introduction

The frequency of rainfall-induced landslides is increasing at an alarming rate in mountainous terrains [1]. The Himalayas contributes to a major share of landslides in the world [2]. Among the different methods of disaster risk reduction, providing early warning is one of the best options. The people at risk can be warned about the possible failure using different monitoring techniques and rainfall forecasts [3]. The most widely followed method for forecasting landslides is the use of rainfall thresholds [4–8]. Statistical analysis of historical data is an effective way to forecast the landslides in the future [9, 10]. For large areas, extending more than hundreds of square kilometers, this is the best method as it surpasses the requirements of site-specific data [11–14]. When the area is small, forecasting can be done using process-based methods as well [12, 15].

Process-based or physically based models require precise measurement of physical parameters at the field. They consider the effect of the climatic, hydrological, and morphological reasons for triggering of landslides by evaluating the physical processes [16–18]. Rainfall thresholds are rainfall conditions beyond which landslides can be triggered in the region. Process-based models can be used for enhancing the performance of rainfall thresholds [19, 20]. The conventional thresholds are defined based on the rainfall parameters, viz. event, duration, and intensity. These parameters are used to define a rainfall event quantitatively. Event is the total amount of precipitation, and duration is the time span of continuous precipitation till the time of failure. Intensity is the ratio of event to duration, which is the average rate

N. Satyam (✉) · M. T. Abraham
Department of Civil Engineering, Indian Institute of Technology Indore, Indore, Madhya Pradesh 453552, India
e-mail: neelima.satyam@iiti.ac.in

M. T. Abraham
e-mail: phd1901204011@iiti.ac.in

S. Kolathayar et al. (eds.), *Civil Engineering for Disaster Risk Reduction*, Springer Tracts in Civil Engineering, https://doi.org/10.1007/978-981-16-5312-4_11

of precipitation. Such thresholds can be defined on local, regional, or global scales. Local and regional scales thresholds are to be defined specifically for an area and cannot be used for another area.

The conventional rainfall thresholds can be categorized into two based on their objectives [21, 22]. They can be minimum thresholds (predicting the minimum level to trigger slope failures) or warning thresholds (with a balance in the number of true and false predictions). Minimum thresholds are conservative in nature, and no landslide is expected to happen below this threshold condition. The major limitation of such thresholds is the higher number of false predictions. Antecedent soil moisture conditions can be used to overcome this limitation, as it plays a critical role in the process of slope failure [23]. Examples of the role of antecedent soil moisture on initiation of landslides are pointed out by many researchers [19, 24, 25]. The moisture content data can be made useful to enhance the performance of statistical thresholds [26–28]. For measuring soil moisture at site, field-based monitoring instruments are required [29, 30], which is often a costly approach. Hence hydrological models can be used, to calculate soil moisture suing rainfall and surface characteristics [31–33].

This study presents two different approaches for enhancing the performance of conventional rainfall thresholds in a small town in Darjeeling Himalayas, called Kalimpong. First, the rainfall thresholds are modified by using soil moisture data using a hydrological model called Système Hydrologique Européen Transport (SHETRAN) [33, 34]. A probabilistic approach using Bayes' theorem is used for the purpose. Finally, the details of field-based monitoring technique and its potential use in improving the rainfall thresholds are also described, using microelectromechanical systems (MEMS) tilt sensors. This has been done using an algorithm-based approach. The study area is a hilly town in West Bengal state of India, which is highly susceptible to landslides.

11.2 Details of Study Area

Kalimpong town in West Bengal (Fig. 11.1) is a hill station and well-known tourist destination. This hilly terrain is suffering from slope stability issues in monsoon seasons. The landslide and rainfall data from 2010 to 2016 has been used to derive rainfall thresholds for the region. Heavy rainfall, infiltration, and highly erosive mountain rivulets (called jhoras) are the reason for the frequent landslides in the region [35]. Due to the increase in number of rainfall-induced landslides, the town has become a focal point of research in the recent times [35–37]. Several rainfall thresholds are established for this region, using empirical and probabilistic approaches. This study is an attempt to improve the performance of the event–duration (ED) thresholds derived for Kalimpong, using antecedent soil moisture and real-time field monitoring data.

The town is hemmed between rivers Teesta and Relli in the west and east sides, respectively. The western face, toward Teesta is steep in nature while the eastern

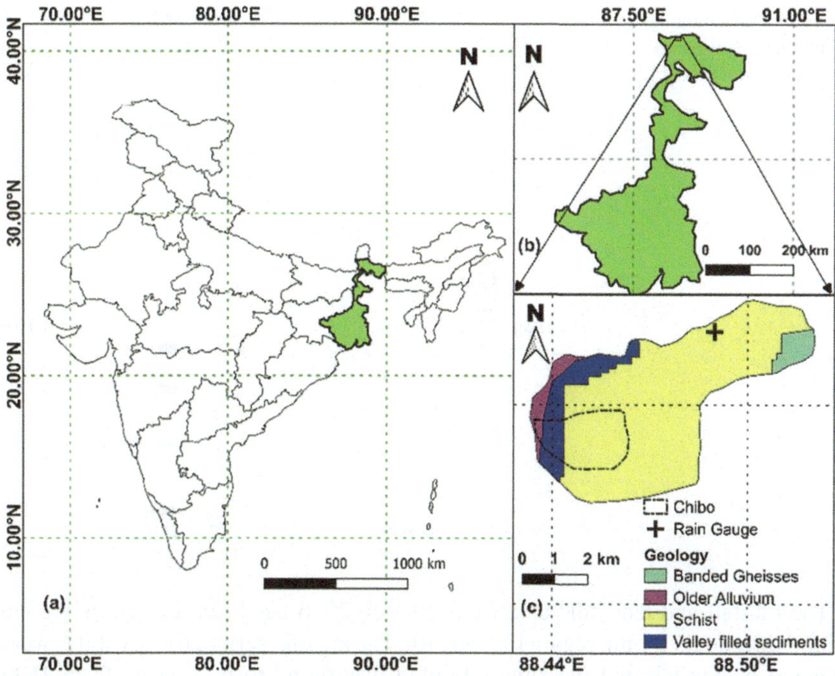

Fig. 11.1 Location details of study area [38]

slopes are gentler. The reddish soil with older debris deposits along the steep slopes is highly prone to slope failures.

Agriculture and tourism are the primary income sources of people of Kalimpong. Landslides affect both these sectors adversely by disrupting transportation facilities and destroying agricultural lands. Increase in population and the resulting demand for more buildings have also increased the risk. Due to the rise in population, construction activities have become inevitable even at steeper slopes, increasing the risk.

The geological composition of Kalimpong town is dominated by Schist, along with banded gneisses and older alluvium (Fig. 11.1) [39]. The rocks are cracked and jointed, allowing easy disintegration and decomposition. The presence of micaceous minerals and the foliations in schist controls the slope failures in the area.

As observed from the drainage map of Kalimpong (Fig. 11.2), the mountain rivulets descend radically. These rivulets are called jhoras in the local dialect. During monsoons, the jhoras cause very intense scouring. The down does not have a planned drain system or urban expansion, which is resulting in excess surface runoff during monsoons. The jhoras with high velocity cause lateral erosion, which results in slope failures.

The landslides in Kalimpong are mostly shallow, but there are cases of continuous subsidence due to deep seated movements. Typical landslides types are rock, or debris slides and debris flows.

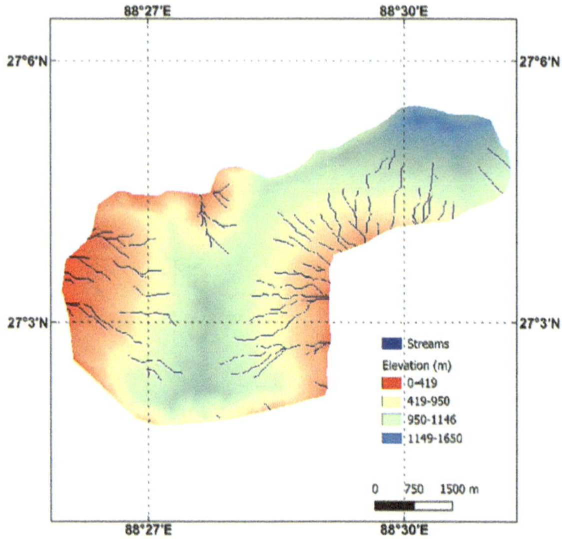

Fig. 11.2 Drainage map of Kalimpong, overlaid on digital elevation model [20]

The rainfall and landslide data from 2010 to 2016 has been used to derive the rainfall thresholds for the region [35, 36, 40]. During this period, 61 landslides were triggered by rainfall and the annual rainfall was found to be varying from 1623 to 2061 mm. Due to heavy rains, the particles lose their shear strength, and they disintegrate and slides down. Weathered rocks and older debris deposits often get transported downslope during rains. The rainfall data has been collected from the station maintained at Tirpai by Save The Hills [41]. From Fig. 11.3, it can be observed that there is a direct relationship between the amount of rainfall and occurrence of

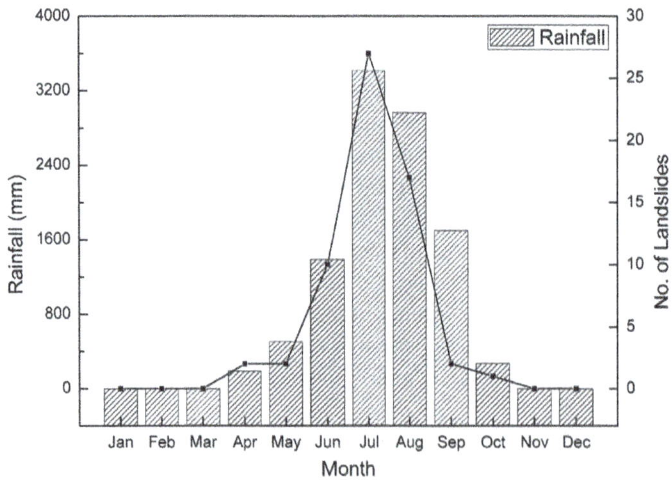

Fig. 11.3 Variation of monthly rainfall and landslide occurrence from 2010 to 2016 [20]

Table 11.1 Monthly rainfall data (mm) during monsoon season in Kalimpong town (2010–2016) [41]

Month	2010	2011	2012	2013	2014	2015	2016
June	316.8	337.0	354.9	248.0	396.4	568.0	327.2
July	665.4	678.0	433.1	424.6	371.2	534.4	869.8
August	425.3	525.6	250.8	401.0	571.8	242.3	262.6
September	268.2	384.1	467.9	113.0	265.4	331.2	366.8

landslides in Kalimpong [42–47]. Maximum rainfall and number of landslides have happened in the month of July, during the study period.

During the study period (2010–2016), the monsoon months (June–September) have contributed around 87% of the total rainfall. The monthly rainfall data used for deriving the rainfall thresholds are mentioned in Table 11.1.

11.3 Rainfall Thresholds, Hydrological Model, and Field Monitoring Data

11.3.1 Rainfall Thresholds

The conventional rainfall thresholds for Kalimpong are defined on both ID and ED planes. In this study, the ED thresholds defined using two different approaches are discussed. The method of derivation of thresholds is the same in both the cases, but the method of identifying responsible rainfall event is different. In the first approach, the nearest rain gauge and closest rainfall event were considered, and the selection was done manually. In the second approach, an algorithm was used to identify the responsible rainfall event. The algorithm considers the spatial distribution of rainfall and landslide events and discards the landslides which do not have a rain gauge within a user-defined buffer distance. 61 landslides were during the study period was used for both the analysis, but in the second method, only 36 landslides were considered for the analysis. The frequentist method for deriving the thresholds can be described as follows:

Once the rainfall events are identified, they are plotted on an event versus duration plot, on logarithmic scale. The best fit line for the data is then derived using the method of least squares. The data can be of varying orders, and hence, logarithmic plot should be used for representing the data. The shift of each data point on y-axis from the best fit line is then calculated. The data follows normal distribution (Fig. 11.4), and hence by using the standard Gaussian distribution, thresholds with different exceedance probability can be defined (Fig. 11.5).

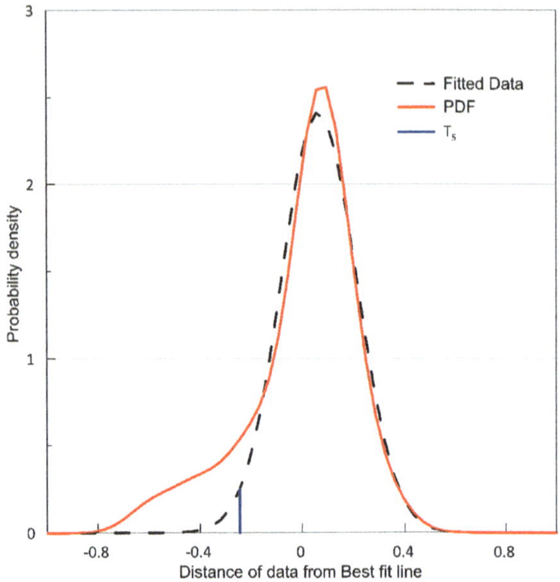

Fig. 11.4 Probability density function of δE distribution [20]

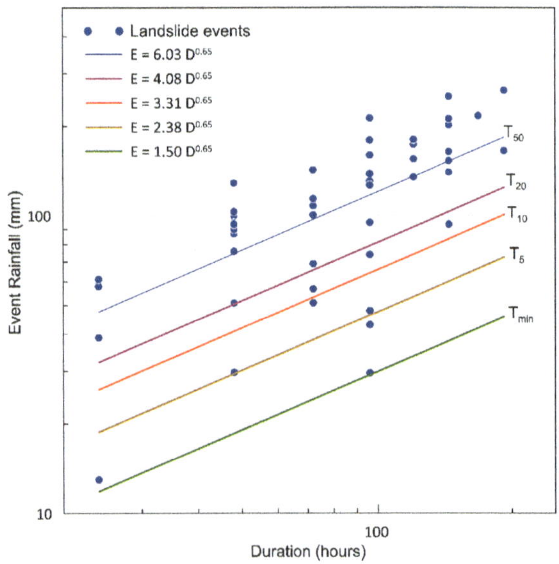

Fig. 11.5 Event–duration thresholds derived using nearest rain gauge approach [20]

Thresholds of different exceedance probability were defined in order to use the thresholds for a probabilistic model, combined with the antecedent moisture conditions. The exceedance probability indicates the percentage of events expected to fall below the line; i.e., 5% of the total landslide events may occur below the T_5 line.

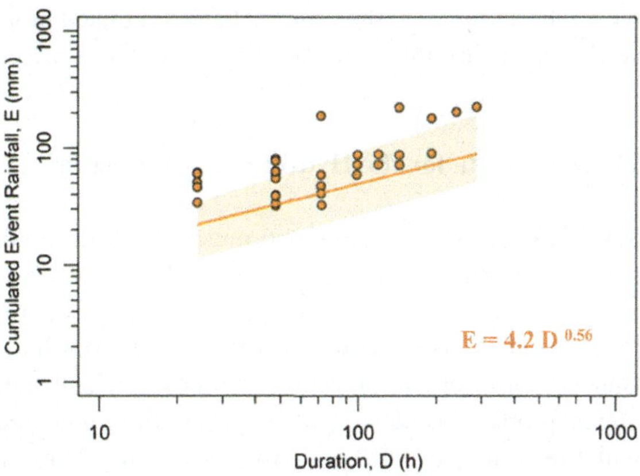

Fig. 11.6 Event–duration thresholds defined using algorithm-based approach [36]

The equations of the minimum threshold line and the best fit line are $E = 1.50D^{0.65}$ and $E = 6.03D^{0.65}$, respectively.

The major limitation of this approach is the uncertainties in data and the spatial distribution of landslide events. The automatic algorithm [48] overcomes this limitation and derives the thresholds by considering the spatial distribution of rain gauges and landslides. The ED threshold for Kalimpong, derived using CTRL-T [36] is plotted in Fig. 11.6.

11.3.2 SHETRAN Model

SHETRAN is a hydrological model to simulate drainage pathways and processes in river basins, which is being used extensively [15, 34, 43, 49]. The model has three components, and this study focuses only on the water flow component of SHETRAN. The partial differential equations governing the flow are solved using the spatially distributed input data. The input data is in the form of a three-dimensional grid, formed using the digital elevation model (DEM).

In this study, the meteorological conditions across the study area have been considered to be uniform across the space, at any time. SHETRAN uses the Penman–Monteith equation, Saint–Venant equations, Rutter equation [50], and the Variably saturated flow Eq. (3D) [51] for understanding the hydrological processes of evapotranspiration, overland flow, channel flow, canopy drip, and subsurface flow.

The model has been calibrated using the Nash–Sutcliffe Efficiency (NSE) [52, 53], defined as:

$$R^2 = 1 - \frac{\sum_{i=1}^{n}\left(w_{\mathrm{i}}^{\mathrm{o}} - w_{\mathrm{i}}^{\mathrm{m}}\right)^2}{\sum_{i=1}^{n}\left(w_{\mathrm{i}}^{\mathrm{o}} - \overline{w_{\mathrm{i}}^{\mathrm{o}}}\right)^2} \tag{11.1}$$

where w_i^o and w_i^m are the observed and simulated moisture content at each time step and $\overline{w_i^o}$ is the arithmetic mean of observed moisture content.

11.3.2.1 Data for the Hydrological Simulation

SHETRAN model requires two meteorological inputs: the rainfall and evapotranspiration. Rainfall data was collected from the nearest available rain gauge, and evapotranspiration was downloaded from the Copernicus Climate Change Service [54]. The DEM used for the study was of 30 m resolution, the Cartosat DEM from the Bhuvan portal of National Remote Sensing Centre, India [55]. While the meteorological properties change with respect to time, catchment properties remain constant with time. Since the area of study is less than 32 km^2, the catchment properties were assumed to be uniform across the space. Spatial variations within this small area may lead to overfitting of data.

The data from 2010 to 2016 was used for calibration, and the soil moisture data from Modern Era Retrospective Analysis for Research and Applications (MERRA) using Goddard Earth Observing System Data Assimilation System Version 5 (GEOS 5) [56] was compared with the simulated values, for calibration.

11.3.2.2 Soil Moisture Content

The rainfall data is used to find the phreatic line, based on the elevation and catchment properties. Thus, the subsurface is divided into unsaturated and saturated zones. For zones above phreatic line, moisture contents are then calculated. These zones at shallow depths are often susceptible to shallow landslides. The water flow modeling in SHETRAN is done using the van Genuchten parameters. To overcome the limitations in modeling, the model has been calibrated and an R^2 value of 0.84 was obtained. The catchment properties after calibration are listed in Table 11.2.

Table 11.2 Catchment properties calibrated using SHETRAN [20]

Parameters	Calibrated value
Strickler overland flow coefficient	0.50 $m^{1/3}$ s^{-1}
Maximum rooting depth	1.6 m
AE/PE at field capacity	1
Canopy storage	5 mm
Saturated water content	0.40
Saturated hydraulic conductivity	1.14 m/day
vanGenuchten-n	1.17
Residual water content	0.08
Leaf area index	1
vanGenuchten-alpha	0.03 cm^{-1}

The calibrated parameters are in good agreement with the parameters predicted using ROSETTA Lite module [57] for Kalimpong, using the particle size distribution of soil [13]. The obtained soil moisture data was then combined with rainfall thresholds, using a Bayesian approach.

11.3.3 Probabilistic Approach Using Rainfall Thresholds and Antecedent Soil Moisture

Bayes' theorem was used to define the conditional probability of occurrence of landslides, using two parameters. The rainfall thresholds were used to classify the rainfall events based on its severity into 6 classes, and the soil moisture data was used to define the soil wetness. The soil moisture data was converted to a scale of 0–1, and the scale value is termed as soil wetness. Based on the value, soil wetness was divided into 5 equal classes. Thus, a 30 cell (5 × 6 matrix) was formed for the two-dimensional analysis. The equations used for calculating the probability are as follows:

$$P(A|B, C) = \frac{P(B, C|A)*P(A)}{P(B, C)} \tag{11.2}$$

Considering our case, *A* is a landslide event, *B* is the soil wetness, and *C* is the rainfall severity. '*B*,*C*' indicates the condition in which the occurrence of *B* and *C* together, within a specific range. This can be called as a cell condition. 30 different cell conditions can be formed in our case, with 6 classes of rainfall severity and 5 classes of soil wetness. The other terms in the equation are:

P(*B*,*C*|*A*) is the probability of occurrence of a certain cell condition when a landslide occurs or the conditional probability or the likelihood of *B*, *C* given *A*.

P(*A*) is the prior probability of the occurrence of a landslide, regardless of *B* and *C*.

P(*B*,*C*) is called the marginal probability of occurrence of a certain cell condition with or without the occurrence of *A*.

P(*A*|*B*,*C*) is the conditional probability or the posterior probability of occurrence of *A* when a '*B*,*C*' is satisfied.

All these values can be estimated in the terms of relative frequencies.

$$P(A) \approx \frac{N_{\mathrm{A}}}{N_{\mathrm{R}}} \tag{11.3}$$

where

N_{A} = The total number of landslide events.

N_{R} = The total number of rainfall events.

$N_{B,C}$ = The number of events in each cell condition.

$N_{(B,C|A)}$ = The number of rainfall events that resulted in landslides while satisfying a cell condition.

Thresholds are classified as less than $T_{\min}$, $T_{\min}$—T_5, T_5–T_{10}, T_{10}–T_{20}, T_{20}–T_{50} and greater than T_{50}. The classification of soil wetness values is 0–0.2, 0.2–0.4, 0.4–0.6, 0.6–0.8, and 0.8–1.0. The 2D probabilistic thresholds are plotted in Fig. 11.7.

It can be observed from Fig. 11.7 that the maximum probabilities are not always associated with the extreme conditions. The maximum probability of 1 was obtained in three cases:

(a) the rainfall severity greater than T_{50} and the soil wetness between 0.4 and 0.6
(b) the rainfall severity between T_{20} and T_{50} and the soil wetness between 0.6 and 0.8
(c) the rainfall severity event between T_5 and T_{10} and the soil wetness between 0.6 and 0.8

From the thresholds, it can be inferred that when the soil moisture is less, only severe rainfalls can trigger landslide but when it the soil moisture is high, even a less

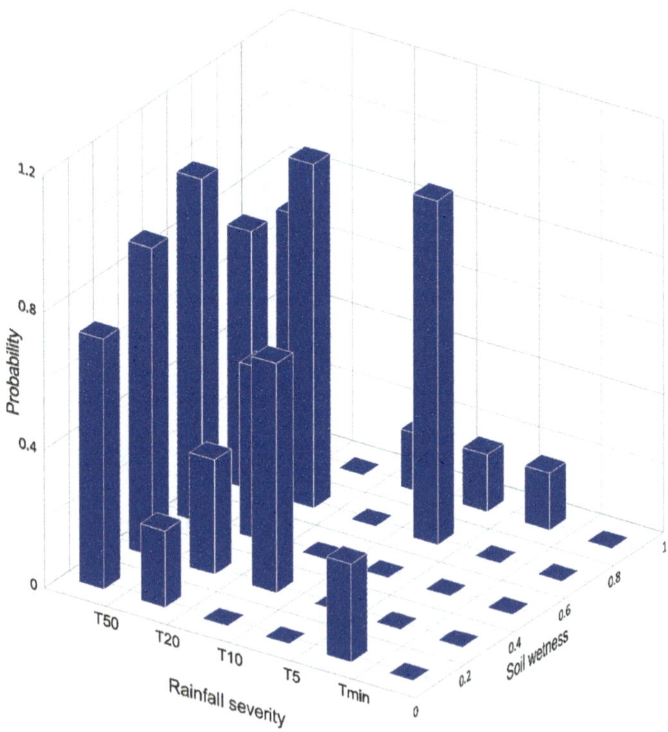

Fig. 11.7 Histogram of conditional probability of occurrence of landslides based on rainfall severity and antecedent soil moisture [20]

severe rainfall can trigger landslides in the town. The results were validated using the rainfall, soil moisture, and landslide data of 2017.

11.3.4 Real-Time Field Monitoring

11.3.4.1 MEMS Tilt Sensor and Volumetric Moisture Content Sensor

The field monitoring data collected from three MEMS sensors located in the Chibo village of Kalimpong was used to enhance the performance of rainfall thresholds defined using the second approach. The sensors were installed on the banks of a jhora, called the Pyarieni jhora, and the slopes are being monitored since July 2017. The sensors consist of an MEMS unit, moisture content sensor, and a control unit to transfer the data in real time (Fig. 11.8).

The tilt sensor measures the angle of tilt in both x- and y-directions. One parallel to the slope and the other, perpendicular to slope, along the surface. The voltage values obtained are converted to the tilt angles using a conversion formula. The control unit transfers the real-time data to a data logger, located in between the sensors and from data logger, it is transferred in real time, via Internet. The MEMS sensor units for this study have been purchased from Chuo Kaihatsu Corporation, Japan. The low-cost sensors can be used as an efficient tool for real-time field monitoring.

Fig. 11.8 Schematic arrangement of sensor unit [38]

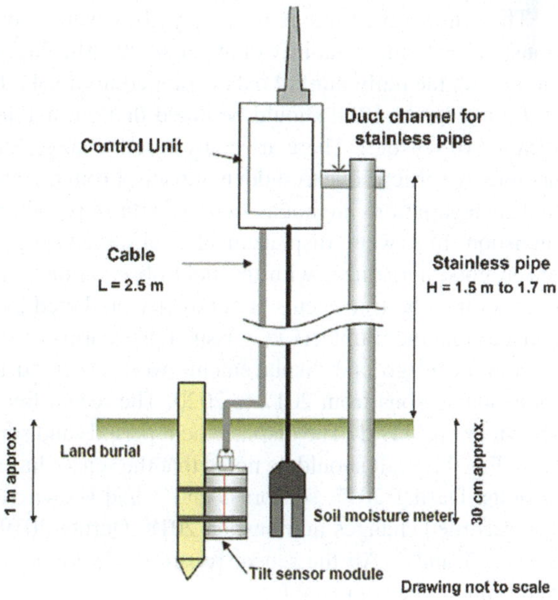

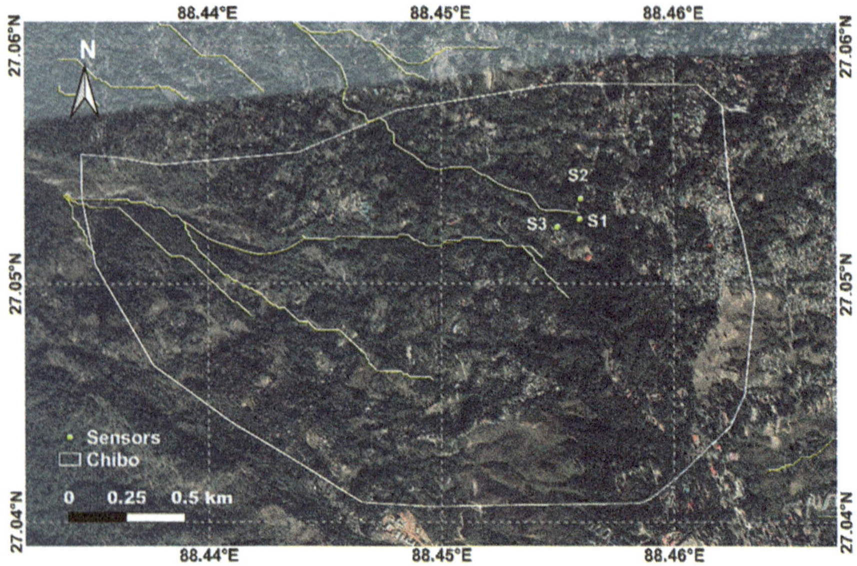

Fig. 11.9 Location of sensors in Chibo, Kalimpong [38]

The site of installing sensors was selected after detailed field surveys and discussion with Geological Survey of India. The locations of sensors and the drainage map of Chibo village are shown in Fig. 11.9.

The sensor observations from July 2017 were compared with the field observations, to verify the reliability of the data [29, 30]. The observations from the sensors, along with the daily rainfall data is plotted in Fig. 11.10.

From Fig. 11.10, it should be noted that the maximum tilt has happened in the location of sensor 2. There are many abrupt changes observed in sensor data, which are mostly vertical, due to sudden impacts. From field observations, it has been noted that such variations are not associated with slope failures all the time. Bu during the monsoon time, when displacement is observed on ground, the tiltmeter variations are in good agreement with the field observations. Hence, it should be noted that even though the true events are correctly predicted by the tiltmeters, there are high chances that the tiltmeters may issue false alarms as well.

Significant ground displacements were observed near sensor 2 during all four monsoon seasons from 2017 to 2020. The recent field observations from 2020 are shown in Fig. 11.11. The displacement periods and sensor readings can be observed from Fig. 11.10. It should be noted that the sensor has tilted during the displacement periods. During 2017, sensors 2 and 3 had shown variations while only sensor 2 had recorded changes in monsoon 2018. During 2019, variations were observed in sensors 1 and 2. All the sensor recordings were compared with field observations and found to be comparable.

The hourly tilt rate in this study has been calculated in a different way than the conventional approach. In this study, the slope of tilt angle vs time has been calculated

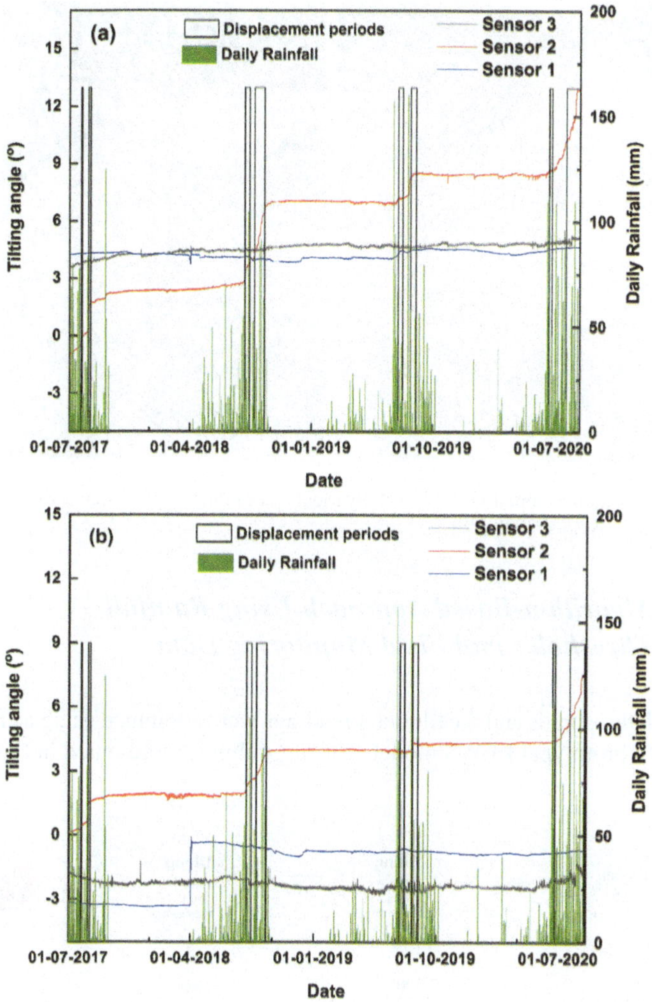

Fig. 11.10 Readings from MEMS tilt sensors along with daily rainfall data and displacement periods. **a** Tilting angle in the direction parallel to slope, and **b** tilting angle in the direction perpendicular to slope [38]

for fixed intervals of time, instead of fixed intervals of tilt angle in the conventional approach. Thus, a threshold was set using the historical cases as 0.03°/h. If the value crosses 0.1°/h, the rate is very critical. The threshold is said to be crossed when the specified value is exceeded continuously, at least for an hour.

Fig. 11.11 Cracks developed in road near Pyarienii jhora during 2020 monsoon. **a** On 27 July 2020, and **b** on 6 August 2020 [38]

11.3.5 Algorithm-Based Approach Using Rainfall Thresholds and Field Monitoring Data

The rainfall thresholds and the tiltmeter readings were combined using an algorithm, to develop a prototypal LEWS model. The algorithm is represented in Fig. 11.12.

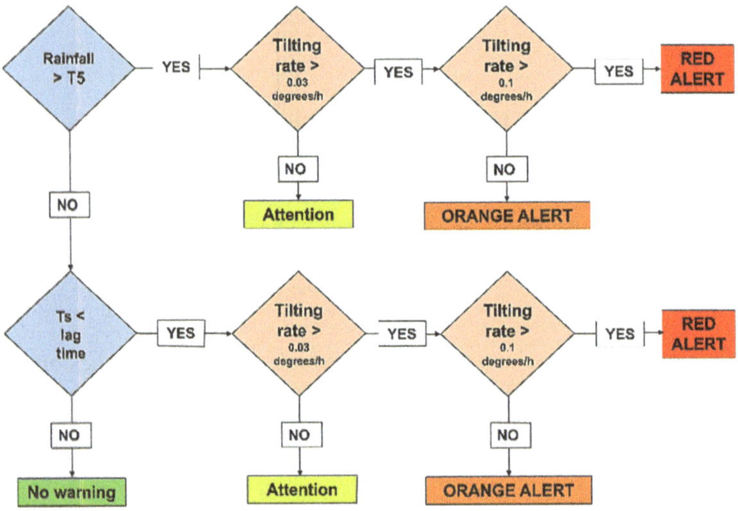

Fig. 11.12 Proposed algorithm for issuing slope failure alert in Chibo area, Kalimpong [38]

The algorithm first checks the occurrence of the precursor event (rainfall) in the study area. If the rainfall is having a severity greater than T_5, then it checks if the tilt threshold is crossed or not. If the threshold of 0.03°/h is crossed, then orange alert is issued and if it is not crossed, only attention is required. If the tilt rate is greater than 0.1°/h, the alert is issued is Red, representing critical condition. If the first check of rainfall severity is failed, algorithm checks if any rainfall event has occurred within a specified lag time (72 h in this study). If such an event has happened, the checks for tilt rates are continued again. Thus, the algorithm takes care of rainfall events happening instantaneously and the effect of immediately preceding rainfall events. The values of tilt rates and lag time for Kalimpong have been calibrated using the tilt sensor data from 2017 to 2020, using a trial-and-error approach.

11.4 Validation

Both the derived thresholds were validated using statistical skill scores [21]. The procedure of validation is required to understand the effectiveness of the proposed models. The process is done with the help of conventional approach of confusion matrix. By using a confusion matrix, the forecasts issued by the model were classified into four categories. They are True Positives (TP), False Positives (FP), True Negatives (TN), and False Negatives (FN). When the model forecasts landslides on a day, and landslide is actually recorded, it is termed as TP. If no landslide happens on the day, it is FP. Similarly, if a landslide happens on a day without actual forecast, it is FN and if no landslide happens on a day without landslide forecast, the prediction is TN. TP and TN are correct predictions and FP, FN are wrong predictions.

Using these fours attributes, some other scores can be derived, to evaluate the overall performance of the model. One such term is called efficiency, which is the ratio of total correct prediction to the total number of all predictions. Two other terms which are indicators to the positive and prediction performances are sensitivity and specificity. Sensitivity is the ratio of landslides correctly predicted to the total number of landslides and specificity is the ratio of non-landslides correctly predicted to the total number of non-landslide days. For an ideal model, both these values should be 1. The ratio of sensitivity to 1-specificity is called the likelihood ratio, another term which can be used to evaluate the overall performance of the model.

11.4.1 Probabilistic Approach Using Rainfall Thresholds and Antecedent Soil Moisture

For the first approach, validation was done using the rainfall, landslide, and soil wetness data of 2017. The results of statistical comparison of the rainfall thresholds and the combined approach are listed in Table 11.3.

Table 11.3 Results of validation of statistical and probabilistic thresholds [20]

	Statistical attributes	TP	FP	FN	TN	Efficiency	Sensitivity = TP/(TP + FN)	Specificity = TN/(FP + TN)	Likelihood ratio = sensitivity/(1−specificity)
ED only	*T*min	5	67	2	291	0.81	0.71	0.81	3.82
	T5	2	47	5	311	0.86	0.29	0.87	2.18
	T10	1	37	6	321	0.88	0.14	0.90	1.38
	T20	1	27	6	331	0.91	0.14	0.92	1.89
	T50	1	15	6	343	0.94	0.14	0.96	3.41
Probabilistic thresholds	P > 0.1	6	41	1	317	0.88	0.86	0.89	7.48
	P > 0.2	2	27	5	331	0.91	0.29	0.92	3.79
	P > 0.4	2	26	5	332	0.92	0.29	0.93	3.93
	P > 0.6	2	25	5	333	0.92	0.29	0.93	4.09
	P > 0.8	2	17	5	341	0.94	0.29	0.95	6.02

Table 11.4 Prediction performance of algorithm-based approach, using rainfall thresholds, and field monitoring data [38]

Statistical attributes	ED + Tilt	ED only	Tilt only
True positives	66	50	86
False positives	63	118	273
False negatives	32	48	12
True negatives	997	942	787
Efficiency	0.92	0.86	0.75
Specificity	0.94	0.89	0.74
Sensitivity	0.67	0.51	0.88
Likelihood ratio	11.33	4.58	3.41

The sensitivity is the highest for when the critical probability is 0.1. For conventional thresholds, the sensitivity is the highest for $T_{\min}$ threshold. In both cases, increasing the threshold level results in a decrease in sensitivity and increase in specificity. The more the likelihood ratio, the better is the model. The likelihood ratio is also the highest for the minimum thresholds considered in both cases. The relative values of likelihood ratio for two-dimensional probabilistic model are higher than those of a conventional model in all the cases, implying the performance is enhanced through the two-dimensional approach.

11.4.2 Algorithm-Based Approach Using Rainfall Thresholds and Field Monitoring Data

The rainfall, field monitoring, and landslide data for Kalimpong from July 2017 to August 2020 were used for evaluation of the performance of the ED threshold, tiltmeters, and the combined approach. The results of validation are listed in Table 11.4.

From Table 11.4, it can be understood that the algorithm-based approach in which the rainfall thresholds and tiltmeters are combined has a better performance than both the methods. The number of false positives was considerably reduced, resulting in an increase in the likelihood ratio and efficiency.

11.5 Conclusions

An LEWS is the result of reliable data collection, analysis, scientific interpretation, and integration of different modules. One of the most widely followed approaches in landslide forecasting is the derivation of conventional rainfall thresholds. This study evaluates the performance of rainfall thresholds and its conceptual improvement, by integrating the thresholds with other parameters, for a hilly town in the Darjeeling Himalayas.

The thresholds were combined with soil moisture estimates derived using a hydrological model called SHETRAN, by means of a 2D Bayesian approach. Also, the thresholds were combined with field-based monitoring data using an algorithm-based approach. Both the methods were then validated using a confusion matrix. The statistical attributes like efficiency and likelihood ratio were used for the purpose of comparison, to evaluate the performance enhancement quantitively.

From the results, it was observed that both the 2D Bayesian and the algorithm-based approaches have improved the performance of conventional thresholds considerably. The algorithm-based approach has the best performance among all, with an efficiency of 0.92 and likelihood ratio of 11.33. The study shows the major disadvantage associated with the conventional rainfall thresholds is the higher number of false alarms. Incorporating in-situ conditions for forecasting (moisture content and tiltmeter readings) can help in reducing the number of false alarms and improve the performance of conventional thresholds. The proposed methods are prototypal and should need further fine tunings before considering them for an operational LEWS.

Acknowledgements The author acknowledges the financial support of the Department of Science and Technology (DST), New Delhi, for funding the research project Grant No. (NRDMS/02/31/015(G)) and the support from Save The Hills (NGO) for carrying out the research.

References

1. Panagoulia D, Zarris D, Maggina K (2004) An assessment of the interaction between storm events and sediment transport. In: Lastra DG, de J, Martínez PV (eds) Proceedings of the 5th international symposium on ecohydraulics, aguatic habitats: analysis and restoration, Madrid, Spain, pp 281–286
2. Froude MJ, Petley DN (2018) Global fatal landslide occurrence from 2004 to 2016. Nat Hazards Earth Syst Sci 18:2161–2181. https://doi.org/10.5194/nhess-18-2161-2018
3. Dick GJ, Eberhardt E, Cabrejo-Liévano AG, Stead D, Rose ND (2015) Development of an early-warning time-of-failure analysis methodology for open-pit mine slopes utilizing ground-based slope stability radar monitoring data. Can Geotech J 52:515–529. https://doi.org/10.1139/cgj-2014-0028
4. Caine N (1980) The rainfall intensity-duration control of shallow landslides and debris flows: an update. Geogr Ann Ser A Phys Geogr 62(1–2):23–27
5. Guzzetti F, Peruccacci S, Rossi M, Stark CP (2008) The rainfall intensity-duration control of shallow landslides and debris flows: an update. Landslides 5:3–17. https://doi.org/10.1007/s10346-007-0112-1
6. Melillo M, Brunetti MT, Peruccacci S, Gariano SL, Guzzetti F (2018) TXT-tool 2.039-1.5: an algorithm for the objective reconstruction of rainfall events responsible for landslides. In: Landslide dynamics: ISDR-ICL landslide interactive teaching tools. Springer International Publishing, Cham, pp 433–447. https://doi.org/10.1007/978-3-319-57774-6_33
7. Abraham MT, Pothuraju D, Satyam N (2019) Rainfall thresholds for prediction of landslides in Idukki, India: an empirical approach. Water 11:2113. https://doi.org/10.3390/w11102113

8. Guzzetti F, Peruccacci S, Rossi M, Stark CP (2007) Rainfall thresholds for the initiation of landslides in central and Southern Europe. Meteorol Atmos Phys 98:239–267. https://doi.org/10.1007/s00703-007-0262-7
9. Martelloni G, Segoni S, Fanti R, Catani F (2012) Rainfall thresholds for the forecasting of landslide occurrence at regional scale. Landslides 9:485–495. https://doi.org/10.1007/s10346-011-0308-2
10. Abraham MT, Satyam N, Kushal S, Rosi A, Pradhan B, Segoni S (2020) Rainfall threshold estimation and landslide forecasting for Kalimpong. India Using SIGMA Model Water 12:1195. https://doi.org/10.3390/w12041195
11. Fusco F, De Vita P, Mirus BB, Baum RL, Allocca V, Tufano R, Clemente ED, Calcaterra D (2019) Physically based estimation of rainfall thresholds triggering shallow landslides in volcanic slopes of Southern Italy. Water 11:1–24. https://doi.org/10.3390/w11091915
12. Baum RL, Savage WZ, Godt JW (2008) TRIGRS—a fortran program for transient rainfall infiltration and grid-based regional slope stability analysis
13. Dikshit A, Satyam N, Pradhan B (2019) Estimation of rainfall—induced landslides using the TRIGRS model. Earth Syst Environ. https://doi.org/10.1007/s41748-019-00125-w
14. Bordoni M, Corradini B, Lucchelli L, Valentino R, Bittelli M, Vivaldi V, Meisina C (2019) Empirical and physically based thresholds for the occurrence of shallow landslides in a prone area of Northern Italian Apennines. Water 11:2653. https://doi.org/10.3390/w11122653
15. Birkinshaw SJ (2008) Physically-based modelling of double-peak discharge responses at Slapton wood catchment. Hydrol Process 22:1419–1430. https://doi.org/10.1002/hyp.6694
16. Crozier MJ (2010) Deciphering the effect of climate change on landslide activity: a review. Geomorphology 124:260–267. https://doi.org/10.1016/j.geomorph.2010.04.009
17. Lin Q, Wang Y (2018) Spatial and temporal analysis of a fatal landslide inventory in China from 1950 to 2016. Landslides 15:2357–2372. https://doi.org/10.1007/s10346-018-1037-6
18. Kirschbaum D, Stanley T, Zhou Y (2015) Spatial and temporal analysis of a global landslide catalog. Geomorphology 249:4–15. https://doi.org/10.1016/j.geomorph.2015.03.016
19. Abraham MT, Satyam N, Rosi A, Pradhan B, Segoni S (2021) Usage of antecedent soil moisture for improving the performance of rainfall thresholds for landslide early warning. Catena 200, 105147. https://doi.org/10.1016/j.catena.2021.105147
20. Abraham MT, Satyam N, Pradhan B, Alamri AM (2020) Forecasting of landslides using rainfall severity and soil wetness: a probabilistic approach for Darjeeling Himalayas. Water 12:0804. https://doi.org/10.3390/w12030804
21. Lagomarsino D, Segoni S, Rosi A, Rossi G, Battistini A, Catani F, Casagli N (2015) Quantitative comparison between two different methodologies to define rainfall thresholds for landslide forecasting. Nat Hazards Earth Syst Sci 15:2413–2423. https://doi.org/10.5194/nhess-15-2413-2015
22. Segoni S, Piciullo L, Gariano SL (2018) A review of the recent literature on rainfall thresholds for landslide occurrence. Landslides 15:1483–1501. https://doi.org/10.1007/s10346-018-0966-4
23. Campbell RH (1974) Debris flows originating from soil slips during rainstorms in Southern California. Q J Eng Geol 7:339–349. https://doi.org/10.1144/GSL.QJEG.1974.007.04.04
24. Crosta GB, Frattini P (2001) Rainfall thresholds for soil slip and debris flow triggering. In: proceeding 2nd EGS plinius conference mediterranean storms, pp 463–487
25. Song S, Wang W (2019) Impacts of antecedent soil moisture on the rainfall–runoff transformation process based on high-resolution observations in soil tank experiments. Water 11:0296. https://doi.org/10.3390/w11020296
26. Segoni S, Rosi A, Lagomarsino D, Fanti R, Casagli N (2018) Brief communication: using averaged soil moisture estimates to improve the performances of a regional-scale landslide early warning system. Nat Hazards Earth Syst Sci 18:807–812. https://doi.org/10.5194/nhess-18-807-2018
27. Segoni S, Rosi A, Fanti R, Gallucci A, Monni A, Casagli N (2018) A regional-scale landslide warning system based on 20 years of operational experience. Water 10:1297. https://doi.org/10.3390/w10101297

28. Zhao B, Dai Q, Han D, Dai H, Mao J, Zhuo L (2019) Probabilistic thresholds for landslides warning by integrating soil moisture conditions with rainfall thresholds. J Hydrol 574:276–287. https://doi.org/10.1016/j.jhydrol.2019.04.062
29. Dikshit A, Satyam DN, Towhata I (2018) Early warning system using tilt sensors in Chibo, Kalimpong, Darjeeling Himalayas. India Nat Hazards 94:727–741. https://doi.org/10.1007/s11069-018-3417-6
30. Abraham MT, Satyam N, Pradhan B, Alamri AM (2020) IoT-based geotechnical monitoring of unstable slopes for landslide early warning in the Darjeeling Himalayas. Sensors 20:2611. https://doi.org/10.3390/s20092611
31. Raffelli G, Previati M, Canone D, Gisolo D, Bevilacqua I, Capello G, Biddoccu M, Cavallo E, Deiana R, Cassiani G, Ferraris S (2017) Local and plot-scale measurements of soil moisture: time and spatially resolved field techniques in plain, hill and mountain sites. Water 9:0706. https://doi.org/10.3390/w9090706
32. Panagoulia D, Dimou G (1996) Sensitivities of groundwater-streamflow interaction to global climate change. Hydrol Sci J 41:781–796. https://doi.org/10.1080/02626669609491545
33. Parkin G (1995) SHETRAN water flow component, equations and algorithms. School of civil engineering and geosciences, Newcastle University, 1–55
34. Birkinshaw SJ, Ewen J (2000) Nitrogen transformation component for SHETRAN catchment nitrate transport modelling. J Hydrol 230:1–17. https://doi.org/10.1016/S0022-1694(00)00174-8
35. Dikshit A, Satyam DN (2018) Estimation of rainfall thresholds for landslide occurrences in Kalimpong, India. Innov Infrastruct Solut 3. https://doi.org/10.1007/s41062-018-0132-9
36. Teja TS, Dikshit A, Satyam N (2019) Determination of rainfall thresholds for landslide prediction using an algorithm-based approach: case study in the Darjeeling Himalayas India. Geosciences 9:302. https://doi.org/10.3390/geosciences9070302
37. Dikshit A, Sarkar R, Satyam N (2018) Probabilistic approach toward Darjeeling Himalayas landslides—a case study. Cogent Eng 5:1–11. https://doi.org/10.1080/23311916.2018.1537539
38. Abraham MT, Satyam N, Bulzinetti MA, Pradhan B, Pham BT, Segoni S (2020) Using field-based monitoring to enhance the performance of rainfall thresholds for landslide warning. Water 12:3453. https://doi.org/10.3390/w12123453
39. Chakraborty I, Ghosh, S, Bhattacharya D, Bora A (2011) Earthquake induced landslides in the Sikkim-Darjeeling Himalayas—an aftermath of the 18th September 2011 Sikkim earthquake. 1–8
40. Satyam N, Abraham MT (2021) Development of landslide early warning using rainfall thresholds and field monitoring: a case study from Kalimpong. In: Garg A, Solanki CH, Bogireddy C, Liu J (eds) Proceedings of the 1st Indo-China research series in geotechnical and geoenvironmental engineering. Springer, Singapore, pp 153–173. https://doi.org/10.1007/978-981-33-4324-5_11
41. Save The Hills Blog (2020) http://savethehills.blogspot.com/. Last accessed 2020/09/15
42. Kim SW, Chun KW, Otsuki K, Shinohara Y, Kim MII, Kim MS, Lee DK, Seo JII, Choi BK (2015) Heavy rain types for triggering shallow landslides in South Korea. J Fac Agric Kyushu Univ 60:243–249
43. Zhang R, Santos CAG, Moreira M, Freire PKMM, Corte-Real J (2013) Automatic calibration of the SHETRAN hydrological modelling system using MSCE. Water Resour Manag 27:4053–4068. https://doi.org/10.1007/s11269-013-0395-z
44. Zhang K, Wang S, Bao H, Zhao X (2019) Characteristics and influencing factors of rainfall-induced landslide and debris flow hazards in Shaanxi province, China. Nat Hazards Earth Syst Sci 19:93–105. https://doi.org/10.5194/nhess-19-93-2019
45. Marc O, Stumpf A, Malet JP, Gosset M, Uchida T, Chiang SH (2018) Initial insights from a global database of rainfall-induced landslide inventories: the weak influence of slope and strong influence of total storm rainfall. Earth Surf Dyn 6:903–922. https://doi.org/10.5194/esurf-6-903-2018
46. De Vita P, Reichenbach P, Bathurst JC, Borga M, Crozier GM, Glade T, Guzzetti F, Hansen A, Wasowski J (1998) Rainfall-triggered landslides: a reference list. Environ Geol 35:219–233. https://doi.org/10.1007/s002540050308

47. Lazzari M, Piccarreta M (2018) Landslide disasters triggered by extreme rainfall events: the case of montescaglioso (Basilicata, Southern Italy). Geosciences 8. https://doi.org/10.3390/geosciences8100377
48. Melillo M, Brunetti MT, Peruccacci S, Gariano SL, Roccati A, Guzzetti F (2018) A tool for the automatic calculation of rainfall thresholds for landslide occurrence. Environ Model Softw 105:230–243. https://doi.org/10.1016/j.envsoft.2018.03.024
49. Banis YN, Bathurst JC, Walling DE (2004) Use of caesium-137 data to evaluate SHETRAN simulated long-term erosion patterns in arable lands. Hydrol Process 18:1795–1809. https://doi.org/10.1002/hyp.1447
50. Abbott MB, Bathurst JC, Cunge JA, O'Connell PE, Rasmussen J (1986) An introduction to the European hydrological system—systeme hydrologique European, "SHE", 2: structure of a physically-based, distributed modelling system. J Hydrol 87:45–59
51. Parkin G (1996) A three-dimensional variably-saturated subsurface modelling system for river basins
52. Nash JE, Sutcliffe IV (1970) River flow forecasting through conceptual models part I—a discussion of principles. J Hydrol 10:282–290
53. Garambois PA, Roux H, Larnier K, Labat D, Dartus D (2015) Caractérisation de comportements de bassins versants et sélection de pluies pour la calibration de modèles hydrologiques dans le cas de crues éclair : bassins de l'est des Pyrénées. Hydrol Sci J 60:424–447. https://doi.org/10.1080/02626667.2014.909596
54. Copernicus Climate Change Service European Union's earth observation programme. https://climate.copernicus.eu/
55. CartoDEM A national digital elevation model from Cartosat-1 stereo data. https://bhuvan.nrsc.gov.in/
56. MERRA data modern-era retrospective analysis for research and applications: national aeronautics and space administration, Goddard Space Flight Center. https://gmao.gsfc.nasa.gov/reanalysis/MERRA/
57. Schaap MG, Leij FJ, Van Genuchten MT (2001) Rosetta: a computer program for estimating soil hydraulic parameters with hierarchical pedotransfer functions. J Hydrol 251:163–176. https://doi.org/10.1016/S0022-1694(01)00466-8

Chapter 12
Real-Time Monitoring System Based on Wireless Sensor Networks and Remote Sensing Techniques for Landslide-Prone Areas in the Northern Region of Thailand

Jessada Karnjana, Suthum Keerativittayanun, Kittikom Sangrit, Pitisit Dillon, Asadang Tanatipuknon, Pakinee Aimmanee, and Ken T. Murata

12.1 Introduction

The northern region of Thailand, such as Chiang Mai, Chiang Rai, Lamphun, and Mae Hong Sorn, is mountainous. According to geological surveys by the Department of Mineral Resources, Ministry of Natural Resource and Environment, many areas in the north are marked as landslide-prone areas. Most landslides were caused by intense rainfall. In 2007, the Department of Disaster Prevention and Mitigation, Ministry of Interior, and National Electronics and Computer Technology Center (NECTEC) have collaborated in the development of a landslide monitoring project based on a network of weather stations. To expand that project in new directions, in 2018, NECTEC has started a series of collaborative research projects under the theme of landslide-related disaster management. These projects aim to explore new approaches based on information and communication technologies (ICT) toward landslide-related disaster management. Research partners include government organizations and universities

J. Karnjana (✉) · S. Keerativittayanun
NECTEC, National Science and Technology Development Agency, 112 Thailand Science Park, Khlong Luang Pathum Thani 12120, Thailand
e-mail: jessada.karnjana@nectec.or.th

K. Sangrit · P. Dillon · A. Tanatipuknon · P. Aimmanee
Sirindhorn International Institute of Technology, Thammasat University, 131 Moo 5, Tiwanon Rd., Bangkadi, Muang, Pathum Thani 12000, Thailand

K. T. Murata
National Institute of Information and Communications Technology, 4-2-1 Nukui-Kitamachi, Koganei, Tokyo 184-8795, Japan

S. Kolathayar et al. (eds.), *Civil Engineering for Disaster Risk Reduction*, Springer Tracts in Civil Engineering, https://doi.org/10.1007/978-981-16-5312-4_12

in Thailand and research institutes in Japan and Southeast Asia.[1] This paper reports the current statuses, activities that have been done so far, and contributions of these projects.

12.2 Landslide Monitoring System Based on Wireless Sensor Network and Visual IoT Camera System

This section reports our implementation of a wireless sensor network (WSN) for collecting environmental parameters in real time at a landslide-prone area in the northern region of Thailand. In the current phase, we are interested in comparing the robustness of two wireless communication protocols, namely ZigBee and LoRa. The evaluation has been done in terms of communication packet losses, which will be discussed in the following subsection. In addition, the WSN works with an image transmission system, so called *Visual IoT camera system*, developed by the National Institute of Information and Communications Technology (NICT), Japan [19]. In the second following subsection, we discuss an image processing technique that can be used to analyze images captured from the Visual IoT system to obtain pieces of information in images captured at different times as many as possible.

12.2.1 WSN-Based Landslide Monitoring System

Recently, a simple WSN has been developed and installed at a landslide-prone area in Doi Pui village in Chiang Mai province, Thailand. Its structure is illustrated in Fig. 12.1. Figure 12.2 shows images of the site where the system operates. It consists of two types of nodes: a sensor node and a gateway. The sensor node is equipped with sensors, such as a temperature sensor and humidity sensor. Each sensor node has two wireless communication modules: ZigBee and LoRa. This work aims to compare the number of packet losses between these two protocols when they work under actual conditions. The sensor node reads sensor data and forwards them to the gateway. The gateway collects data from both channels and connects to the Internet via Wi-Fi and a 3G/4G network. The collected data are transmitted to a database for further analysis.

[1] Research partners include the Department of Disaster Prevention and Mitigation (DDPM) of the Ministry of Interior, the Department of Mineral Resources (DMR) of the Ministry of Natural Ressource and Environment, Sirindhorn International Institue of Technology (SIIT), Thammasat University, Mahidol University, Chiang Mai University, Asain Institute of Technology, and Kasetsart University, Thailand, Technology Computer and Electronic Institute, the National University of Laos, Lao PDR, Philippines Insitute of Volcanology and Seismology (PHIVOLCS), Philippines, National Institute of Information and Communications Technology (NICT), Gunma University, and Teikyo Heisei University, Japan, Thuyloi University, and Transport and Communication University of Vietnam, Vietnam.

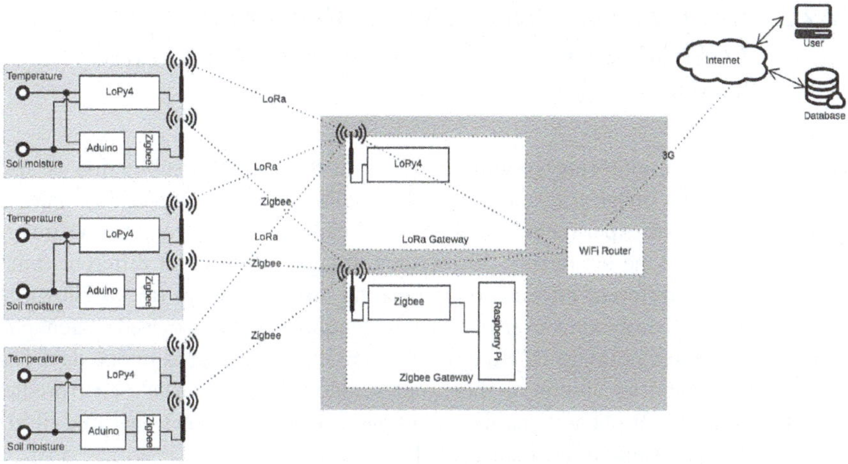

Fig. 12.1 Structure of the wireless sensor network installed at the experimental site

Fig. 12.2 Network installation (left) and experimental site in Chiang Mai province, Thailand (right)

According to data collected every minute from April 4 to May 7 of 2019, the humidity data had a strong correlation with the rain precipitation. The packet loss rate of ZigBee was about 50% less than that of LoRa. Thus, ZigBee might be more suitable in this situation. In addition, the number of packet losses depends mainly on the rain quantity measured as rain depth. However, these collected data show that the received signal strength indication (RSSI) did not correlate with the rain precipitation. Therefore, we hypothesize that the packet loss might be due to the stability of the 3G/4G network when it is raining. At the moment, this issue is under investigation.

12.2.2 *Image Fusion Method Used in Monitoring Landslide-Prone Areas with the Visual IoT Camera System*

The WSN-based monitoring system described in the previous subsection is also integrated with an image transmission system called a NICT's Visual IoT camera system. In our field experiment, the Visual IoT camera system captures an image of the landslide-prone area every 30 minutes. Images captured at different times of a day contain different information depending on the incident ray, as shown in Fig. 12.3a–c. This study aims to investigate techniques that can be used to generate a single representative image of a day from many images captured in one day. The analysis of a chronological series of these representative images can be used to predict some land-movement phenomena. For example, it has been shown that tree tilting relates to slope movement mechanisms [3].

From this study, we propose an image fusion method based on human visual perception and multi-exposure fusion for a landslide-prone area monitoring system [16]. The proposed method consists of two steps. First, all captured images are enhanced by a scaling function based on human visual perception. Second, all original images and enhanced images are fused by applying Gaussian and Laplacian pyramid-based blending.

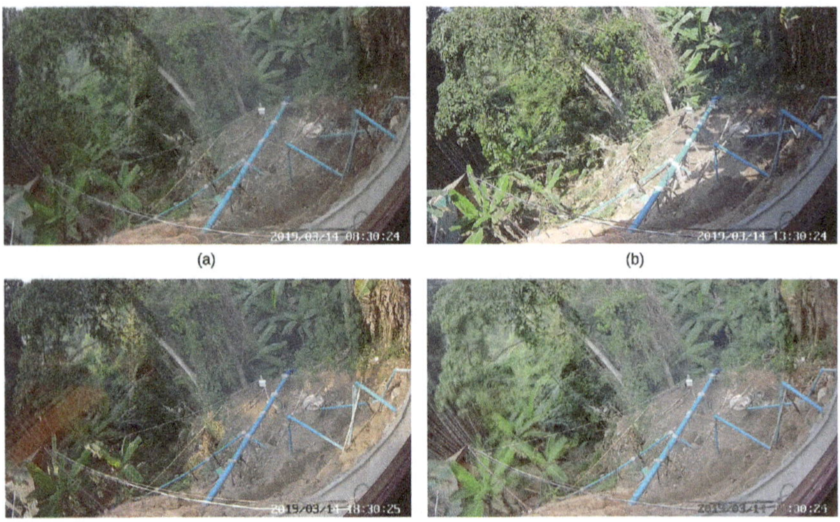

Fig. 12.3 **a**–**c** Input images captured by a Visual IoT camera installed at a landslide-prone area in Doi Pui village of Chiang Mai province, Thailand, and **d** the representative image obtained from the proposed method

An example of the representative image is shown in Fig. 12.3d. The proposed method can generate a representative image with satisfying objective evaluation scores, including image enhancement metric [13], universal image quality index [25], and perceptual similarity index [21]. However, the proposed method introduces some ghosting artifacts in the resulting image since the input sequence is dynamic.

12.3 Landslide Detection Based on Multi-temporal Satellite Imagery Analysis

The landslide monitoring system can be done locally [12] and remotely [2], as discussed in the previous section. The remote monitoring system includes satellite imagery analysis or image processing on data obtained from the unmanned aerial vehicle (UAV) attached with cameras [20]. Each system has its own advantages and disadvantages. For example, the local-sensor-based system can accurately acquire data, get, and analyze data in (near) real time. However, the local system can work with only some limited areas depending on the WSN scale. For the remote sensing system, the coverage area is larger, but the accuracy is lower.

This section reports two activities concerning satellite imagery analysis investigated by our team. First, we demonstrate that, given two multi-temporal satellite images, changes introduced by landslides in those images can be detected automatically and computational time effectively compared to a conventional method based on principal component analysis (PCA) [23]. Second, we show an example of applications of deep-learning approaches to landslide detection.

12.3.1 Change Detection in Multi-temporal Satellite Images Based on Structural Patch Decomposition and k-means Clustering

Fundamentally, change detection analyzes change data, which can be detected from many methods, for example, image differencing, normalized difference vegetation indexing, change vector analysis, principal component analysis (PCA), and image rationing [7]. Among these methods, the PCA-based method is experimentally proven to be better than the others when the trade-off between the computational cost and detection performance is considered. Inspired by this finding, we propose a simpler but more robust one based on structural patch decomposition (SPD) [17] and adaptive k-means clustering [23].

The proposed method is sketched in Fig. 12.4. Two satellite images from different times are taken as the inputs. Both are analyzed based on SPD to determine a structural consistency map [17]. This is done by bitwise comparison of satellite images binarized from its k-clustered images. Note that the value of k is adjusted

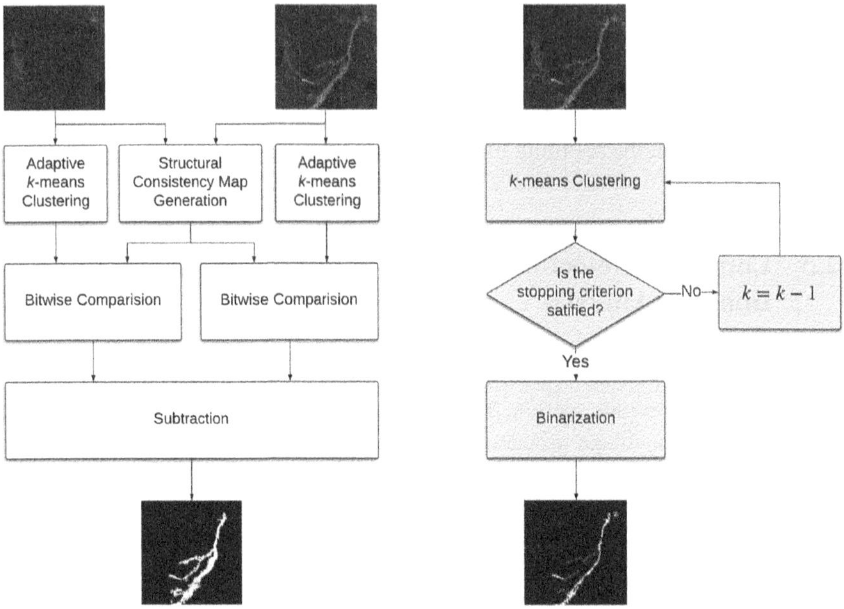

Fig. 12.4 Proposed method (left) and the proposed adaptive k-means clustering algorithm (right) [23]

automatically by the proposed algorithm [23]. Finally, the output, called a change map, is drawn from the difference between two binary images obtained from the bitwise comparison.

The experimental results show that the pixel error rate, correctness, and robustness against Gaussian and speckle noise of the proposed method [23] are slightly better than those of the PCA-based method [7]. However, both methods are comparable in terms of completeness. Besides, the computational time of the proposed method is about 1.84 times faster than the PCA-based method. Note that the proposed method can still detect changes effectively when two input images are of different resolutions, as shown in Fig. 12.5, whereas the PCA-based method performs poorly.

12.3.2 Landslide Dectection Based on Faster R-CNN Model

This study aims to investigate the possibility of detecting landslides in satellite imagery by using a variation of convolutional neural networks (CNNs), namely faster R-CNN (a faster version of regions with CNN), of which its architecture is shown in Fig. 12.6. To test the faster R-CNN's detection performance, we evaluate it with an open-access dataset from the TripleSat satellite (770 images) [14]. The

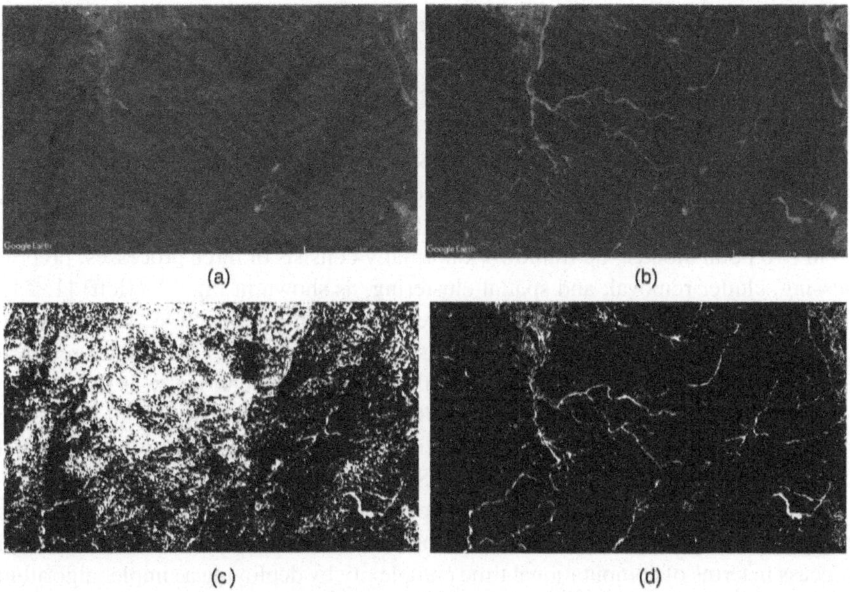

Fig. 12.5 **a–b** Multi-temporal input images, and **c–d** change maps obtained from the PCA-based method [7] and the proposed method [23], respectively

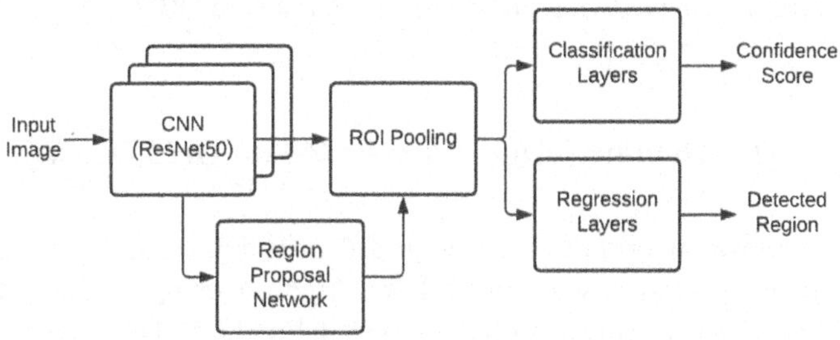

Fig. 12.6 Faster R-CNN's architecture. It takes an RGB satellite image as the input and returns a rectangular region with a confidence score as the output

accuracy, precision, recall, F-measure, and intersection-over-union (IoU) [22] score are 0.97, 0.85, 0.64, 0.71, and 0.61, respectively.

Even though the recall, F-measure, and IoU score are less than 80%, the faster R-CNN model provides promising results in terms of accuracy and precision. The model can be improved further by integrating information from other sources, such as digital elevation models (DEMs).

12.4 Rockfall Detection from LiDAR Point Clouds

This section describes our activities relating to the analysis of terrestrial LiDAR point clouds for rockfall detection, which is considered the fastest type of landslide and causes damages to properties and lives [4]. One of the approaches used to study rockfall recently is analyzing data obtained from a terrestrial laser scanner (TLS) because it provides high-resolution data [1, 5, 18, 24]. To detect rockfall events from two point clouds, the framework normally consists of three processes: preprocessing, clutter removal, and spatial clustering, as shown in Fig. 12.7 (left) [1, 24]. The preprocessing includes point-cloud registration and subtraction. This could be done using any available software such as CloudCompare [8]. The clutter-removal process is for removing noise data from the dataset. Conventionally, it is based on the nearest neighbor clutter-removal (NNCR) algorithm [6]. The spatial-clustering process is used to determine rockfall events and is based on density-based spatial clustering of applications with noise (DBSCAN) algorithm [11].

This section is divided into two subsections. First, as it is known that NNCR is time-consuming, we demonstrate an example of how to improve the clutter-removal process in terms of computational time complexity by deploying a simpler algorithm based on grid density [10]. Second, we show a drawback of the standard DBSCAN, which is a parametric algorithm. Its performance depends on two predefined parameters. This makes predefining these parameters challenging. Thus, we overcome this problem by introducing a divide-and-conquer approach to DBSCAN.

12.4.1 Clutter-Removal Algorithm Based on Grid Density

The clutter-removal algorithm based on grid density [10] can replace NNCR, as shown in Fig. 12.7 (right). It works as follows. After obtaining a surface-difference image from the preprocessing process, as shown in Fig. 12.8c, the surface-difference image is projected into 2D. The area is divided into squares of the same dimension, as shown in Fig. 12.8d. Then, data points are counted in each square to determine the maximum and minimum numbers of data points from all squares. The square with data points less than half of the difference between the maximum and minimum is considered a clutter square, and data points in clutter squares are removed from the surface-difference image, as shown in Fig. 12.8e. This clutter-removed surface-difference image is then fed to DBSCAN to determine rockfall events.

According to our simulations, NNCR is slightly better than the proposed method in terms of the numbers of rockfall events. However, in terms of false-positive detection, the proposed method is better. More interestingly, the proposed method has a better computational time complexity, i.e., for 52,000 data points, the proposed method is about 16 times faster than NNCR [10].

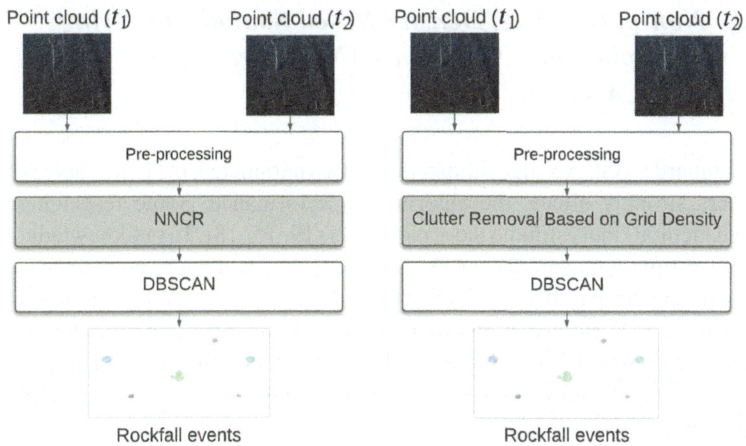

Fig. 12.7 Conventional method with NNCR (left) and the proposed method (right) [10]. Both methods take two point clounds from different times t_1 and t_2 and their inputs and return rockfall-event maps

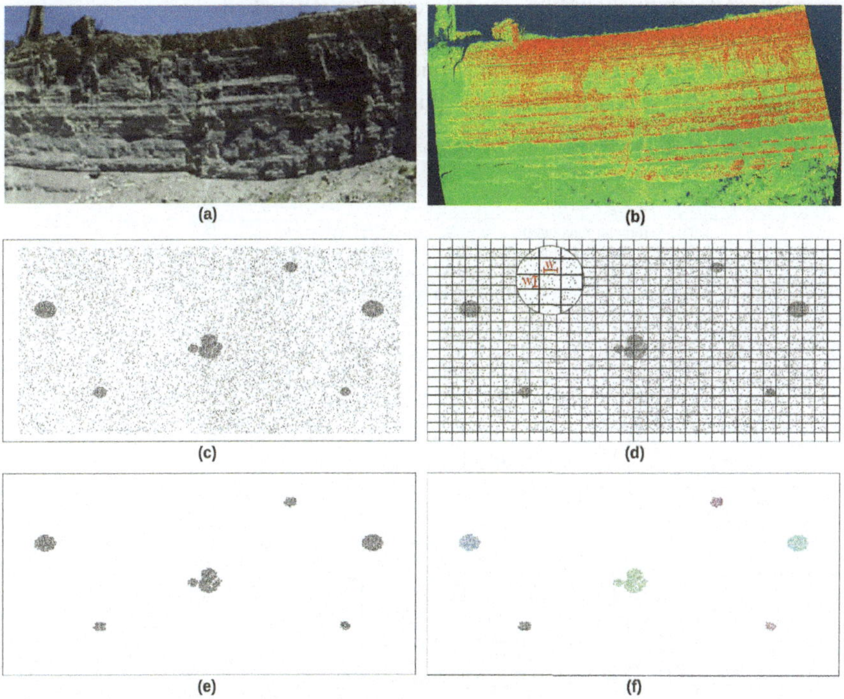

Fig. 12.8 **a** Puigcercos cliff, Catalonia, Spain [1, 24], **b** 3D point cloud dataset [1, 24], **c** surface-difference image, **d** grid of equal squares, **e** clutter-removed surface-difference image obtained from the proposed clutter-removal algorithm, and **f** detected rockfall events obtained by the proposed method [10]

12.4.2 Non-parametric, Density-Based Spatial Clustering of Applications with Noise (Non-parametric DBSCAN)

As the standard DBSCAN algorithm requires two parameters [11], the choice of these parameters strongly affects the algorithm's performance. Some researchers have proposed methods that estimate these parameters [9, 15, 18]. To our knowledge, those parameter-estimation methods can estimate only constant parameters. However, the constant-parameter-based DBSCAN may not be feasible when it is applied to a dataset with a wide variety of data-point densities. To solve this problem, our team has developed a new DBSCAN-based algorithm, called divide-and-conquer-based DBSCAN or DC-DBSCAN, that can cluster data points automatically and effectively even though data density varies.

The proposed framework is illustrated in Fig. 12.9 (left), and it consists of two steps. First, the parameters are estimated roughly by grid analysis. Second, the standard DBSCAN with the estimated parameters is used to cluster the 3D point cloud. Then the DBSCAN with updated, estimated parameters is applied to each cluster separately and iteratively until a stopping criterion is satisfied.

To verify the performance of the proposed DC-DBSCAN, we conduct experiments with the same dataset as mentioned in the previous subsection. The simulation results show that the averaged purity score of DC-DBSCAN (96.22%) is

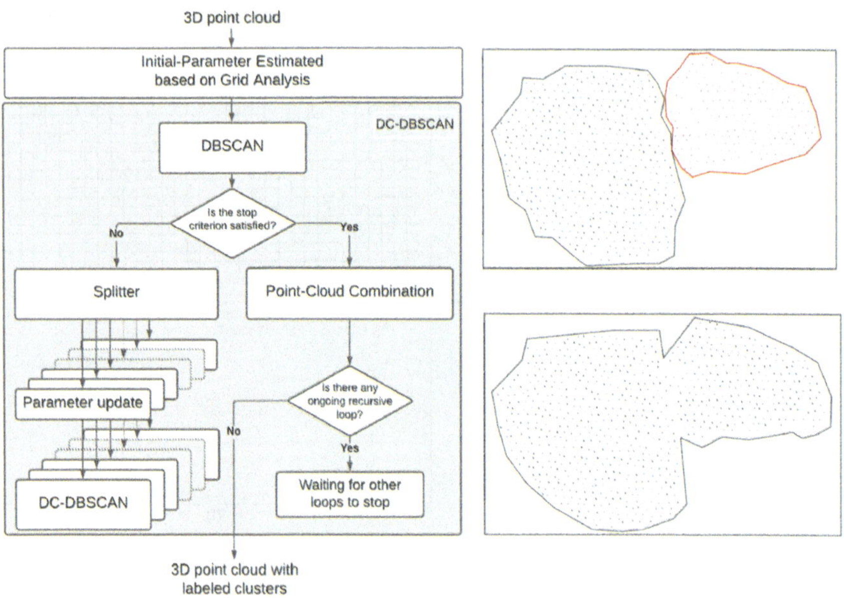

Fig. 12.9 Framework of DC-DBSCAN (left). Note that, in some case, DC-DBSCAN can detect rockfall events (top-right) the standard DBSCAN fails to detect (bottom-right)

considerably greater than that of the standard DBSCAN (91.09%). Interestingly, DC-DBSCAN can detect rockfall events the standard DBSCAN fails to detect, as shown in Fig. 12.9 (right).

12.5 Conclusion

This paper reported current collaborative research activities under projects endorsed by ASEAN COSTI and e-Asia JRP, aiming to investigate potential landslides' environmental parameters and conditions using real-time monitoring based on wireless sensors networks (WSN) and remote sensing techniques. It demonstrated the implementation of a WSN-based monitoring system integrated with NICT's Visual IoT camera system and discussed a technique that can generate a representative image of a day by applying an image enhancement technique and a multi-exposure fusion technique. Also, remote sensing techniques used to detect changes in satellite imagery and 3D point clouds were discussed: change detection in multi-temporal satellite images based on structural patch decomposition and k-means clustering, landslide detection based on a faster R-CNN model, clutter-removal algorithm based on grid density for rockfall detection, non-parametric, and density-based spatial clustering of applications with noise algorithm. The integration of these approaches is expected to provide useful information for landslide-related disaster management.

Acknowledgements This work has been partially supported by the ASEAN Committee on Science, Technology and Innovation (COSTI) under the ASEAN Plan of Action on Science, Technology and Innovation (APASTI) funding scheme and by e-Asia JRP funding scheme.

Also, it is supported by Thailand Advanced Institute of Science and Technology (TAIST), National Science and Technology Development Agency (NSTDA), and Tokyo Institute of Technology under the TAIST-Tokyo Tech program.

References

1. Abellán A, Calvet J, Vilaplana JM, Blanchard J (2010) Detection and spatial prediction of rockfalls by means of terrestrial laser scanner monitoring. Geomorphology 119(3–4):162–171
2. Amit SNKB, Aoki Y (2017) Disaster detection from aerial imagery with convolutional neural network. In: 2017 international electronics symposium on knowledge creation and intelligent computing (IES-KCIC), pp 239–245. IEEE
3. Ballesteros-Cánovas JA, Márquez-Peñaranda J, Sánchez-Silva M, Díez-Herrero A, Ruiz-Villanueva V, Bodoque J, Eguibar M, Stoffel M (2015) Can tree tilting be used for paleoflood discharge estimations? J Hydrol 529:480–489
4. Bates RL, Jackson JA (1987) Glossary of geology
5. Bitelli G, Dubbini M, Zanutta A (2004) Terrestrial laser scanning and digital photogrammetry techniques to monitor landslide bodies. Int Arch Photogr Rem Sens Spat Inf Sci 35(B5):246–251
6. Byers S, Raftery AE (1998) Nearest-neighbor clutter removal for estimating features in spatial point processes. J Am Stat Assoc 93(442):577–584

7. Celik T (2009) Unsupervised change detection in satellite images using principal component analysis and *k*-means clustering. IEEE Geosci Rem Sens Lett 6(4):772–776
8. CloudCompare Website. https://www.danielgm.net/cc/
9. Darong H, Peng W (2012) Grid-based dbscan algorithm with referential parameters. Phys Procedia 24:1166–1170
10. Dillon P, Karnjana J, Aimmanee P (2020) Rockfall detection from terrestrial lidar point clouds by using dbscan with clutterremoval based on grid density. Technical report, EasyChair
11. Ester M, Kriegel HP, Sander J, Xu X et al (1996) A density-based algorithm for discovering clusters in large spatial databases with noise. Kdd 96:226–231
12. Intrieri E, Gigli G, Mugnai F, Fanti R, Casagli N (2012) Design and implementation of a landslide early warning system. Eng Geol 147:124–136
13. Jaya V, Gopikakumari R (2013) IEM: a new image enhancement metric for contrast and sharpness measurements. Int J Comput Appl 79(9)
14. Ji S, Yu D, Shen C, Li W, Xu Q (2020) Landslide detection from an open satellite imagery and digital elevation model dataset using attention boosted convolutional neural networks. Landslides pp 1–16
15. Karami A, Johansson R (2014) Choosing dbscan parameters automatically using differential evolution. Int J Comput Appl 91(7):1–11
16. Keerativittayanun S, Sangrit K, Srisukanun P, Dillon P, Karnjana J (2019) Image enhancement technique based on human visual perception and multi-exposure fusion for a landslide-prone area monitoring system. In: 2019 58th annual conference of the Society of Instrument and Control Engineers of Japan (SICE), pp 382–387. IEEE
17. Ma K, Li H, Yong H, Wang Z, Meng D, Zhang L (2017) Robust multi-exposure image fusion: a structural patch decomposition approach. IEEE Trans Image Process 26(5):2519–2532
18. Micheletti N, Tonini M, Lane SN (2017) Geomorphological activity at a rock glacier front detected with a 3d density-based clustering algorithm. Geomorphology 278:287–297
19. Murata KT, Mizuhara T, Pavarangkoon P, Yamamoto K, Muranaga K, Aoki T (2018) Design and development of real-time video transmission system using visual iot device. In: International conference on multi-disciplinary trends in artificial intelligence, pp 263–269. Springer, Berlin
20. Niethammer U, James M, Rothmund S, Travelletti J, Joswig M (2012) Uav-based remote sensing of the super-sauze landslide: evaluation and results. Eng Geol 128:2–11
21. Reisenhofer R, Bosse S, Kutyniok G, Wiegand T (2018) A haar wavelet-based perceptual similarity index for image quality assessment. Signal Process Image Commun 61:33–43
22. Rezatofighi H, Tsoi N, Gwak J, Sadeghian A, Reid I, Savarese S (2019) Generalized intersection over union: a metric and a loss for bounding box regression. In: Proceedings of the IEEE conference on computer vision and pattern recognition, pp 658–666
23. Tanatipuknon A, Aimmanee P, Keerativittayanun S, Karnjana J (2020) Unsupervised change detection in multi-temporal satellite images based on structural patch decomposition and k-means clustering for landslide monitoring. In: International symposium on integrated uncertainty in knowledge modelling and decision making, pp 259–269. Springer, Berlin
24. Tonini M, Abellan A (2014) Rockfall detection from terrestrial lidar point clouds: a clustering approach using r. J Spat Inf Sci 2014(8):95–110
25. Wang Z, Bovik AC (2002) A universal image quality index. IEEE Signal Process Lett 9(3):81–84

Chapter 13
Perturbation of Earth Surface Process by Geophysical and Meteorological Process in the Nepal Himalaya

Basanta Raj Adhikari

13.1 Introduction

Himalaya is a product of collision between Indian and Eurasian plates [2, 12, 14] and seismically active due to reoccurrences of earthquakes. The Himalayan arc extends 2400 km from Namcha Barwa (7756 m) in the east to Nanga Parbat (8138 m) in the west with 230–350 km width [9]. There are difference slices in the Himalaya along the principal thrusts/faults and stacked one over the other propagating southward and building the architecture of the Himalaya [12]. Nepal Himalaya lies in the central Himalayan arc and occupies nearly 800 km extending from Mahakali River in the west to Mechi River in the east. These tectonic units divided longitudinally into five major tectonic zones from south to north separated by major thrusts (Fig. 13.1). These tectonic units are: Indo-Gangetic plain, Siwaliks, Lesser Himalayan Zone, Higher Himalayan Zone and Tibetan Tethys Zone.

These tectonic units are playing an important role to carve the landscape coupled with forming different physiographic units. These physiographic units are divided on the basis of elevation and climate [8]. They are: Terai (below 200 masl), Siwaliks (100–2000 masl), hill (200–3500 masl), middle mountain (700–4100 masl) and high mountain (1800–8888.86 masl) (Fig. 13.2). The Terai region receives lots of sediments from the Siwaliks, hills and mountains transported by the major river systems in Nepal. These sediments deposited in the foothills of the Himalaya created lots of flooding problems every year, which is one of the major hazards in Nepal killing 303 people in 2020 [11]. The Siwaliks (Chure) consists of sedimentary rocks composed of sandstone, mudstone and conglomerate. Therefore, many landslides and debris

B. R. Adhikari (✉)
Department of Civil Engineering, Pulchowk Campus, Institute of Engineering, Tribhuvan University, Lalitpur, Nepal
e-mail: bradhikari@ioe.edu.np

S. Kolathayar et al. (eds.), *Civil Engineering for Disaster Risk Reduction*, Springer Tracts in Civil Engineering, https://doi.org/10.1007/978-981-16-5312-4_13

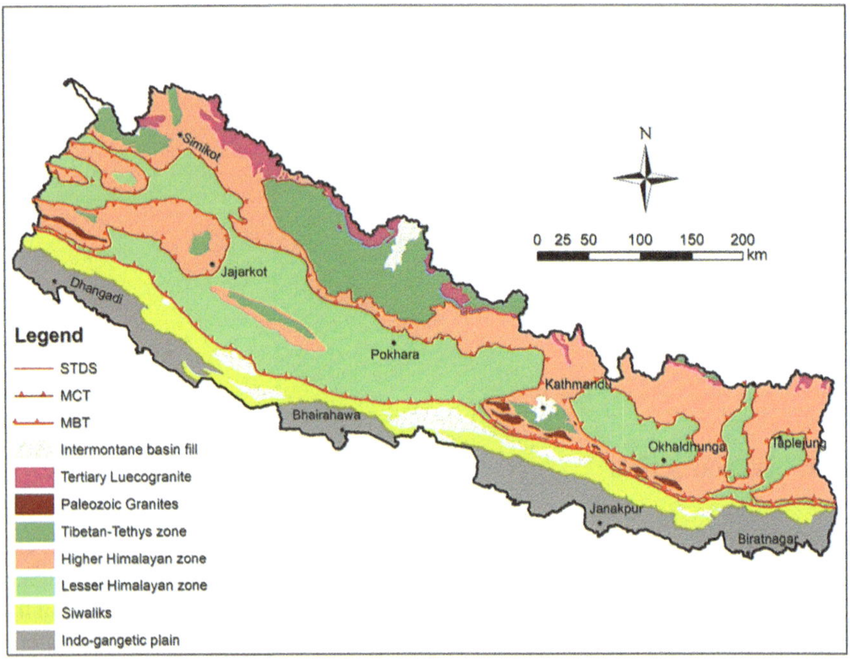

Fig. 13.1 Geological subdivision of the Nepal Himalaya (Modified after Amatya and Janawali [1]). MBT: Main Boundary Thrust, MCT: Main Central Thrust, STDS: South Tibetan Detachment System

flow events occur in this region destroying life and properties. The hills and mountains occupy more than 80% of the country, experiencing different tectonic activities. These tectonic activities coupled with monsoon and anthropogenic interventions trigger many slope failures. Landslides, debris flows, flooding and glacial lake outburst floods (GlOFs) are major hazards in the Nepal Himalaya.

13.2 Large-Scale Mass Movement

The Nepal Himalaya experienced many large-scale mass movements in the past and completely changed the landscape (Table 13.1). These mass movements were triggered by seismotectonic movement of the Himalaya and meteorological causes. The Tsergo Ri landslide (109 m^3) (Langtang Himal, Nepal) is one of the largest rock slides in the Nepal Himalaya that occurred in prehistoric period (fission track age 'some' 104 years) [20]. This landslide melted the sliding surface and formed hyalomylonites in its primary sliding planes.

[15]. Rockslide deposits (30–40 ka) around Ringmao village dammed the Phoksundo lake in the western part of Nepal [21]. Similarly, the Kalopani landslide,

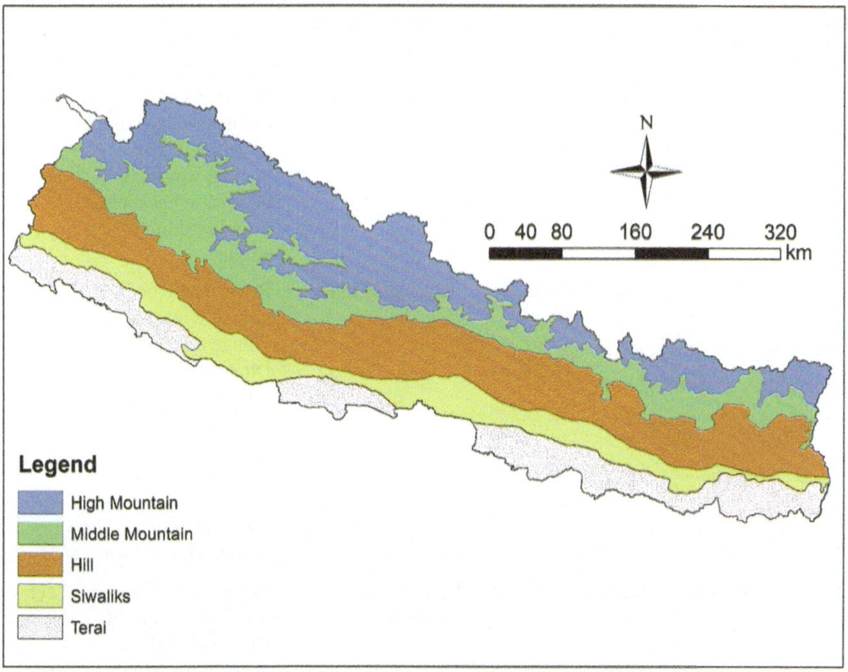

Fig. 13.2 Physiographic subdivision of the Nepal Himalaya

Dhukur Pokhara landslide and Latamarang landslide [21] occurred in the post-glacial period in the Nepal Himalaya. There are many other landslides and debris flow events in the modern history. The Pokhara valley was formed by the transported materials along the Seti River due to collapse of Sabche Cirque in the Annapurna region. This collapse was occurred due to three large-scale earthquakes dates in 1100, 1255 and 1344 [16]. This valley experienced a similar kind of debris flow in 2012 but in a smaller scale, which killed more than 72 people and destroyed lots of property [13]. The Bhotekoshi River in the central Nepal has experienced many large-scale glacial lake outburst lake floods in 1982 [7] and in 2016 [3]. The Jure landslide, one of the biggest landslides in Asia, occurred in 2014 in this valley that killed more than 156 people and blocked the Bhotekoshi River for few days [19]. Similarly, many landslides and debris flow events occurred in the central Nepal in 1993 due to cloud burst [5], which was one of the deadliest cloud burst events in the modern history of Nepal.

Table 13.1 Large-scale mass movement in the Nepal Himalaya

S. no.	Date	Name	Location	Description	Source
1	(10^4 years)	Tsergo Ri	Langtang, Rasuwa	The effect of prehistoric landslide of about 10^9 m^3 volume was huge because it has melted even its sliding surface	[20]
2	30–40 ka	Ringmo rockslide	Phoksundo Lake	Rockslide deposit around Ringmo village affected Dhaulagiri limestones of the Tibetan Tethys sediments, and the detritus materials are covered by loess in which intercalations of reddish soil layers	[21]
3	Post-glacial	Kalopani landslide	Kaligandaki, Mustang	The rockslide masses originated south of Dhaulagiri Himal deposited near the Kalopani village and dammed the Kaligandaki River covering of 10 km^2	[21]
4	Post-glacial	Dukur Pokhara	Marsyangdi Valley, Manang	The source of the landslide materials is from the southeastern ridge of Naur Himal at elevations of 4700–4900 m. The Tibetan Tethys sediment was slide down due to steep slope and folded structures	[21]
5	Post-glacial	Latamarang	Marsyangdi Valley, Manang	The landslide material deposit stretches over 4 km along the river from the village of Thanchauk passing Latamarang	[21]

(continued)

Table 13.1 (continued)

S. no.	Date	Name	Location	Description	Source
6	1100–1344	Pokhara Valley	Kaski	Pokhara Valley was formed due to three major earthquake materials transported from the Sabche Cirque in the Annapurna region	[16]
7	1962	Darbang	Myagdi	Buried the Darbhanga Bazaar and killed 500 people	[22]
8	1976	Madi River	Kaski	Landslide killed about 75 people	[18]
9	1982	Bhotekoshi	Sindhupalchowk	GLOFs hazard and destroyed lots of properties	[7]
10	1993	Phedi Gau	Makawanpur	Large-scale landslides and debris flow in central Nepal and destroyed property. Phedigaon debris flow destroyed 52 houses and 52 death	[5]
11	2012	Seti River	Kaski	Flood has killed more than 70 people and destroyed properties	[13]
12	2014	Jure landslide	Sindhupalchowk	Landslide has buried a village with 156 death and blocked the Sunkoshi River forming 55-m-high dam	[19]
13	2016	Bhotekoshi	Sindhupalchowk	GLOFs hazard which destroyed lots of property	[3]

13.3 Spatiotemporal Variation of Landslide

The occurrences of landslides in the Nepal Himalaya are mainly controlled by geological structures such as faults, thrusts, folds and joints [17]. The landslide generation in this terrain is related to the coupling effect of geophysical causes and Asian monsoon. Besides, rapid urbanization and haphazard infrastructure construction are also playing an important role to make more susceptible for landslides. The total number of fatal landslide is 4844 from 1971 to 2020 which killed 5740 people (Fig. 13.3) [4, 11]. The reported number of landslide and fatality has significantly

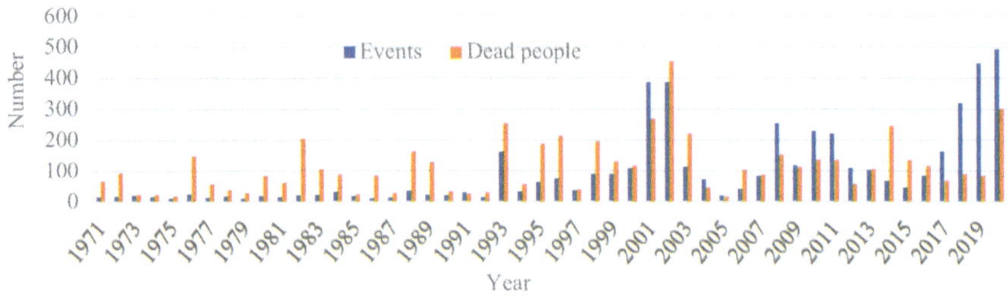

Fig. 13.3 Temporal variation of landslides from 1971 to 2020 [4, 11]

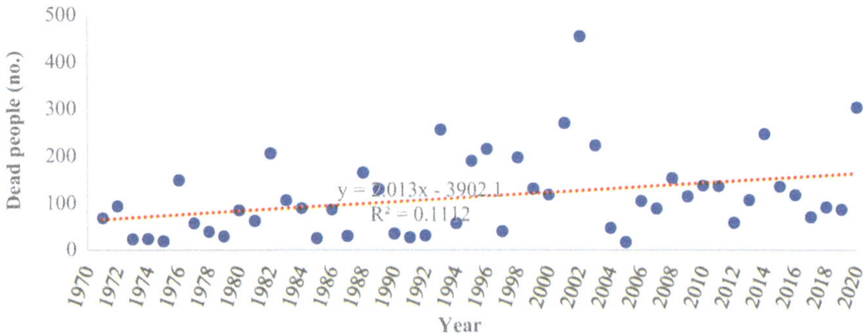

Fig. 13.4 Landslide trend analysis from 1971 to 2020 [4, 11]

increased from 1993. The increasing number of landslides is related to the non-engineered village road construction. A recent study [10] shows that the rainfall-triggered landslides are more likely to occur within 100 m of a road than the landslide generated by seismic events. However, the improvement of reporting systems such as telephone, Internet and mobile phones (Fig. 13.4) has also helped to document and report the landslide events and fatality.

This might be related to internal migration because most of the poor people who migrated from rural area to urban area are living in the vulnerable zone and faced different kinds of disaster.

Gorkha earthquake 2015 has triggered more than 19,000 co-seismic landslides [6] which are reactivated in subsequent years by monsoon. The analysis shows that almost 85% of the fatal landslide occur between July to September. The spatial distribution of landslide shows that most of the landslides are concentrated in the mid-hills of Nepal (Fig. 13.5). This is mainly due to the non-engineered road construction, improper agricultural practices and excessive rainfall during the monsoon season. The districts of central and western Nepal experienced lots of landslide due to concentrated rainfalls and anthropogenic activities with high population density.

The monsoon was very active in the year of 2020 and affected central and western part of Nepal killing 303 people. Most of the landslide occurred between July and August in the following districts: Sindhupalchowk, Myagdi, Parbat, Jajarkot,

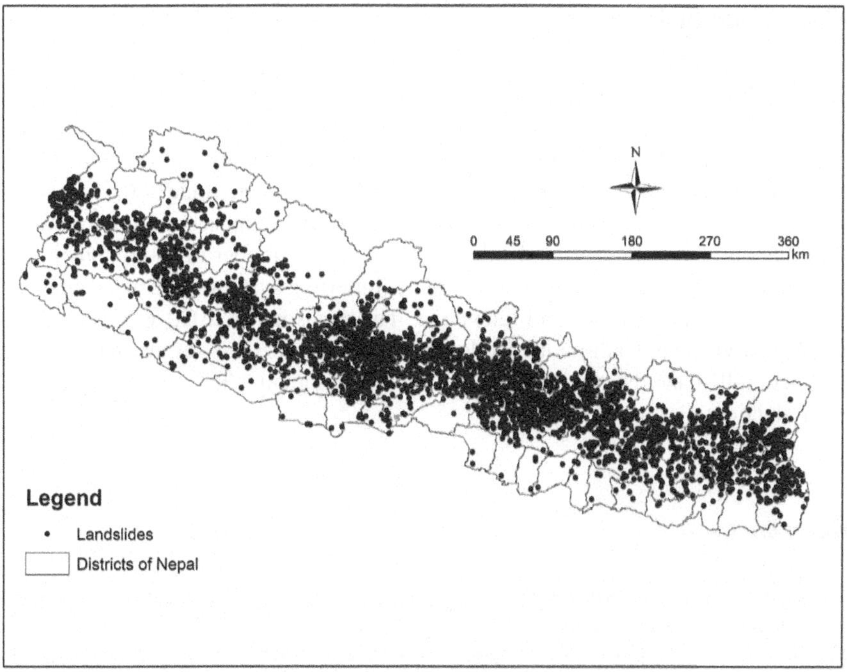

Fig. 13.5 Spatial distribution of fatal landslides from 1971 to 2020 [4, 11]

Syangja, Kaski, Bajhang, Accham and Baglung. A large landslide was occurred on 14 August at Lidi, Sindhupalchowk, that took the lives of 27 people and destroyed many houses. The landslide was occurred due to prolonged rainfall in the weak geomorphological slope created by Gorkha earthquake 2015. The ground cracks before the landslide suggest that this area was gradually moving away from the geotechnically equilibrium state. Another big landslide occurred at Barekot, Jajarkot, on July 11, 2020, which killed 11 people and 5 houses were destroyed. The most remarkable debris flow event was occurred at Ramaroshan, Accham, on 23 August, and at least 6 people were dead.

These events show that Nepal Himalaya is very vulnerable for the landslide and debris flow due to high rainfall events and weak geology. Therefore, proper engineering geological investigation before road construction, watershed management and adequate drainage management should be applied to reduce the disaster risk. The newly established National Disaster Risk Reduction and Management Authority (NDRRMA) is working closely with the concerned departments and institution for the risk reduction; however, proper investigation of multi-hazard risk should be forefront of the planning process.

13.4 Conclusion

Nepal Himalaya is seismically active due to the occurrences of different scales of earthquakes. These activities weaken the earth's surface to create different kinds of mass movement. The coupling effect of weak geology and monsoon is responsible for triggering many landslides, debris flow and floods every year. Some of the past large-scale mass movements in the Nepal Himalaya changed the landscape completely. The spatiotemporal variation of landslides shows that the number of fatal landslides is increasing every year. It might be due to better reporting systems with the availability of mobile phones, Internet and telephone. The monsoon in 2020 was very active and triggered many shallow landslides and debris flows. It is utmost necessary to conduct multi-hazard risk assessment along this mountain chain for the sustainable risk reduction.

References

1. Amatya KM, Janawali BM (1994) Geological map of Nepal. scale: 1: 1,000,000. Department of Mines and Geology, Nepal
2. Bilham R (2019) Himalayan earthquakes: a review of historical seismicity and early 21st century slip potential. Geol Soc Lond Spec Publ 483(SP483):16
3. Cook KL, Andermann C, Gimbert F, Adhikari BR, Hovius N (2018) Glacial lake outburst floods as drivers of fluvial erosion in the Himalaya. Science 362(6410):53–57
4. Desinventar (2016) Country Profile of Nepal. UNDRR
5. Dhital MR, Khanal N, Thapa KB (1993) The rold of extreme weather events, mass movements, and landuse changes in increasing natural hazards. In: Workhop on causes of recent damage incurred in south-central nepal
6. Gnyawali KR, Adhikari BR (2017) Spatial relations of earthquake induced landslides triggered by 2015 Gorkha Earthquake Mw= 7.8. Workshop on world landslide forum
7. Hewitt K (1982) Natural dams and outburst floods of the Karakoram Himalaya. IAHS 138:259–269
8. LRMP (1986) Land resources mapping project. Survey Department, Government of Nepal and Kenting Earth Sciences
9. Le Fort P (1996) Evolution of the Himalaya. In: Yin A, Harrison TM (eds) The tectonic evolution of Asia. Cambridge University Press, Cambridge
10. McAdoo BG, Quak M, Gnyawali KR, Adhikari BR, Devkota S, Rajbhandari PL, Sudmeier-Rieux K (2018) Roads and landslides in Nepal: how development affects environmental risk. Nat Hazard 18(12):3203–3210
11. MoHA (2021) Nepal disaster risk reduction portal
12. Molnar P (1984) Structure and tectonics of the Himalaya: Constraints and implications of geophysical data. Annu Rev Earth Planet Sci 12(1):489–516
13. Oi H, Daisuke HIGAKI, Hiroshi YAGI, Nobuhiro USUKI, Kousuke YOSHINO (2014) Report of the investigation of the flood disaster that occurred on May 5, 2012 along the Seti River in Nepal. Int J Erosion Control Eng 7(4):111–117
14. Sapkota SN, Bollinger L, Klinger Y, Tapponnier P, Gaudemer Y, Tiwari D (2013) Primary surface ruptures of the great Himalayan earthquakes in 1934 and 1255. Nat Geosci 6(1):71–76
15. Schramm J-M, Weidinger JT, Ibetsberger HJ (1998) Petrologic and structural controls on geomorphology of prehistoric Tsergo Ri slope failure, Langtang Himal, Nepal. Geomorphology 26(1–3):107–121

16. Schwanghart W, Bernhardt A, Stolle A, Hoelzmann P, Adhikari BR, Andermann C, Tofelde S, Merchel S, Rugel G, Fort M (2016) Repeated catastrophic valley infill following medieval earthquakes in the Nepal Himalaya. Science 351(6269):147–150
17. Upreti BN (2001) The physiography and geology of nepal and landslide hazards. In: Chalise SR, Upreti BN, Tianchi L (eds) Landslide problem mitigation to the Hindu-kush Himalayas, ICIMOD, p 312
18. Upreti BN, Dhital MR (1996) Landslide studies and management in Nepal, International Centre for Integrated Mountain Development. Kathmandu, Nepal, p 94
19. Van der Geest K (2018) Landslide loss and damage in Sindhupalchok District, Nepal: comparing income groups with implications for compensation and relief. Int J Disaster Risk Sci 9(2):157–166
20. Weidinger JT, Schramm J-M, Surenian R (1996) On preparatory causal factors, initiating the prehistoric Tsergo Ri landslide (Langthang Himal, Nepal). Tectonophysics 260(1–3):95–107
21. Weidinger JT (2006) Ladslide dams in the high mountains of India, Nepal and China stability and life span of their dammed lakes. Italian J Eng Geol Environ 1
22. Yagi H, Maruo Y, Saijo K, Nakamura S (1990) The September 1988 large landslide in the vicinity of MCT, Darbang, Nepal. J Jpn Geol Soc 26(4):45–49

Chapter 14
Post-earthquake Reconnaissance: Theories Versus Observations

Siau Chen Chian

14.1 Introduction

Earthquake remains one of the most destructive natural phenomena to date, leading to catastrophic damage to buildings and lifelines. Dependent on the hypocentre characteristics, propagation path effects and effect of the sediments overlying the bedrock, the outcome of the ground motion can vary between different locations. An earthquake ground motion time history can be described based on three features: the amplitude, frequency content and duration of shaking. These features have significant effects to the displacement of the building which causes damage to rigid structures.

Apart from earthquake shaking characteristics, the reinforcement detailing and construction sequence of the building are also keys to the consequence of the shaking. Insufficient or inappropriate placement of shear links, in particular at columns, forms the weak spot of the structure during shaking, leading to buckling column or, in severe cases, soft-storey collapse. The sequence of construction causing weak joints may also lead to soft-storey collapses at higher levels of buildings rather than at the typical ground level. Earthquake shaking also affects the ground condition, causing landslides, settlement of ground and foundations. Liquefaction has also been observed in several earthquake events which led to upheaval of suspended ground slabs as well as uplift of underground utilities manholes and pipelines.

Case studies of recent major earthquakes would be used to demonstrate the theories with observations. Reference would be made to earthquake reports co-written by the author.

S. C. Chian (✉)
National University of Singapore, Singapore, Singapore
e-mail: sc.chian@nus.edu.sg

S. Kolathayar et al. (eds.), *Civil Engineering for Disaster Risk Reduction*, Springer Tracts in Civil Engineering, https://doi.org/10.1007/978-981-16-5312-4_14

14.2 Ground Motion Characteristics

Earthquakes are violent ground shaking which are typically recorded and measured by a network of seismic stations globally. A seismograph is a device for measuring the movement of the ground. This device consists of a seismometer which acts as a ground motion detector and operates based on the principle of inertia. Seismometers are highly sensitive to ground movements and can detect and record motions in the east–west, north–south and up–down directions. A simple seismometer can be understood by visualising a weight hanging on a spring. In the event of an earthquake, the relative motion between the weight and the ground provides a measurement of the ground motion. A schematic diagram of a seismometer can be illustrated in Fig. 14.1.

The graphical output of a seismometer is a seismogram, which shows the earthquake time history of the earthquake. From the time history, the amplitude, frequency content and duration of the shaking can be determined. Earthquake engineers typically convert the time history information to an acceleration versus period plot, known as a response spectra, to observe the dominant frequency contents of the shaking. Frequency of a wave refers to the number of waves that passes through a point in one second. Period is the amount of time it takes for one wave cycle to pass the given point. Period is hence a reciprocal of frequency. Figure 14.2 shows the response spectra for the 2009 Padang and 2011 Tohoku earthquakes, respectively.

In order to correlate these response spectra to building damage, the effect of building resonance is to be discussed. All buildings have a natural period, or resonance, which is the number of seconds it takes for the building to naturally oscillate with greater amplitude at some frequencies than at others. Resonant frequency of a building is the frequency at which the maximum amplitude oscillation occurs,

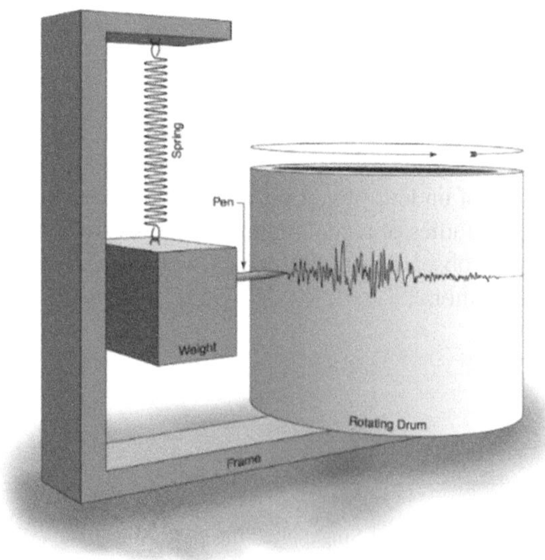

Fig. 14.1 Principles of a seismometer [14]

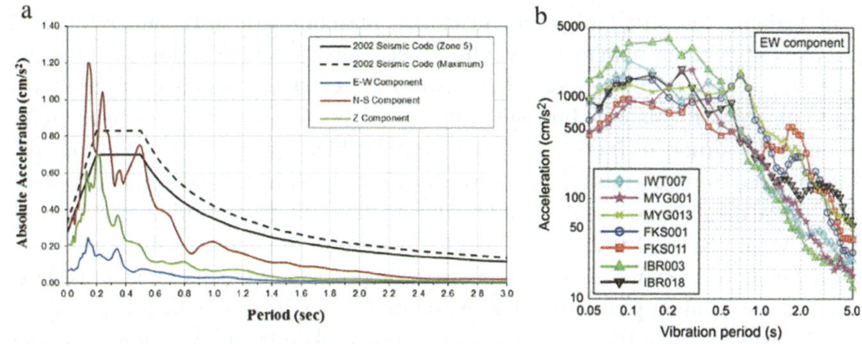

Fig. 14.2 Response spectra of 2009 Padang earthquake response spectra vs. uniform hazard response spectra of Indonesian seismic code (left, A) and elastic response spectra of the recorded mainshock ground motions, east–west component at the seven K-NET station (right, B) [9, 10]

leading to development of cracks if it exceeds the ductility of the building. The ground also has a specific resonant frequency. Hard bedrock has higher frequencies than softer sediments. If the frequency of ground motion coincides with the resonant frequency of the building, it will undergo the largest oscillations and suffer the greatest damage. Low-rise buildings are more affected by short and frequent waves (i.e. high frequency shaking), while high-rise buildings are affected by long and slow shaking (i.e. low frequency shaking). A higher amplitude shaking in units of acceleration often also implies stronger shaking with larger displacement,hence expected to generate more significant damage to structures. A longer duration of shaking has the same outcome of greater damage owing to more oscillations and opportunities for the cracks to further develop and propagate, leading to a potential catastrophic collapse of a building.

14.2.1 The 2009 Padang Earthquake

On 30 September 2009, a major earthquake registering a magnitude Mw7.6 occurred off the coast of Padang in West Sumatra in Indonesia, followed by an inland earthquake measuring Mw6.6 the next day morning. The main shock was triggered by oblique thrust faulting near a long undersea subduction fault interface between the Australian and Sunda Plates, while the inland earthquake occurred due to a dextral (right lateral) strike-slip movement at the proximal Great Sumatran Fault [24, 25]. The United States Geological Survey (USGS) estimated the focus of the Mw7.6 earthquake to be located at a depth of 81 km and a distance of 60 km west–north-west of the city of Padang. A tsunami warning was issued but later withdrawn. Significant damage to buildings due to ground shaking and large-scale landslides was resulted. Details of the earthquake and distribution of damages can be found in [2, 16].

In Padang, the structures were primarily masonry structures and reinforced concrete (RC) frames with masonry infill. A vast majority of the buildings in Padang were one- or two-storey residential and two- to three-storey commercial structures due to planning regulations limiting building heights to 6 storeys prior to the relocation of the airport in 2005. Only a few steel buildings were observed in the city. Unreinforced masonry infill panels performed poorly with extensive cracking, indicating large displacement of the building during the earthquake. Buildings over three storeys were observed to be most affected, owing to the depth of the earthquake where layers of soil filtered high frequency components of the earthquake. BMKG recorded a maximum peak ground acceleration (PGA) of 0.3 g, which lasted about 20 s. Considering that the acceleration record was taken from a location with relatively stiff soil according to geological map [12], the earthquake ground shaking was expected to have amplified at Padang city which lies on relatively thick alluvium deposits with an expected PGA between 0.4 and 0.6 g across the city [9].

Based on the synthetic earthquake time history in Fig. 2a, dominant frequencies were between 2 and 5 Hz. This range of frequencies is more likely to coincide with natural frequencies of 2- to 5-storey building, which is substantiated with observations where RC buildings below six storeys suffered structural damage (see Fig. 14.3). The 1-storey masonry residential building, 3-storey RC hotel building and 4-storey RC commercial building shown in the figure demonstrate this point. In contrast, multi-storey RC hotel and shopping centres, including one with glass façade as shown in Fig. 3f, did not suffer as extensive structural damage. These patterns of damage indicate strong evidences of significance of theory of resonance on expected structural performance of buildings subject to strong earthquake shaking.

14.2.2 The 2011 Tohoku Earthquake

On 11 March 2011, a Mw9.0 earthquake occurred off the coast of Tohoku in north-east of Honshu, Japan, at the Japan Trench where the lithosphere of the Pacific Plate subducts under the Okhotsk Plate. The rupture of the lithosphere was about 500 km long and 200 km wide [1]. Ground shaking was felt across the entire Japan and lasted for almost 4 min (220 s). A large tsunami along the Tohoku coast was also generated, which damaged hundreds of kilometres of sea defences and inundated coastal towns and cities along the east coast. At seismic measuring stations in Miyagi and Ibaraki prefectures, high peak ground acceleration exceeding 1 g was recorded, partly due to proximity to localised strong motion generation areas along the coastline. Details can be found in [2, 10].

At the affected north-eastern region of Honshu, the main island of Japan, the typology of residential houses in Japan was predominantly timber frame structures, steel frame or RC construction. Commercial and industrial buildings were typically steel frame or composite steel and RC construction. Continuous improvement in building regulations could have hence resulted in majority of the buildings suffering minimal damage from the earthquake shaking. Another reason for the low damage

Fig. 14.3 Structural performance of buildings in Padang city. **a** 1-storey masonry residential building, **b** 3-storey RC hotel building, **c** 4-storey RC commercial buildings, **d** multi-storey RC hotel building, **e**, **f** multi-storey shopping centres

despite the strong shaking was partly due to the non-coincidence of building natural frequency and the frequency content of the shaking which was dominant between 3 and 10 Hz, as shown in Fig. 2b. Damage to wooden structures was observed in Sendai, Sukagawa and Shirakawa, which indicated the susceptibility of wooden structures against strong earthquake shaking in these areas [13]. Figure 14.4 shows the performance of wooden and RC buildings in Sendai, the closest city to the earthquake epicentre.

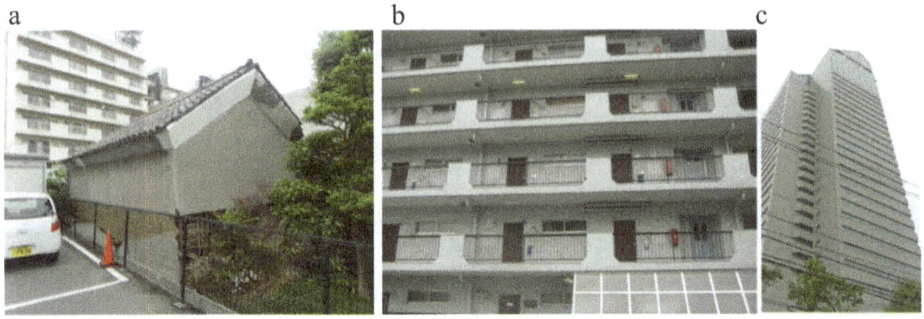

Fig. 14.4 Performance of buildings in Sendai. **a** Damage to wooden house, **b** diagonal shear cracks at RC building wall [10], **c** no visible damage to Lion Tower Kotodai [20]

14.2.3 The 2016 Muisne Earthquake

In the evening of 16 April 2016, a megathrust earthquake measuring Mw7.8 shook Ecuador's Pacific Coast. The hypocentre of the earthquake was located approximately 29 km SSE of Muisne and 168 km from the country's capital Quito. The earthquake occurred at the interface where the Nazca Plate subducts eastwards beneath the South American Plate at a rate of about 61 mm/year [24] and a dipping angle of about 16° [18]. Peak ground accelerations recorded at strong motion stations by the Instituto Geofisico ranged from 0.51 g in Portoviejo to 1.55 g in Pedernales. According to geological records, several of these towns are sited on young Quaternary sediment deposits [6].

Figure 14.5 presents the spectral accelerations recorded at seismic stations in Pedernales, Manta and Portoviejo, overlaid with the average damage grades of RC

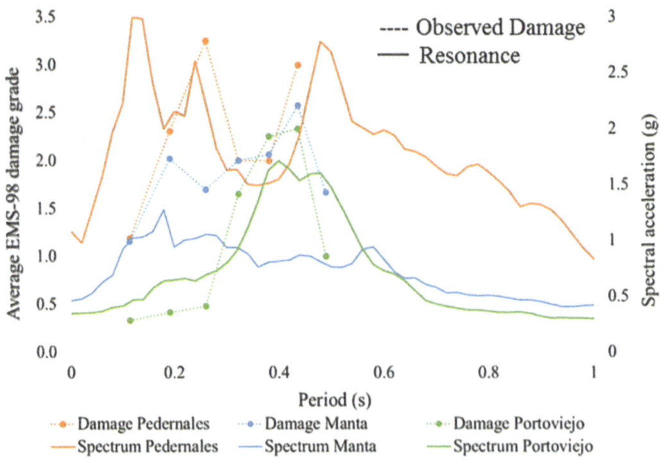

Fig. 14.5 Response spectra from ground motion records and observed building damage in the affected areas in Ecuador [2]

a b c

Fig. 14.6 Performance of buildings affected by the Muisne earthquake. **a** RC frame with masonry infill building at Pedernales, **b** RC building at Manta, **c** RC building at Portoviejo

and timber/bamboo structures of different fundamental periods estimated based on their typology and height. It can be observed that higher grades of damage generally coincide with higher spectral acceleration of the shaking in Pedernales, Manta and Portoviejo at similar periods, thereby confirming damage attributed to structural resonance. In specific, significant damage was observed for two- to five-storey buildings at Pedernales which coincides with the dominant periods of 0.2 and 0.5 s. Similarly, two- to four-storey buildings were observed to be most heavily damaged at Manta and Portoviejo, respectively, which coincides with ground motion periods of young Quaternary sediments at those locations. Figure 14.6 shows the performance of buildings in the affected areas.

14.3 Deficiencies in Reinforcement Detailing

Steel reinforcement detailing is also crucial in ensuring structural integrity when a building is subject to strong earthquake shaking. Over the years, the design of these steel reinforcements has evolved and updated with more experiences of earthquake events around the world. As a result, buildings constructed with older versions of the seismic building codes tend to be more susceptible to damage in the event of an earthquake, which necessitate retrofitting to enhance structural integrity of buildings where suitable. In severe cases, failure of key structural members such as column can lead to catastrophic soft-storey collapses. Practices of developing countries extending the height of buildings by leaving reinforcement bars sticking out over several years can cause poor bonding of these rebars with concrete.

14.3.1 The 2009 Padang Earthquake

At Padang in Indonesia, majority of small structures were very poorly constructed with low-quality building materials. Of the more recent and larger engineered structures, the quality of construction was reasonable, with occasional poor quality

Fig. 14.7 Deficiency in reinforcement and typical collapse mode. **a** Smooth plain bars used as main reinforcements (found in a collapsed shophouse) at Padang city, **b** soft-storey collapse of the Public Works Building at Padang city, **c** lack of transverse reinforcement in beam–column connection

observed. During the field survey by the UK Earthquake Engineering Field Investigation Team (EEFIT), very large spacing between confining links, particularly at the beam–column connections and joints, was commonly observed in collapsed RC buildings. The use of smooth bars for links and main reinforcement instead of ribbed bars was commonly observed in collapsed buildings as shown in Fig. 14.7. Such smooth bars have lesser bonding with the surrounding concrete and therefore lower performance of composite action against significant strain of the structural members of the RC building. Soft-storey collapse was the major mode of failure of several engineered structures.

The first Indonesian national building code (PPIUG) was developed in 1983 [7]. It was adapted from the New Zealand seismic standard which applied uniform hazard concept with 200 years' design earthquake return period. This Indonesian code SNI-1726:2002 [21] was revised several times since then, however, the mandatory requirement to adhere to the code started only in 2002. The most recent versions adopted approaches in design similar to the American code [23]. This may explain the seemingly inadequate design and construction of buildings prior to year 2002.

14.3.2 The 2011 Tohoku Earthquake

Several amendments to the building code were made over the years, which led to older structures generally less seismically resistant as compared to modern structures. Notable amendments include the 1981 implementation which incorporated experiences from the 1978 Miyagi-Oki earthquake, which introduced the two-level approach (serviceability and ultimate limit states) and stricter requirements for tall buildings. The latest amendments to the code were in 2000 following the 1995 Kobe earthquake, where seismic performance requirements and verification based on earthquake response spectra at bedrock with soil amplification factors were specified [16].

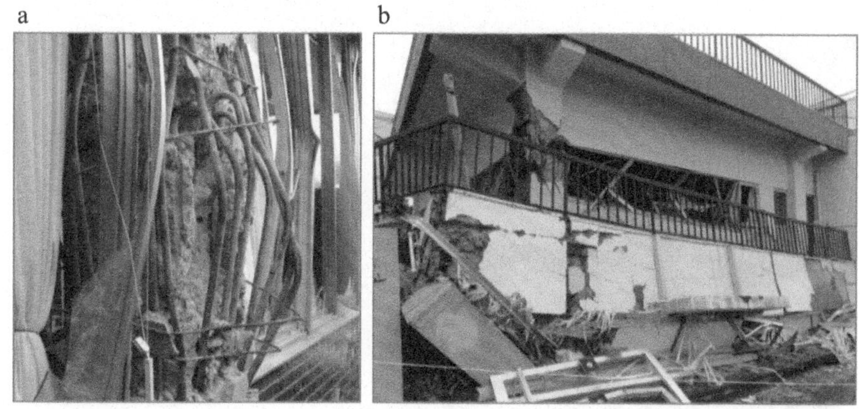

Fig. 14.8 Failure of RC buildings following the Tohoku earthquake. a Buckling at mid-span of column of Sukagawa city office building and b soft-storey collapse of RC building at Sendai

Observations of some RC residential apartments used insufficient hook length and angles which were exposed after those structural members buckled failures following the strong earthquake. Other RC buildings such as the Sukagawa city office, constructed in 1970, showed similar failures (see Fig. 8a). The buckling and shear failure of columns were also largely due to the construction adhering to an older version of the code during construction, with emphasis placed on providing more reinforcements at the joints at that time. Hence, failures tend to occur towards the middle section of the column where there was less provision of reinforcement. In contrast to the frequent observations of soft-storey failures in many reconnaissance missions such as the 2009 Padang earthquake, soft-storey collapse was not a common observation in the Tohoku earthquake event. An exception was a three-storey RC building in Sendai (Fig. 8b), where columns failed by shear due to insufficient confining reinforcement stirrups. This building was constructed in 1969, in which the design followed the older version of the building code.

14.3.3 The 2016 Muisne Earthquake

Following the 2016 Muisne earthquake, column failure due to insufficient and poor shear design and detailing was also observed, some with poor lapping of link without 135° bend (see Fig. 9a, c). In making matter worst, concrete cover was insufficient or missing (0–20 mm) in some cases, which encouraged corrosion to the steel in the structural member. Most seismic-induced soft-storey collapse failures typically occur at the ground level where there is concentration of high loads from the upper storeys. Ground floor columns with minimal infill walls can also lead to the building being more vulnerable to such collapse pattern. Such soft-storey failures, alike to those observed at Padang and Tohoku, were observed as shown in Fig. 9d.

Fig. 14.9 Failures caused by inadequate design of infills and reinforcement detailing following the Muisne earthquake. **a** School building with short column failure, **b** column head failure showing insufficient shear design, **c** insufficient lapping of column link with no 135°-bend, **d** inadequate design/detailing of RC frame and infills at Portoviejo, **e**, **f** exposed reinforcement for future building extension in Ecuador.

However, observations of upper floor soft-storey failures in a number of buildings were observed following this earthquake event. Change in stiffness of the storey or plan was one of the key factors leading to such occurrence of upper storey collapses. Figure 9b, c shows examples of discontinuity of stiffness observed that likely caused the collapse of the Magistrate building and a commercial building in Portoviejo and Manta, respectively. There were however occasions where upper soft-storey collapse occurred on buildings without obvious change in structural stiffness, plan or storey height. One example is presented in Fig. 9d. Such failure could have been aggravated by construction practice in these towns where extension of the building takes place long after the bottom floors were completed (see Fig. 9e, f). Several other seismically active countries such as Turkey and Greece have similar practices, which can be structural concerns as demonstrated in past earthquake events [8, 12]. The joints between new and old members were typically poorly executed with insufficient lapping of the rebar. This results in weak and brittle connection with local reduction in capacity, therefore posing a weakness and potential failure at these areas. Reinforcement bars left out for many years can experience surface corrosion which weaken their bond to the new concrete during casting. Another reinforcement detailing shortcoming for such extension of building also means that reinforcement bar would be lapped at the same height where the extension commences, resulting in the use of the same length bars in all columns. The increased axial loads would also cause more severe problems for the bottom floors, should earlier design of the bottom floors fail to anticipate the actual loadings of the extended section of the building.

Similar to practices in other countries, Ecuador's first adoption of seismic design regulations and subsequent updates have often followed major earthquakes in the country or elsewhere. The first building code CEC-51 in Ecuador that included seismic design requirements was introduced in 1951 following the 1949 Mw6.9 earthquake in Ambato. Following its introduction, it was deemed mandatory for the whole country in 1952; however, these were not applied by all contractors and government entities [15]. No seismic hazard zonation was included in this code. An update of this code was published in 1977, following damaging earthquakes in 1976 in Esmeraldas and Cotopaxi. The seismic loads in this version of the code were defined as a function of the building's characteristics, local geology and an importance factor. Despite the code not having a seismic hazard component, a practical design guide [15], published around the same time as a companion to the code, provided good practice seismic design and detailing for reinforced concrete buildings. For instance, avoidance of plan and vertical irregularities was encouraged; closed 135°-bend shear links were specified; lap splices needed to be avoided at column ends; etc. These requirements were however applied only to buildings up to 3 storeys at that juncture. Although there were earthquakes in the 1980s in Ecuador such as the 1987 Mw7.2 earthquake, revision to the seismic code was not published until 2001 following the 1998 Mw7.2 earthquake in Bahía de Caráquez [4]. This Ecuadorian Construction Code CPE INEN 5:2001 (also referred to as CEC-2000) incorporated a seismic hazard zonation, with values of bedrock PGA ranging from 0.15 to 0.40 g (zones 1–4) for a 475-year return period. The highest seismicity (zone 4) corresponded to the coastal areas in the west, and it decreased towards the east of the country, with the exception of a central area that was also assigned to zone 4. This update adopted several design guidelines from the American UBC-97 and adapted them to local Ecuadorian practice. The most recent version of the Ecuador seismic code was published in 2011 and made official in 2015 [17]. This code NEC-SEDS included six zones, with values of bedrock PGA ranging from 0.15 to > 0.50 g for a 475-year return period. It also includes the option of adopting the displacement-based design approach as an alternative of the typical force-based approach that was followed in most design codes in the world. The code also included a seismic risk, assessment and retrofit section (NEC-SE-RE) that requires essential and special structures to undergo a seismic performance check using a nonlinear analysis. Procedures for seismic risk and loss assessments of buildings and at a regional scale were also suggested with guidance on rapid assessments of buildings following the American standards.

14.4 Soil Liquefaction

The effects of soil liquefaction have been studied extensively since the 1964 Niigata and 1964 Alaska earthquakes. This phenomenon remains a pressing issue in earthquake geotechnical engineering to date with soil liquefaction largely responsible for extensive damage to residential properties and underground utilities in Christchurch, New Zealand, following major earthquakes in 2010 and 2011 [3]. On 28 September 2018, soil liquefaction buried the suburb of Balaroa and Petobo village in 3-m-deep mud following a strong earthquake in Central Sulawesi Province of Indonesia [22].

Soil liquefaction occurs when a saturated or partially saturated soil substantially loses strength and stiffness in response to an applied cyclic shaking or vibration such as during an earthquake. The soil which is ordinarily a solid behaves like a liquid in this state. The loss of strength and stiffness is attributed to the generation of excess pore water pressure in the soil which pushes the cohesionless granular soil grains apart, leading to these grains suspended in the water phase and therefore offering little resistance when a stress is applied onto it. Conditions encouraging soil liquefaction to occur are: (1) the soil is saturated, (2) the arrangement of soil particles is loose, and (3) the soil is cohesionless, such as sand. Soil saturation often below the groundwater table is necessary to invoke generation of excess pore water pressure within the voids in the soil matrix. Loose soil tends to compress when subject to an applied load. This collapsing void within the loose soil matrix leads to the pore water being compressed which develops excess pore water pressure. Finally, fine sandy soil is most susceptible to liquefaction mainly due to its cohesionless nature where the strength of the soil derives from the contact friction between grains, along with its lower permeability range as compared to larger sand grains and gravels. When water is less likely to be dissipated from the collapsing void, the greater the time permitted for build-up of high excess pore pressure to liquefy the soil.

14.4.1 The 2009 Padang Earthquake

In some areas of Padang, soil liquefaction occurred largely attributed to the high sand content and water table present along its low coastal strip plain. The tension cracks on the road and near the beach were the consequence of lateral spreading as demonstrated in Fig. 10a. Landwards into the Padang city centre, the Mere Amelie Church

Fig. 14.10 Consequence of soil liquefaction following the Padang, Tohoku and Muisne earthquakes. **a** Tension cracks arising from lateral spreading at Padang city. **b** Sand boils in adjoining building of Mere Amelie Church, Padang. **c**, **d** The well that was blocked with sand due to liquefaction at Siteba Village, Padang. **e** Upheaval of floor slabs of village house due to liquefaction at Perumdam Village, Padang. **f**, **g** Uplift of manhole and lateral spreading at Shirakawa. **h** Ground settlement on pavement at Pedernales. **i** Uplift of manhole close to the observed ground settlement at Pedernales

suffered effects of liquefaction as supported with observations of sand boils in the adjacent school building. These sand boils (Fig. 10b) were found to be predominantly silty sand which came from the soil deep beneath the topsoil, thereby confirming occurrence of soil liquefaction. In low-lying land regions, soil liquefaction was also found as evident from a well in Siteba Village which had been choked by the rising sand brought up by the high pore water pressure during the earthquake as shown in Fig. 10c, d. The excavated soil was investigated and found to be uniformly graded fine sand, confirming liquefaction susceptibility. In nearby Perumdam Village, two adjacent single-storey houses near the river had suffered from the effects of liquefaction. The consequence was the upheaval of suspended ground slabs caused by excess pore water pressure building up beneath these slabs as shown in Fig. 10e. Once the cracks in the floor slabs had opened, the water could flow out and the high excess pore pressure was relieved, allowing the floor slabs to settle back.

14.4.2 The 2011 Tohoku and 2016 Muisne Earthquakes

In Japan following the Tohoku earthquake, liquefaction-induced uplift of manholes and lateral spreading were also observed in Shirakawa and Sendai (Fig. 10e, f). They were located adjacent to agriculture fields where high water table was present and encouraged liquefaction to occur. The cause of the uplift of manholes is attributed to the lower submerged unit weight of the manhole as compared to the surrounding soil [3]. When the soil liquefies, the uplift buoyancy force subject to Archimedes principle exceeds the self-weight of the manhole and frictional resistance at the soil–manhole wall interface, hence leading to the uplift failure to occur. In Ecuador, evidence of earthquake-induced soil liquefaction, such as ground settlement and uplift of underground pipes and manholes, was observed at several locations visited in the Manabí Province. Along the seafront in Pedernales, settlements of 100 mm were observed on interlocking block pavements (Fig. 10g) with manholes protruding by some 30 mm (Fig. 10h). The former was attributed to the compaction of loose soil deposits under strong shaking, while the latter caused similarly by buoyancy of the underground utilities.

14.5 Conclusion

Following the observations presented from the three earthquakes over the past two decades, structural resonance remains to be one of the factors leading to heavy damage to buildings and should be taken into consideration in design carefully. The observation of few number of near-collapse buildings despite the strong shaking but non-coinciding ground shaking versus building frequencies at Sendai city near the epicentre of the Tohoku earthquake is another evidence of importance of resonance effect to buildings. Insufficient and poor reinforcement detailing was observed in severely damaged buildings in all three earthquakes, most of which were designed based on older version of building codes. The minor or no damage to recently constructed buildings in these three countries also showed that modern building codes adopted are largely able to offer sufficient seismic resistance against severe damage if followed diligently.

Soil liquefaction was also discussed based on the findings of these earthquake events, which confirms theories to observations of damage. In comparison with structural failure, consequences of soil liquefaction such as foundation and ground settlements, lateral spreading and uplifting of underground utilities have shown to be less lethal in the urban environment. However, the inconvenience resulted can be rather troubling as observed in the 2010 and 2011 Christchurch earthquakes and 2011 Tohoku earthquake where several low-rise residential buildings had to be demolished due to severe settlement. It should also be noted that in rural areas, the impact could be more significant such as slope stability as reported in the 2018 Sulawesi earthquake, which buried a large community.

Acknowledgements The author would like to express his thanks to the comradeship of fellow 2009 Padang, 2011 Tohoku and 2016 Muisne earthquakes' field investigation team members as well as the Engineering and Physical Sciences Research Council (EPSRC) [Grant No.: EP/H043616/1, EP/I01778X/1 and EP/P025951/1], AIR Worldwide, Cambridge Architectural Research Ltd, Global Earthquake Model, Institution of Civil Engineers (QUEST Travel Award), Sasakawa Foundation, Sellafield Ltd., Willis Research Network, Guy Carpenter, Centre for Urban Sustainability and Resilience at University College London, School of Technology at University of Cambridge, Queens College, Cambridge, and Arup, London, for their financial support to the missions. The background support from the Institution of Structural Engineers (IStructE) and the committee and staff of Earthquake Engineering Field Investigation Team (EEFIT) for the reconnaissance mission is also greatly appreciated. The authors are also grateful to the organisations and individuals whom have also helped in the missions in one way or another.

References

1. Ammon CJ, Lay T, Kanaore H, Cleveland M (2011) A rupture model of the great 2011 Tohoku earthquake. Earth Planets Space 63:693–696. https://doi.org/10.5047/eps.2011.05.015
2. Chian SC, Wilkinson SM, Whittle JK, Mulyani R, Alarcon JE, Pomonis A, Saito K, Fraser S, Goda K, Macabuag J, Offord M, Hunt-Raby AC, Sammonds P, Franco G, Stone H, Ahmed B, Hughes FE, Jirouskova NK, Kaminski S, Lopez J (2019) Lessons learnt from the 2009 padang Indonesia, 2011 Tohoku Japan and 2016 muisne ecuador earthquakes. Front Built Environ 5(73). https://doi.org/10.3389/fbuil.2019.00073
3. Chian SC, Tokimatsu K, Madabhushi SPG (2014) Soil liquefaction-induced uplift of underground structures: physical and numerical modeling. J Geotech Geoenviron Eng ASCE 140(10):04014057. https://doi.org/10.1061/(ASCE)GT.1943-5606.0001159
4. Chlieh M, Mothes PA, Nocquet J-M, Jarrin P, Charvis P, Cisneros D, Font Y, Collot J-Y, Villegas- J-C, Rolandone F, Vallée M, Regnier M, Segovia M, Martin X, Yepes H (2014) Distribution of discrete seismic asperities and aseismic slip along the Ecuadorian megathrust. Earth Planet Sci Lett 400:292–301
5. Código de Práctica Ecuatoriano (CPE) (2001) Ecuadorian building code, general design specifications. Ecuadorian Standardization Institute CPE INEN 5:2001, Quito
6. Dirección de Geologiì a yMinas (DGGM) (1982).Mapa Geológico Actualizado del Ecuador, escala 1:1'000.000. Instituto Ecuatoriano de Mineria, Quito
7. Direktorat Penyelidikan Masalah Bangunan (1983) Peraturan Pembebanan Indonesia Untuk Gedung. PPIUG, Bandung
8. Earthquake Engineering Field Investigation Team (EEFIT) (2003) The Kocaeli, Turkey Earthquake of 17 August 1999: Field Report, The Institution of Structural Engineers. Available online at: http://www.istructe.org/resources-centre/technical-topic-areas/eefit/eefit-reports. Accessed 6 Feb 2021

9. Earthquake Engineering Field Investigation Team (EEFIT) (2009) The Padang, Sumatra—Indonesia Earthquake of 30 September 2009: Field Report, The Institution of Structural Engineers. Available online at: http://www.istructe.org/resources-centre/technical-topic-areas/eefit/eefit-reports. Accessed 6 Feb 2021
10. Earthquake Engineering Field Investigation Team (EEFIT) (2011) The Tohoku Earthquake and Tsunami of 11 March 2011: Field Report, The Institution of Structural Engineers. Available online at: http://www.istructe.org/resourcescentre/technical-topic-areas/eefit/eefit-reports. Accessed 6 Feb 2021
11. Earthquake Engineering Field Investigation Team (EEFIT) (2016) The Muisne, Ecuador Earthquake of 16 April 2016: Field Report. The Institution of Structural Engineers. Available online at: http://www.istructe.org/resourcescentre/technical-topic-areas/eefit/eefit-reports. Accessed 6 Feb 2021
12. Geological Agency-Department of Energy and Mineral Resources of Republic of Indonesia (ESDM) (2009) Geological map of Padang City. Ministry of Energy and Mineral Resources of the Republic of Indonesia
13. Goda K, Pomonis A, Chian SC, Offord M, Saito K, Sammonds P (2012) Ground motion characteristics and shaking damage of the 11th March 2011 Mw9.0 Tohoku Earthquake. Bull. Earthquake Eng. 11:141–170. https://doi.org/10.1007/s10518-012-9371-x
14. Incorporated Research Institutions for Seismology (IRIS) (2020) How does a seismometer work? Education and Outreach Series, No. 7, Washington DC. Available online at: https://www.iris.edu/hq/sis/resources/seismometers. Accessed on 6 Feb 2021
15. Instituto Ecuatoriano de Normalizaci6n (INEN) (1976) Guia popular de construccion sismo resistente. Instituto Ecuatoriano de Normalizaci6n, Quito, Ecuador
16. Midorikawa M, Iiba M, Koshika N (2004) Seismic performance evaluation of seismically isolated buildings introduced to the building code of Japan. J Press Vessel Technol 126:18–24. https://doi.org/10.1115/1.1638390
17. Norma Ecuatoriana de la Construcci6n (2015) Ecuadorian Building Code, Seismic Loads, Earthquake-resistant Design,Ministry of Urban Development and Housing. Norma Ecuatoriana de la Construcci6n, Quito
18. Parra H, Benito MB, Gaspar- JM (2016) Seismic hazard assessment in continental Ecuador. Bull Earthquake Eng 14:2129–2159. https://doi.org/10.1007/s10518-016-9906-7
19. Pomonis A, Kappos A, Karababa F, Panagopoulos G (2011) Seismic vulnerability and collapse probability assessment of buildings in Greece. In: Spence R, So E, Scawthorn C (eds) Human casualties in earthquakes: progress in modelling and mitigation (Advances in Natural and Technological Hazards Research). Springer, London
20. Structural Engineers Association Washington (SEAW) (2011) 2011 Great East Japan (Tohoku) Earthquake & Tsunami. Earthquake Reconnaissance Observations, Structural Engineers Association Washington, June 2011. Available online at: https://www.seaw.org/s/seaw-2011japaneqreport.pdf. Accessed 6 Feb 2021
21. Standar Nasional Indonesia (2002) Tata Cara Perencanaan Ketahanan Gempa Untuk Rumah dan Gedung. Jakarta: SNI-1726-2002
22. The Washington Post (2018) 'Not a straightforward event': how multiple disasters stunned experts and ravaged a corner of Indonesia. Available online at: https://www.washingtonpost.com/world/asia_pacific/not-a-straightforward-event-how-multiples-disasters-stunned-experts-and-ravaged-a-corner-of-indonesia/2018/10/06/4394e20c-c967-11e8-b2b5-79270f9cce17_story.html. Accessed 6 Feb 2021
23. Uniform Building Code (1997) International conference of building officials. Uniform Building Code, Whittier, CA

24. United States Geological Survey (USGS) (2010a) Magnitude 7.6—Southern Sumatra, Indonesia. Available online at: https://earthquake.usgs.gov/earthquakes/eventpage/usp000h237/executive. Accessed 6 Feb 2021
25. United States Geological Survey (USGS) (2010b) Magnitude 6.6—Southern Sumatra, Indonesia. Available online at: https://earthquake.usgs.gov/earthquakes/eventpage/usp000h24c/executive. Accessed 6 Feb 2021
26. United States Geological Survey (USGS) (2016) General Summary of the M7.8 Muisne Event. USGS, Reston

Chapter 15
Estimation of Local Site Effects in Indian Scenario: Lessons from Past Earthquakes, Current Practices, and Future Trends

C. Shreyasvi and K. Venkataramana

15.1 Introduction

The ground-shaking or tremors is the manifestation of the stress accumulated underneath due to the movement of the tectonic plates. The ground motion recorded at the surface is a product of the complex amalgamation of the seismic source, wave propagation path, and local geological characteristics. The local geological and geotechnical conditions are capable of modifying the amplitude, frequency content, and duration of ground motion. This phenomenon is commonly known as local site effects. The seismic waves usually tend to amplify as they pass through the surficial layers of soil and can be observed in sites with irregular topography, deeper basins, and soils with poor mechanical properties. This phenomenon is deemed to be the reason for the difference in shaking observed at a different location during any given earthquake.

There has been early evidence of spatial variation in the intensity of shaking due to the amplification of unconsolidated sediments during the 1906 San Francisco Earthquake. The impact of local geology on the ground motion was observed during earthquakes in Japan in the early 1930s by Ishimoto [21] and Sezawa [57]. The ideology of local site amplification gained attention after the 1985 Guerreros Michoacán and 1989 Loma Prieta earthquakes. The earthquakes that occurred in the early 90s such as 1994 Northridge and 1995 Kobe brought forward more evidence strengthening the concept of local site effects. In the Indian scenario, the country has witnessed some great earthquakes demonstrating that more than 50% of the land area is prone to seismic activity [23]. India hosts multiple tectonic regimes such as active

C. Shreyasvi (✉)
GEM Foundation, Pavia, Italy
e-mail: shreyasvic@globalquakemodel.org

K. Venkataramana
Department of Civil Engineering, National Institute of Technology Karnataka, Surathkal, India
e-mail: katta@nitk.edu.in

S. Kolathayar et al. (eds.), *Civil Engineering for Disaster Risk Reduction*, Springer Tracts in Civil Engineering, https://doi.org/10.1007/978-981-16-5312-4_15

(Himalayan region), stable (Peninsular India), and active crustal (Gujarat) region. The main cause of seismic activity in the country is due to subduction of the Indian plate below the Eurasian plate. The earthquake catalog compiled by Oldham [47] reveals the history of seismic activity in the early 1800s originating from the west coast (Kutch earthquake, 1819) and the Assam earthquake (1897). In recent years, there have been more devastating earthquakes such as 2001 Bhuj, 2004 Sumatra–Andaman, 2015 Nepal, 2017 Uttarakhand earthquake to list a few. The northeast part of the country, the Himalayas, and the Kutch region of Gujarat have been witnessing earthquake events of low to moderate size ($M_W < 5$) more frequently. The second author had the privilege to be part of the reconnaissance team after the Bhuj earthquake, and the geotechnical damage observed post the event has been shown in Fig. 15.1.

Since local site effects have a direct impact on the damage caused to the built environment, characterization of the local site becomes an essential component in hazard mitigation. The initial studies in quantifying the site amplification aimed at developing a correlation between site property and its amplification. The studies mostly focused on 1D wave propagation of S-wave assuming that the geological structure underneath the site is homogenous. In this regard, V_S was found to be a suitable parameter for correlating with the damage [56], Kitagawa and Matsushima

a Settlement of ground due to liquefaction. **b** Sand boils observed near Kandla port

c The collapse of telephone poll due to loss of strength in the supporting soil.

Fig. 15.1 Photographs of the geotechnical damage observed at Kandla port during Bhuj 2001 earthquake

[26]. The site characteristics vary spatially, and the lateral homogeneity of the underlying geological structure is usually confined to valley or basin type of geological structure. The seismic waves undergo multiple reflections along the basin edges which are termed basin-induced surface waves and are capable of causing extensive damage. The first observation of basin effects was reported during the 1971 San Fernando earthquake by Hanks [15]. Later studies [33, 67] confirmed that the long-period ground motions observed in San Fernando and Los Angeles basins were due to dispersion and Rayleigh wave trajectories.

The longer duration of ground motions recorded during the 1985 Guerreros Michoacán earthquake also implicates basin effects. The one-dimensional wave propagation assumption cannot explain the complex patterns undertaken by the waves in causing sustained amplification. This paved the way toward 2D and 3D simulation of site response. It has been well established that the basin effects influence the ground motion in a certain frequency range depending on the dynamic and geometrical properties of the site. Many researchers are working on theoretical aspects to explain the phenomena as well as devise experiments to replicate the observations.

The idea behind site characterization is to integrate the site factor in seismic hazard estimation. The hazard is usually estimated for a reference site condition of mostly rock or stiff site with $V_S > 700\ \mathrm{ms}^{-1}$. The site factor is individually estimated using various methods depending on the feasibility to procure the necessary site data and the importance level of the infrastructure. The site factor is combined with the previously computed hazard values to obtain the hazard estimates concerning the surface site conditions. The level of sophistication necessary in estimating the site factor and its integration with hazard depends on various criteria of the project. The readers are advised to refer to Barani and Spallarossa [4] for more information on site-specific hazard analysis. The present article is an attempt to outline the current practices in the country concerning the estimation of local site effects. The article presents a detailed review of various methods but is mostly limited to site response simulation and empirical amplification functions. The 2D and 3D response simulation has been left out from the article as the available works in this area are quite limited. The authors intend to provide a comprehensive review of the methods to understand the phenomena of local site effects and aid in choosing a method for a related study.

15.2 Computational Methods to Estimate Local Site Effects

The modification of the seismic wave traveling toward the surface from a depth can be due to a variety of site factors such as its geometry, presence of structural irregularities such as cavities, ground water-bearing strata, and topography. The amplification or de-amplification in the amplitude of the seismic wave is usually an outcome of one or more of these factors. The phenomena associated with each of these factors contributing toward site effects are different and so are the estimation techniques. The methods available for estimating site effects due to each of these factors are outlined in the present section.

15.2.1 Topographic Site Amplification

Topography has a significant impact on the amplification that can be observed in the transmitted ground motion. The surface topography affects the amplitude and frequency content of the ground motion. The topography is broadly categorized into convex such as the peak of the mountainous terrain, ridges, and concave such as valleys and foothills. The amplification observed at the hilltops is higher as compared to its base. These observations were made during the 1976 Fruili, 1980 Iripinia, and 1985 Chile earthquakes. Theoretical and experimental studies have revealed that convex surfaces cause amplification and concave surface causes de-amplification. The extent of these effects depends on numerous parameters such as incidence angle, wave type, i.e., body or surface wave, and azimuth angle. Various topographic proxies are used to correlate the feature with the amount of amplification. The proxies are derived from the 30 and 15-arcsecond digital elevation models (DEMs). These models are usually inclusive of topography as well as bathymetry data. The DEMs of varying resolutions are freely available for the entire globe and can be accessed through OPENDEM (https://www.opendem.info/link_dem.html). The topographic proxies derived from the DEM are slope, topographic position index (TPI), roughness, and terrain ruggedness index (TRI). These terrain data are presented in the form of grids divided by the uniformly spaced latitudes and longitudes and referenced according to the World Geodetic System (WGS84). A typical example of the topographical map derived from the DEM data is presented in Fig. 15.2.

Upon deriving the terrain parameters from the DEM, the next step is to correlate them to site effects. However, the approach is less straightforward than said, and numerous studies have been carried to understand the wave propagation behavior in the case of strong topographical features. In the Indian scenario, very few studies have considered site effects associated with topographic slopes. Khan et al. [25] studied the influence of topography during the 2005 Kashmir earthquake using the

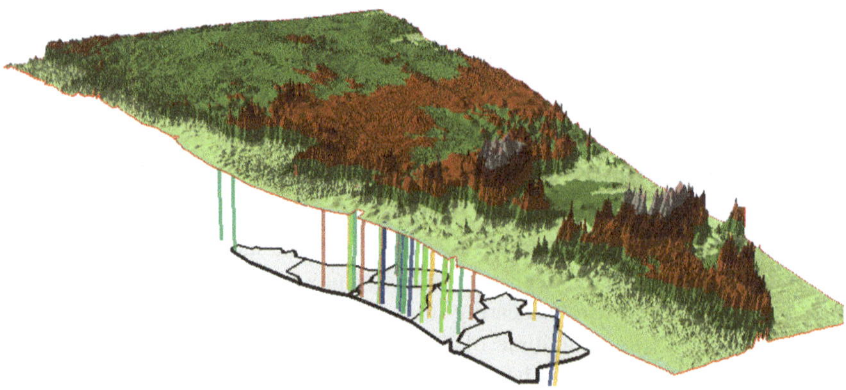

Fig. 15.2 Topographic map of the South Western part of India generated using the DEM [63]

spectral finite element method. The study highlights that the topographical amplification contributed to the damage observed in 98% of the affected area. On the other hand, there have been studies indirectly correlating the topographic slope to the commonly referred site proxy, i.e., $V_{S(30)}$. Wald and Allen [71] proposed a correlation between the slope and $V_{S(30)}$ as a function of the tectonic regime, and as a result, a coarser $V_{S(30)}$ map can be derived for any region in the world. The study was based on the premise that the more competent material with higher $V_{S(30)}$ constitutes the topographical features with steeper slopes. These correlations were further validated by [34] suggesting that these correlations can be used for first-order site characterization used in generating regional- or national-level seismic hazard maps. These findings were utilized by Shreyasvi et al. [59] in their regional hazard assessment, and surface hazard maps were created for the South West region of India. The procedure for deriving the topographic gradient from the DEM and generation of $V_{S(30)}$ map and the hazard map can be referred from Sitharam [65], Shreyasvi et al. [61]. A sample of the $V_{S(30)}$ map derived from topographic slope has been presented in Fig. 15.3. The slope map and $V_{S(30)}$ map for the whole of India have been generated using topographic slope as a proxy for quantifying site effects and generating surface-level hazard maps.

Apart from the studies at a larger scale, the effect of topography on site amplification can be studied at a site-specific scale by 3D finite element modeling of the site. However, there have been very limited experiments in this regard especially in India.

15.2.2 Site Response Simulation

The previous earthquake experiences have revealed that the intensity of ground shaking is stronger on soft soils when compared to bedrock outcrop. Soft soils are common in valleys and most of the cities are located on such sedimented river valleys. Srinagar, Guwahati, Aizawl, Pithoragarh, Uttarkashi, Nainital, and Bhagalpur are some of the cities with high seismic hazard and risk according to the report prepared by the National Disaster Management Authority (NDMA). Estimating site effects through topographic slope for a populated region is not accurate enough and also hazard assessment in urban environment demands site-specific investigations. Hence, a detailed analysis of the complex phenomena of wave propagation in a non-uniform medium is necessary. The non-uniform medium implies the spatial variation in the site in terms of soil type, depth to bedrock, and the difference in soil properties such as density and water content. A simplified illustration of the seismic wave propagation from the source to the site through various geological strata has been presented in Fig. 15.4. The computational methods for estimating site effects in such cases deal with the simulation of wave propagation through the complex media. There are numerous studies of varying complexity considering wave propagation in 1, 2, and 3 dimension. The simplest and the easiest way to compute the site response is through a 1D simulation. There are few open-source software available such as DEEPSOIL,

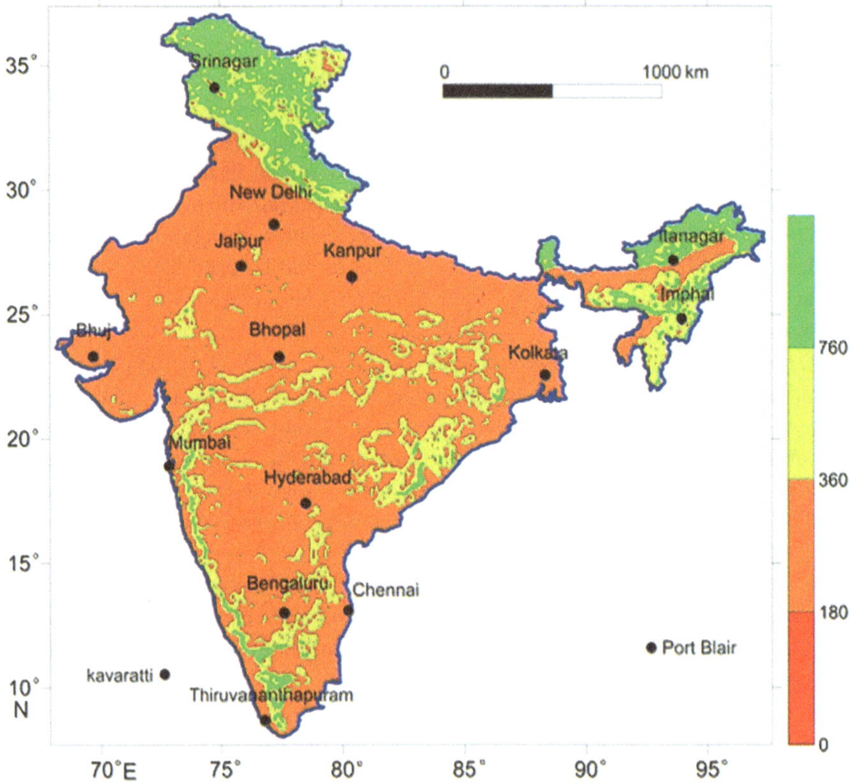

Fig. 15.3 A contour map depicting the spatial variation of $V_{S(30)}$ derived from the topographical attributes [65]

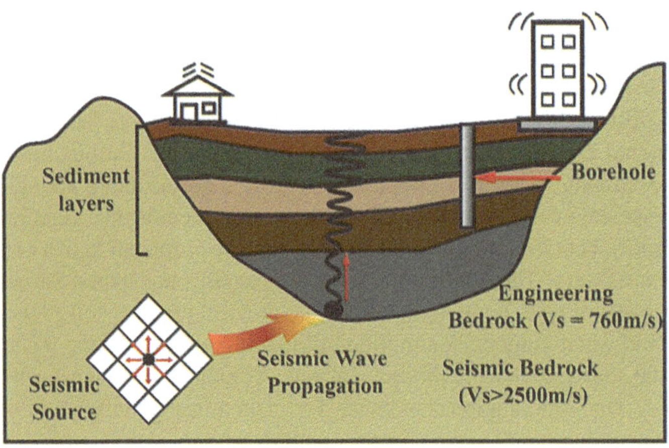

Fig. 15.4 Illustration of the seismic wave propagation from the source to site through various rock and soil strata [58]

Table 15.1 Summary of computational codes available for estimating site response

Dimensions	Equivalent linear	Nonlinear
1	ProShake, SHAKE2000, EERA, STRATA, CYCLIC 1D, DEEP SOIL	CyberQuake, DEEP SOIL, NERA, FLAC, STRATA
2 or 3	QUAKE/W, SASSI2000	FLAC, PLAXIS

STRATA, NERA, and commercial packages such as SHAKE2000. Additionally, a user-friendly Excel program was developed by Iglesia and Stiady [19] giving better insight into the concept and the computation involved in site response computations. The available software for computing site responses is summarized in Table 15.1. Two- and three-dimensional site responses can be computed using finite element software such as ANSYS, FLAC3D, and PLAXIS. Additionally, a MATLAB-based code for computing equivalent linear site response in the frequency domain has been recently developed by Kumar and Mondal [31].

The site response simulation requires two major inputs that can be categorized as soil modeling and input motion suite. The site characteristics such as shear wave velocity (V_S), unit weight (γ), modulus reduction (M-R), and damping (D) curves as a function of strain and thickness of various soil layers constitute the soil modeling aspect of input. The input motion that is passed through the layered soil media is usually a suite of scaled/unscaled ground motions recorded near the site or elsewhere. A schematic representation of the one-dimensional site response is depicted in Fig. 15.5.

The soil or site characteristics are determined through in-situ tests such as standard penetration test (SPT), dynamic cone penetration test (DCPT), P-S logging, and multichannel analysis of surface wave (MASW). The dynamic characteristics of the soil, i.e., M-R and D curves, for the site response computation can be generated from laboratory tests such as cyclic tri-axial, bender element, and resonant column tests

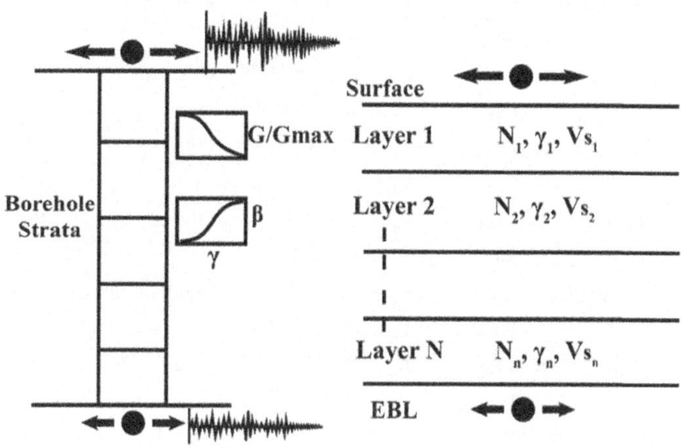

Fig. 15.5 Schematic representation of the one-dimensional site response [58]

[24, 53] or adopt the standard curves for different soil types [13, 55, 70]. However, some of the in-situ tests are expensive, demand skilled labor, and may not be feasible at all times. Hence, in such cases, there have been numerous correlations developed to link the commonly available SPT 'N' value with the $V_S$profile for a given site [1, 8, 9, 68] as given in Table 15.2. There have been studies, wherein the site response was estimated for sites with only SPT 'N' values. Studies have explored the use of SPT—V_S correlations developed for other regions such as Japan, Taiwan, Bangladesh, and different parts of India [52, 66]. A typical site consisting of layers of different geomaterial and the estimated average Vs profile from the correlation listed in Table 15.2 is depicted in Fig. 15.6.

The idea of the computation methods is to accurately represent the actual site using mathematical models, so that the response estimated can be very close to reality. The improvement in the computational capacity and quality of the site data has encouraged researchers to explore a variety of techniques for estimating site response. However, a certain amount of uncertainty is involved in the estimates of site characteristics and input motions. This uncertainty can be reduced by considering multiple

Table 15.2 List of the commonly used V_s–N correlations for site response study in India

S. No.	Author	Soil type	Correlation	Region	Nomenclature
1	Dikmen [12]	Clay	$V_s = 44\ N^{0.48}$	Turkey	UD(09)
		Sand	$V_s = 73\ N^{0.33}$		
2	Kumar et al. [32]	Clay	$V_s = 89.31\ N^{0.358}$	Lucknow	KAS(10)
		Sand	$V_s = 100.53\ N^{0.265}$		
3	Tsiambaos and Sabatakakis [67]	Clay	$V_s = 140.1\ N^{0.29}$	Greece	TS(11)
		Sand	$V_s = 92\ N^{0.341}$		
4	Chatterjee and Choudhury [9]	Clay	$V_s = 77.11\ N^{0.3925}$	Kolkata	CC(13)
		Sand	$V_s = 54.82\ N^{0.5255}$		
5	Kirar et al. [26]	Clay	$V_s = 94.4\ N^{0.379}$	Roorkee	TS(11)
		Sand	$V_s = 100.3\ N^{0.348}$		
6	Uma Maheswari et al. [68]	Clay	$V_s = 89.31\ N^{0.358}$	Chennai	UM(10)
		Sand	$V_s = 100.53\ N^{0.265}$		
7	Lee [33]	Clay	$V_s = 114\ N^{0.31}$	USA	LEE(90)
		Sand	$V_s = 57\ N^{0.49}$		
8	Pitilakis et al. [51]	Clay	$V_s = 114\ N^{0.31}$	Greece	P(92)
		Sand	$V_s = 57\ N^{0.49}$		
9	Jafari et al. [22]	Clay	$V_s = 27\ N^{0.73}$	South of Tehran	J(02)
		Sand	$V_s = 80\ N^{0.33}$		
10	Imai [20]	Clay	$V_s = 80.2\ N^{0.292}$	Japan	IM(07)
		Sand	$V_s = 80.6\ N^{0.331}$		
11	Hasancebi and Ulusay [16]	Clay	$V_s = 97.9\ N^{0.269}$	Turkey	HU(07)
		Sand	$V_s = 90.8\ N^{0.319}$		

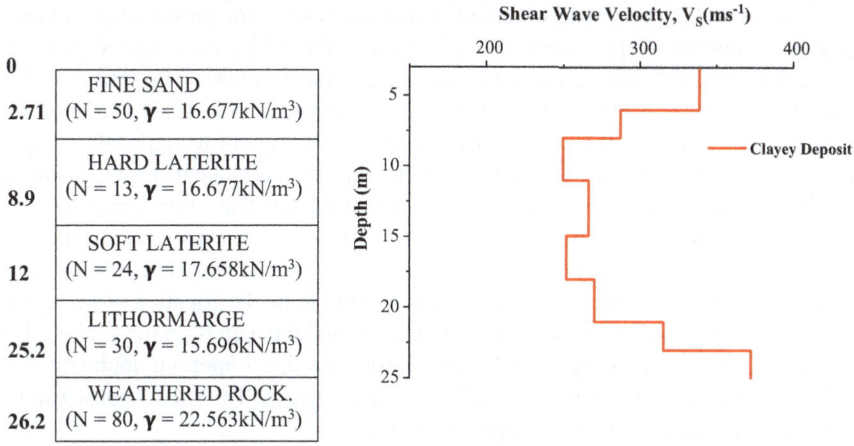

Fig. 15.6 Typical soil stratigraphy and estimated average Vs profile for a clay site

probable soil profiles instead of a single profile for a given site. Further multiple input motions are also considered as they can collectively represent the ground motions that can be expected at the site. The question arises about the threshold value for the number of realizations of soil profiles as well as the input motion. In this regard, a study was carried out for a site located in Calicut (Kerala, Southern part of India) to define a threshold value for obtaining stable estimates by Shreyasvi et al. [62]. The study suggests that a minimum of 20 different realizations of soil profiles and ground motion suits consisting of 7–10 records must be sufficient in making a rationale estimate of site response.

The next question that needs to be addressed upon successfully compiling all the input data necessary for computation is the type of site response model. The model can be linear or nonlinear describing the behavior of the constituent material under the influence of dynamic motion. The linear and the nonlinear models behave similarly at lower strains, i.e., ground motions with lower values of intensity measures and/or the soil sites are relatively stiff. At stronger ground motions with PGA > 0.1 g, the soil tends to behave nonlinearly [58]. Nonlinear models are capable of accurately capturing the dynamic response of sites at higher strain levels using a more advanced constitutive modeling of soil behavior. However, for nonlinear analysis, the previously explained M-R and D curves are not applicable. Kondner [28] proposed a pressure-dependent hyperbolic model that was later modified by Matasovic [39] to capture the nonlinear soil behavior. Phillips and Hashash [50] developed a procedure wherein the hyperbolic model is modified to accommodate the M-R and D curves, so that the final model can accommodate both small and large strain problems. The soils are subjected to a series of reloading and unloading cycles and this can be modeled by extending the Masing rules [17].

The nonlinear models are computationally expensive; hence, an interim model known as the equivalent linear model was proposed that has the capabilities of a nonlinear model but operates at the cost of a linear model. The proposed model uses an iterative scheme to approximate the nonlinear soil behavior and computes average shear modulus (G_{sec}) over an entire loading cycle of the hysteresis loop. The model is efficient and can be used for cases, wherein the strains are within 1–2% and intensity measure values are less than 0.4 g [30]. The only limitation of the equivalent linear method is that they cannot be used for calculating the permanent displacement as the model assumes elastic behavior of the soil implying that upon completion of loading, the strain will become zero. Further, pore pressure fluctuation in the soil cannot be modeled as the model operates on total stress analysis. These limitations can be overcome by using nonlinear models that integrate the equation of motion at small time intervals. The accuracy of a nonlinear model depends on the constitutive model derived through field and laboratory tests. Another important application of nonlinear models is the assessment of liquefaction hazards. The site response can be computed in both the time and frequency domain.

The choice of the type of site response analysis (SRA) depends on the domain used for calculation, the sophistication of the constitutive model, incorporation of pore water pressure, and dimension of the medium considered for simulating wave propagation. A detailed classification of types of SRA has been presented in Fig. 15.7. The initial attempts for characterizing the local site effects were through microzonation projects. Some examples of such initiatives are Mohanty et al. [41] for Delhi, Mahajan et al. [37] for Dehradun city, Anbazhagan and Sitharam [2] for Bangalore, and Boominathan et al. [6] for Chennai. The shear wave velocity (V_S) was determined through geophysical methods such as MASW and 1D SRA which was computed using SHAKE2000. Most of the available correlations between SPT 'N' value and V_S were developed as a part of these projects. The major outcomes of these projects were a V_S map and maps of multiple intensities for a study region.

The choice of input motion plays a major role in estimating response parameters such as amplification spectrum and surface-level response spectrum. Amplification (Y) is defined as the ratio of the spectral quantity at a depth (X) to that measured at

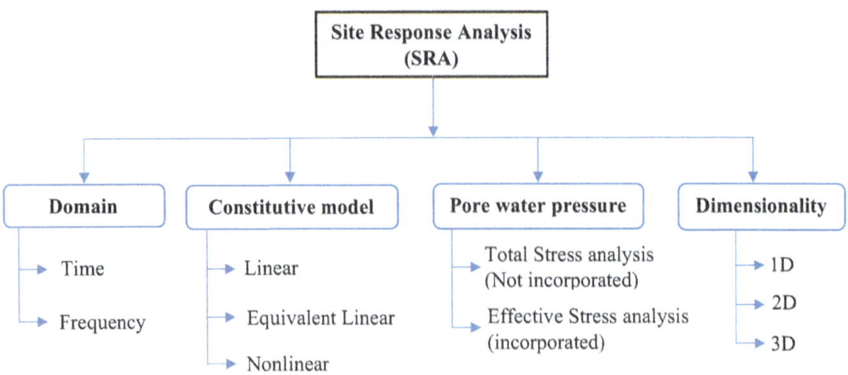

Fig. 15.7 Classification of site response computational methods

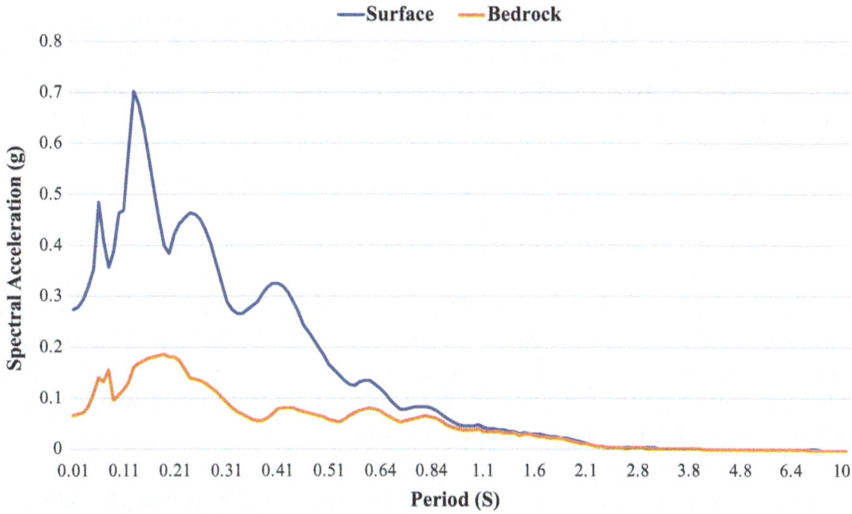

Fig. 15.8 Spectral acceleration observed at the bedrock and the surface level

the surface level (Z). A typical example of the variation of spectral quantity at the bedrock level and the surface level has been presented in Fig. 15.8. The local site effects (higher amplification) is usually predominant at short spectral period range unless the natural site period is greater than 1 s.

$$Y = \frac{Z}{X} \text{ or } \ln(Y) = \ln(Z) - \ln(X) \tag{15.1}$$

Certain studies use regionally recorded ground motions from Uttarakashi (1991), Chamoli (1999), Bhuj (2001), Sikkum (2011) such as Choudhury et al. [11, 52], Parihar and Anbazhagan [48] as input for SRA. Few studies use a combination of regional (Bhuj 2001) and global ground motions (Kobe 1995; Loma Prieta 1989) [49]. Studies have attempted to generate synthetic ground motion for regions with no available data and performed SRA (Govindaraju and Bhattacharya [14] (WAVEGEN). Nampally et al. [43], (EXSIM) Anbazhagan and Sitharam [3] (SMSIM)). Recently, a study by Shreyasvi et al. [63] attempted to understand the impact of ground motions recorded in different tectonic regimes in estimating the site response.

The success of numerous experiments with the input parameters encouraged the scientists to apply the technique for manmade sites such as municipal solid waste (MSW) landfills. The seismic behavior of the landfill located in Mavallipura, Bangalore, was examined through field and laboratory tests, and the site response was simulated in DEEP SOIL [46]. In addition to MSW landfills, pond ash embankments have also been studied with the aid of 2D modeling through OPENSEES [40]. The study was more oriented toward the liquefaction susceptibility of the site, and hence, it was concluded that the embankment is not safe against liquefaction under fully saturated conditions. Recently, attempts toward capturing the 2D variation in soil properties and wave propagation have been investigated by Chandran and Anbazhagan [7]. Six sites from the Southern part of India were chosen, and 2D MASW was carried out for these sites. The 2D nonlinear site response has been computed using FLAC-2D. The amplification factors from the study have been provided in two categories based on the period of vibration, i.e., short- and long-period amplification factor. Also, the study points out that the heterogeneity in the subsurface profile influences the low amplitude of the input motion.

The site response computation procedure described so far depends on multiple input motions as well. However, it may not always be possible to obtain a suite of ground motions for a given site. Hence, Kottke and Rathje [29] proposed an alternative scheme based on random vibration theory (RVT). Instead of the iterative procedure, the site response can be computed in a single simulation based on extreme value statistics using the Fourier spectrum and duration of the input motion. In the Indian scenario, RVT based site response has not been attempted so far as per the knowledge of the authors.

Integration of the local site response into the hazard results is crucial, and there are various techniques with different levels of complexity. The easiest and straight forward approach is to multiply the amplification factor obtained from the SRA with the computed hazard at the bedrock level or any reference site condition. This approach is known as deterministic approach. Under the probabilistic approach, there are two approaches such as modifying the ground motion model of the seismic hazard analysis and convolution approach [5]. The former method is explained systematically through Eqs. 15.2 and 15.3.

$$\ln Y(T) = a + b * \ln(X_{\mathrm{ref}}(T) + c) + \varepsilon_{\ln Y(T)} * \sigma_{\ln Y(T)} \quad (15.2)$$

$$\ln Z(T) = \ln X_{ref}(T) + \ln Y(T) \quad (15.3)$$

The amplification factor (Y) and spectral quantity at a reference site condition (X_{ref}) obtained from seismic hazard analysis are function of spectral period/frequency. Hence, the Y is regressed in nonlinear domain over X_{ref} at each spectral period with a, b, and c as the regression coefficients.

15.3 Experimental Methods

The experimental methods essentially involve macroseismic observations, strong motion and/or weak motion data, microtremor measurements. Some of these techniques have found application even before the advent of computational models. Each of these techniques is briefly discussed to give an insight to the readers and serve as a motivation for further reading. A detailed review of the experimental methods is out of the scope of the present article.

15.3.1 Macroseismic Observations

There have been cases wherein the sites have previously experienced large destructive earthquakes and a detailed macroseismic observation is already available. In such cases, the macroseismic intensity maps can be overlapped with the regional tectonic as well as the geological map to identify the zones with different damage levels. In the earlier days, when the seismometers were still new and sparsely located, the macroseismic observations have been used to estimate the intensity measures such as PGA, SD, and duration of shaking. However, one must be cautious while using macroseismic data as they are highly subjective and demands expertise in quantifying the observed damage. One such attempt was done by Hough and Roger [18], wherein they used the previously published data related to intensity and geodetic information of the Shillong (1897), Kangra (1905), and Nepal (1934) earthquakes. These data were used to investigate the rupture zones as well as the amplification of the ground motion in the Indo-Gangetic plains.

15.3.2 Strong/Weak Motion Recordings

The installation of seismometers in severe seismic zones of the country such as the Himalayan belt and the North East part of India has given ample opportunity to researchers to understand the phenomena of site effects. Site response in the Sikkim Himalayas was analyzed using the strong motion recording from the seven stations located in the vicinity [45]. Further, the study hints at nonlinearity in the estimated amplification factors. Records from digital seismometers installed in and around Delhi were used to understand the site-to-site variability component in the amplification by Nath et al. [44]. Weak motion data was used for the generalized inversion method of estimating site effects. Bhuj (2001) aftershocks were used to assess the site response in Kachchh region of Gujarat by Mandal et al. [38] by calculating the H/V spectral ratios. The study categorizes the region into different zones based on the geologic age, and each category is associated with the amplification factor. Recently, there have been studies, wherein 395 ground motion records from 94 earthquakes recorded at 90 sites of the Indian strong motion network were used for the classification of the sites [10]. A map of the stations installed and maintained by the National Center for Seismology (https://seismo.gov.in/) has been given in Fig. 15.9 presenting the extent of the spatial coverage of the network.

15.3.3 Microtremor Survey

These surveys are conducted to understand the wave propagation path in a given medium with the aid of ambient noise. This technique is popularly known as the

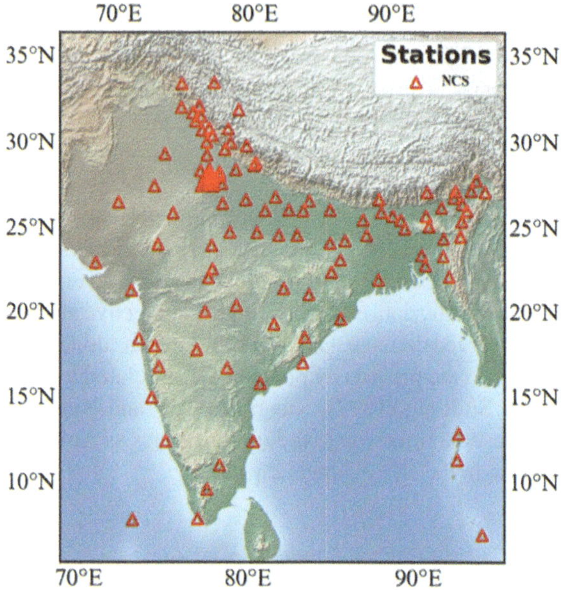

Fig. 15.9 Map depicting the geographical locations of the seismic stations in India

Nakamura method for estimating the resonant frequency and amplification spectrum of the ground motion due to the presence of surface layers. The readers are encouraged to refer to Nakamura [42] and other associated articles for more information about the method. The spectral features of the microtremor spectra are closely associated with the local geological conditions. For instance, the short predominant period implies that the site is stiff composed of weather rocks to stiff soils, and on the other hand, the long period implies loose soil deposit. Microtremor studies have been conducted in different parts of the country for characterizing the site response. A few examples of such studies are Singh et al. [64] for Surat (Gujarat), Putti and Satyam [54] for Vishakhapatnam, [36] for Jammu city.

15.4 Recommendation and Scope for Future Work

i. In the absence of site-specific data, and for the first level of site, characterization using the geological maps and DEM data for estimating site amplification is both feasible and economical. These studies aid in choosing sites with a higher potential for damage for site-specific investigation.

ii. There has been a constant debate about the assumption of 1D wave propagation in SRA and the need for 2D and 3D modeling of the sites. One must be aware of the complexity in the simulation as well as in obtaining the site data before making a conscious decision to use the higher dimensional SRA.

iii. The concept and the methodology for linear, equivalent linear, and nonlinear is quite well established at least for 1D SRA. This gives plenty of opportunity to develop codes rather than relying on the existing computer programs.
iv. Certain soil types such as laterite (which is commonly found near the coastal parts of the country) or manmade sites are difficult to model as there are no standard M-R and D curves. With the availability of the necessary experimental setups, standard curves for unconventional materials can be generated.
v. The RVT based SRA has not found wide applications owing to its drawbacks in certain aspects of estimating the site response parameters. However, given that the method has not been investigated in the Indian scenario, this gives ample opportunity to the readers to study and improvise the method.
vi. Empirical amplification functions can be developed for any given site if the ground motion recordings are available. Further, these functions can be integrated with the ground motion models in seismic hazard analysis for estimating site-specific hazard estimates.
vii. Irrespective of the method chosen for determining the site response, there is scope for developing design spectra for a given site or region [59]. This particular parameter has not been estimated by most of the site response studies.

References

1. Anbazhagan P, Kumar A, Sitharam TG (2013) Seismic site classification and correlation between standard penetration test N value and shear wave velocity for Lucknow City in Indo-Gangetic Basin. Pure Appl Geophys 170(3):299–318
2. Anbazhagan P, Sitharam TG (2008) Seismic microzonation of Bangalore, India. J Earth Syst Sci 117(2):833–852
3. Anbazhagan P, Sitharam TG (2020) Site characterization and site response studies using shear wave velocity. J Seismol Earthq Eng 10(2):53–67
4. Barani S, Spallarossa D (2017) Soil amplification in probabilistic ground motion hazard analysis. Bull Earthq Eng. https://doi.org/10.1007/s10518-016-9971-y
5. Bazzurro P, Cornell CA (2004) Nonlinear soil-site effects in probabilistic seismic-hazard analysis. Bull Seismol Soc Am 94(6):2110–2123
6. Boominathan A, Dodagoudar GR, Suganthi A, Maheswari RU (2008) Seismic hazard assessment of Chennai city considering local site effects. J Earth Syst Sci 117(2):853–863
7. Chandran D, Anbazhagan P (2020) 2D nonlinear site response analysis of typical stiff and soft soil sites at shallow bedrock region with low to medium seismicity. J Appl Geophys 179:104087
8. Chatterjee K, Choudhury D (2013) Variations in shear wave velocity and soil site class in Kolkata city using regression and sensitivity analysis. Nat Hazards 69(3):2057–2082
9. Chatterjee K, Choudhury D (2013) Variations in and soil site class in Kolkata city using regression and sensitivity analysis. Nat Hazards 2057–2082
10. Chopra S, Kumar V, Choudhury P, Yadav RBS (2018) Site classification of Indian strong motion network using response spectra ratios. J Seismolog 22(2):419–438
11. Choudhury D, Phanikanth VS, Mhaske SY, Phule RR, Chatterjee K (2015) Seismic liquefaction hazard and site response for design of piles in Mumbai city. Indian Geotech J 45(1):62–78
12. Dikmen U (2009) Statistical correlations of and penetration resistance for soils. J Geophys Eng 6:61–72

13. Gazetas G, Dakoulas P (1992) Seismic analysis and design of rockfill dams: state-of-the-art. Soil Dyn Earthq Eng 11(1):27–61
14. Govindaraju L, Bhattacharya S (2012) Site-specific earthquake response study for hazard assessment in Kolkata city, India. Nat Hazards 61(3):943–965
15. Hanks TC (1975) Strong ground motion of the San Fernando, California, earthquake: ground displacements. Bull Seismol Soc Am 65(1):193–225
16. Hasancebi N, Ulusay R (2007) Empirical correlations between and penetration resistance for ground shaking assessments. Bull Eng Geo Environ 66:203–213
17. Hashash YMA, Phillips C, Groholski DR (2010) Recent advances in non-linear site response analysis. In: Proceedings of the 5th international conference on recent advances in geotechnical
18. Hough SE, Roger B (2008) Site response of the Ganges basin inferred from re-evaluated macroseismic observations from the 1897 Shillong, 1905 Kangra, and 1934 Nepal earthquakes. J Earth Syst Sci 117(2):773–782
19. Iglesia GR, Stiady JL (2010) Seismic site response analysis using spreadsheets. In: Proceedings of International Conferences on Recent Advances in Geotechnical Earthquake Engineering and Soil Dynamics. Paper no. 10
20. Imai T (1977) P- and S-wave velocities of the ground in Japan. In: Proceedings of 9th international conference on soil mechanics and foundation engineering, vol 2, pp 127–132
21. Ishimoto M (1932) I2. Comparaison accélérométrique des secousses sismiques dans deux parties de la ville de Tôkyô
22. Jafari MK, Asghari A, Rahmani I (1997) Empirical correlation between (Vs) and SPT-N value for south of Tehran soils. In: Proceedings of the 4th international conference on civil engineering, Tehran, Iran (in Persian)
23. Jain SK (1998) Indian earthquakes: an overview. Indian Concr J 72:555–562
24. Jaya V, Dodagoudar GR, Boominathan A (2012) Modulus reduction and damping curves for sand of south-east coast of India. J Earthq Tsunami 6(04):1250016
25. Khan S, Meijde MVD, Werff HVD, Shafique M (2020) The impact of topography on seismic amplification during the 2005 Kashmir earthquake. Nat Hazard 20(2):399–411
26. Kirar B, Maheshwari BK, Muley P (2016) Correlation between shear wave velocity (vs) and SPT resistance (N) for Roorkee region. Int J Geosynth Gr Eng 2:9 https://doi.org/10.1007/s40891-016-0047-5
27. Kitagawa Y, Matsushima Y (1984) Evaluation of dynamic ground characteristics and seismic microzoning. Bull N Z Soc Earthq Eng 17(1):15–23
28. Kondner RL (1963) A hyperbolic stress-strain formulation for sands. In: Proceedings of 2nd Panamerican conference on soil mechanics and foundation engineering, Brazil, vol 1, pp 289–324
29. Kottke AR, Rathje EM (2013) Comparison of time series and random-vibration theory site-response methods. Bull Seismol Soc Am 103(3):2111–2127
30. Kramer SL, Paulsen SB (2004, March) Practical use of geotechnical site response models. In: Proceedings of international workshop on uncertainties in nonlinear soil properties and their impact on modeling dynamic soil response, p 10. University of California, Berkeley
31. Kumar A, Mondal JK (2017) Newly developed MATLAB based code for equivalent linear site response analysis. Geotech Geol Eng 35(5):2303–2325
32. Kumar A, Anbazhagan P, Sitharam TG (2010, December) Shear wave velocity profiling for Lucknow urban centre using seismic refraction survey. In: Indian Geotechnical Conference, GEO. Trendz pp 167–170
33. Lee SHH (1990) Regression models of shear wave velocities. J Chin Inst Eng 13:519–532
34. Lemoine A, Douglas J, Cotton F (2012) Testing the applicability of correlations between topographic slope and VS 30 for Europe. Bull Seismol Soc Am 102(6):2585–2599
35. Liu HL, Heaton T (1984) Array analysis of the ground velocities and accelerations from the 1971 San Fernando, California, earthquake. Bull Seismol Soc Am 74(5):1951–1968
36. Mahajan AK, Mundepi AK, Chauhan N, Jasrotia AS, Rai N, Gachhayat TK (2012) Active seismic and passive microtremor HVSR for assessing site effects in Jammu city, NW Himalaya, India—a case study. J Appl Geophys 77:51–62

37. Mahajan AK, Slob S, Ranjan R, Sporry R, van Westen CJ (2007) Seismic microzonation of Dehradun City using geophysical and geotechnical characteristics in the upper 30 m of soil column. J Seismolog 11(4):355–370
38. Mandal P, Chadha RK, Satyamurty C, Raju IP, Kumar N (2005) Estimation of site response in Kachchh, Gujarat, India, region using H/V spectral ratios of aftershocks of the 2001 Mw 7.7 Bhuj earthquake. Pure Appl Geophys 162(12):2479–2504
39. Matasovic N (1993) Seismic response of composite horizontally-layered soil deposits, Los Angeles, Ph.D. thesis. University of California at Los Angeles, California, p 449
40. Mohanty S, Patra NR (2016) Dynamic response analysis of Talcher pond ash embankment in India. Soil Dyn Earthq Eng 84:238–250
41. Mohanty WK, Walling MY, Nath SK, Pal I (2007) First order seismic microzonation of Delhi, India using geographic information system (GIS). Nat Hazards 40(2):245–260
42. Nakamura Y (2019) What is the Nakamura method? Seismol Res Lett 90(4):1437–1443
43. Nampally S, Padhy S, Trupti S, Prasad PP, Seshunarayana T (2018) Evaluation of site effects on ground motions based on equivalent linear site response analysis and liquefaction potential in Chennai, south India. J Seismolog 22(4):1075–1093
44. Nath SK, Sengupta P, Srivastav SK, Bhattacharya SN, Dattatrayam RS, Prakash R, Gupta HV (2003) Estimation of S-wave site response in and around Delhi region from weak motion data. J Earth Syst Sci 112(3):441–462
45. Nath SK, Sengupta P, Sengupta S, Chakrabarti A (2000) Site response estimation using strong motion network: a step towards microzonation of the Sikkim Himalayas. Current Sci 1316–1326
46. Naveen BP, Sitharam TG, Sivapullaiah PV (2019) Seismic behavior and dynamic site response of municipal solid waste landfill in India. In: Recent challenges and advances in geotechnical earthquake engineering. IGI Global, pp 168–196
47. Oldham RD (1899) Report of the great earthquake of 12th June, 1897. Office of the Geological Survey
48. Parihar A, Anbazhagan P (2020) Site response study and amplification factor for shallow bedrock sites. Indian Geotech J 50(5):726–738
49. Phanikanth VS, Choudhury D, Reddy GR (2011) Equivalent-linear seismic ground response analysis of some typical sites in Mumbai. Geotech Geol Eng 29(6):1109
50. Phillips C, Hashash YM (2009) Damping formulation for nonlinear 1D site response analyses. Soil Dyn Earthq Eng 29(7):1143–1158
51. Pitilakis KD, Anastasiadis A, Raptakis D (1992) Field and laboratory determination of dynamic properties of natural soil deposits. In: Proceedings of 10th world conference earthquake engineering, Rotterdam, pp 1275–1280
52. Puri N, Jain A, Mohanty P, Bhattacharya S (2018) Earthquake response analysis of sites in state of Haryana using DEEPSOIL software. Procedia Comput Sci 125:357–366
53. Puri N, Jain A, Nikitas G, Dammala PK, Bhattacharya S (2020) Dynamic soil properties and seismic ground response analysis for North Indian seismic belt subjected to the great Himalayan earthquakes. Nat Hazards 103:447–478
54. Putti SP, Satyam N (2020) Evaluation of site effects using HVSR microtremor measurements in Vishakhapatnam (India). Earth Syst Environ 4:439–454
55. Seed HB, Sun JH (1989) Implication of site effects in the Mexico city earthquake of September 19, 1985 for earthquake-resistance-design criteria in the San Francisco Bay Area of California. University of California, Berkeley
56. Seed HB, Whitman RV, Dezfulian H, Dobry R, Idriss IM (1972) Soil conditions and building damage in 1967 Caracas earthquake. J Soil Mechan Found Div 98(8):787–806
57. Sezawa K (1930) Possibility of the free-oscillations of the surface-layer excited by the seismic-waves. 東京帝國大學地震研究所彙報= Bull Earthq Res Inst 8(1):1–11
58. Shreyasvi C, Venkataramana K, Chopra S (2019) Local site effect incorporation in probabilistic seismic hazard analysis—a case study from southern peninsular India, an intraplate region. Soil Dyn Earthq Eng 123:381–398

59. Shreyasvi C, Venkataramana K, Chopra S, Rout MM (2019) Probabilistic seismic hazard assessment of Mangalore and its adjoining regions, a part of Indian Peninsular: an intraplate region. Pure Appl Geophys 176(6):2263–2297
60. Shreyasvi C (2019) Probabilistic seismic hazard assessment and site characterisation of Southwest India. Doctoral dissertation, National Institute of Technology Karnataka, Surathkal
61. Shreyasvi C, Venkataramana K (2020a) Seismic hazard estimation for Southwest India. In: Advances in computer methods and geomechanics. Springer, Singapore, pp 207–220
62. Shreyasvi C, Rahmath NB, Venkataramana K (2020b) Influence of variabilities of input parameters on seismic site response analysis. In: Advances in computer methods and geomechanics. Springer, Singapore, pp 233–244
63. Shreyasvi C, Venkataramana K, Chopra S (2020c) Influence of tectonic regime in the selection of ground motions for seismic site response analysis. In: Proceedings of 17th world conference on earthquake engineering, 17WCEE, Sendai, Japan, September 13th to 18th 2020. Paper ID 1d-0104
64. Singh AP, Parmar A, Chopra S (2017) Microtremor study for evaluating the site response characteristics in the Surat city of western India. Nat Hazards 89(3):1145–1166
65. Sitharam TG, Kolathayar S, James N (2015) Probabilistic assessment of surface level seismic hazard in India using topographic gradient as a proxy for site condition. Geosci Front 6(6):847–859
66. Thaker TP, Rao KS, Gupta KK (2010) Ground response and site amplification studies for coastal soil, Kutch, Gujarat: a case study. Int J Earth Sci Eng 3(4):543–553
67. Tsiambaos G, Sabatakakis N (2010) Empirical estimation of from in situ tests on soil formations in Greece. Bull Eng Geol Environ 291–297
68. Uma Maheswari R, Boominathan D, Dodagoudar GR (2010) Use of surface waves in statistical correlations of and penetration resistance of Chennai soils. Geotech Geol Eng 28:119–137
69. Vidale JE, Helmberger DV (1988) Elastic finite-difference modeling of the 1971 San Fernando, California earthquake. Bull Seismol Soc Am 78(1):122–141
70. Vucetic M, Dobry R (1991) Effect of soil plasticity on cyclic response. J Geotech Eng 117(1):89–107
71. Wald DJ, Allen TI (2007) Topographic slope as a proxy for seismic site conditions and amplification. Bull Seismol Soc Am 97(5):1379–1395

Part III
Resilient Infrastructures

Chapter 16
Disaster Resilient Properties: Built Environment Discourse

N. A. C. Shavindree, D. Amaratunga, and R. Haigh

16.1 Introduction

Over the last ten years, the natural hazards and its after-effects have been responsible for the loss of 1.23 million human lives, victims amounting to 4.03 billion, and reported direct economic losses valued at US$ 2.97 trillion [45]. A disaster which is followed up by a natural hazard is defined as a solemn disruption to the societal functions resulting in extensive human, economic, and environmental losses to the system which is ideally beyond the capacity of the affected community to handle by themselves [49]. Disasters are not only a result of natural environment consequences but rather a mix of such consequences and the vulnerability of the communities in the affected areas [34]. The vital necessity of mainstreaming disaster risk management into the development planning has been highlighted [37, 41, 50]. Bosher [4] identifies two significant concepts related to disaster risk management as built-in resilience and disaster risk reduction. He further defines the built-in resilience as a quality of the built environment ability to adapt to existing and emergent threats, and this ability varies through physical, institutional, economic, and social dimensions [4, 15]. Several other studies state that improving the overall disaster resilience of a given context, built environment plays a critical role as it is extremely vulnerable to the effects of natural hazards and their consequences [17, 26, 44]. The resilience level of the community and ultimately the national resilience depends on the ability of the built environment to maintain the functionality during and aftermath of a hazardous event [26]. Therefore, in-depth understanding of the expertise knowledge and the know-how of mitigating the effects of hazards and threats is vital in designing and constructing a resilient built environment [5, 22].

N. A. C. Shavindree (✉) · D. Amaratunga · R. Haigh
Global Disaster Resilience Centre, School of Applied Sciences, University of Huddersfield, Huddersfield, UK
e-mail: Chrishani.Shavindree@hud.ac.uk

S. Kolathayar et al. (eds.), *Civil Engineering for Disaster Risk Reduction*, Springer Tracts in Civil Engineering, https://doi.org/10.1007/978-981-16-5312-4_16

Within this context, this study explores the underlying characteristics of the concept of disaster resilience, which is instrumental in developing the built-in resilience of the built environment together with the multiple roles of the built environment towards societal resilience. Several disaster resilience studies highlight and discuss disaster resilient properties for different contexts such as infrastructure resilience and transport sector resilience. In science, property refers to a characteristic or trait which is used to illustrate a matter through means of observation, measurement, or combination. Furthermore, the same resilient properties are being identified as resilience dimensions and resilience principles in some other studies. Despite these different classifications, a clear set of disaster resilient properties or basic underlying characteristics for the context of the built environment has not been discussed in detail as yet. In the light of this, the Chapter discusses what are the different underlying characteristics or properties relating to the concept of disaster resilience in the built environment.

16.2 Literature Review

16.2.1 The Concept of Disaster Resilience

Resilience term widely known today has been mentioned as a design principle embedded in the traditional construction knowledge before the nineteenth century which has been achieved through characteristics such as oversizing components and spaces, redundancy, and reparability [18]. However, these concepts were taken over by the modern engineering definition of resilience which is aimed at optimizing structural safety while reducing material consumption. However, the definition of resilience changes according to the context to which it is applied in. In the context of disaster risk management, the resilience aspect is known as disaster resilience and has received the attention of scholars and many studies globally. Disaster resilience can be defined as the capability of a system and community to resist, absorb, accommodate, and recover from the hazard in a timely and efficient manner [40]. The main elements of disaster resilience are the context or whose resilience is aimed to build, what kind of disturbances is resisted, capacity to respond, and the reaction to the disasters [9].

Chang et al. [7] suggest that resilience can be conceptualized along four interrelated dimensions: technical, organizational, social, and economic (Fig. 16.1). In further elaborating on these different dimensions, technical dimension refers to the capability of physical systems inclusive of all their subcomponents and connections to deliver the desired output levels. The organization dimension covers the capacity of organizations or institutional units that are handling the essential facilities. These must possess the ability to carry out the critical disaster-related functions effectively. Social dimension strategies are to mitigate the negative societal impact on the communities and governmental jurisdictions facing the hazardous situation. The

Fig. 16.1 Resilience dimensions adapted and recreated based on the Bruneau et al. [6]

economic dimension aims at reducing both the direct and indirect economic losses from a hazardous situation [6]. Due to this, multidimensional nature of the disaster resilience concept developing resilience measures which are concise, quantifiable, and simple is challenging [8].

The target outcomes of adopting resilience measures are reduced failure probabilities, reduced negative consequences in terms of loss of lives and economic and societal damages, and reduced restoration times to the normal operational capacities [6]. In line with this background, the next section explores the role of the built environment towards enhancing disaster resilience.

16.2.2 Role of the Built Environment

The increasing risk of natural hazards due to urbanization, population growth, and climate changing patterns [36] are surging the necessity for enhancing the disaster resilience of the built environment. Specifically, in the developing countries, the rapid unregulated urbanization has created a backlog of unsafe developments and exposed communities amounting to 325 million by 2030 [46]. Built environment is made up of man-made structures which often constitutes the weakest links and will turn out to be the most dangerous places during disasters. Furthermore, the impact of disasters is largely felt by inhabitants, communities, and countries mainly because the infrastructure still remains complicated, interdependent, and connected [19]. Investing in physical and societal resilience complements to reinforce financial resilience as well, and this is reflected in the Sendai Framework for Disaster Risk Reduction which promotes to invest in DRR and disaster preparedness via structural and non-structural measures [24].

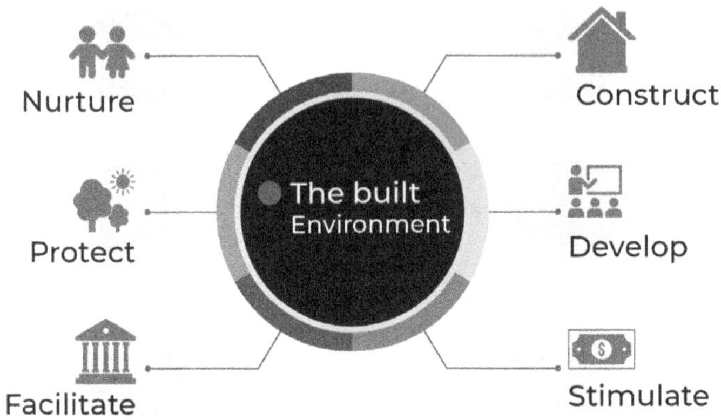

Fig. 16.2 Role of the built environment towards societal resilience, Haigh and Amaratunga [2]

In facing the potential hazards and withstanding them, the built environment needs to be equipped with an effective level of resilience [20]. In achieving a sustainable future for all these urban areas, the disaster risk needs to be embedded into the national development agenda and also focus on managing disaster risk rather than managing emergency situations due to the consequences of natural disasters [1]. The composition of the built environment varies through single buildings to large-scale infrastructure projects and urban development activities.

Significance of the built environment in the event of disaster risk realization has been evident; for instance, the tremendous losses occurred as a result of Wenchuan earthquake were largely due to the defects of built environment, which also disturbed local emergency response and external aid after the hazard. Furthermore, this highlighted the disparity in dwelling destruction amidst the rural and urban context owing to the socially deprived conditions making the rural built environment more vulnerable during the hazardous situation [48]. In the long run, vulnerability could be decreased by imposing DRR measures and increasing the preparedness levels of the built environment as promoted by Build Back Better Framework [12]. Haigh and Amaratunga [2] identifies six multiple roles built environment plays in embedding societal resilience.

The physical assets in the built environment need to be equipped with the structural features that will enhance the resilience of the built environment which is covered under the role of construct. It is vital to reduce the risk by the use of hazard resilient designs, specifications, construction methods, materials, technologies, and construction of protective infrastructure [25]. Furthermore, proper skills and technical know-how need to be developed among the built environment stakeholders also which is reflected by the developing role for resilience. The disaster risk research relating to the different aspects of the built environment needs to be conducted so that the best practices could be developed based on these scientific findings. The next role of the built environment is to stimulate resilience approaches, and this could be done via an incentive-based resilience approach which promotes and supports the protection, maintenance, recovery, and sustainability of the built environment. Provision of incentives through reduced consent fees, tax deductibles, low insurance premiums, reduced interest rate loans, reduced permit fees, and fee waivers will stimulate the built environment stakeholders, specially building owners and investors to adopt resilience practices [1]. The incentive schemes could be linked with these basic disaster resilient characteristics in the projects so that the built-in resilience of the systems could be improved.

Facilitating resilience is another key role of the built environment. This role covers the public, social, and security infrastructures such as health facilities, civil defence, and security services which are particularly critical to the survival during and after a hazardous occurrence [17]. Accordingly, the built environment needs to facilitate accessibility of health facilities, medical care capacity, density of principal roads, and the capacity of emergency treatment centres and rescue missions following a disaster which contributes to the resilient properties in the built environment [39]. Furthermore, since resilience is a transferable skill, efforts to facilitate resilient development will support students to transit, adapt, and thrive in the demanding construction industry as well [42]. With reference to the protection role of the built environment, the protective characteristics of the built environment offer an important means by which the risk posed by hazards could be reduced, thereby preventing a disaster. Conversely, post-disaster, the loss of critical buildings, and infrastructure can greatly increase a community's vulnerability to hazards in future [17]; therefore, protecting critical infrastructure availability during and aftermath of a hazard occurrence is vital. The attempt to reduce injuries inducing blast debris and the development of glazing materials that do not contribute to the explosion-induced projectile hazards and have enhanced security application are few of the ground-level strategies adopted [17].

When discussing the nurturing role of the built environment, social resilience refers to the ability to start, nurture, and maintain positive relationships even in the face of threats or unpleasant eventualities or mishaps [35]. Promoting space and designing the built environment in a manner that promotes social cohesion with a sense of well-being in their residential areas will further nurture individual resilience in the circumstance of a hazardous situation. It has been proven that residents who believe and trust their neighbours and community managers always show resilient behaviours [39]. The coping capacity can be nurtured through a focus on three transformative attributes: mindfulness and re-connection to natural (water, food, energy, and air) and human-made elements (governance, infrastructures, and economic) supporting people's well-being; sense of agency, in both individuals and groups; and strong social and community cohesion [38]. A resilient built environment could be developed by nurturing creativity by providing time and resources for experimentation, rewarded innovation, tolerance for failure, and an atmosphere in which employees felt safe to share new ideas [3]. Also, the factor of nurturing diversity for resilience requires understanding of the specifics of the particular piece of heritage and the diversity of the built environment within which it exists which requires understanding the conditions of resilience or basic underlying characteristics of disaster resilience [11]. Accordingly, the next step is to investigate what are the basic underlying characteristics or properties that will govern these resilience roles.

16.2.3 Disaster Resilient Properties

The terminology disaster resilient properties can be found in numerous disaster resilience studies; however, the most commonly referred disaster resilient properties are the four resilient properties highlighted by Bruneau et al. [6]. They define four resilience properties in relation to both physical and social systems and their resilience. These four resilience properties are as follows:

- **Robustness**: strength, or the ability of elements, systems, and other units of analysis to withstand a given level of stress or demand without suffering degradation or loss of function.
- **Redundancy**: the extent to which elements, systems, or other units of analysis exist that are substitutable, i.e. capable of satisfying functional requirements in the event of disruption, degradation, or loss of functionality.
- **Resourcefulness**: the capacity to recognize difficulties, identify priorities, and mobilize resources when conditions exist that threaten to disrupt some element, system, or other unit of analysis. Resourcefulness can be further conceptualized as consisting of the ability to apply material (i.e. monetary, physical, technological, and informational) and human resources to meet established priorities and achieve goals.
- **Rapidity**: the capacity to meet priorities and achieve goals in a timely manner in order to contain losses and avoid future disruption.

Zhou et al. [52] have highlighted that disaster resilience possesses two properties of inherent resilience and adaptive resilience. The inherent capacity is defined for the routine operations, determined by the structure and functions of system before, during, and after disaster, until the structure and functions are destroyed. The adaptive resilience is the ability to resist and recover from the disaster basically after the hazardous event has taken place [52]. Lorenz [23] has identified that three properties of social systems which constitute resilience are highlighted to be adaptive capacity, participative capacity, and coping capacity. The adaptive capacity is the property of a system in which structures are adapted in a style to prevent or minimize future disasters, whereas the coping capacity is the system's property of dealing with devastating processes that occurred in the past. Adaptive capacity is divided into themes of governance, policy and leadership, and social and community engagement [33]. Adaptive capacity is the arrangements and processes that enable adjustment through learning, adaptation, and transformation. The participative capacity is a measure of the system's ability to adapt its own structures in face of the interventions by other systems, decreasing the system's resilience [23]. Coping capacity is the means by which people or organizations use available resources, skills, and opportunities to face adverse consequences that could lead to a disaster. Coping capacity is categorized under the themes of social character, economic capital, infrastructure and planning, emergency services, community capital, and information and engagement.

In summarizing the factors that have been highlighted as disaster resilient properties in the existing studies, they are adaptive resilience, inherent resilience, robustness, redundancy, resourcefulness, rapidity, participative capacity, adaptive capacity, and coping capacity. These same factors have been referred by various other terminologies in other studies. Since this paper is focused on developing a set of resilient properties for the context of the built environment, it is necessary to study where else these individual factors have been used.

The four factors robustness, redundancy, resourcefulness, and rapidity identified by Bruneau et al. [6] are shortly known as 4 Rs named as attributes or qualities of resilience as well in another study by Parker [32]. Norris et al. [30] suggest three attributes of disaster resilience are robustness: the capacity must be resistant to a wide variety of dangers; redundancy: elements could be substitutable or replaced in the event of disruption or degradation (e.g. social networks need numerous interconnections in order that communication is not stalled by the removal of a single network member); and rapidity: how quickly any needed resources can be accessed and used. The redundancy factor has been highlighted as one of the key principles relating to disaster resilience in Wildavsky's principles of resilient systems in 1986 [18].

As aforementioned, the disaster resilient properties or underlying characteristics in the built environment context have not been studied in detail so far; however, resilient characteristics relating to different arenas of built environment have been already discussed. Such a study reveals nine principles of resilience that needs to be embedded across the multiple areas in the transport sector [28]. The nine basic resilient principles identified are information flows, robustness, capacity to learn, resourcefulness, flexibility, responsiveness, redundancy, safe failure, and good governance. These nine principles include the three of the disaster resilient properties, robustness, responsiveness, and redundancy, which were categorized by Bruneau et al. [6] as resilient properties. A similar study done in relation to transport network resilience [27] identifies a slightly different set of resilience dimensions as redundancy, strength, collaboration, mobility, autonomous components, ability to recover quickly, efficiency, adaptability, safety, and diversity.

Francis and Bekera [14] conclude the factors that affect resilience are concluded as robustness, resourcefulness, redundancy, rapidity or ability to return to normal operating capacity in a timely manner, interconnectedness, cross-functional stakeholders, anticipative capacity, stakeholders' cooperation, capacity to recognize threats, evaluation of the model used to obtain and retain competence, capacity to prepare for future protection efforts, and ability to reduce likely risks. Twigger-Ross et al. [43] have presented key dimensions of resilience along disaster cycle. Accordingly, the core dimensions of resilience are robustness: the maintenance of desired characteristics of the system in spite of changes to the system, and adaptability: the capacity of actors in the system to influence resilience [47] or similarly the capacity of actors in a system to manage resilience in the face of uncertainty and surprise (Gunderson and

Holling 2001). It is related to learning and flexibility to change and transformability: the capacity of people to create a fundamentally new socioecological system when existing conditions (ecological, social, political, or economic) are affected. However, despite these multiple definitions and conceptual frameworks, there is still no single definition of disaster resilience in the field of disaster management that researchers and practitioners agree upon [10, 13, 51]. The following table presents a collection of the factors discussed in these existing studies relating to the theme of resilient properties (Table 16.1).

The analysis of this literature findings was conducted in order to develop a theoretical framework relating the different terminologies and concepts attached specifically under the theme of disaster resilient properties.

16.3 Methodology

In this study, the existing literature and studies relating to disaster resilient features, properties, and characteristics relating to the built environment context are studied and synthesized with the aim of conceptualizing various dimensions of disaster resilience in the built environment into a single framework which can be an instrumental tool for policymakers, practitioners, and researchers in provisioning a solid basis for disaster resilience-related work. The method adopted for this study is a mapping review which is alternatively named as descriptive review, and it is mostly used for the studies based on literature surveys. This specific methodology is mainly adopted to map out and categorize existing literature on a selected area and contextualizing in-depth systematic reviews and identification of gaps in the evidence-based research [16]. This methodology was further developed and refined by Evidence for Policy and Practice Information and Co-ordinating Centre (EPPI-Centre), Institute of Education, London.

Furthermore, this methodological approach determines the extent to which a body of the literature on the selected area disaster resilient properties reveals any interpretable pattern to the pre-existing propositions, theories, methodologies, and findings [21]. As a result of drawing overall conclusions about the existing conceptualizations, propositions, methods, and findings at the end of this descriptive review, the study will derive the state of the art in the particular domain of disaster resilient properties and associated terminologies [31]. Finally, the resilient properties are linked with the multiple roles of the built environment towards enhancing societal resilience which could be further developed through empirical research as a basis for practitioners and decision-makers in disaster risk.

Table 16.1 Disaster resilient properties and associated factors

Source	Disaster resilience-related terminology		
[6]	*Resilience properties*		
	Robustness		
	Redundancy		
	Resourcefulness		
	Rapidity		
[52]	*Properties of resilience*		
	Inherent resilience		
	Adaptive resilience		
[23]	*Resilient properties*		
	Adaptive capacity		
	Participative capacity		
	Coping capacity		
[32]	*Attributes or qualities of resilience*		
	Robustness		
	Redundancy		
	Rapidity		
[30]	*Attributes of resilience*		
	Robustness		
	Redundancy		
	Rapidity		
[28]	*Basic resilient principles*		
	Information flows		
	Robustness		
	Capacity to learn		
	Resourcefulness		
	Flexibility		
	Responsiveness		
	Redundancy		
	Safe failure		
	Good governance		
[27]	*Basic resilient dimensions*		
	Redundancy		
	Strength		
	Collaboration		
	Mobility		
	Autonomous components		

(continued)

Table 16.1 (continued)

Source	Disaster resilience-related terminology		
	Ability to recover quickly		
	Efficiency		
	Adaptability		
	Safety		
	Diversity		
[14]	*Factors that affect resilience*		
	Robustness		
	Resourcefulness		
	Redundancy		
	Rapidity		
	Interconnectedness		
	Cross-functional stakeholders		
	Anticipative capacity		
	Stakeholders' cooperation		
	Capacity to recognize threats		
	Evaluation of the model used to obtain and retain competence		
	Capacity to prepare for future protection efforts		
	Ability to reduce likely risks		
[43]	*Resilient dimensions*		
	Level 1	*Level 2*	*Level 3*
	Robustness	Resistance	Diversity
	Adaptability	Learning capacity	Creativity
	Transformability	Innovation	Memory
		Resourcefulness	Cohesion
		Flexibility	Strength
		Redundancy	Individual capacity
			Corporation
			Transformability

16.4 Discussion

The discussion mainly comprises two sections, and the first is around the theme of developing a set of resilient properties. In the next section, these resilient properties are further elaborated along the six roles the built environment for developing societal resilience: construct, develop, facilitate, stimulate, nurture, and protect.

16.4.1 Resilient Properties

The factors that have been highlighted as disaster resilient properties in the existing studies can be summarized as: adaptive resilience, inherent resilience, robustness, redundancy, resourcefulness, rapidity, participative capacity, adaptive capacity, and coping capacity.

The adaptive resilience and the inherent resilience have been named as two properties of resilience, and this classification is mostly based in relation to the timing of the disaster as inherent resilience is relating to the capacity of the system to withhold its functioning during and after the disaster until the system is damaged or affected and the adaptive resilience is the capacity of the system to resist and recover from the disaster basically after the hazardous event has taken place. Therefore, inherent resilience and adaptive resilience sound more like two types of resilience that needs to be accomplished in for a system to be disaster resilient. However, Bruneau et al. [6] directly identify four factors as resilient properties, robustness, redundancy, resourcefulness, and rapidity, and Lorenz [23] identifies participative capacity, adaptive capacity, and coping capacity as the basic properties of a resilient social system. These factors have been cross-referred to in various studies mostly the four properties by Bruneau et al. [6], but they have been categorized under different terminologies such as resilience attributes, principles, dimensions, and factors affecting resilience.

The participative capacity refers to the system's ability to change its own structures with regard to interventions by other systems. This capacity could be linked with the flexibility highlighted by Nakat and Salim [28] and Twigger-Ross et al. [43] in their studies.

Coping capacity is the means by which people or organizations use available resources, skills, and opportunities to face adverse consequences that could lead to a disaster. Coping capacity covers social character, economic capital, infrastructure and planning, emergency services, community capital, and information and engagement. This relates closely to the disaster resilient property resourcefulness defined by Bruneau et al. [6] as it is the capacity to identify problems, establish priorities, and mobilize resources when conditions exist that threaten to disrupt some element, system, or other unit of analysis. Resourcefulness can be further conceptualized as consisting of the ability to apply material (i.e. monetary, physical, technological, and informational) and human resources to meet established priorities and achieve goals. Furthermore, resourcefulness has been highlighted by Nakat and Salim [28] and Twigger-Ross et al. [43] in their studies as factors relating to disaster resilience.

The adaptive capacity is the property of a system in which structures are modified to prevent future disasters. Adaptive capacity is the arrangements and processes that enable adjustment through learning, adaptation, and transformation. Adaptive capacity could be linked with the key dimensions of resilience in the disaster cycle highlighted by Twigger-Ross et al. [43], learning capacity, adaptability, and the transformability; capacity of people to create a fundamentally new socioecological system when existing conditions (ecological, social, political, or economic) are affected as it covers the transformation aspect as well. Furthermore, capacity to prepare for future protection efforts and capacity to recognize threats highlighted by Francis and Bekera [14] under factors that affect resilience are also covered by the adaptive capacity.

Rapidity, which refers to the capacity to meet priorities and achieve goals in a timely manner in order to contain losses and avoid future disruption, links with the ability to recover quickly and ability to reduce likely risks which are highlighted as factors relating to resilience dimensions. Redundancy which refers to the extent to which elements, systems, or other units of analysis exist which are substitutable and capable of satisfying functional requirements in the event of disruption, degradation, or loss of functionality closely links with efficiency, safe failure, substitutability, and corporation as well. Robustness refers to the ability to maintain desired characteristics of the system in spite of changes to the system [29]. This again relates to safety, strength, and responsiveness factors which are other attributes highlighted in other studies. Autonomous components and self-reliance could be linked with each other. Memory and information flows could be linked together. Collaboration and corporation could be linked together.

Following the synthesis on the disaster resilient properties, the resilient properties have been included and Fig. 16.3 has been developed. The inherent resilience and adaptive resilience by their definition refers to the different stages of resilience therefore has not been included as a basic underlying quality which is the definition based on which this study was conducted. As aforementioned, the resourcefulness property highlighted by Bruneau et al. [6] is inclusive under the coping capacity; therefore, the theoretical model developed for disaster resilient properties contains six resilient properties.

These are robustness, redundancy, rapidity, participative capacity, adaptive capacity, and coping capacity which have clear definitions have been placed at the outer circle.

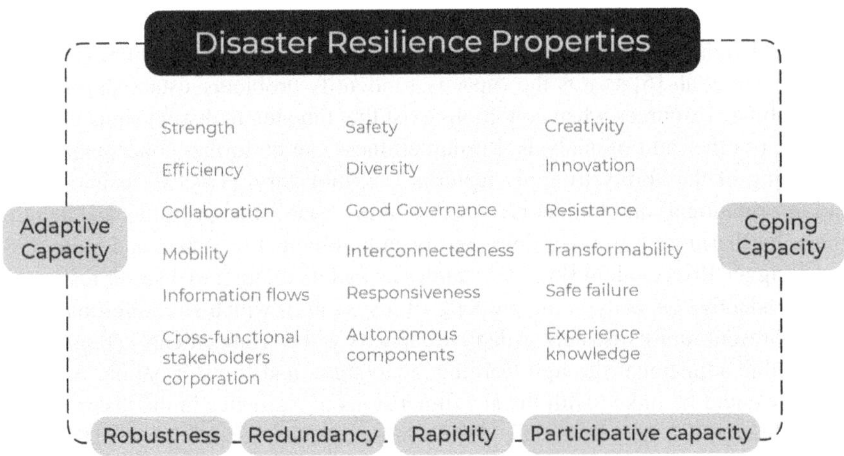

Fig. 16.3 Disaster resilient property framework (*source* Authors)

Table 16.2 Definitions of disaster resilient properties

Terminology	Definition
Participative capacity	System's ability to change its own structures with regard to interventions by other systems
The adaptive capacity	Property of a system in which structures are modified to prevent future disasters
Coping capacity	The capacity to identify problems, establish priorities, and mobilize resources when conditions exist that threaten to disrupt some element, system, or other unit of analysis
Robustness	The ability to maintain desired characteristics of the system in spite of changes to the system
Redundancy	The extent to which elements, systems, or other units of analysis exist that are substitutable and capable of satisfying functional requirements in the event of disruption, degradation, or loss of functionality
Rapidity	The capacity to meet priorities and achieve goals in a timely manner in order to contain losses and avoid future disruption

Following the descriptive mapping review on the different elements relating to disaster resilient properties, Fig. 16.3 is developed. These factors by their definitions have a depth and cover considerably larger scopes in relation to the other factors in the studies. However as aforementioned, these properties have been mixed up and named under different categories such as attributes, principles, dimensions, and factors affecting resilience in various resilience models. In these models since these properties are placed with several other factors, it cannot be clearly demarcated under which category they should be placed without proper empirical evidence. However, some of these factors are clearly overlapping or covered by the umbrella definitions of the disaster resilient properties, whereas few others are linked with the resilient properties but due to the lack of proper definitions they cannot be eliminated without further studies.

In summary, there are mainly six disaster resilient properties which are participative capacity, adaptive capacity, coping capacity, robustness, redundancy, and rapidity. The working definitions synthesized from the literature study are presented in Table 16.2.

16.4.2 *Role of the Built Environment in Embedding Disaster Resilient Properties*

The next part of the discussion is built around applying these resilient properties within the built environment discourse which is mainly done through the six multiple roles of the built environment towards societal resilience.

Under the construction role of the built environment, there is a vital necessity to embed the aforementioned resilience properties to the new constructions as well as existing buildings and infrastructure in a manner that they could even function during a hazard with the minimal disruption to the societal functions. The physical assets in the built environment need to be equipped with the structural features that will enhance the participative capacity, adaptive capacity, coping capacity, robustness, redundancy, and rapidity which is mainly covered under the role of construct.

As a part of the developing role of the built environment, stakeholders have to seek ways to upgrade and develop the building technology so that the resilient properties could be visible or included in the built-in structures and the built environment components. For instance, the construction knowledge needs to be developed on how the physical assets could be enriched with participative capacity, adaptive capacity, coping capacity, robustness, redundancy, and rapidity.

The next role, which is stimulation, could be implemented by the provision of incentives for the investors, building owners, and project implementers on the basis of including the resilient properties in the construction works. The incentive schemes could be linked with the resilient characteristics or properties, and this could be used as a rational or basis for evaluating the resilience in the built environment.

In facilitating public social and security infrastructures such as health facilities, civil defence, and security services, the built environment needs to facilitate accessibility to health facilities, medical care capacity, density of principal roads, and the capacity of emergency treatment centres and rescue work after a hazard [39]. This could be mainly achieved through embedding resilient properties such as redundancy where even if one access is disrupted the people have access to the most critical services via anther route.

With reference to the protection role of the built environment, the protective characteristics of the built environment could be achieved via the adoption of the resilient practices based on the identified resilient properties. There has been significant investigation aimed at expanding knowledge that will enable the construction of a generation of buildings that are more robust and safer which ultimately reflects the initial disaster resilient properties highlighted in this study.

In discussing the nurturing role of the built environment, social resilience refers to the ability to start, nurture, and maintain positive relationships even in the face of threats or unpleasant eventualities or mishaps [35]. The coping capacity can be nurtured through a focus on three transformative attributes: mindfulness and re-connection to natural (water, food, energy, and air) and human-made elements (governance, infrastructures, and economic) supporting people's well-being; sense of agency, in both individuals and groups; and strong social and community cohesion [38]. A resilient built environment could be developed by nurturing creativity by providing time and resources for experimentation, rewarded innovation, tolerance for failure, and an atmosphere in which employees feel safe to share new ideas [3]. Further, the factor of nurturing diversity for resilience requires understanding of the specifics of the particular piece of heritage and the diversity of the built environment within which it exists which requires understanding the conditions of redundancy and obsolescence which are the basic underlying disaster resilient properties.

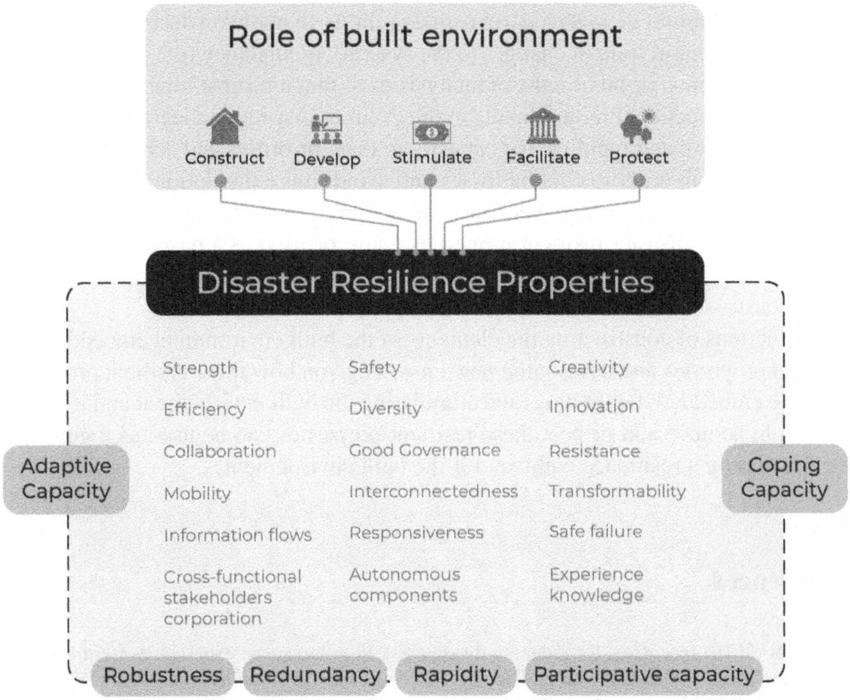

Fig. 16.4 Resilience pathway for the built environment (*source* Authors)

This discussion clearly demonstrates that the multiple resilience roles of the built environment are closely linked and could be built on the six disaster resilient properties identified through this study. Therefore, a resilience pathway for the built environment could be developed based on these findings, as presented in Fig. 16.4.

16.5 Conclusion

This study was focused on synthesizing the concepts and terminologies relating to disaster resilient properties and underlying characteristics in the built environment discourse.

In summarizing all the resilient properties and related dimensions, six disaster resilient properties were summarized as participative capacity (system's ability to change its own structures with regard to interventions by other systems), adaptive capacity (property of a system in which structures are modified to prevent future disasters), coping capacity (capacity to identify problems, establish priorities, and mobilize resources when conditions exist that threaten to disrupt some element, system, or other unit of analysis), robustness (ability to maintain desired characteristics of the system in spite of changes to the system), redundancy (the extent to which elements, systems, or other units of analysis exist that are substitutable and capable of satisfying functional requirements in the event of disruption, degradation, or loss of functionality), and rapidity (the capacity to meet priorities and achieve goals in a timely manner in order to contain losses and avoid future disruption).

In addressing the critical necessity of making the built environment disaster resilient. These disaster resilient properties could be used as a basis for developing a resilience pathway for the built environment, governed by its multiple resilience roles: construct, develop, stimulate, facilitate, protect, and nurture.

The actions of constructing the elements of the built environment embedding the resilient properties and developing new knowledge on how these resilient properties could be embedded, facilitating, and stimulating the built environment and its stakeholders to achieve and protect these resilient properties can be used as a guideline for developing a resilience pathway for the built environment.

References

1. ADB (2016) Incentives for reducing disaster risk in urban areas: experiences from Da Nang (Viet Nam), Kathmandu Valley (Nepal), and Naga City (Philippines) (978-92-9257-478-9). Retrieved from https://www.adb.org/sites/default/files/publication/185616/disaster-risk-urban.pdf
2. Amaratunga D (2016) Developing a resilient built environment. Global Disaster Resilience Centre, University of Huddersfield, UK Sêr Cymru NRN-LCEE & Royal Society of Biology Workshop on Exploring Resilience
3. Barasa E, Mbau R, Gilson L (2018) What is resilience and how can it be nurtured? A systematic review of empirical literature on organizational resilience. Int J Health Policy Manag 7(6):491
4. Bosher L (2014) Built-in resilience through disaster risk reduction: operational issues. Build Res Inform 42:240–254. https://doi.org/10.1080/09613218.2014.858203
5. Bosher L, Carrillo P, Dainty A, Glass J, Price A (2007) Realising a resilient and sustainable built environment: towards a strategic agenda for the United Kingdom. Disasters 31(3):236–255. https://doi.org/10.1111/j.1467-7717.2007.01007.x
6. Bruneau M, Chang S, Eguchi R, Lee G, O'Rourke T, Reinhorn A et al (2003) A framework to quantitatively assess and enhance the seismic resilience of communities. Earthq Spectra. https://doi.org/10.1193/1.1623497
7. Chang SE, McDaniels T, Fox J, Dhariwal R, Longstaff H (2014) Toward disaster-resilient cities: Characterizing resilience of infrastructure systems with expert judgments. Risk Anal 34(3):416–434
8. Chang SE, Shinozuka M (2004) Measuring improvements in the disaster resilience of communities. Earthq Spectra 20(3):739–755
9. Combaz E (2014) Disaster resilience: topic guide. Retrieved from Birmingham, UK
10. Cutter SL, Barnes L, Berry M, Burton C, Evans E, Tate E, Webb J (2008) A place-based model for understanding community resilience to natural disasters. Glob Environ Chang 18(4):598–606
11. de Bruijn K, Buurman J, Mens M, Dahm R, Klijn F (2017) Resilience in practice: five principles to enable societies to cope with extreme weather events. Environ Sci Policy 70:21–30. https://doi.org/10.1016/j.envsci.2017.02.001

12. de Ruiter MC, Couasnon A, van den Homberg MJ, Daniell JE, Gill JC, Ward PJ (2020) Why we can no longer ignore consecutive disasters. Earth's Fut 8(3):e2019EF001425
13. Fekete A, Hufschmidt G, Kruse S (2014) Benefits and challenges of resilience and vulnerability for disaster risk management. Int J Disaster Risk Sci 5(1):3–20. https://doi.org/10.1007/s13753-014-0008-3
14. Francis R, Bekera B (2014) A metric and frameworks for resilience analysis of engineered and infrastructure systems. Reliab Eng Syst Saf 121:90–103. https://doi.org/10.1016/j.ress.2013.07.004
15. Gaillard JC (2011) Hazards and the built environment: attaining built-in resilience. Disaster Prev Manage Int J 20(2):215–216. https://doi.org/10.1108/09653561111126148
16. Grant MJ, Booth A (2009) A typology of reviews: an analysis of 14 review types and associated methodologies. Health Info Libr J 26(2):91–108
17. Haigh R, Amaratunga D (2010) An integrative review of the built environment discipline's role in the development of society's resilience to disasters. Int J Disaster Resilience Built Environ
18. Hassler U, Kohler N (2014) Resilience in the built environment. Build Res Inform 42(2):119–129. https://doi.org/10.1080/09613218.2014.873593
19. Havko J, Mitašová V, Pavlenko T, Titko M, Kováčová J (2017) Financing the disaster resilient city in the Slovak Republic. Procedia Eng 192:301–306. https://doi.org/10.1016/j.proeng.2017.06.052
20. Jacob KH (2011) Hazards and the built environment: attaining built-in resilience. Struct Infrastruct Eng 7:729–731. https://doi.org/10.1080/15732479.2010.532633
21. Lau F, Kuziemsky C (2016) Handbook of eHealth evaluation: an evidence-based approach
22. Lorch R (2005) What lessons must be learned from the tsunami? Build Res Inf 33(3):209–211. https://doi.org/10.1080/09613210500070219
23. Lorenz DF (2013) The diversity of resilience: contributions from a social science perspective. Nat Hazards 67(1):7–24. https://doi.org/10.1007/s11069-010-9654-y
24. Mahul O, Signer BL, Hamada H, Gamper CD, Xu R, Anne H (2019) Boosting financial resilience to disaster shocks: good practices and new frontiers—World Bank Technical Contribution to the 2019 G20 Finance Ministers' and Central Bank Governors' Meeting (English). Retrieved from Washington, D.C.
25. Malalgoda C, Amaratunga D, Haigh R (2014) Challenges in creating a disaster resilient built environment. Procedia Econ Finance 18:736–744. https://doi.org/10.1016/S2212-5671(14)00997-6
26. McAllister T (2013) Developing guidelines and standards for disaster resilience of the built environment: a research needs assessment. Retrieved from https://nvlpubs.nist.gov/nistpubs/TechnicalNotes/NIST.TN.1795.pdf
27. Murray-Tuite P (2006) A comparison of transportation network resilience under simulated system optimum and user equilibrium conditions
28. Nakat E, Salim Z (2015) Disaster risk management in the transport sector: a review of concepts and international case studies (English). Retrieved from Washington, D.C.: http://documents.worldbank.org/curated/en/524081468188378328/Disaster-risk-management-in-the-transport-sector-a-review-of-concepts-and-international-case-studies
29. Noor KBM (2008) Case study: a strategic research methodology. Am J Appl Sci 5(11):1602–1604. https://doi.org/10.3844/ajassp.2008.1602.1604
30. Norris FH, Stevens SP, Pfefferbaum B, Wyche KF, Pfefferbaum RL (2008) Community resilience as a metaphor, theory, set of capacities, and strategy for disaster readiness. Am J Community Psychol 41(1–2):127–150
31. Paré G, Kitsiou S (2017) Methods for literature reviews. In: Handbook of eHealth evaluation: an evidence-based approach [Internet]. University of Victoria
32. Parker DJ (2020) Disaster resilience—a challenged science. Environ Hazards 19(1):1–9. https://doi.org/10.1080/17477891.2019.1694857
33. Parsons M, Glavac S, Hastings P, Marshall G, McGregor J, McNeill J et al (2016) Top-down assessment of disaster resilience: a conceptual framework using coping and adaptive capacities. Int J Disaster Risk Reduct 19:1–11. https://doi.org/10.1016/j.ijdrr.2016.07.005

34. Peng T, Lemay L, Cody B (2013) Concrete building systems: disaster resilient solutions for safer communities. In: Paper presented at the 1st Residential Building Design & Construction Conference, Sands Casino Resort, Bethlehem
35. Perera S, Adeniyi O, Babatunde SO (2017) Analysing community needs and skills for enhancing disaster resilience in the built environment. Int J Disaster Resilience Built Environ
36. Phillips E, Fraser SA, Murnane RJ, Paul N (2018) Aftershocks: remodeling the past for a resilient future. Retrieved from Washington, D.C.
37. Prabhakar S, Srinivasan A, Shaw R (2009) Climate change and local level disaster risk reduction planning: need, opportunities and challenges. Mitig Adapt Strat Glob Change 14(1):7
38. Romero- P, Gnatz DM, Wilhelmi O, Hayden M (2016) Urban sustainability and resilience: from theory to practice. Sustainability 8(12):1224
39. Song J, Huang B, Li R, Pandey R (2020) Construction of the scale-specific resilience index to facilitate multiscale decision making in disaster management: a case study of the 2015 Nepal earthquake. Soc Indic Res 148(1):189–223
40. Tanner T, Lovell E, Wilkinson E, Ghesquiere F, Reid R, Rajput S (2015) Why all development finance should be risk-informed. Retrieved from London
41. Trujillo M, Baas S, Ricoy A, Battista F, Herold J, Vantwout T (2015) Mainstreaming disaster risk reduction in agriculture: an assessment of progress made against the Hyogo framework for action. In: Food and Agriculture Organization of the United Nations (FAO). Final report, USA, 96p
42. Turner M, Scott-Young C, Holdsworth S (2019) Resilience and well-being: a multi-country exploration of construction management students. Int J Construct Manage, pp 1–12
43. Twigger-Ross C, Kashefi W, Brooks K, Deeming H, Forrest S et al (2014) Flood resilience community pathfinder evaluation rapid evidence assessment
44. UNISDR (2011) Global assessment report on disaster risk reduction. Retrieved from Geneva, Switzerland. https://www.preventionweb.net/english/hyogo/gar/2011/en/bgdocs/GAR-2011/GAR2011_Report_FrontCover.pdf
45. UNISDR (2019) The human cost of disasters: an overview of the last 20 years 2000–2019
46. Urbano De Bettencourt M, Ebinger TS, Olga J, Fay M, Ghesquiere F, Gitay H, Krausing J, Kull DW, Mccall K, Reid RCJ, Simpson AL (2013) Building resilience: integrating climate and disaster risk into development—the World Bank Group experience: Main report (English). Retrieved from Washington DC. http://documents.worldbank.org/curated/en/762871468148506173/Main-report
47. Walker B, Holling CS, Carpenter SR, Kinzig A (2004) Resilience, adaptability and transformability in social–ecological systems. Ecol Soc 9(2). Retrieved from www.jstor.org/stable/26267673
48. Wang S, Tang W, Qi D, Li J, Wang E, Lin Z, Duffield CF (2017) Understanding the role of built environment resilience to natural disasters: lessons learned from the Wenchuan earthquake. J Perform Constr Facil 31(5):04017058
49. Wannous C, Velasquez G (2017) United Nations Office for Disaster Risk Reduction (UNISDR)—UNISDR's Contribution to Science and Technology for Disaster Risk Reduction and the Role of the International Consortium on Landslides (ICL), Cham
50. Weekes C, Bello OD (2019) Mainstreaming disaster risk management strategies in development instruments (II): policy briefs for Barbados, Guyana, Saint Lucia, Suriname, and Trinidad and Tobago
51. Yoon DK, Kang JE, Brody SD (2016) A measurement of community disaster resilience in Korea. J Environ Plann Manage 59(3):436–460
52. Zhou H, Zhou H, Wang JA, Wang JA, Wan J, Wan J et al (2010) Resilience to natural hazards: a geographic perspective. Nat Hazards 53(1):21–41. https://doi.org/10.1007/s11069-009-9407-y

Chapter 17
Multiple Benefits of Blue-Green Infrastructure and the Reduction of Environmental Risks: Case Study of Ecosystem Services Provided by a SUDS Pond

Vladimir Krivtsov, Steve Birkinshaw, Valerie Olive, Janeé Lomax, Derek Christie, and Scott Arthur

17.1 Introduction

Sustainable Urban Drainage System (SUDS) are indispensable for the proper functioning and sustainable development of the built environment [1–9]. SUDS ponds have been constructed in many UK cities to help reduce the risk of urban flooding that has been exacerbated by the expansion of impervious areas [10]. These ponds are an essential part of Green and Blue-Green infrastructure (GI/BGI) and provide a number of additional benefits, including water quality improvement, biodiversity, amenity and recreation values among others [11–13]. Here we present a case study of Granton pond located in Edinburgh (Scotland, UK—see Fig. 17.1) and discuss the relevance of its ecosystem services to the alleviation of multiple ecological, environmental and socio-economic risks.

V. Krivtsov (✉)
Royal Botanic Garden Edinburgh, Edinburgh, UK
e-mail: vkrivtsov@rbge.org.uk

S. Birkinshaw
Newcastle University, Newcastle upon Tyne, UK

V. Olive
SUERC, University of Glasgow, Glasgow, UK

J. Lomax · S. Arthur
Heriot Watt University, Edinburgh, UK

D. Christie
Botanical Society of Scotland, Edinburgh, UK

V. Krivtsov
University of Edinburgh, Edinburgh, UK

S. Kolathayar et al. (eds.), *Civil Engineering for Disaster Risk Reduction*, Springer Tracts in Civil Engineering, https://doi.org/10.1007/978-981-16-5312-4_17

Fig. 17.1 The location of the Granton pond

17.1.1 Site Description

The pond (Figs. 17.1 and 17.2) was established in 2005 and is situated in the Forthquarter Park, close to a supermarket and a college and has an area of approximately 2600 m^2. The site appears to be managed by Capita Symonds/National Grid [14]. Adjacent to the pond on its eastern side is the Caroline Burn watercourse, which emerges at this point from the culverted section. The burn is slow flowing in this section, with wetlands on both sides. Analysis of historical and aerial maps reveals that this wetland was re-established in 2003 following de-culverting of its lower sections.

It should be noted, however, that contrary to its initial appearance, there is no outflow from Granton pond on its Eastern side into Caroline Burn and the surrounding wetlands. The outflow is situated on the Northern side and then follows the Scottish Water pipework towards the Firth of Forth.

A hydrographic survey carried out as part of this research provided an estimate for the pond's volume of 1904 m^3 for the normal operating conditions. However, during flooded conditions (also observed during this research) the pond's volume increases to *circa* 2545 m^3.

The catchment area draining into the pond is 0.082 km^2 (82,000 m^2) and it consists of two main parts (Fig. 17.2). The western section of the catchment consists of a new housing development (Waterfront Park) established at the same time as the pond

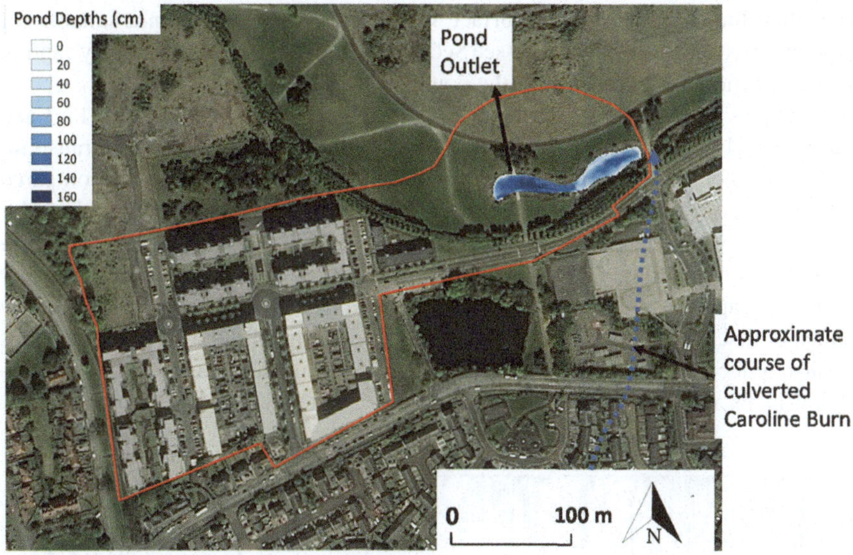

Fig. 17.2 Granton pond catchment

with Scottish Water stormwater drains flowing directly into the south-west corner of the pond. The eastern section of the catchment is predominately grassland that drains naturally into the pond. Two metre resolution LIDAR data of the catchment are available and provided the elevations in both models. Daily rainfall data were supplied by a Scottish Environmental Protection Agency (SEPA) rain gauge situated at the Royal Botanic Garden located 3 km south-east of the pond.

17.2 Hydrology

The water catchment of the pond (Fig. 17.2) was delineated using details of the stormwater drains obtained from Scottish Water and the elevations from the LIDAR data. Hydrology of the pond and its catchment have subsequently been assessed using the well-established and physically-based distributed hydrological model SHETRAN [15]. Details of the precipitation, potential evaporation and such catchment qualities as soil and land cover were applied to the model. The simulation was subsequently run for 19 months from 1/1/2018 to 31/7/2019, corresponding to the period when the hydrochemical and hydrobiological sampling was undertaken. The output from SHETRAN contains simulations of pond discharge, surface and subsurface runoff, as well as time series of evapotranspiration, free surface evaporation and soil moisture dynamics, among others. It should be noted that the surface water component within SHETRAN is simulated with the diffusive wave approximation of the full 2D shallow water equations (rather than the full hydrodynamic solution), so the model is unable to capture the full dynamics of surface water flow. It is therefore not suitable for simulating extreme rainfall events (see Sect. 17.2.1), which need to accurately capture the propagation of flood waves in the pond.

Examples of the results obtained using SHETRAN are shown in Fig. 17.3 and Table 17.1. Higher precipitation in winter combined with low rates of evapotranspiration and evaporation from free surface leads to higher values of discharge compared with the rest of the year, although there is variation between the years with Jan-Mar 2018 being wetter than Jan-Mar 2019. In contrast, lower levels of rainfall during early Autumn and Summer months combined with higher temperatures and higher values of evapotranspiration and evaporation from free surface lead to relatively low values of discharge (Fig. 17.3). These seasonal patterns of the discharge dynamics are also reflected in the increase of water residence time during the summer months (Table 17.1), with a maximum residence time of 19.8 days from July to Sept 2018.

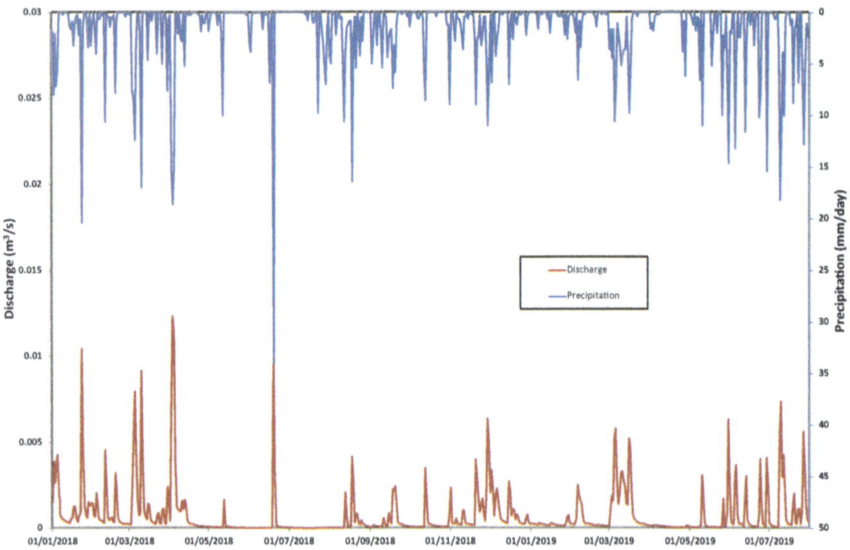

Fig. 17.3 Pond discharges simulated using Shetran

Table 17.1 Granton pond simulated pond nominal residence times using Shetran

Month	Residence time (days)
Jan–Mar 2018	4.1
Apr–Jun 2018	7.8
July–Sept 2018	19.8
Oct–Dec 2018	11.3
Jan–Mar 2019	10.3
Apr–Jun 2019	7.7

This has implications both for biogeochemical cycling and ecological functioning, as well as its overall performance as a SUDS feature.

17.2.1 Simulation of Extreme Events

The hydrodynamic model CityCAT [16] was used to simulate extreme events within the catchment. The model includes the full hydrodynamic solution of the 2D shallow water equations so it is able to capture the full dynamics of surface water flow to the pond and within the pond. Evaporation or subsurface flows are not incorporated into the model so it is not suitable for long-term simulations. The catchment area is divided into buildings, impervious areas and green areas, with infiltration considered in the 'green areas'.

A number of extreme precipitation events were simulated using CityCAT to provide insight into potential flooding problems. Figure 17.4 shows an example of maximum water depths for a one hour, one in 100 year rainfall event for 'with pond' (i.e. the current situation) and 'with no pond' (i.e. what would have happened if the pond was not there). It should be noted that in the latter scenario the extreme precipitation event results in flooding (up to a depth of 0.4 m) and the outflow from the temporary water body thus created at the depression around the present position of the pond is to the East into the Caroline Burn (Fig. 17.4a). However, in the current situation (i.e. 'with pond' scenario) the flooding problem is minimized; the pond fills to its maximum depth (*circa* 1.6 m) and the outflow from the pond is to the North via the Scottish Water pipework (see Fig. 17.4b). As expected, water depths are the same in both scenarios away from the pond in the western part of the catchment.

CityCAT results (Table 17.2) also show that the presence of the pond delays and reduces the peak discharge after extreme precipitation events. Among the simulated scenarios, the maximum alleviation of the peak discharge is expected for the 30 min and 1 h events with a reduction in the peak discharge of 0.18 m^3/s. The largest relative reduction is for the 15 min event, with the peak reduced by 42%. The maximum delay for the 6 h event (Table 17.2) with the peak 8 min later when the pond is present. For less extreme events, the relative reduction in peak discharge is expected to be greater and the exact realized values will depend on a combination of factors, including the state of the catchment saturation, rainfall duration and intensity and the conditions of the drains. The state of the catchment saturation at the start of a rainfall event can be simulated using SHETRAN and these results are used as initial conditions in CityCAT; this is particularly important in the 'green' parts of the catchment. Thus, coupling of SHETRAN and CityCAT models improves the understanding of the flood resilience provided by the pond under a range of scenarios.

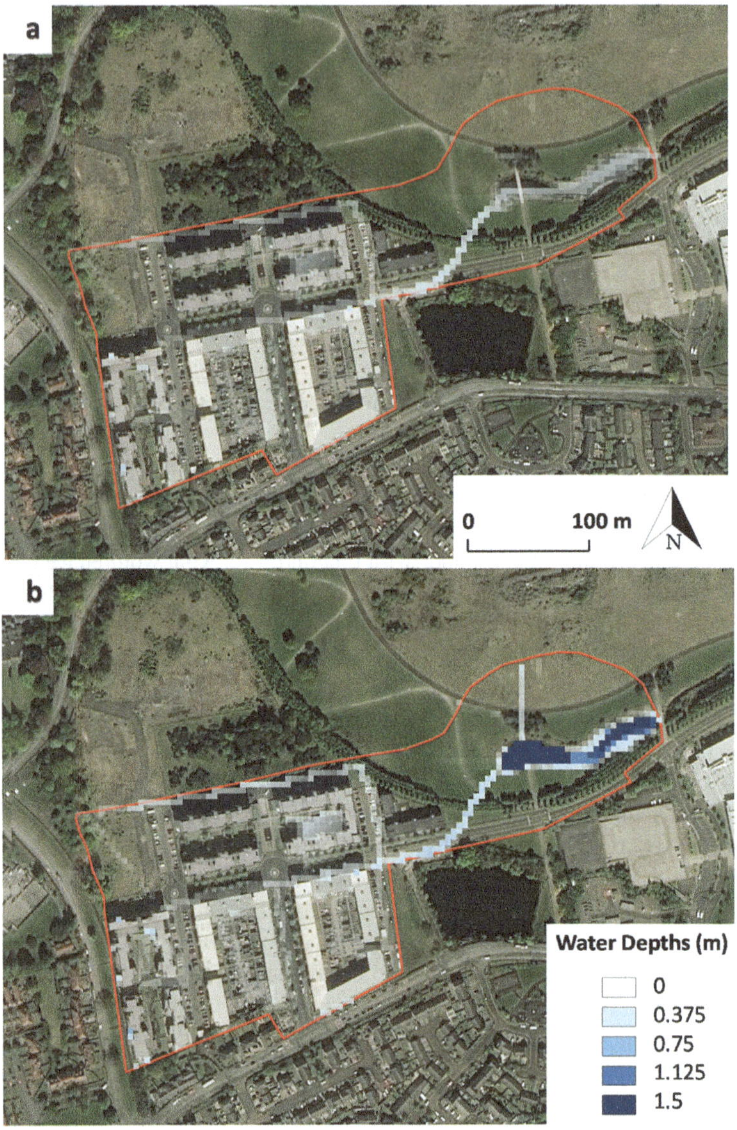

Fig. 17.4 CityCAT simulated maximum water depths for a one hour, one in 100 year rainfall event **a** no pond, **b** with pond

17.3 Water Chemistry

Consideration of water chemistry is indispensable for understanding ecological patterns and the overall functioning of aquatic systems [17–27]. In the present study, concentrations of chemical elements in the surface water samples (collected between

Table 17.2 Granton peak discharges and delayed time to peak (compared with no pond for a one in 100 year event of different durations

Storm duration	Total rainfall (mm)	Peak rainfall rate (mm/h)	No Pond	With Pond	
			Peak discharge (m^3/s)	Peak discharge (m^3/s)	Delay (min)
15 min	24	227	0.36	0.21	7
30 min	31	182	0.46	0.28	6
1 h	38	136	0.50	0.32	6
2 h	46	94	0.47	0.33	6
3 h	52	74	0.43	0.31	7
6 h	65	47	0.31	0.25	8

Apr 2018 and May 2019) have been measured using ICP MS. Figure 17.5 shows the boxplots for selected chemicals of specific interest. The results revealed that for the majority of analyzed elements, the values at Granton pond are similar or do not considerably exceed the typical values for other urban ponds in Edinburgh area. This is true for such potentially problematic pollutants as As and P which show no significant differences, whilst levels of Zn appear to be significantly lower (Kruskal–Wallis test) than the typical values in other ponds. The levels of Si, a biogenic element important for diatom growth, were also not significantly different from the other ponds.

In some cases, however, chemical concentrations in Granton surface water appear to be significantly higher (revealed by Kruskal–Wallis test) compared to the levels observed at other ponds. Relatively small exceedances have been observed for, e.g. U, Li, Sr, Sb, Se, whilst for B, Mg and in particular Rb and K the exceedances were rather more pronounced. Intermediate exceedances were observed for Na and Ca. It should be noted, however, that the error bars for chemical levels at Granton are typically tighter than those for combined distributions at other ponds. This indicates that, except K and Rb, for all the other chemicals the levels at Granton are never the highest, although the analysis of the detailed results (data not shown) reveals that it has the third highest levels for B and Mg and the second highest for Ca, Na, Sb and Se. The median values of K and Rb in Granton surface water are the highest among the ponds studied in Edinburgh area, although the measurements at Granton are exceeded by the outliers from elsewhere.

The pathways for all these elements are not certain, but it is likely that road runoff is a contributor [28, 29]. However, it is also possible that a considerable part of the inputs might have been attributable to the air-borne dust accumulating on the leaves of the surrounding deciduous trees, which end up in the pond after being shed and have been observed in large quantities during the hydrobiological sampling. Correlations between the water concentrations of chemical elements (Table 17.3) provide important clues for the tentative interpretation of their sources and pathways. For example, there is evidence of positive association between P and As, which might be indicative of the As adsorption on the organic P-rich detritus. The association of

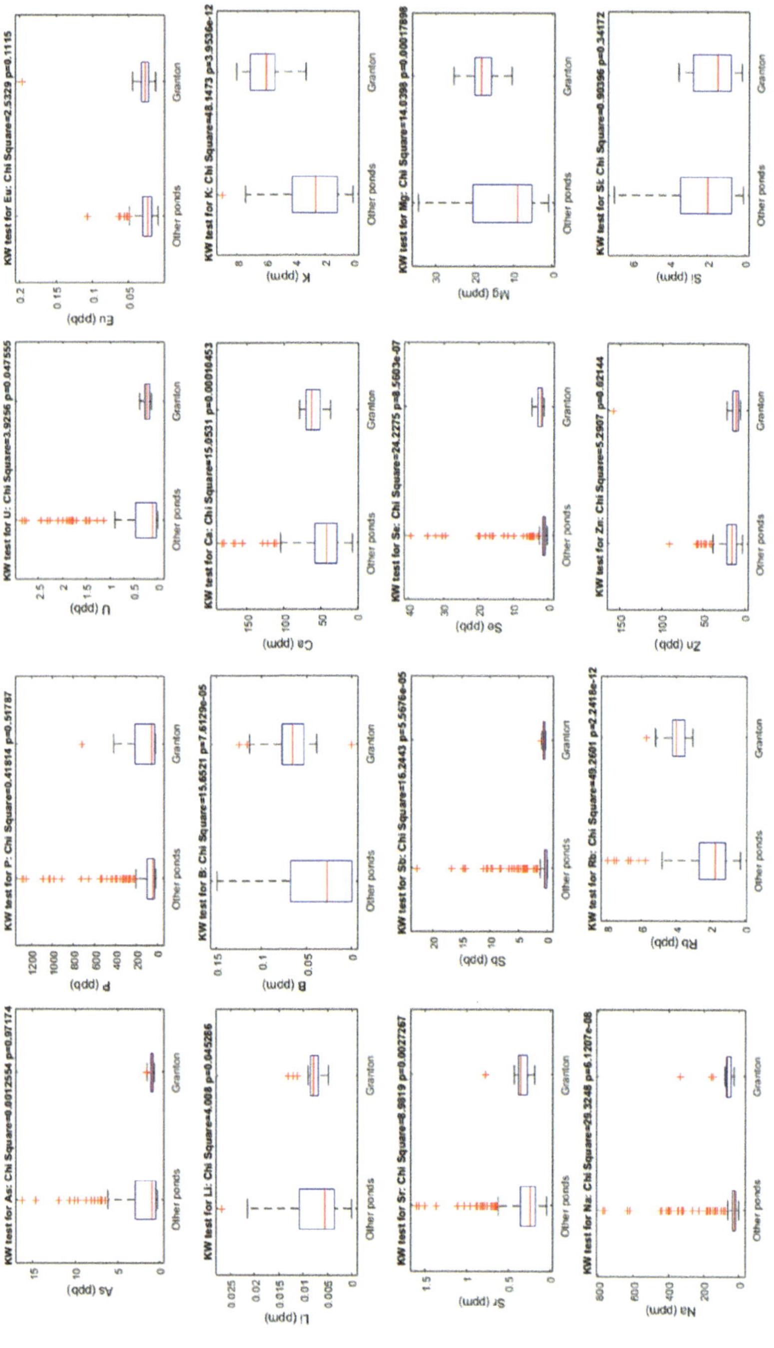

Fig. 17.5 Surface water concentrations of selected chemicals in Granton SUDS pond in comparison to other urban ponds in the area. 'KW' = Kruskal–Wallis test

Table 17.3 Correlations between the surface water concentrations of chemical elements

Variables	As	P	U	Eu	Li	B	Ca	K	Sr	Sb	Se	Mg	Na	Rb	Zn	Si
As	1	0.61	0.15						0.28		0.33		0.3	0.16		
P	0.61	1				0.13	-0.19		-0.15	-0.14			-0.14			
U	0.15		1		0.27	0.19	0.14		0.44							
Eu				1	0.34	0.14	0.51	0.36	0.46		0.38	0.35	0.36	0.35	0.52	0.22
Li			0.27	0.34	1	0.29	0.66	0.35	0.61		0.31	0.68	0.29	0.28		0.54
B		0.13	0.19	0.14	0.29	1	0.15	0.52	0.13	-0.25		0.47	-0.13			0.34
Ca		-0.19	0.14	0.51	0.66	0.15	1	0.49	0.81	0.16	0.73	0.72	0.7	0.43	0.13	0.47
K				0.36	0.35	0.52	0.49	1	0.42		0.3	0.58	0.31	0.72		0.42
Sr	0.28	-0.15	0.44	0.46	0.61	0.13	0.81	0.42	1	0.27	0.79	0.44	0.81	0.4		0.21
Sb		-0.14				-0.25	0.16		0.27	1	0.44		0.52		0.13	
Se	0.33			0.38	0.31		0.73	0.3	0.79	0.44	1	0.31	0.95	0.3	0.15	
Mg				0.35	0.68	0.47	0.72	0.58	0.44		0.31	1	0.3	0.27		0.63
Na	0.3	-0.14		0.36	0.29	-0.13	0.7	0.31	0.81	0.52	0.95	0.3	1	0.32	0.13	
Rb	0.16			0.35	0.28		0.43	0.72	0.4		0.3	0.27	0.32	1		0.26
Zn				0.52			0.13			0.13	0.15		0.13		1	
Si				0.22	0.54	0.34	0.47	0.42	0.21			0.63		0.26		1

Only significant (95% probability levels) correlations are listed

Zn and Eu might be indicative of their source from discarded electronics. In addition, there are a number of positive correlations among Si and the metals belonging to the first two groups of the periodic table (e.g. K, Li, Na, Ca, Mg, Sr). One possible interpretation is that these elements were, to a considerable degree, sourced from the weathering of natural silicates and building materials. However, the correlations observed (for instance a strong positive relationship between K and Rb) would, in part, have resulted from the similarity of chemical properties. Consideration of the temporal changes in surface water concentrations (data not shown) also provides complementary clues for the interpretation of patterns observed. For example, the levels of Na show peaks in Winter and Spring indicating its partial supply from the application of road salt, whilst the strong correlation of Na with Se may be indicative of the presence of Se impurities in the commercial road salt products.

17.4 Hydrobiological Patterns and Biological Water Quality

The dynamics of hydrobiological community was monitored in detail between Apr 2018 and May 2019; preliminary samples were also taken since Nov 2017. Figure 17.6 gives examples of the organisms encountered in the samples. There was a relatively biodiverse bloom of phytoplankton between Jul and Sep. *Spirogyra* was frequent in most planktonic samples throughout the warm period and was observed in periphyton throughout the year. However, it was not encountered in planktonic form between Oct 2018 and Mar 2019. Spirogyra reappeared in the plankton after that and was registered as frequent in April and as abundant in May 2019 samples.

Cyanobacteria (mainly *Oscillatoria* but also *Nostoc*) were present between Jun and Sep but were never abundant. *Eudorina* and *Desmodesmus quadricauda* occurred in variable (sometimes elevated) quantities, whilst *Mougeotia* was frequent in Jun and Sep samples. Diatoms in plankton were variously represented by a number of species (e.g. *Navicula, Craticula, Stauroneis, Synedra, Nitzschia* spp, *Cymatopleura solea*) whilst dinoflagellates (mainly *Peridinium,* also *Gymnodinium*) were encountered from Jul onwards. *Crucigenia* was frequent or abundant late in the year, whilst *Synura* was registered in Jun, Aug and Sep samples.

Protozoa were observed throughout the investigation period, with a particularly diverse community in mid summer. Rotifers were also present throughout, with a particularly diverse and abundant community (*Brachionus*, *Polyarthra*, *Keratella*, *Filinia*, *Synchaeta*) observed during the colder period of the year. Interestingly, *Daphnia* were only registered in May 2018, whilst *Bosmina cornuta* was frequent in May 2019. Copepoda (mainly *Diaptomus*) were observed from Sep onwards.

Chemicals dissolved in the water affect plankton [21, 26, 30] and aquatic invertebrates [31–33]. In this study, sweep sampling of freshwater invertebrates was carried out to assess the biological water quality. The community was characterized by the dominance of animals tolerant of a wide range of conditions, including *Gamarus*

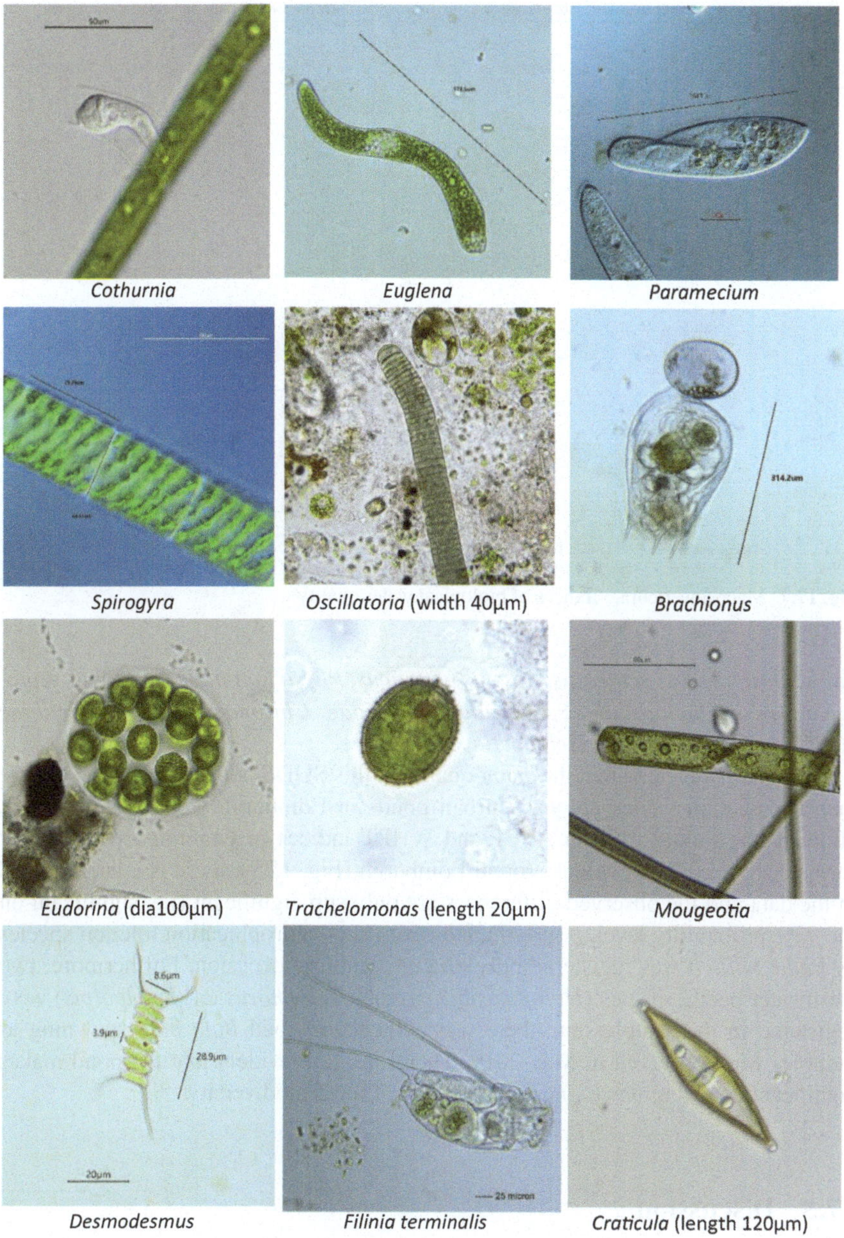

Fig. 17.6 Examples of planktonic genera encountered in Granton Pond

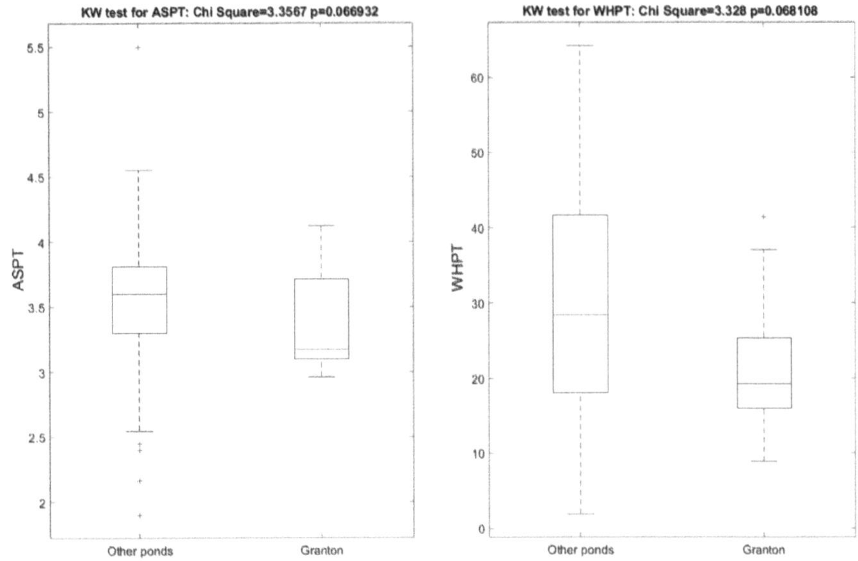

Fig. 17.7 Macroinvertebrate indices of biological water quality

pulex, *Gasterosteus aculeatus*, *Asselus aquaticus*, *Physa fontinalis*, as well as representatives of *Planorbidae*, *Sphaeriidae*, *Corixidae*, *Chironomidae*, *Erpobdellidae* and *Glossiphoniidae* (e.g. *Helobdella stagnalis*).

The results revealed that the water quality in this SUDS pond appears to be worse than the median values among 9 urban ponds in Edinburgh area [28]. However, although the values of both ASPT and WHPT indices in Granton are somewhat lower than the median values around Edinburgh (Fig. 17.7), there is a large spread in the data and the observed differences are only sub-significant (i.e. significant on the 93% probability level, $p < 0.07$). However, such eutrophication tolerant species as *Radix baltica* was registered only on one sampling occasion. Furthermore, two freshwater beetle species (*Haliplus sibiricus* and *Helophorus* c.f. *brevipalpis*) were registered in the samples and there was evidence of toad *Bufo bufo* spawning as tadpoles were observed in May 2019. Therefore, it is evident that the pond makes significant contributions to the enhancement of local biodiversity.

17.5 Discussion

Ponds are an important part of urban Blue-Green Infrastructure (BGI) and contribute to a number of ecosystem services, including alleviation of flood risk and improvement of runoff water quality, as well as amenity and biodiversity. Multiple benefits associated with urban ponds have only recently started to be fully appreciated and the examples of in-depth interdisciplinary insights are still rare. Our results highlight the value of ecosystem services provided by BGI in general [34], and by SUDS ponds in particular. This study is focused on hydrology, water chemistry, hydroecology and biological water quality, which are an integral part of multiple benefits associated with BGI. These benefits also contribute to the BGI recreational capacity and its amenity and biodiversity values. Consequently, the locality is enjoyed every day by many visitors and is particularly popular with joggers, dog walkers and families.

The results presented here unambiguously contribute to the evidence from other research concluding that urban ponds provide alleviation of flood risk [35]. It should be noted, however, that there are a number of other relevant environmental risks, which SUDS ponds help to alleviate, including those related to, e.g. water and air quality, biodiversity and the decrease in wellbeing due to the loss of people connecting with green space. Although we have not monitored the inputs of chemicals to the pond, the comparison of surface water concentrations with those observed in other similar ponds suggests that this SUDS site is providing adequate water quality improvement. This also links well with the recent results documenting the accumulation of pollutants in sediments [29]. It should also be noted that although there is a concern in regards to the risk for public health due to the potential development of cyanobacteria, they do not appear to be problematic for this SUDS pond. Furthermore, the ICP MS analysis of water chemistry, the observed planktonic community and the biological water quality estimated by macroinvertebrate sampling are all broadly similar to those in other ponds.

It is noteworthy that the overall functioning of SUDS ponds and other green and blue-green infrastructure assets is best understood by the simultaneous consideration of all their components and an in-depth consideration of their interlinkages [11, 13]. For example, the detailed consideration of the site's hydrology, catchment properties and water chemistry is indispensable for the correct understanding of the characteristics of its biological community and the hydroecological patterns observed. Nonetheless the resulting biological community provides a good integral measure of the overall water quality and could give important clues in regard to pollutant inputs and their involvement in biogeochemical cycling.

It should also be noted that aquatic ecosystems are characterized by a plethora of indirect interrelations [36–38] and many of those would be beyond the scope of this manuscript. For instance, biodiversity of the terrestrial vegetation is important for the increase in local amenity and biodiversity values, for improvements in air quality and for the understanding of pollutant inputs, including those associated with airborne particles [39]. Furthermore, both the aquatic and the adjacent terrestrial habitats available at SUDS sites are particularly important for provision of wildlife corridors connecting with other suitable habitats scattered throughout the urban landscapes.

Finally, it is worth mentioning that although SUDS ponds have many benefits, in some areas there is a reluctance of residents to adopt them [11]. Parents may fear for their children's safety if a pond is put in, some people fear that a SUDS pond will attract more insects or other pests and others simply don't like the idea of a SUDS pond. This is a very interesting phenomenon because if there is a lake or "water feature" in a community then its citizens don't seem to have those same concerns, yet when something is labelled as a SUDS feature suddenly there are community-wide concerns. SUDS advocates and designers sometimes need to come up with creative solutions to overcome these prejudices against SUDS features. It is also important to educate the public on the positives and low amount of danger that comes with having a SUDS pond nearby [40–43] and consequently engage them in citizen science projects [40–43]. The research on multiple benefits presented here is very relevant in that respect.

17.6 Conclusion

This study highlights flood risk reduction and the value of multiple benefits and ecosystem services provided by BGI, which are relevant for a wide range of environmental, economic and social risks. It also contributes to the accumulating evidence of interlinkages among hydrology, ecology, biogeochemistry and biological water quality at the SUDS ponds sites, thus underlying the need for a comprehensive simultaneous consideration of their subsystems. These aspects are vital for planning future developments and nature-based solutions, as well as for the management of the existing SUDS assets. Consequently, this research should prove useful for studies elsewhere and for further refinements of GI and BGI management. These refinements, based on a better understanding of the underlying science, are indispensable for reducing the risk of negative environmental effects [44] and for enhancing the public's engagement and participation in their use and maintenance [11, 12, 43].

Acknowledgements Data collection for this study was, in part, supported by the EPSRC funding for the UFR project. Adrian Sumner, Heather Forbes, Alejandro Sevilla, Alice Masip, Achiraya Kraiphet, Yamina Monteiro and Cesare Pertusi are kindly thanked for their various contributions to fieldwork, data processing and identification/biological recording. Help of Garth Foster was invaluable in identifying water beetles.

References

1. Cotterill S, Bracken LJ (2020) Assessing the effectiveness of sustainable drainage systems (SuDS): interventions, impacts and challenges. Water 12(11):3160
2. Esmail BA, Suleiman L (2020) Analyzing evidence of sustainable urban water management systems: a review through the lenses of sociotechnical transitions. Sustainability 12(11):1–46
3. Gimenez-Maranges M, Breuste J, Hof A (2020) Sustainable drainage systems for transitioning to sustainable urban flood management in the European Union: a review. J Clean Prod 255:120191
4. Green A (2019) Sustainable drainage systems (SuDS) in the UK. Urban stormwater and flood management. Springer, Berlin, pp 69–101
5. Lashford C, Rubinato M, Cai Y, Hou J, Abolfathi S, Coupe S et al (2019) SuDS & sponge cities: a comparative analysis of the implementation of pluvial flood management in the UK and China. Sustainability 11(1):213
6. McDonald RK (2018) Sustainable urban drainage systems (SUDS) in Scotland: assessment of monitoring and maintenance within local authorities and Scottish Water

7. Melville-Shreeve P, Cotterill S, Grant L, Arahuetes A, Stovin V, Farmani R et al (2018) State of SuDS delivery in the United Kingdom. Water Environ J 32(1):9–16
8. Rodak CM, Moore TL, David R, Jayakaran AD, Vogel JR (2019) Urban stormwater characterization, control, and treatment. Water Environ Res 91(10):1034–1060
9. Spence K, Bridge J, McLuckie D, Kandasamy J (2019) Urban stormwater and flood management: urban stormwater and flood management. Springer, Berlin, pp 139–157
10. Woods-Ballard B, Kellagher R, Martin R, Jefferies C, Bray R, Shaffer P (2007) The SuDS manual. CIRIA C697, London
11. CIRIA (2019) Blue-green infrastructure—perspectives on planning, evaluation and collaboration. CIRIA C780a, London
12. CIRIA (2019) Blue-green infrastructure—perspectives on water quality benefits. CIRIA C780b, London
13. O'Donnell E, Thorne C, Ahilan S, Arthur S, Birkinshaw S, Butler D et al (2020) The blue-green path to urban flood resilience. Blue-Green Syst 2(1):28–45
14. Jarvie J, Arthur S, Beevers L (2017) Valuing multiple benefits, and the public perception of SUDS ponds. Water 9(2):128
15. Ewen J, Parkin G, O'Connell PE (2000) SHETRAN: distributed river basin flow and transport modeling system. J Hydrol Eng 5(3):250–258
16. Glenis V, Kutija V, Kilsby CG (2018) A fully hydrodynamic urban flood modelling system representing buildings, green space and interventions. Environ Model Softw 109:272–292
17. Krivtsov V, Howarth M, Jones S, Souza A, Jago C (2008) Monitoring and modelling of the Irish Sea and Liverpool Bay: an overview and an SPM case study. Ecol Model 212(1–2):37–52
18. Krivtsov V (2001) Study of cause-and-effect relationships in the formation of biocenoses: their use for the control of eutrophication. Russ J Ecol 32(4):230–234
19. Krivtsov V, Bellinger E, Sigee D (2000) Incorporation of the intracellular elemental correlation pattern into simulation models of phytoplankton uptake and population dynamics. J Appl Phycol 12(3–5):453–459
20. Krivtsov V, Bellinger E, Sigee D (2002) Water and nutrient budgeting of Rostherne Mere, Cheshire, UK. Nordic Hydrol 33(5):391–414
21. Krivtsov V, Bellinger E, Sigee D (2003) Ecological study of Stephanodiscus rotula during a spring diatom bloom: dynamics of intracellular elemental concentrations and correlations in relation to water chemistry, and implications for overall geochemical cycling in a temperate lake. Acta Oecologica-Int J Ecol 24(5–6):265–274
22. Krivtsov V, Bellinger EG, Sigee DC (2005) Elemental composition of Microcystis aeruginosa under conditions of lake nutrient depletion. Aquat Ecol 39(2):123–134
23. Krivtsov V, Sigee D, Bellinger E (2001) A one-year study of the Rostherne Mere ecosystem: seasonal dynamics of water chemistry, plankton, internal nutrient release, and implications for long-term trophic status and overall functioning of the lake. Hydrol Process 15(8):1489–1506
24. Krivtsov V, Sigee DC, Bellinger EG, Porteous G (2001) Determination of P release from Rostherne Mere sediment cores. Acta Hydrochim Hydrobiol 29(2–3):111–117
25. Krivtsov V, Tien C, Sigee D, Bellinger E (1999) X-ray microanalytical study of the protozoan Ceratium hirundinella from Rostherne Mere (Cheshire, UK): dynamics of intracellular elemental concentrations, correlations and implications for overall ecosystem functioning. Netherlands J Zool 49(4):263–274
26. Krivtsov V, Sigee DC (2005) Importance of biological and abiotic factors for geochemical cycling in a freshwater eutrophic lake. Biogeochemistry 74(2):205–230
27. Krivtsov V, Bellinger E, Sigee D (1999) Modelling of elemental associations in Anabaena. Hydrobiologia 414:75–81
28. Krivtsov V, Arthur S, Buckman J, Bischoff J, Christie D, Birkinshaw S et al (2019) Monitoring and Modelling SUDS Retention Ponds: Case Studies from Scotland ICONHIC, Chania, Greece. http://www.urbanfloodresilience.ac.uk/documents/krivtsov-et-al.-iconhic-2019b.pdf
29. Krivtsov V, Arthur S, Buckman J, Kraiphet A, Needham T, Gu W et al (2020) Characterisation of suspended and sedimented particulate matter in blue-green infrastructure ponds. Blue-Green Syst 2(1):214–236

30. Krivtsov V, Sigee D, Bellinger E (2002) Elemental concentrations and correlations in winter micropopulations of Stephanodiscus rotula: an autecological study over a period of cell size reduction and restoration. Eur J Phycol 37(1):27–35
31. Bloor MC, Banks CJ, Krivtsov V (2005) Acute and sublethal toxicity tests to monitor the impact of leachate on an aquatic environment. Environ Int 31(2):269–273
32. Bloor MC, Banks CJ, Krivtsov V (2006) Population dynamics in Asellus aquaticus as modified by chronic leachate stress. Eng Geol 85(1–2):9–13
33. Bloor MC, Banks CJ, Krivtsov V (2006) How leachate stress modified the population dynamics of two macro invertebrates: Water quality improvements by constructed wetlands. Int Assoc Theor Appl Limnol 29:1507–10
34. Fenner R, O'Donnell E, Ahilan S, Dawson D, Kapetas L, Krivtsov V et al (2019) Achieving urban flood resilience in an uncertain future. Water 11(5)
35. Krivtsov V, Birkinshaw S, Arthur S, Knott D, Monfries R, Wilson K et al (2020) Flood resilience, amenity and biodiversity benefits of an historic urban pond. Phil Trans R Soc A 378(2168):20190389
36. Krivtsov V (2004) Investigations of indirect relationships in ecology and environmental sciences: a review and the implications for comparative theoretical ecosystem analysis. Ecol Model 174(1–2):37–54
37. Krivtsov V (2008) Indirect effects in ecology. In: Jorgensen SE, Fath BD (eds) Encyclopedia of ecology, Newnes pp 1948–58
38. Krivtsov V, Howarth M, Jones S (2009) Characterising observed patterns of suspended particulate matter and relationships with oceanographic and meteorological variables: studies in Liverpool Bay. Environ Model Softw 24(6):677–685
39. Krivtsov V, Birkinshaw S, Forbes H, Olive V, Chamberlain D, Lomax J et al (2020) Hydrology, ecology and water chemistry of two suds ponds: detailed analysis of ecosystem services provided by blue-green infrastructure. WIT Trans Built Environ 194:167–178
40. Li L, Collins AM, Cheshmehzangi A, Chan FKS (2020) Identifying enablers and barriers to the implementation of the Green Infrastructure for urban flood management: a comparative analysis of the UK and China. Urban For Urban Greening 54:126770
41. Mukhtarov F, Dieperink C, Driessen P, Riley J (2019) Collaborative learning for policy innovations: sustainable urban drainage systems in Leicester, England. J Environ Plann Policy Manage 21(3):288–301
42. Rae M, Miró A, Hall J, O'Brien K, O'Brien D (2019) Evaluating the validity of a simple citizen science index for assessing the ecological status of urban drainage ponds. Ecol Ind 98:1–8
43. Williams J, Jose R, Moobela C, Hutchinson D, Wise R, Gaterell M (2019) Residents' perceptions of sustainable drainage systems as highly functional blue green infrastructure. Landscape Urban Plann 190:103610
44. Ahilan S, Guan MF, Wright N, Sleigh A, Allen D, Arthur S et al (2019) Modelling the long-term suspended sedimentological effects on stormwater pond performance in an urban catchment. J Hydrol 571:805–818

Chapter 18
Static and Seismic Assessment of Soil Arching in Piled Embankments

Sanjay Nimbalkar and Naveen Kumar Meena

Notations and Abbreviations

Notations

[C]	Damping matrix (dimensionless)
[M]	Mass matrix (dimensionless)
$[K']$	Stiffness matrix (dimensionless)
ξ_0	Damping ratio (dimensionless)
ω_i and ω_j	Circular frequencies (rad/s)
α and β	Damping coefficients (dimensionless)
λ	Logarithmic hardening constant (dimensionless)
γ	Unit weight (kN/m^3)
v	Poisson's ratio (dimensionless)
ψ	Dilation angle (degree)
ϕ'	Effective friction angle (degree)
c'	Effective cohesion (kN/m^2)
p_o'	Effective overburden pressure (kN/m^2)
σ_h	Horizontal stress (kN/m^2)
σ_p	Vertical stress on pile (kN/m^2)
σ_s	Vertical stress on subsoil (kN/m^2)
σ_v	Vertical stress (kN/m^2)
E_{em}	Embankment modulus (MPa)
E_p	Pile modulus (GPa)
F_d	Design equivalent dynamic load (kN)
F_s	Static wheel load (kN)

S. Nimbalkar (✉) · N. K. Meena
School of Civil and Environmental Engineering, University of Technology Sydney, Ultimo, NSW 2007, Australia
e-mail: Sanjay.Nimbalkar@uts.edu.au

S. Kolathayar et al. (eds.), *Civil Engineering for Disaster Risk Reduction*, Springer Tracts in Civil Engineering, https://doi.org/10.1007/978-981-16-5312-4_18

K_o	Lateral stress coefficient at rest state (dimensionless)
K_p	Lateral stress coefficient at passive state (dimensionless)
N_{em}	Normalized embankment height (dimensionless)
N_{vs}	Normalized vertical stress (dimensionless)
P_{es}	Plane of equal settlement (m)
a_o	Initial yield surface size (kN/m^2)
e_o	Initial void ratio (dimensionless)
e_1	Void ratio at unit pressure (dimensionless)
D	Pile diameter (m)
E	Modulus of elasticity (MPa)
K	Lateral stress coefficient (dimensionless)
M	Critical-state stress ratio (dimensionless)
SAR	Soil arching ratio (dimensionless)
d	Pile wall width (m)
h	Embankment height (m)
k	Logarithmic bulk modulus (kN/m^2)
q	Surcharge on the embankment top (kN)
s	Pile spacing (m)
a, b, c, a', b', and c'	Relationship constants (dimensionless)

Abbreviations

2D	Two dimensional
3D	Three dimensional
CINPE4	Four-node plane strain linear infinite element
CPE8R	Eight-node plane strain element with reduced integration
DAF	Dynamic amplification factor
EA	Equivalent area
FEA	Finite element analysis
MC	Mohr–Coulomb
MCC	Modified cam clay
ORE	Office of research and experiments
PGA	Peak ground acceleration

18.1 Introduction

Across the globe, increase in population and demand of associated roads, railways, and public transport infrastructure has stimulated professionals to use soft compressible soils, which otherwise deemed conventionally unsuitable foundation material.

Since, the weak soil has less bearing capacity and high compressibility, piled embankment is often used as a suitable engineering alternative providing benefits in terms of rapid construction, lesser differential settlement, and elevating the ground level for construction of transport corridors on such weak soils.

In the pile-supported railway embankment, most of the imposed load is transferred to the rigid pile through a shearing stress mechanism named as "soil arching" [1]. In the recent decade, several studies [2–5] focused on the mechanism of soil arching. Han and Gabr [2] performed numerical investigations to assess the soil arching by considering three major influence factors: the height of the fill, the tensile stiffness of geosynthetic, and the pile elastic modulus under static loading condition. They reported that soil arching is significantly affected by the variation in piled embankment properties. A case study of pile-supported transport embankment is reported by Liu et al. [3]. The case history was back analyzed in finite element analysis and authors concluded that significant load was transferred on the pile top due to the soil arching. Wu et al. [4] reported a full-scale experiment to investigate the performance of piled embankment. A numerical analysis was also performed in two-dimensional (2D) plane strain condition to compare applicability of the different numerical approaches. The numerical model adopting equivalent area (EA) approach was in good agreement with experimental results. In addition, modified cam clay (MCC) soil constitutive model yielded good prediction with the experimental results. Almeida et al. [5] performed numerical and analytical modeling of a series of centrifuge tests, where static surcharge was considered on the embankment top. Results of the numerical modeling yielded good agreement with European guidelines.

Most of the past studies investigated soil arching under the static loading condition on the piled embankment. Due to traffic loading and seismic excitation, the complex behavior of soil arching in the pile-supported railway embankment is largely unknown. Heitz et al. [6] reported model tests to investigate the vertical stress distribution on the pile and subsoil under the cyclic loading. Lehn et al. [7] conduced a three-dimensional (3D) numerical simulation under the cyclic loading and observed that the soil arching was stable up to 50 cycles of load. As per the authors' best knowledge, there is no study yet to be reported which deal with the investigation of the seismic assessment of soil arching under a pile-supported railway embankment.

In the present book chapter, results of finite element analysis (FEA) focusing on the static and seismic assessment of soil arching in a pile-supported railway embankment are presented and discussed. The key input parameters of the piled embankments are also identified which are found to largely affect the mobilization of soil arching.

18.2 Finite Element Analysis

The finite element based commercial software ABAQUS 2018 [8] is used to assess the soil arching under the static loading and earthquake condition. The details of the FEA such as analyzed area, soil constitutive model, meshing (mesh size and element type), interaction, boundary condition, loading, and modeling procedure are illustrated in this section.

18.2.1 General Description

A series of FEA are carried out to assess the soil arching in the 2D plane strain condition. The 2D modeling requires less computational effort and time needed for simulation. The central part of a pile-supported railway embankment considered for the numerical simulation. The embankment height (h), pile spacing (s), and diameter (D) are chosen as 3.5–6.5 m, 2–3.5 m, and 1 m, respectively. Subsoil depth is chosen as 8 m, and the pile is socketed into the bedrock. A typical pile-supported railway embankment with modeled unit cell is shown in Fig. 18.1.

Various methods have been reported in the past which report the conversion of a 3D piled embankment into 2D, and it is argued that equivalent area (EA) method is in good agreement with the 3D model [4, 10]. Area replacement ratio of the pile to surrounding soil in both 3D and 2D is preserved in the EA method. However in other methods, pile wall thickness is kept same with equivalent elastic modulus considering the normal and flexural stiffness of pile. During the conversion of 3D to 2D, pile arrangement is assumed in square pattern (i.e., s_x and $s_y = s$). The principle of the EA method to convert a pile diameter into the equivalent wall is explained in Fig. 18.2.

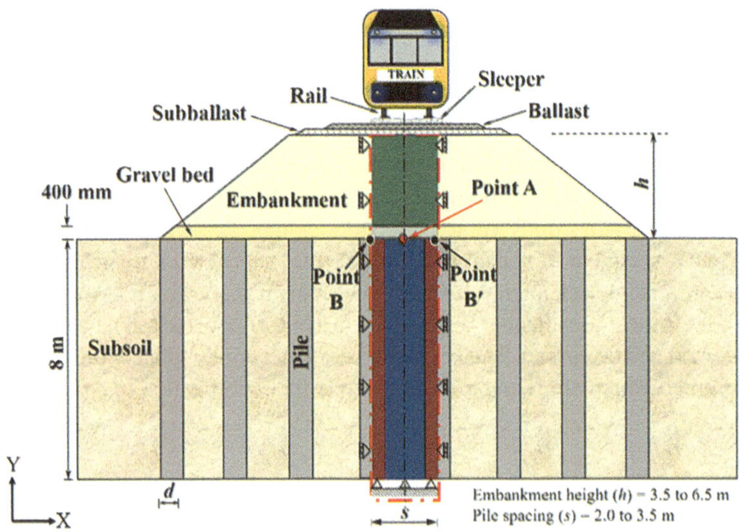

Fig. 18.1 Typical pile-supported railway embankment with modeled unit cell (modified from Meena et al. [9])

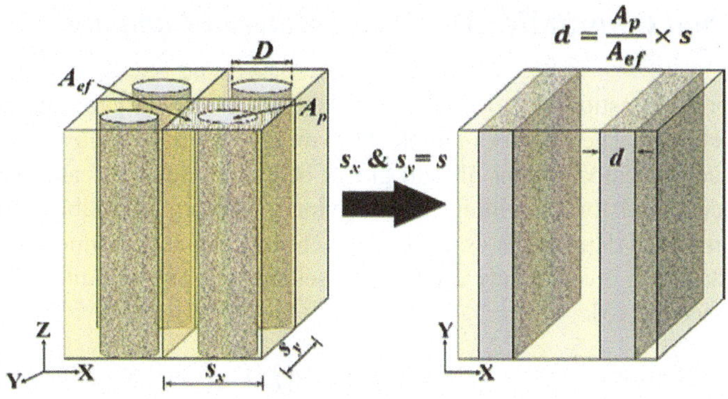

Fig. 18.2 Typical illustration of the principle of EA method (modified from Zhang et al. [10])

18.2.2 Boundary Condition, Meshing, Interaction and Seismic Input

Vertical boundaries of the numerical model are laterally restrained, while the base boundary is fully fixed. The adoption of infinite element boundary can reduce the wave reflection from the model boundaries during the seismic analysis. The FEA results can be affected by the element type and mesh size. The eight-node plane strain element with reduced integration (CPE8R) is used for this study. The four-node plane strain linear infinite element (CINPE4) is used for infinite boundary under the seismic analysis [11]. In addition, the interaction between the pile and surrounding soil is stimulated using basic coulomb friction model [12]. Based on the accuracy in results with less computation time, the mesh size is chosen.

The seismic input of Christchurch 2011 earthquake is used for the seismic assessment of the soil arching [13]. The magnitude on Richter scale is 6.3 M_L for 30 s time duration. The peak ground acceleration (PGA) is taken as 0.34 g. The seismic parameters of the Christchurch 2011 earthquake are similar with a few major Australian earthquakes as summarized in Table 18.1.

Table 18.1 Few major earthquakes in Australian territories (data sourced from [14])

Year	States of Australia	Earthquake magnitude (M_L)	Peak ground acceleration (PGA) (g)
2016	Northern Territory	6.1	0.49
1997	Western Australia	6.2	0.50
1979	Western Australia	6.1	0.20
1968	Western Australia	6.5	0.20
1941	Western Australia	6.3	0.20

18.2.3 Soil Constitutive Model and Material Damping

An appropriate constitutive model of soil in the FEA is necessary to simulate precise material behavior. The Mohr–Coulomb (MC) model is commonly used for granular material such as embankment fill and gravel bed. The modified cam clay (MCC) is well recognized for describing the behavior of soft compressible soil such as clay. The material damping is essential characteristics of the seismic analysis. The Rayleigh damping is commonly used for the seismic analysis in finite elements and it is expressed as [15]:

$$[C] = \alpha[M] + \beta[K^{'}] \tag{18.1}$$

where $[C]$ is damping matrix, $[M]$ is mass matrix, $[K']$ is stiffness matrix, α and β are the damping coefficients calculated as:

$$\alpha = 2 \times \left(\frac{\omega_i \cdot \omega_j}{\omega_i + \omega_j} \right) \times \xi_0 \tag{18.2}$$

$$\beta = \left(\frac{2}{\omega_i + \omega_j} \right) \times \xi_0 \tag{18.3}$$

The ω_i and ω_j are circular frequencies of the model, and ξ_0 is the damping ratio. The damping ratio is considered as 3%. Based on Eqs. 18.2 and 18.3, the damping coefficients α and β are calculated as 0.032 and 0.0017.

All the material properties are summarized in Table 18.2. This study is performed assuming the fully drained condition consequently the development of excess pore water pressure is neglected. For the subsoil simulated as MCC material, initial yield surface size (a_o) is calculated as [3]:

$$a_o = \frac{1}{2}\exp[\frac{(1+e_o)(e_1 - e_o.k.\ln p_o^{'})}{(-k)}] \tag{18.4}$$

where e_o is initial void ratio; e_1 is void ratio at unit pressure; k is logarithmic bulk modulus; p_o' is effective overburden pressure, and λ is logarithmic hardening constant.

18.2.4 Loading on the Embankment Top

In the present book chapter, an equivalent dynamic load is applied on the top of the modeled unit cell to mimic the load of rail track with moving train. The equivalent dynamic load is defined as [16, 17]:

Table 18.2 Material properties used in FEA [3, 9]

Material properties	Embankment fill	Gravel bed	Subsoil
Constitutive model	MC	MC	MCC
Unit weight, γ (kN/m^3)	20	21	19.7
Young's modulus, E (MPa)	20	25	–
Poisson's ratio, v	0.25	0.25	0.35
Effective cohesion c′ (kPa)	0.1	0.1	–
Effective friction angle, φ' (degree)	30	35	–
Effective dilation angle, Ψ (degree)	0	5	–
Critical-state stress ratio, M	–	–	1.2
Logarithmic hardening constant, λ	–	–	0.06
Logarithmic bulk modulus, k	–	–	0.012
Initial yield surface size, a_o (kPa)	–	–	103*
Void ratio at unit pressure, e_1	–	–	0.87
Initial void ratio, e_o	–	–	0.45

Note *Calculated from Eq. (18.4)

$$F_d = DAF.F_s \tag{18.5}$$

where F_d is design equivalent dynamic load (kN); DAF is dynamic amplification factor (dimensionless), and F_s is static wheel load (kN). Office of research and experiments (ORE) and 2:1 methods are adopted to calculate the DAF and equivalent dynamic load on the embankment top, respectively. In this study, heavy haul freight train with axle load of 35 tone moving at a speed of 40 km/h is considered. More details on the variation in train-induced dynamic loads against the nominal range of train speeds (40 to 160 km/h) are provided in Table 18.3.

Table 18.3 Train-induced equivalent dynamic load on the embankment top (data sourced from [9])

Train speed, V (km/h)	Static wheel load, F_s (kN)	Dynamic amplification factor	Equivalent dynamic load, F_d (kN)
40	81.23	1.29	105
80	81.65	1.31	107
120	80.94	1.36	110
160	81.17	1.45	118

18.2.5 Modeling Procedure

The FEA is carried out in three steps. First, initial stress and predefined void ratio are established in the subsoil using geostatic step. The rigid pile is activated in the next step, then the embankment including the gravel bed is constructed in the stages. After achieving full embankment height in 4 days, an equivalent dynamic load of railway track including the moving train load is applied on the embankment top. Subsequently, seismic excitation appropriate to the Christchurch 2011 earthquake is applied on the bottom of model adopting the dynamic implicit stepping scheme.

18.3 Results and Discussions

18.3.1 Vertical Stress Distribution in the Embankment Fill

The soil arching is assessed in terms of the vertical stress distribution from the embankment top to base at point A (refer Fig. 18.1). The effect of embankment height (h) and pile spacing (s) on vertical stress distribution in the embankment fill from top to point A is shown in Figs. 18.3 and 18.4, respectively. The vertical stress and embankment height are normalized for the sake of general applicability. The pile spacing (s) is taken 2.5 m to investigate the effect of embankment height on the soil arching. It is evident from Fig. 18.3 that normalized vertical stress ($N_{\mathrm{vs}} = \sigma_{\mathrm{s}}/\{\gamma.(s$

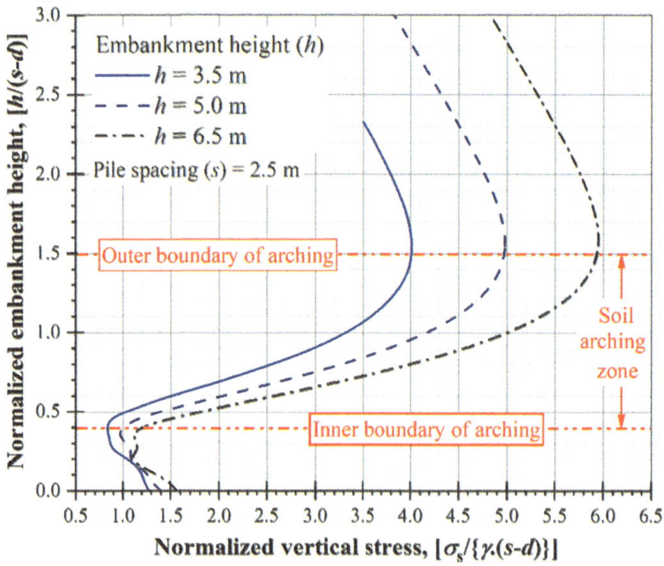

Fig. 18.3 Vertical stress over the point A in embankment fill for different embankment height

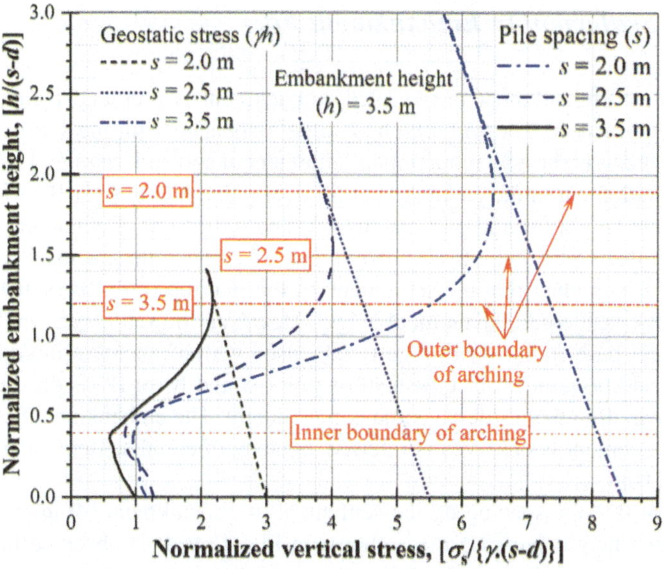

Fig. 18.4 Vertical stress over the point A in embankment fill for different pile spacing

$- d)\}$) follows the depth-wise linearly increasing trend of geostatic stress from the embankment top to the depth attaining 1.5 fold normalized embankment height ($N_{em} = h/(s - d)$). This embankment height represents the outer boundary of soil arching for all considered embankment height (i.e., 3.5, 5.0, and 6.5 m). Below the outer boundary, the N_{vs} shows decrease up to the $0.4 \times N_{em}$ which is represented as the inner boundary of soil arching. The vertical depth located between the upper and inner boundaries of soil arching can be defined as soil arching zone and the majority of vertical stress is transferred to the pile top (on point B or B′) at the inner boundary of soil arching. The N_{vs} again increases following the linearly increasing trend due to the self-weight of embankment fill underneath the inner boundary of soil arching. The same trend of vertical stress has been observed elsewhere in the literature [3]. It is worth noting that the soil arching zone is not effected by the embankment height.

In Fig. 18.4, the embankment height (h) is chosen 3.5 m, and it is evident that for the given range of pile spacing, the N_{vs} trend follows the same trend as illustrated in Fig. 18.3. The inner boundary is the same for all pile spacing. However, the upper boundary varies with pile spacing. The upper boundary lies on the $1.9 \times N_{em}$, $1.5 \times N_{em}$ and $1.3 \times N_{em}$ for the pile spacing of 2.0, 2.5 and 3.5 m, respectively. It implies from Fig. 18.4 that the soil arching zone increases with a decrease in pile spacing. Therefore, it is crucial to provide an optimum pile spacing, allowing for the installation tolerances of rigid piles.

18.3.2 Settlement in Embankment Fill

Settlement of embankment in the piled embankment is a crucial factor from the serviceability aspect. The soil arching can be associated with the serviceability of piled embankment through reduction in settlement of soft soil. Figures 18.5 and 18.6 show the settlement in the embankment fill over point A and B (or B′) for different embankment height (h) and pile spacing (s). The settlement of embankment fill for different embankment heights is shown in Fig. 18.5. The pile spacing is considered as 2.5 m. It is evident that the settlement on the pile top (point B or B′) is nearly zero for all considered embankment height. However, the settlement on the mid of embankment base (point A) is slightly increased with an increase in embankment height. In addition, uniform settlement on point A and B (or B′) is observed above the 1.9 × N_{em} for all considered embankment height. The embankment height when uniform settlement is observed is referred as plane of equal settlement (P_{es}) in the literature [18].

Effect of the pile spacing on the settlement of embankment fill under the fixed embankment height (i.e., 3.5 m) is shown in Fig. 18.6. It is observed that for the considered range of pile spacing, the settlement follows the same trend as Fig. 18.5. However, the P_{es} varies with the pile spacing. The P_{es} occurred at 2.4 × N_{em}, 1.9 × N_{em} for the pile spacing of 2 and 2.5 m, respectively. The P_{es} does not seem to exist for the pile spacing 3.5 m. From Figs. 18.4 and 18.6, it is concluded that the outer boundary of soil arching develops below the P_{es}.

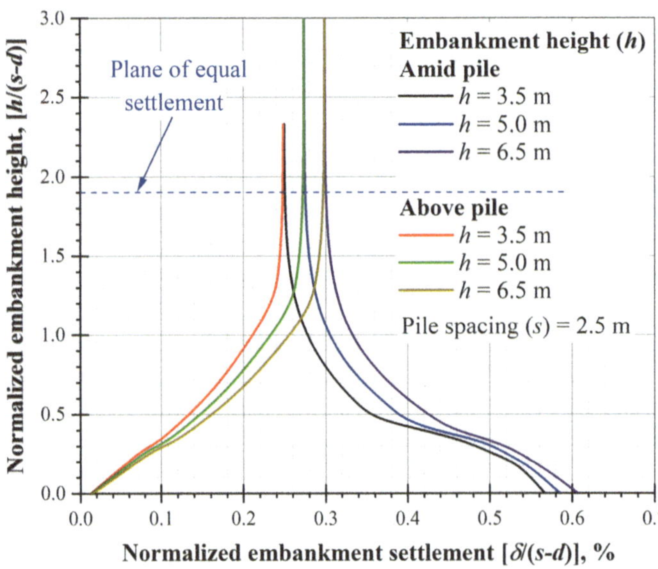

Fig. 18.5 Settlement over points A and B (or B′) in embankment fill for different embankment height

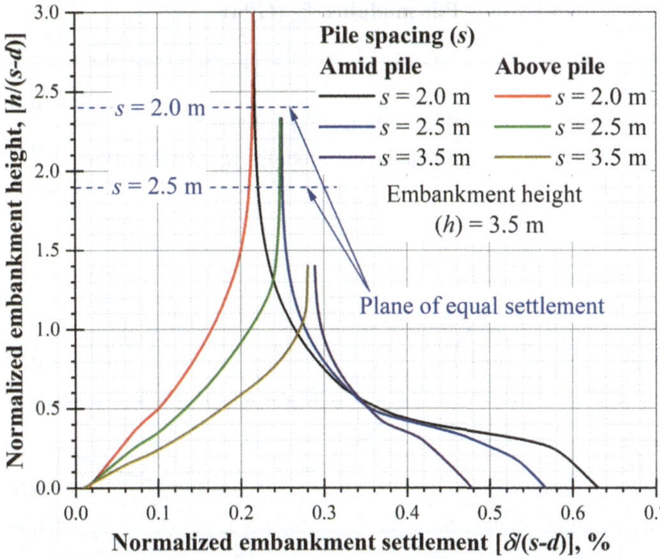

Fig. 18.6 Settlement over points A and B (or B′) in embankment fill for different pile spacing

18.3.3 Effect of Key Parameters of Piled Embankment on Soil Arching Ratio

The degree of soil arching in a piled embankment can be described by the soil arching ratio (SAR). The ratio of vertical stress on point A (σ_s) to geostatic stress including surcharge ($\gamma.h + q$) is known as soil arching ratio [2]:

$$\text{SAR} = \frac{\sigma_s}{(\gamma.h + q)} \tag{18.6}$$

where σ_s is vertical stress on point A; γ is unit weight of embankment fill; h is embankment height, and q is surcharge. The SAR lies in the range of 0 to 1, where 0 denotes full mobilization of soil arching and 1 represents no soil arching.

The effect of pile and embankment moduli on SAR is illustrated in Fig. 18.7. It shows that the SAR decrease up to 25% with an increase in embankment modulus (E_{em}) from 15 to 30 MPa. However, the pile modulus (E_p) has a negligible effect on the SAR due to higher modulus of elasticity of rigid pile. The E_p is considered fixed at 20 GPa for the SAR corresponding to the E_{em}. In contrast, the E_{em} is fixed at 20 MPa. The effect of E_{em} on the SAR is confirmed in Han and Gabr [2].

Figure 18.8 shows the effect of effective friction and dilation angle on SAR. It is observed that the SAR decreases up to 30% with an increase in effective friction angle (ϕ') from 30° to 45°. In addition, the SAR decreases up to 7% with an increase in dilation angle (Ψ) from 0° to 15°. The value of Ψ is fixed at 0° for the SAR corresponding to the varying ϕ'. In contrast, ϕ' is fixed at 30°. It implies that effective

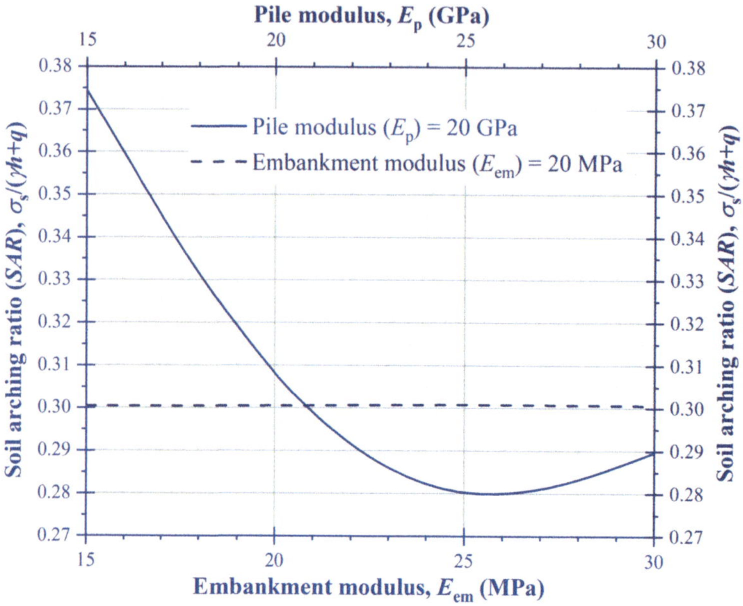

Fig. 18.7 Effect of pile and embankment modulus on the soil arching ratio

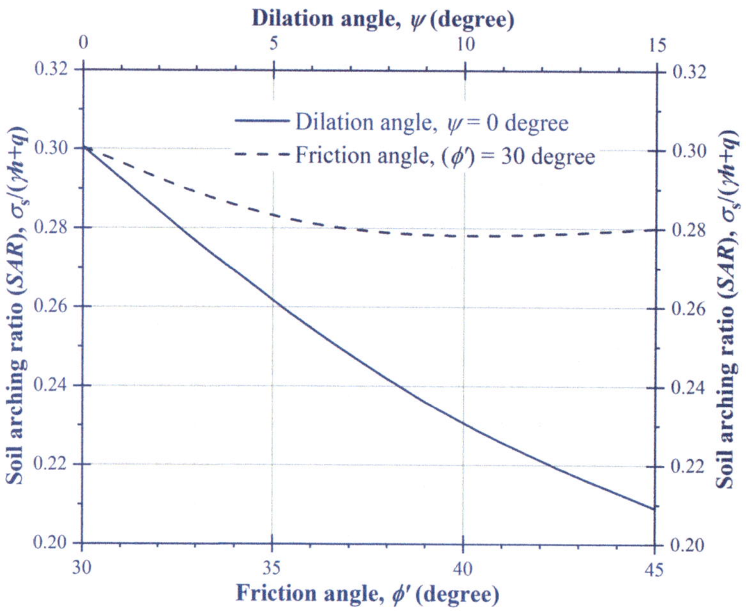

Fig. 18.8 Effect of friction and dilation angle on the soil arching ratio

friction angle of embankment fill is an important parameter compared to dilation angle. Thus, Figs. 18.7 and 18.8 conclude that the embankment modulus and friction angle is more prominent parameters of the piled embankment and higher values must be maintained for achieving full mobilization of the soil arching.

18.3.4 Effect of the Earthquake on Soil Arching

The piled embankment with 3.5 m embankment height and 2.5 m pile spacing is considered to study the effect of earthquake on soil arching. Figure 18.9 illustrates the N_{vs} distribution in the embankment fill over point A with and without earthquake condition. In the absence of earthquake, the N_{vs} distribution follows the same trend as shown in Fig. 18.3. The N_{vs} is consistent with the geostatic stress from the embankment top to 1.5 N_{em}. Subsequently it decreases up to $0.4 \times N_{em}$ and beneath this height, the N_{vs} again increases due to the self-weight of embankment fill. In contrast, the N_{vs} consistently increase from the embankment top to bottom during the earthquake. However, a marginal increment of the N_{vs} is observed on the pile top (point B or B′). It implies that soil arching is not mobilized properly under the earthquake condition.

Figure 18.10 shows the lateral stress coefficient (K) in embankment over the point A with and without earthquake. The lateral stress coefficient is the ratio of horizontal stress (σ_h) to vertical stress (σ_v). In the absence of earthquake, the lateral stress coefficient exists at rest ($K_o = 0.5$) after the normalized embankment height (N_{em})

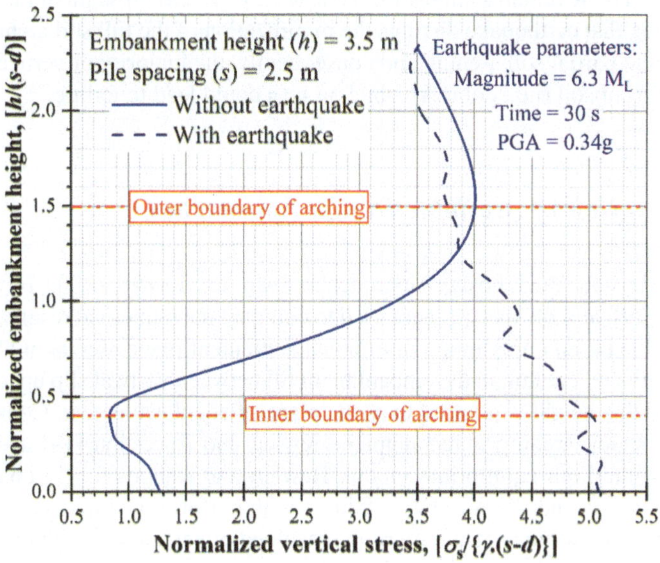

Fig. 18.9 Effect of earthquake on vertical stress over the point A in embankment fill

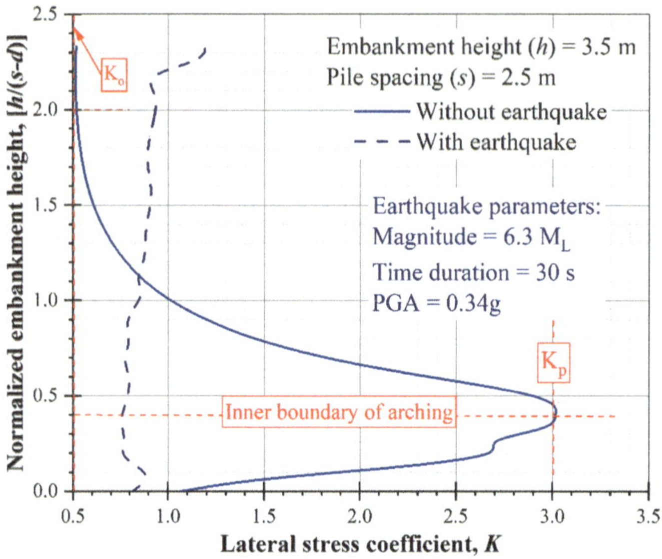

Fig. 18.10 Effect of earthquake on lateral stress coefficient over the point A in embankment fill

of 2.0. This embankment height is referred to as the plane of equal settlement as discussed earlier in Fig. 18.5. Subsequently, the K tends to approach the passive state ($K_p = 3.0$) at the N_{em} of 0.4, which represent the inner boundary of soil arching as explained in Fig. 18.3. In contrast, the K does not follow the same trend during the earthquake. The K remains almost the same (i.e., $K = 0.8$) throughout the N_{em}. It is worth noting that earthquake disrupts the mobilization of the full soil arching, while achieving the partial soil arching and consequently, the majority of vertical stress is imparted on subsoil (i.e., point A) which in turn could lead to failure.

18.3.5 Review of Available Design Approaches

Three well-established design approaches such as Terzaghi [1], Hewlett and Randolph [19], and BS8006 [20] are used to compare the numerical results. Terzaghi method [1] is based on assumption of equal settlement plane. Terzaghi [1] defined an imaginary soil column in the embankment fill between vertical slip up to a certain height, and this height is referred to as the plane of equal settlement. Consequently, shape of the soil arching is rectangular. Hewlett and Randolph [19] and BS8006 [20] follow the limit equilibrium model. Hewlett and Randolph [19] observed the semicircular arch and a hemispherical dome with uniform thickness corresponding to 2D and 3D conditions. The failure is presumed either on the crown of soil arch (i.e., semicircular arch and a hemispherical dome) or on the pile top. It is confirmed that the pile top is critical area in the plane strain condition while crown of soil arch is more critical in 3D condition [19]. The failure theory adopted in Hewlett and Randolph [19] is shown in Fig. 18.11.

The BS8006 [20] altered these empirical models using plane of equal settlement. The BS8006 [20] reported two conditions for soil arching; (i) partial aching, when 0.7 $(s\text{-}D) \leq h \leq 1.4$ $(s\text{-}D)$ and (ii) full arching, when $h \geq 1.4$ $(s\text{-}D)$. In this chapter, the range of embankment height (h) is taken as 3.5–6.5 m. Also, full soil arching formula [20] is considered.

Table 18.4 is illustrated the different equations used to estimate the SAR in these approaches. Figure 18.12 shows the variation in the results even for the same piled embankment condition. The effect of seismic excitation is not considered in these approaches. BS8006 [20] shows the higher value of the SAR followed by the Terzaghi

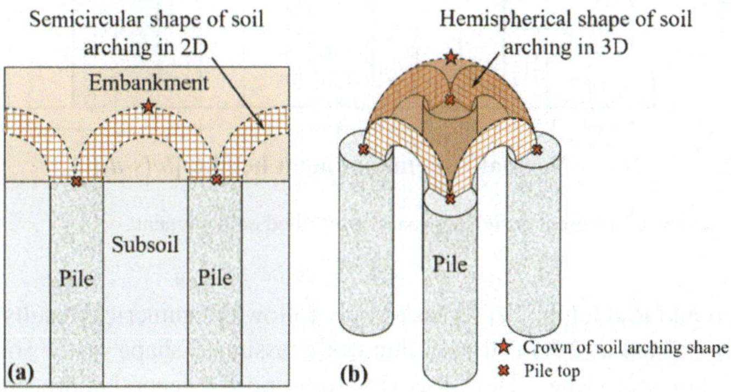

Fig. 18.11 Failure theory for **a** 2D plane strain and **b** 3D condition (modified from Hewlett and Randolph [19])

Table 18.4 Review of different design approaches

Design method	Soil arching ratio (SAR)
Terzaghi [1]	$\mathrm{SAR} = \frac{(s^2-d^2)}{4.h.d.K.\tan\phi'}\left(1 - e^{\left(\frac{-4.h.d.K.\tan\phi'}{s^2-d^2}\right)}\right)$ where $K = (1 - \sin\phi')$
Hewlett and Randolph [19]	*At the pile top*
	$\mathrm{SAR} = \left(\frac{1}{\left(\frac{2K_p}{K_p+1}\right)\left[\left(1-\frac{d}{s}\right)^{(1-K_p)}-\left(1-\frac{d}{s}\right)\left(1+\frac{d}{s}K_p\right)\right]+\left(1-\frac{d^2}{s^2}\right)}\right)$ where, $K_p = \frac{(1+\sin')}{(1-\sin')}$
BS8006 [20]	*For full arching*
	$\mathrm{SAR} = \frac{2.8s}{(s+d)^2h}\left[s^2 - d^2\left(\frac{\sigma_p}{\gamma h}\right)\right]$ where, σ_p is vertical stress on the pile

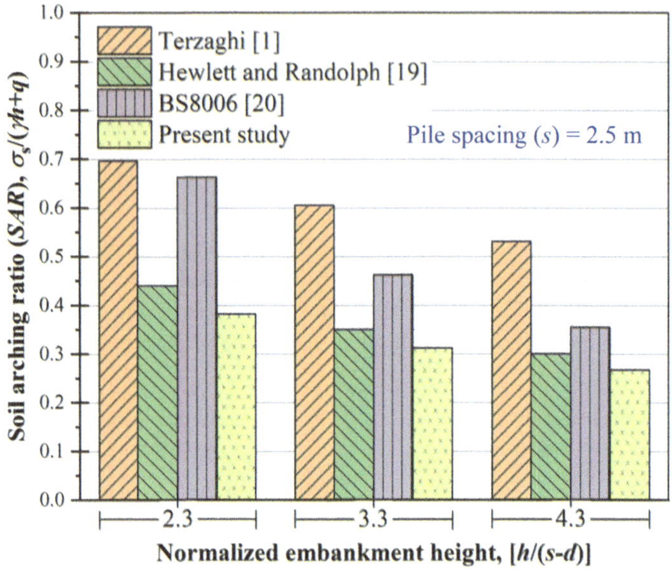

Fig. 18.12 Review of different design approaches for piled embankment

[1]. Hewlett and Randolph [19] shows closely follow the numerical results. The pile spacing (s) is fixed at 2.5 m. It may due to the assumed shape of the soil arching in these design approaches. Thus, it is vital to propose a general approach for piled embankment design.

18.4 Practical Implications

Followings are the practical implications of this study:

(i) It is evident that soil arching is associated with the minimum embankment height. In this study, the minimum embankment height is associated with the clear pile spacing ($s - d$). However, results also infer that piled embankment parameters have a significant effect on the soil arching. FEA results show that embankment modulus and effective soil friction angle are the key parameters for the mobilization of soil arching. The relationship between soil arching ratio (SAR), embankment modulus (E_{em}), and effective friction angle (ϕ') is given as:

$$\text{SAR} = a.(E_{em})^2 - b.(E_{em}) + c \quad (18.7)$$

$$\text{SAR} = a'.\left(\phi'\right)^2 - b'.\phi' + c' \quad (18.8)$$

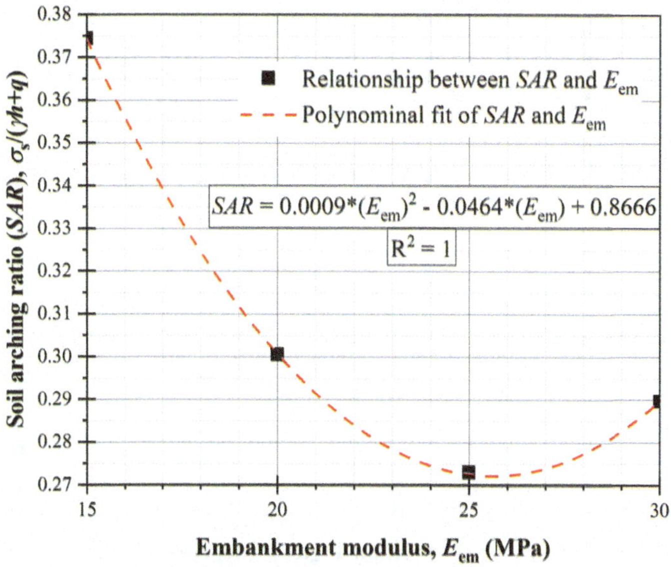

Fig. 18.13 Relationship between SAR and E_{em}

where a, b, c, a', b', and c' are empirical parameters and their values are determined as 9×10^{-4}, 4.64×10^{-2}, 8.67×10^{-1}, 2×10^{-4}, 2.16×10^{-2}, and 7.63×10^{-1}, respectively. The graphical representation of these relationships is shown in Figs. 18.13 and 18.14.

(ii) This study reports the effect of seismic excitation on the soil arching which is a crucial aspect for the piled embankment design in earthquake-prone areas, and there is need of a general design approached for the piled embankment. The mobilization partial soil arching, if not at all, is evident during seismic excitation as revealed from the numerical investigations.

Therefore, this study has practical implications to identify the key parameters of piled embankment. In addition, seismic assessment of soil arching may lead to prevent the potential embankment failure in earthquake-prone area.

18.5 Summary

In this study, static and seismic assessment of the soil arching mechanism in a piled embankment has been investigated under the plane strain condition. Based on the results and discussions, following important findings may be summarized:

(i) The pile spacing significantly influences to the soil arching zone. The soil arching zone expands with less pile spacing. Thus, the pile spacing should be optimized to achieve the full use of soil arching.

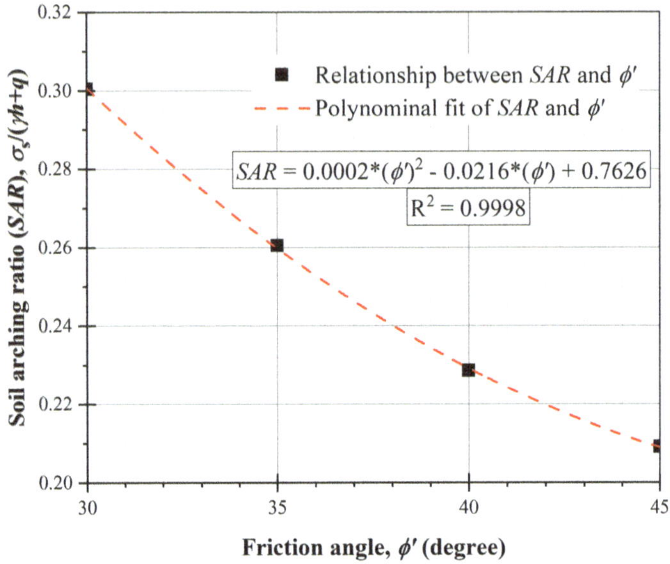

Fig. 18.14 Relationship between SAR and ϕ'

(ii) The embankment modulus and effective soil friction angle are the key parameters for soil arching. The soil arching increases with an increase in embankment modulus and effective friction angle. Thus, the values of embankment modulus and effective friction angle should be maintained large enough to attain the full mobilization of soil arching.

(iii) The seismic excitation significantly affects the soil arching which may lead to failure of piled embankment. Thus, the seismic assessment of soil arching should be considered in the design of piled embankment.

(iv) The comparison with different design approaches reveals that current design approaches need to be improved including the seismic considerations in the piled embankment design.

References

1. Terzaghi K (1943) Theoretical soil mechanics. Johnwiley & sons. New York, pp 11–15
2. Han J, Gabr MA (2002) Numerical analysis of geosynthetic-reinforced and pile-supported earth platforms over soft soil. J Geotech Geoenviron Eng 128(1):44–53
3. Liu HL, Ng CW, Fei K (2007) Performance of a geogrid-reinforced and pile-supported highway embankment over soft clay: case study. J Geotech Geoenviron Eng 133(12):1483–1493
4. Wu L, Jiang G, Ju N (2019) Behavior and numerical evaluation of cement-fly ash-gravel pile-supported embankments over completely decomposed granite soils. Int J Geomech 19(6):04019048

5. Almeida MSS, Fagundes DF, Thorel L, Blanc M (2020) Geosynthetic-reinforced pile-embankments: numerical, analytical and centrifuge modelling. Geosynth Int 27(3):301–314
6. Heitz C, Lüking J, Kempfert HG (2008) Geosynthetic reinforced and pile supported embankments under static and cyclic loading. Strain, 1:1–5
7. Lehn J, Moormann C, Aschrafi J (2016) Numerical investigations on the load distribution over the geogrid of a basal reinforced piled embankment under cyclic loading. Procedia Eng 143:435–444
8. Dassault Systèmes (2018) ABAQUS® User Manual (2018), SIMULIA, a division of Dassault Systèmes, Providence, RI, USA
9. Meena NK, Nimbalkar S, Fatahi B, Yang G (2020) Effects of soil arching on behavior of pile-supported railway embankment: 2D FEM approach. Comput Geotech 123:103601
10. Zhang Z, Han J, Ye G (2014) Numerical investigation on factors for deep-seated slope stability of stone column-supported embankments over soft clay. Eng Geol 168:104–113
11. Pham HV, Dias D (2019) 3D numerical modeling of a piled embankment under cyclic loading. Int J Geomech 19(4):04019010
12. Potyondy JG (1961) Skin friction between various soils and construction materials. Geotechnique 11(4):339–353
13. Meena NK, Nimbalkar S (2019) Effect of water drawdown and dynamic loads on piled raft: two-dimensional finite element approach. Infrastructures 4(4):75
14. USGS Earthquake Hazards Program, https://earthquake.usgs.gov/earthquakes/search/, Last accessed 06/01/2021
15. Nimbalkar S, Indraratna B, Dash SK, Christie D (2012) Improved performance of railway ballast under impact loads using shock mats. J Geotech Geoenviron Eng 138(3):281–294
16. Nimbalkar S, Indraratna B (2016) Improved performance of ballasted rail track using geosynthetics and rubber shockmat. J Geotech Geoenviron Eng 142(8):04016031
17. Doyle NF (1980) Railway track design: a review of current practice. Australian Government Publishing Service, Canberra
18. Jenck O, Dias D, Kastner R (2007) Two-dimensional physical and numerical modeling of a pile-supported earth platform over soft soil. J Geotech Geoenviron Eng 133(3):295–305
19. Hewlett WJ, Randolph MF (1988) Analysis of piled embankments. Int J Rock Mechan Min Sci Geomech Abstracts 25(6):297–298
20. BIS (2010) BS8006-1: (2010): Code of practice for strengthened/reinforced soils and other fills. Design of embankments with reinforced soil foundations on poor ground

Chapter 19
RETRACTED CHAPTER: Vulnerability of Interspersed Railway Tracks Exposed to Flood and Washaway Conditions

Sakdirat Kaewunruen, Yuki Nishinomiya, and Mitsuru Hosoda

19.1 Introduction

In practice, locally sourced materials are commonly chosen for construction and maintenance of railway lines such as steel rails, sleepers, fasteners, ballast, formation (capping layer over compacted soil), subgrade and foundation. Generally, train and track interactions usually induce the dynamic loading burdens acting on railway tracks stemming from either commuter and/or freight trains. These can incite dynamic behaviours (which are the amplified phenomena above simple static behaviour) of a compliant railway track. The dynamic behaviour is dominant and can be noticed when a train travels above 60 km/h. It is crucial to consider the track dynamic responses under various loading regimes [1] because irregular behaviours can lead to detrimental train derailments. It is important to note that the dynamic loading burdens from the large impact loads due to wheel/rail irregularities (e.g. wheel flats, out-of-round wheels, etc.) can cause structural cracks in brittle sleepers, densify and pulverise ballast support. For instance, a traditional transient waveform pattern of wheel impacts due to a dipped joint can be seen in Fig. 19.1. Apparently, the amplitude of the impact loads can vary from 200 to 400 kN while the duration can be from 2 to 10 ms. Based on a transient pulse concept (i.e. Duhamel's integral), these impact

---

The original version of this chapter was retracted. The retraction note to this chapter is available at https://doi.org/10.1007/978-981-16-5312-4_32

S. Kaewunruen (✉)
Department of Civil Engineering, University of Birmingham, Birmingham, UK
e-mail: s.kaewunruen@bham.ac.uk

Y. Nishinomiya · M. Hosoda
Railway Technical Research Institute, Tokyo, Japan
e-mail: nishinomiya.yuki.78@rtri.or.jp

M. Hosoda
e-mail: hosoda.mitsuru.75@rtri.or.jp

S. Kolathayar et al. (eds.), *Civil Engineering for Disaster Risk Reduction*, Springer Tracts in Civil Engineering, https://doi.org/10.1007/978-981-16-5312-4_19

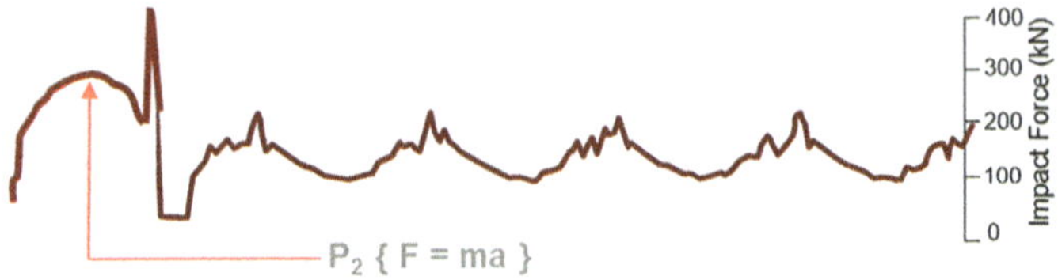

Fig. 19.1 Example of dynamic impact loading pattern

pulses are equivalent to the dynamic excitations with a frequency range from 100 to 500 Hz ($f = 1/T$: f is the frequency and T is the period). Note that this frequency range can incite the resonances of track components and result in pre-mature damage and reduction of the durability and service lives of track components. In reality, wheel/rail interaction generates dynamic forces acting on rails and rail seats. Noting that the dynamic load patterns depend on train speed, track geometry, axle load, vehicle type, and wheel/rail defects or irregularities. Railway and track engineers must be aware of the frequency ranges of static and dynamic loadings in order to plan and realise the life cycle asset maintenance and management of railway tracks with respect to critical train speeds and bespoke operational parameters [1–10].

Timber sleepers have been originally installed on railway track systems around the world, especially in North America, Africa and to certain extent in Europe, Australia and Asia. Their life cycle is approximated to be around 10–15 years subject to their usage applications, service exposures, operation parameters, environmental factors and the level of maintenance quality. After certain time, the deteriorated timber sleepers require renewals to ensure safety in operations. Partial replacement or spot replacement of timber sleepers by new pre-stressed concrete sleepers is a concept that has been adopted globally. This temporary solution is to assure track quality and enhance short-term mitigation that could be agile, inexpensive, effective and relatively fast. This type of spot replacement is usually implemented on the second- or third-class timber tracks or in some countries on the first-class main line. This technique is referred to as "interspersed tracks". In general, restricted train speeds are regularly used when the deteriorated condition of track is under the base operation conditions (BOCs), or a reasonably safe condition. By using the interspersed technique, a full operational speed can still be attained. Moreover, this approach improves the ability to withstand high train speed operations or to control longitudinal rail forces preventing a track buckling [9–11].

It is understood that the spot replacement of aged, rotten timber sleepers is relatively more economical than a complete track renewal or reconstruction. However, the interspersed track imposes some risks and disadvantages. Generally, the degraded timber sleepers will be removed and then the new stiffer concrete sleepers will be installed onto old and weakened foundation. Note that the foundation tends to be in services for a very long time without major reconstruction. In fact, local track stiffness of the renewed track with spot concrete sleepers tends to be inconsistent as the existing and ageing timber tends to degrade too. The inconsistency of track stiffness and the fluctuating track decay rate can be a reason of uneven settlement and foundation failure [9–13]. Considering the differential track stiffness, deterioration processes, track component durability and operational parameters, many patterns of interspersed railway tracks have been introduced such as 1 in 2, 1 in 3, 1 in 4 and so on (note: this means that there is 1 concrete sleeper in every indicated number of sleeper; for instance, 1 in 4 refers to 1 concrete sleeper in every 4 timber sleepers including the concrete itself). It is crucial to note that this type of railway track mainly exists in a rail network with low operational train speeds. One in four interspersed track is usually seen in practical adoptions and will be the prime focus in this study. A key reason is that this type of track has various flaws derived from how it is built. The long-term performance of interspersed railway tracks can be impaired as shown in Fig. 19.2 [13]. Figure 19.2 shows the example of interspersed railway tracks in low-speed operation (< 25 km/h). Note that the tracks in Fig. 19.2 have been commissioned between 2006 and 2008 and have served as a main high-speed link to the maintenance junctions.

Serviceability state of a railway track is one of the governing criteria for sleepers manufactured using different material types in the existing aged track systems. It is noteworthy that a general recommendation (e.g. by Australian Office of Transport Safety Investigations) is to carry out concrete sleeper installation only 'in-face' (i.e. the practice of installing the same sleeper type continuously rather than interspersed with other sleepers in between, also referred to as 'on face') [11–13]. The in-face method is generally advised to improve vulnerability of the track systems. However, cost and time constraints discourage the in-face installation. Many railway companies have employed on-face installation (spot replacement of concrete sleepers) to support operational demands without disruption from deteriorations of track components.

Fig. 19.2 Example of one in four interspersed tracks (one concrete sleeper after three timber sleepers—a set of four)

In addition, complexities of climate change and extreme weather conditions have raised an essential concern of risks and uncertainties for railway operators. Extreme weather conditions can significantly affect railway operations and safety, such as fatalities, injuries and property damage. Note that climate change and extreme weather conditions incur serious challenges to infrastructure systems. Nonetheless, most research (over 200 journal articles annually) have been focussed only on the development of high-level holistic frameworks for risk reduction, crisis responses, systems resilience and top-down infrastructure management. There is very little research that has been conducted to understand the true capacity to identify vulnerability to the transport infrastructures or to implement real actions to prevent and recover the natural crisis. It is widely recognised that there is an urgent need to integrate bottom-up consideration of climate change, its vulnerability, its structural integrity and its extreme weather impacts in policies, design, maintenance and reconstruction of infrastructure systems. In practice, everyday decision makings do not consider the consequences that could affect the new assets and infrastructures in the future. On this ground, this study is essential for railway owners, managers, maintainers and regulators in order to embrace real insights into climate change adaptation and resilience-based measures that mitigate the risks and uncertainty derived from extreme climatic conditions. For instance, the climate in South East Asia (such as in Thailand, Indonesia, Malaysia, Vietnam, etc.) is dominated by two monsoon regimes, namely as northeast monsoon and southwest monsoon. The northeast monsoon occurs during December, January and February, and the period frequently possess the most flooding conditions. Being in the equatorial zone and tropical country, the mean temperature throughout the year is constantly high (e.g. 26 °C) and has a very high humidity due to the high temperature. As a case study, Malaysia, for example, also can have a very heavy rainfall season, which can be more than 2500 mm per year. It is clear that one of the most detrimental natural disasters experienced in many continents (e.g. Europe, Asia, Africa, etc.) are floods and their consequential landslides, as illustrated in Fig. 19.3. These conditions can soften the soil formation underneath the tracks, and can also cause washaway when the ballast under the sleepers have been removed by rainfalls and runoffs. This study will thus pay special attention to the risks associated with heavy rainfall and flood.

This paper aims at investigating the vulnerability of the interspersed railway tracks exposed to flooding conditions. Dynamic responses in terms of displacements and accelerations of the interspersed railway tracks under moving train loads will be considered as the precursor to identify the level of serviceability. Based on critical literature review, this research has never been presented in open literature [14–21]. A group of two-dimensional interspersed track models in three-dimensional space was created using Timoshenko beams in a finite element package, STRAND7. Tensionless ballast property has been implemented. The dynamic displacements have been evaluated to understand the geometric behaviours of rail over sleeper, rail at midspan, cross level and twists. The insight into the interspersed track vulnerability will help rail track engineers to manage risks and uncertainty due to flooding conditions and to enable a truly predictive maintenance and improve the reliability of infrastructure asset maintenance and management.

Fig. 19.3 Washway of railway tracks occurred in Malaysia East Coast Line railway bridge, which cross Nenggiri River in Kemubu, Kelantan, had totally lost due to massive flood in December 2014 (*Courtesy* Malaysian Department of Public Works)

19.2 Methodology and Data

19.2.1 Track Modelling

In this study, non-linear finite element models of interspersed tracks have been adopted from previous studies [21, 22]. The models were established and validated using field measurement data. In the models, a two-dimensional Timoshenko beam model has been constructed in three-dimensional space. Timoshenko beam was found to be one of the most suitable options to simulate rails and concrete sleepers [21]. Using the numerical and experimental modal parameters [22], the finite element models can be fully verified. Figure 19.4 displays the finite element models for an in situ 1:4 interspersed railway track with different types of sleepers. Using a general-purpose finite element software STRAND7, the beam elements can take into account fully shear and flexural deformations, to appropriately represent the sleepers and rails. Note that each sleeper model consists of 60 beam elements and each rail model consists of 200 beam elements. The 60 kg rail cross section and sectional parameters (area: 17,789.9 mm^2; second moment of area: 43.2 × 106 mm^4) have been assigned [21]. The trapezoidal cross-section has been applied to the concrete sleeper elements in accordance with the standard medium duty sleepers (204 mm top-wide × 250 mm bottom-wide × 180 mm deep) [22]. The rectangular cross-section has been applied to the timber sleeper elements in accordance with the standard timber sleepers (230 mm wide × 130 mm deep) used in Australia [22]. The rail pads at rail seats have been

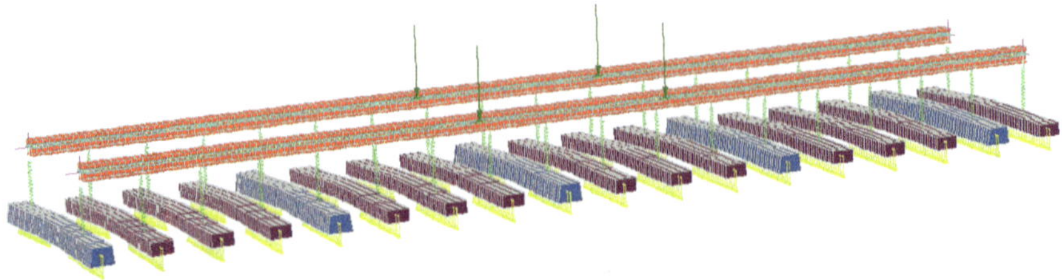

Fig. 19.4 Validated 1:4 interspersed track model (blue: concrete sleepers; and purple: timber sleepers). The model is subjected to a moving train axle (two-wheel sets)

modelled using a series of spring-dashpot elements. The novelty in this study is a more rational representation of the support condition, which has been modelled using the non-linear tensionless beam support feature in STRAND7. This attribute enables the beam to lift over the support while the tensile supporting stiffness is omitted. This feature is critical especially when the support is deteriorated unsymmetrically. The tensionless support option is more appropriate to correctly simulate the ballast characteristics in a railway tracks in real life [21].

19.2.2 Engineering Properties

Engineering properties of each element used in the analyses are given in Table 19.1, which show the geometrical and material properties of the finite element model. All dimensions are given in millimetres. This study adopts the partial support condition, which has been reported to be more suitable for standard gauge tracks. Spring—dashpot model has been used to simulate rail pads. Four separated forces with a constant magnitude of 100 kN have been used to imitate the loading condition of a passenger train bogie (2 per each rail, 2 m apart). Note that this load magnitude has been selected for benchmarking purpose [21–23]. The non-dimensional analyses have then been carried out to investigate the maximum dynamic responses in terms of displacements and cross level (inferring track twists) considering the effects of train speed and frequency domain.

Table 19.1 A summary of engineering parameters in the model

Parameters	Range	Unit		Remarks
Length	$l_r = 10.8$	m		Standard gauge is 1.435 m*
Gauge	$g = 1.5$	m		1.5 m is distance between wheel loads*
Modulus	$E_r = 205$	GPa		
Poisson's ratio	$v_r = 0.25$	–		
Rail pad stiffness	$k_p = 17$	MN/m		Obtained from dynamic modal testing*

19.2.3 Risk Exposures to Flood and Washaway Conditions

When a railway track is exposed to flood and washaway conditions, the formation strength and capacity will be undermined. The severity of strength reduction depends on the duration of rainfalls and runoffs. In most cases, when water ponding exists, total track inspection cannot be adequately conducted, making it a very dangerous situation to operate any train. In an event of heavy rainfall (e.g. 2 h continuously), a flash flood can incur. Any flash flood along railway corridors can weaken the formation, resulting in a very low to nil track modulus. The location with low level of terrains will often suffer this problem and sometimes lead to track mud pumping overtime. In practice, engineers may not be able to observe this problem until the severity and damage scale is large.

In a case that the gradient or vertical slope of railway tracks and corridor is steep, the runoffs can cause erosion of formation and cause ballast washaway. This event will completely eliminate the ballast and track formation that support the track systems. The severity of this incident depends on the volume and the speed of runoff and whether any water-borne debris exists. If the railway corridor has been properly designed (e.g. with a crossfall tapering towards the drainage), the ballast washaway might occur partially (e.g. only half of track support), but the scale of damage might be large (e.g. a large number of sleepers are effected). If the flood condition exists, rail engineers may not be able to observe the affected zone until major damages incur such as land slide, derailments, etc. For instance, land slip could also occur as illustrated in Fig. 19.5.

When a railway track is located in an inclined plane of terrain, cross water runoffs can also cause ballast washway as illustrated in Fig. 19.6. The cross flow can infiltrate the ballast and erode the ballast particles (and potentially formation), causing the ballast washaway (loss of track support), and eventually land slips. When the track system is exposed to a large area with ballast washaway, any operation of a train is reckless.

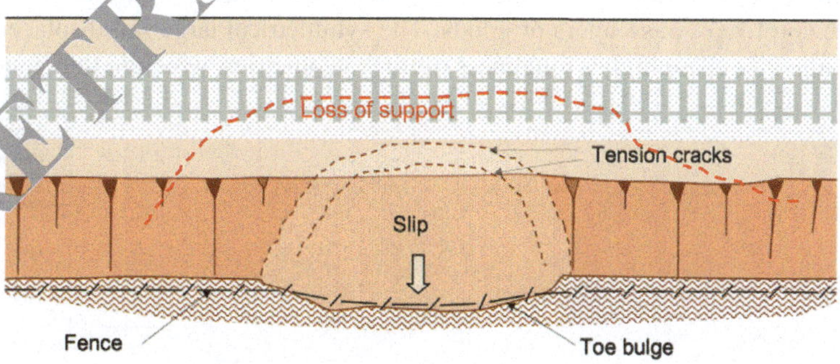

Fig. 19.5 Risks of heavy rainfalls and runoff and flood conditions. Initially, loss of track support will occur, followed by tension cracks and land slips. Track engineers are generally unable to observe or notice occurrences of the loss of track support

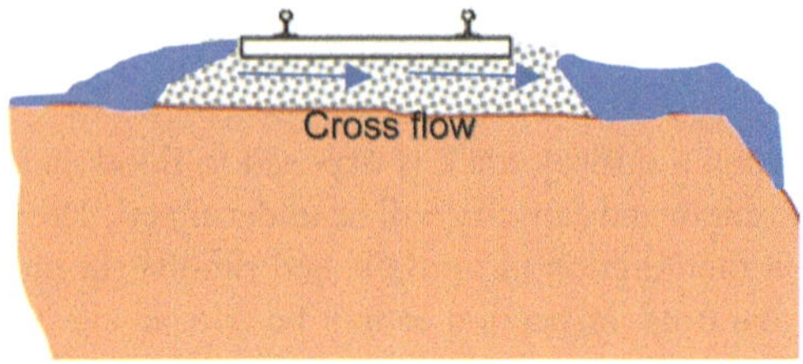

Fig. 19.6 Cross run-off causing ballast washaway

In this study, a special attention to the initial flood condition when it undermines the track support is considered. This is because, under this situation, engineers and operators cannot inspect the track and observe any problem. In some extent, a service train is operated on the flooded track systems. This study will identify the vulnerability and potential risks when the train services are exposed to such conditions. The emphasis is placed on the interspersed railway tracks since these interspersed methods are often adopted in vulnerable railway corridors and networks.

19.3 Results and Discussions

Based on the track models, the dynamic responses of the railway tracks (without any damage) under moving train loads can be seen in Fig. 19.7. It is clear that the train speed influences the dynamic displacements of the track systems. When the train speed increases, the dynamic displacement generally increases. The variance of the dynamic displacement can be observed and is because the dynamic properties or structural periods of track systems can respond differently to different excitation frequencies (i.e. $v = f \times \lambda$ or $f = 1/T$).

For the track systems with a good track support, the symmetry of dynamic displacements on both rails (left and right rails) can be observed. The movement of trains with large rail displacements on interspersed tracks would affect simply the ride comfort of passengers or goods. The symmetrical large rail displacements will commonly cause higher roughness of track geometries, which in turn generally

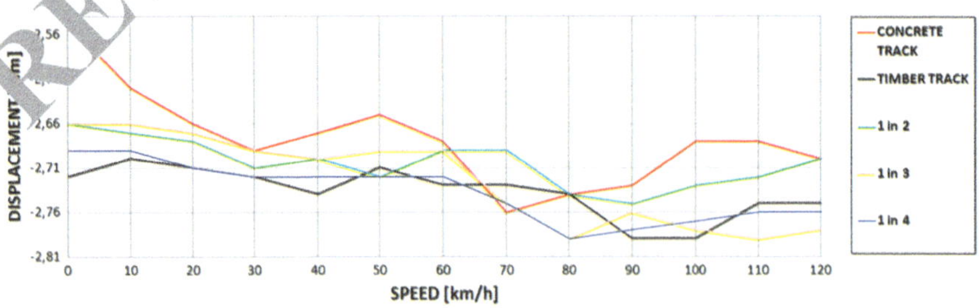

Fig. 19.7 Dynamic displacements of rails subjected to moving train loads (for track systems with a good track support condition)

induce higher vibrations (e.g. on-board vibration), louder noises (e.g. rolling noises) and poorer ride comfort.

The analyses into the vulnerability of the 1:4 interspersed track systems have been conducted in comparison with timber-sleepered track systems. Figure 19.8 illustrates the dynamic response envelopes of track systems exposed to small-scale and large-scale losses of support conditions. In this study, only half of sleeper support is considered for the effect of floods and washaway condition on the loss of support conditions as the case study.

It is clear from Fig. 19.8 that the train loads incur the difference in dynamic rail displacements on left and right rails. This difference at a position is often referred to as 'cross level'. When a train bogie or a train body travels over the different cross levels, the twists in the train body or bogie can incur. These twists can cause train derailments. The twist on train body is often called 'long twist', while the twist on train bogie is called 'short twist'. These twist limits can be illustrated in Fig. 19.9 (adopted from a maintenance standard of Transport for NSW, Australia). If the track twists reach E2 and E1, this situation is at danger and requires emergency actions. The train could derail when travel over E2/E1 conditions.

The dynamic twists of the interspersed track systems considering the losses of support conditions are shown in Table 19.2. The short twist is determined using 2 m cord, while the long twist is based on 14 m cord. The twist results have been correlated with the risk colours shown in Fig. 19.9 (green is normal, light blue is P2, dark blue is P1, yellow is E2, red is E1) .

From Table 19.2, it should be noted that N is normal condition; P3 is a situation needed to repair within 3 months; P2 is a situation needed to repair within 28 days; P1 is a situation needed to repair within 7 h; E2 is a situation needed to repair within 24 h; E1 is a situation needed to repair immediately. This implies that when the 1:4 interspersed track is exposed to large-scale loss of support condition, it could be very dangerous to operate a train above 40 km/h. In fact, it will still be at risk when a train travels at 20 km/h since the long twist defect could derail the train, especially when the train could also have certain defects (e.g. stiff bogies, deflated suspensions, etc.). On this ground, it is clear that rail operators should be very careful in train operations when the railway tracks become vulnerable due to flood and washaway conditions. In order to mitigate this issue, engineers should consider to apply ballast bond solutions to enable free drainage whilst reinforce the ballast particles [24, 25]. This insight will help track engineers develop appropriate climate change adaptation method and policy for operations of interspersed railway tracks facing extreme rainfall and flooding conditions.

19.4 Conclusion

This study identifies the vulnerability in the railway infrastructures exposed to flood and washaway conditions. This study is the world's first to determine the capability of operating trains over vulnerable track systems. A special track system, called

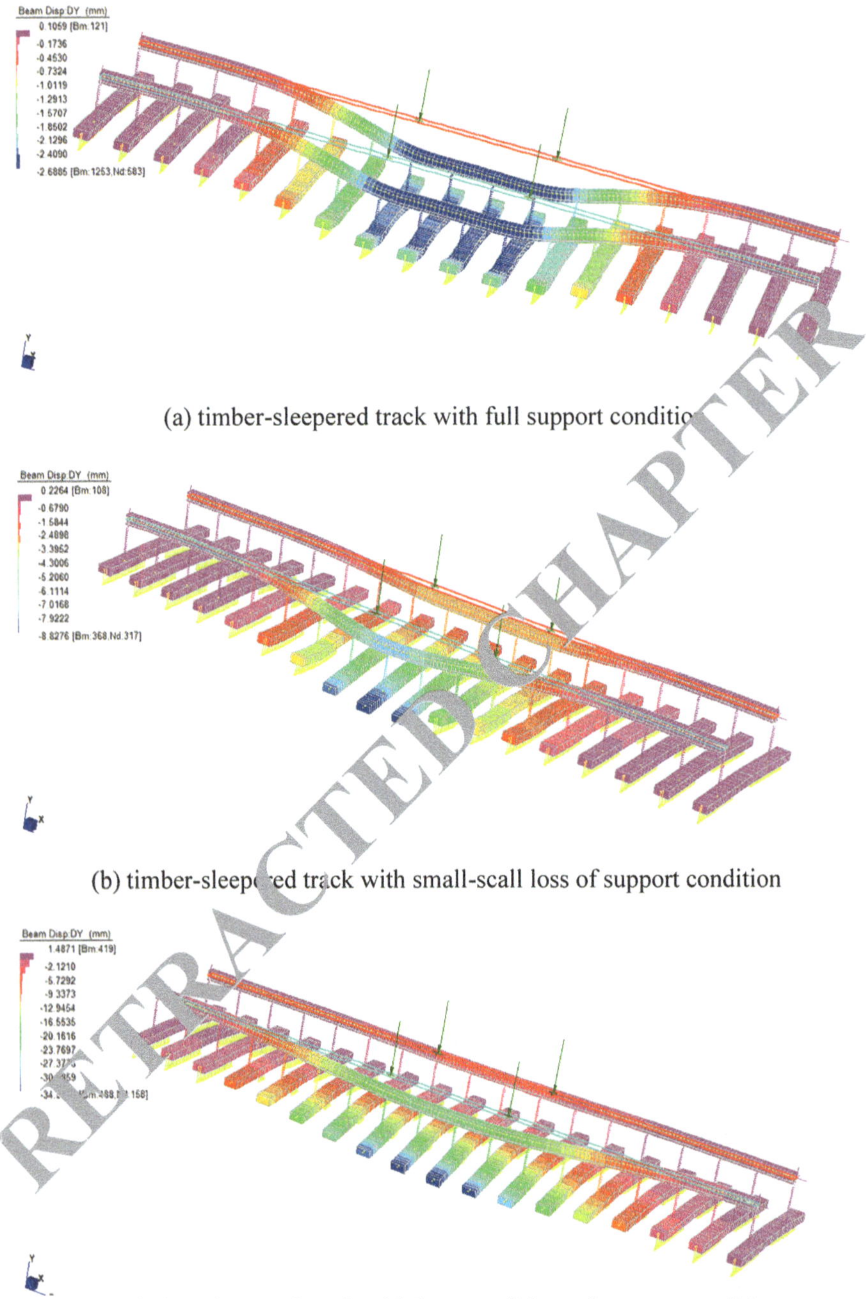

(a) timber-sleepered track with full support condition

(b) timber-sleepered track with small-scall loss of support condition

(c) timber-sleepered track with large-scall loss of support condition

Fig. 19.8 Dynamic responses to 120 km/h moving train loads of track systems exposed to flood and washaway conditions

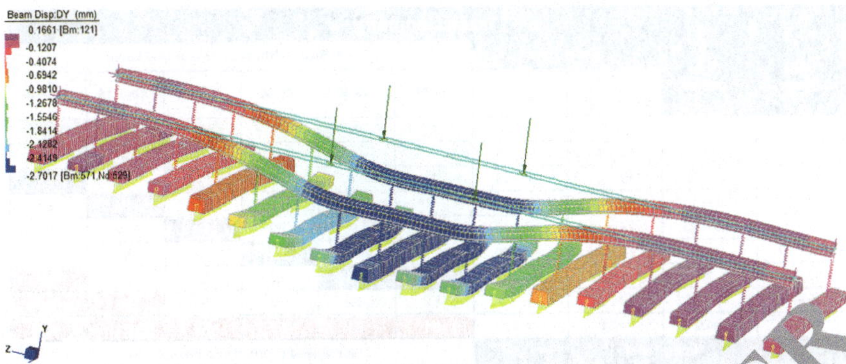

(d) 1:4 interspersed track with full support condition

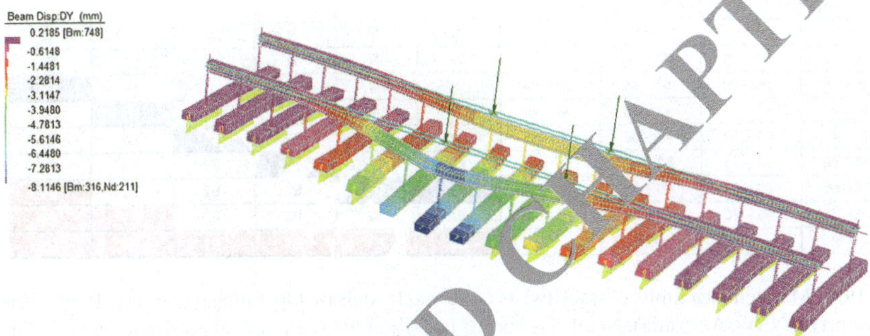

(e) 1:4 interspersed track with small-scall loss of support condition

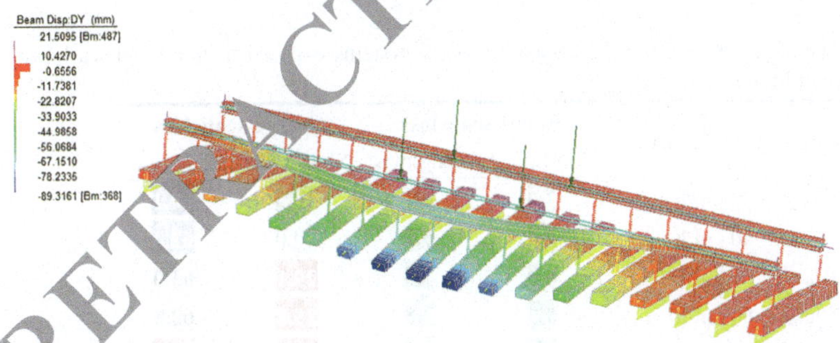

(f) 1:4 interspersed track with large-scall loss of support condition

Fig. 19.8 (continued)

Track Geometry									
Wide Gauge	Tight gauge	Short Twist		Track Speed (Normal / Passenger) km/hr					
		2.7m DO NOT USE	2m	20/20	40/40	60/60	80/90	100/115	115/160
<21	<10	<16	<12	N	N	N	N	N	N
21 – 22	10	16 – 18	12 – 13	N	N	N	N	P3	P2
23 – 26	11 - 12	19 – 21	14 - 15	N	N	N	P3	P2	P1
27 – 28	13 – 14	22 – 23	16	N	N	P3	P2	P1	E2
29 – 30	15 – 16	24 – 25	17 – 18	N	P3	P2	P1	E2	E2
31 – 32	17	26 – 27	19 – 20	P2	P2	P1	E2	E2	E2
33 – 34	18	28 – 29	21 – 22	P1	P1	E2	E2	E2	E1
35 – 37	19 – 20	30 – 31	23	E2	E2	E2	E2	E1	E1
>37	>20	>31	> 23	E1	E1	E1	E1	E1	E1
Long Twist				Track Speed (Normal / Passenger) km/hr					
Not in a Transition		In a Transition							
13.2m DO NOT USE	14m	13.2m DO NOT USE	14m	20/20	40/40	60/60	80/90	100/115	115/160
<29	<31	<32	<34	N	N	N	N	N	N
29 – 33	31 – 35	32 – 36	34 – 38	N	N	N	N	P3	P2
34 – 38	36 – 40	37 – 41	39 – 43	N	N	N	[illegible]	P2	P1
39 – 43	41 – 46	42 – 46	44 – 49	N	N	P3	P2	P1	E2
44 – 49	47 – 52	47 – 52	50 – 55	N	P3	P2	P1	E2	E2
50 – 56	53 – 59	53 – 59	56 – 62	P2	P2	P1	E2	E2	E2
57 – 60	60 – 64	60 – 63	63 – 66	P1	P1	E2	E2	E2	E1
61 – 66	65 – 70	64 – 69	67 – 72	E2	E2	E2	E2	E1	E1
>66	>70	>69	>72	E1	E1	E1	E1	E1	E1

Fig. 19.9 Maintenance limits of track twists (adapted from Base Operating Condition, BOC, from Transport for NSW, Australia). Note: N is normal condition; P3 is a situation needed to repair within 3 months; P2 is a situation needed to repair within 28 days; P1 is a situation needed to repair within 7 h; E2 is a situation needed to repair within 24 h; E1 is a situation needed to repair immediately. Courtesy: Transport for NSW

Table 19.2 Dynamic twists of 1:4 interspersed track systems exposed to flood and washaway conditions (unit in mm)

Train speed (km/h)	Small-scale loss		Large-scale loss	
	Short twist	Long twist	Short twist	Long twist
Class 1: 20 km/h	3.0	6.1	10.5	60.1
Class 2: 40 km/h	3.6	6.9	20.0	62.4
Class 3: 60 km/h	4.0	7.0	29.1	63.0
Class 4: 80 km/h	5.7	7.8	25.8	62.8
Class 5: 100 km/h	6.4	8.0	30.2	65.1
Class 6: 120 km/h	6.6	8.1	31.6	66.0

*Colour backgrounds are correlated with risk profiles defined in Fig. 19.9

the interspersed track, is used as case study. Non-linear finite element analyses of interspersed track systems have been established. A clear novelty in the model is the adoption of tensionless support condition that can mimic the actual ballast condition. It is very important to realistically simulate the actual ballast condition when the track is vulnerable and the asymmetric instabilities occur. This study considers the loss of support conditions as the consequence of flood and washaway conditions stemmed from extreme weather and climatic events.

The dynamic responses of the interspersed track systems exposed to the extreme weather events have demonstrated the vulnerability of the operations. By considering the risk profiles, the dynamic responses can be instrumental in identifying risks with respect to the operations and track conditions. Dynamic track twists can be derived and employed as the catalyst in vulnerability determination. It is clear that track conditions exposed to flood conditions cannot be easily determined from traditional inspections or observations by engineers, maintainers or operators. On this ground, it is at risk to operate a train over vulnerable track systems. Considering the 1:4 interspersed track systems, it is found that a train should not be operated above 40 km/h when it is suspected that the track suffers from flood and washaway conditions. In an emergency, a train might be able to travel at a low speed (e.g. less than 20 km/h) but vigilant monitoring and control is mandatory. Note that low speed trains could derail in a fail-safe situation if careful monitoring and control is set. However, in general, it is not advisable to operate a train over a vulnerable interspersed track, especially when there is no appropriate monitoring and control measures. A temporary solution to mitigate this issue has been proposed. When heavy rainfalls or extreme weather conditions (e.g. storm, hurricane or typhoon) are anticipated, engineers and maintainers should develop a solution to reinforce the support condition, for example, by using ballast bonding agents.

Acknowledgements The authors are sincerely grateful for the financial sponsorships to Japan Society for the Promotion of Science (JSPS) via Grant No. JSPS-L15701, BRIDGE Grant (Collaboration between University of Birmingham and University of Illinois at Urbana Champaign), and European Commission for H2020-MSCA-RISE Project No. 691135 "RISEN: Rail Infrastructure Systems Engineering Network", which enables a global research network that tackles the grand challenge in railway infrastructure resilience and advanced sensing in extreme conditions (www.risen2rail.eu). The valuable discussions and comments from ISO/BSI standard committee for railway sleepers (TC269 WG7 chaired by Dr Makoto Ishida and BSI WG by Neil Gofton) are gratefully acknowledged.

References

1. Remennikov AM, Kaewunruen S (2008) A review on loading conditions for railway track structures due to wheel and rail vertical interactions. Struct Control Health Monit 15(2):207–234
2. Remennikov AM, Kaewunruen S (2005) Determination of dynamic properties of rail pads using instrumented hammer impact technique. Acoust Australia 33(2):63–67

3. Kaewunruen S, Remennikov AM (2009) Influence of ballast conditions on flexural responses of railway concrete sleepers. Concrete Australia J Concrete Inst Australia 35(4):57–62
4. Kaewunruen S, Remennikov AM (2010) Dynamic properties of railway track and its components: Recent finding and future research directions. Insight—Non-Destruct Test Condition Monit 52(1):20–22
5. Ngamkhanong C, Kaewunruen S, Costa BJAJI (2018) State-of-the-art review of railway track resilience monitoring. 3(1): 3
6. Kaewunruen S, Remennikov AM, Akira A, Hirotaka S (2014) Free vibrations of interspersed railway track systems in three-dimensional space. Acoust Australia 42(1):20–26
7. Cai Z (1992) Modelling of rail track dynamics and wheel/rail interaction, Ph.D. Thesis, Department of Civil Engineering, Queen's University, Ontario, Canada
8. Kaewunruen S, Remennikov AM (2007) Effect of improper ballast packing/tamping on dynamic behaviours of on-track railway concrete sleeper. Int J Struct Stab Dyn 7(1):167–177
9. Kaewunruen S, Remennikov AM (2008) Nonlinear transient analysis of railway concrete sleepers in track systems. Int J Struct Stab Dyn 8(3):505–520
10. Kohoutek R (1991) Dynamic and static performance of interspersed railway track. Proc Conf Railway Eng 1:153–159
11. Lake M, Ferreira L, Murray MH (2002) Using simulation to evaluate rail sleeper replacement alternatives. Transp Res Rec 1785:58–63
12. Kaewunruen S, Remennikov AM (2006) Sensitivity analysis of free vibration characteristics of an in-situ railway concrete sleeper to variations of rail pad parameters. J Sound Vib 298(1–2):453–461
13. Kaewunruen S, Remennikov AM (2007) Investigation of free vibrations of voided concrete sleepers in railway track system. In: Proceedings of the Institution of Mechanical Engineers Part F Journal of Rail and Rapid Transit, vol 221, no 4, pp 495–508
14. Binti Sa'adin SL, Kaewunruen S, Jaroszweski D (2016) Operational readiness for climate change of Malaysia high-speed rail. In: Proceedings of the Institution of Civil Engineers-Transport. 2016. Thomas Telford Ltd.
15. Binti Sa'adin SL, Kaewunruen S, Jaroszweski DJC (2016) Risks of climate change with respect to the Singapore-Malaysia High Speed Rail System. Climate 4(4):65
16. Kaewunruen S et al (2018) Vulnerability of structural concrete to extreme climate variances. 6(2):40
17. Binti Sa'adin SL, Kaewunruen S, Jaroszweski D (2016) Heavy rainfall and flood vulnerability of Singapore-Malaysia high speed rail system. Australian J Civil Eng 14(2):123–131
18. Dindar S et al (2018) Bayesian Network-based probability analysis of train derailments caused by various extreme weather patterns on railway turnouts. Saf Sci 110:20–30
19. Kaewunruen S, Sussman JM, Matsumoto A (2016) Grand challenges in transportation and transit systems. 2(1)
20. Kaewunruen S, Sussman JM, Einstein HH (2015) Strategic framework to achieve carbon-efficient construction and maintenance of railway infrastructure systems. 3(6)
21. Kaewunruen S, Lewandrowski T, Chamniprasart K (2018) Dynamic responses of interspersed railway tracks to moving train loads. Int J Struct Stab Dyn 18(01):1850011
22. Kaewunruen S, Ngamkhanong C, Ng J (2019) Influence of time-dependent material degradation on life cycle serviceability of interspersed railway tracks due to moving train loads. Eng Struct 199:109625
23. Ngamkhanong C, Wey CM, Kaewunruen S (2020) Buckling analysis of interspersed railway tracks. Appl Sci 10(9):3091
24. Kaewunruen S (2014) Impact damage mechanism and mitigation by ballast bonding at railway bridge ends. Int J Railway Technol 3(4):1–22
25. Ngamkhanong C, Kaewunruen S, Baniotopoulos C (2021) Nonlinear buckling instabilities of interspersed railway tracks. Comput Struct 249:106516

Chapter 20
Modeling to Support Acceleration of Restoration of a Residential Building System in Southeastern B.C. Due to Riverine Flooding

David N. Bristow and Afia Siddika Ivy

20.1 Introduction

Floods are the most frequent and costliest types of natural hazard events around the globe [1]. In Canada, for example, flooding has been the single most damaging type of disaster in terms of economic and social losses since the beginning of the twentieth century [2]. A major source of impacts is the inundation of many buildings. Recovery of a flooded building is a complex process [3]. It requires a number of expert workers with knowledge on dealing with flood-damaged properties to work together. In addition, an inefficient recovery plan can prolong the recovery process, increasing secondary water damage.

Deniz et al. [1] explain that most of the available damage estimation models for buildings are location specific. Available models mostly focus on Europe and Japan with scarce coverage of North America. The multi-hazard software program Hazus [2, 4] is one of the few in North America and is most commonly used in the United States of America (US). However, it provides loss estimation based on assessed building value rather than actual replacement costs. To overcome this shortcoming, Deniz et al. [1] developed a model using building architectural layouts and inventories of building components for a 2013 Colorado flood and then assessed actual loss. The findings of these papers provide valuable information on damage estimation useful for improving understanding of recovery challenges.

Another valuable finding is the recovery timeline provided by NiyimIT Inc. [5]. In this technical report on the Hazus tsunami model, there are building recovery timelines. Both water depth damage and flow velocity damage to the building are obtained and damages are converted to direct economic losses. The report has a classification of the recovery requirements in three categories, namely, moderate

D. N. Bristow (✉) · A. S. Ivy
Cities and Infrastructure Systems Lab, Department of Civil Engineering, University of Victoria, Victoria, Canada
e-mail: dbristow@uvic.ca

S. Kolathayar et al. (eds.), *Civil Engineering for Disaster Risk Reduction*, Springer Tracts in Civil Engineering, https://doi.org/10.1007/978-981-16-5312-4_20

repair (associated with 5–25% monetary loss), extensive repair (associated with 25–100% monetary loss) and complete repair (associated with 100% monetary loss). However, this report does not base recovery time on a temporal recovery model for time estimation but rather assumes a series of default values of building recovery times to estimate repair and replacement costs.

Similarly, previous work reported in the Hazus flood technical manual [4] also assumes default values for recovery time. A group of experts provided scenarios regarding possible depths of flooding and associated building recovery steps (i.e., clean-up, physical restoration and political delays). The restoration timeline of buildings is generated following the pattern of the Hazus 2013 earthquake model [6]. Remaining recovery steps were included as add-ons (i.e., a constant value for any damage state) to further assist the building repair and replacement cost estimation. Although these expert judgments work as a reference point, they do not indicate potential optimizations in the recovery process.

Among recovery-related challenges, the complex nature of the recovery process is the most important one as recovery time estimation consists of several interacting and interdependent factors, such as the structural characteristics, hazard intensities and the available workforce [3]. This complexity is similar to the challenge in understanding recovery of infrastructure systems. He and Cha [7] state the importance of understanding the interdependencies within a civil infrastructure network as the key to post-disaster recovery planning. Very few recovery-related studies, however, can address the interdependencies residing in a system in a quantitative manner. He and Cha [7] are one of the few who estimate a timeline for recovery in their lifeline recovery model. They consider electric power, telecommunications and water supply as components of a system and use the national and regional commodity-transaction data to create a dynamic input output model. However, the model developed is a hypothetical infrastructure network model and the damage consideration was based on generic assumptions. An opportunity for improvement of this model is the usage of coefficients in the recovery time formulas which have yet to be adequately assessed.

Kammouh and Cimellaro [3] also offer an approach to address the complex nature of recovery modeling. They propose an empirical probabilistic model of restoring lifelines (e.g., power, water, gas and telecom) after an earthquake event. They derive fragility restoration functions based on datasets collected from a wide range of literature focusing on earthquake recovery over the past century. The aim was to introduce a simplified recovery time estimation method. However, the historic database model lacks the ability to create optimized recovery plans that is sensitive to differing resource constraints.

Another consequence of interdependency and complexity in the built environment is the potential for cascading effect of hazards among components. To avoid cascades, it is important for the damaged system to recover quickly or de-couple to limit additional failures. How restoration efforts should be directed is a multi-objective decision making problem that is not only subject to constraints (e.g., the efficient use of time, cost or other resource constraints) but also consideration of system characteristics (i.e., systems resilience). For example, Afrin and Yodo [8] outline an assessment methodology to determine the optimized recovery strategy given a particular network structure of a water distribution system.

For our present purposes at the buildings-scale, however, the literature on recovery models of buildings after a natural hazard is lacking in the following ways: first and foremost, there is a scarcity of empirical restoration data related to flood hazards in the Canadian context; secondarily, there is a need to undertake systems and networked recovery modeling that delineates and analyzes complex dependencies among components; and finally, there is an operational and economic benefit to implementing strategic optimization of the recovery pathways that has yet to be capitalized on. This paper addresses these gaps in the literature by contributing a computational recovery methodology that accounts for complex dependencies within the recovery process to inform system-wide risk analyses for disaster risk reduction and to optimize recovery efforts given workforce availability constraints. The context of this research is the recovery of residential single detached houses that experience flood damage. The aim is to determine a strategic solution for bringing back the building systems to their initial functional capacity in the shortest possible time to reduce further losses. The primary objectives consists of establishing a building restoration simulation method that addresses the complex recovery process.

In summary, a building is composed of systems and components that underlie the functioning of the building. Informed decision-making during damage recovery after a disastrous situation requires understanding the damage as well as probable restoration requirements for components. The novel contribution of this paper is the development and case study analysis of a systems methodology that can enable the computation of complex relationships among component repairs and resource requirements that can be further assessed based on time constraints to determine an optimized pathway.

20.2 Methodology

The methodology developed here uses the Graph Model for Operational Resilience (GMOR) [9]. GMOR is a computational recovery assessment engine that allows for the assessment of recovery times in interdependent systems. Our methodology is the construction of a new model for use in GMOR. GMOR models can delineate the recovery of individual system components which collectively maintain operational functions. Interactions and relations among components are modeled for recovery assessment instead of using aggregated component recovery data that has been a common practice for previous recovery studies as described in the introduction. Assessment of the individual state of entities and the system's overall state vector depends upon the computation of interactions among the entities after a shock and probable entity disruption. For instance, the state of a component is an initial condition that could be estimated based on the height of a component versus the local flood height. GMOR correlates restoration requirements with damage and computes when each component recovers. This capability makes the use of GMOR manageable

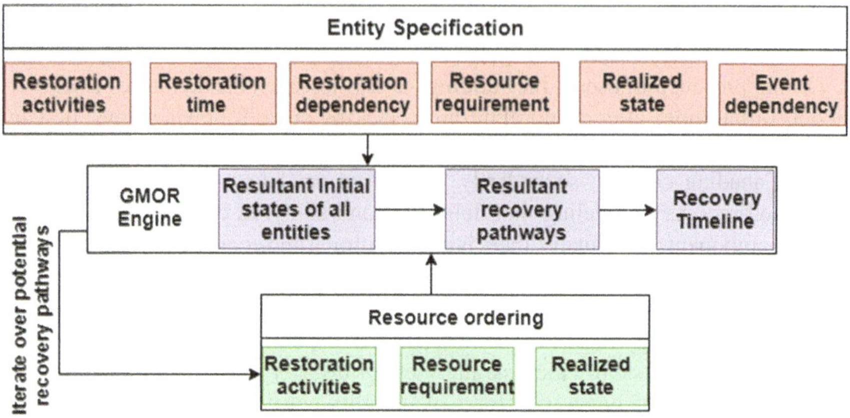

Fig. 20.1 Schematic depiction for determining recovery time of a single building using GMOR

for credibly sized and data-driven assessments. Therefore, it is feasible to perform rapid model creation for recovery time assessment of spatially-explicit systems for multiple types of damage. Moreover, GMOR models are used to analyze the dependencies involved in recovery of a function and are capable of quantifying the impact of the order of restoration steps on recovery [10].

The approach here combines a location-specific damage restoration time estimation method which is identified as a key to recovery design [1] with a means to simulate the complex recovery of a residential building. In particular, GMOR serves as the computational engine to calculate the recovery time. Finally, a permutation technique is used to determine the shortest possible time for recovery which involves the integration of a decision tree and GMOR as presented below. The decision tree enumerates of all possible resource orderings, and the resulting recovery times for these resource orderings are then computed in GMOR.

Figure 20.1 illustrates the methodology to be carried out. Here, the GMOR computational engine serves as the center of all these processes. The first step is to specify a data inventory that quantifies infrastructure component restorations. An entity is a general means to encapsulate all the data for a building component and serves as input to a GMOR model. Characterizing an entity involves specifying six variables:

(1) *restoration activities* refer to a list of discrete tasks such as demolition of a certain component, servicing or installation that would be required to bring the component back to functioning;
(2) *restoration dependency* would specify relationships of precedence among restoration activities;
(3) *restoration time* specifies the time estimates to complete a restoration activity;
(4) *resource requirement* specifies the labor needed to work on the task—both time and resources impose constraints on the recovery pathway optimization problem;

(5) *event dependency* would specify how hazard event states, such as flood depth, maps onto damage states (DS) and in turn to the level of restoration activities required to bring the building back to functioning;
(6) *realized state* is the present condition of the building.

Given these data as input, a GMOR model of the building abstracted to the level of restoration tasks is created. An analysis of the GMOR model provides a number of possible recovery pathways with associated recovery timelines. As the order of restoration activities can be modulated within the constraints of available resources, an optimization technique using a decision tree is proposed to iterate over the possible recovery pathways to find the most efficient in terms of time and labor utilization.

20.2.1 Building Damage Scenarios and Restoration Schemes

Damage to a residential building caused by a flood depends on a number of variables, such as depth of water, velocity of floodwater, duration of flooding, warning time, sediment and effluent content [11]. Most damage curves, however, only use depth of flood as the principle factor when velocity is not a concern. For example, Pistrika et al. [12] demonstrate a step-by-step methodology for devising depth-damage functions using actual flood depths and damage data collected for an actual flood event. Previous flood related studies undertaken in Alberta, Canada, [13] also consider building damage as a function of depth of inundation. The *Canadian Guidelines and Database of Flood Vulnerability Functions* report [14] suggests the use of depth-damage curves in damage estimations.

A review of the literature was conducted to synthesize a complete spectrum of damage scenarios according to hazard level and in turn develop an inventory of restoration tasks related to the damaged states. Table 20.1 lists the literature sources from which depth damage relationships for residential buildings (considering the full range of possible damages) are extracted. All but one source are US-based data; one dataset is based in Canada. Using the findings contained in these sources, we have refined the relationship between flood intensity (i.e., depth of water) and building component failure criteria. The type of flood assumed for our present study is riverine flooding which may last 2–3 days; and building locations subject to relatively negligible flood velocity for demonstration purposes.

Permanent structural and non-structural components of a building are both prone to damage by riverine flooding. These components are responsible for collectively performing different functions of a building. One could also consider internal contents (e.g., furniture, clothing) that are not a permanent part of a building and exterior attributes (e.g., backyard shed) and that, once the building is flooded, are all expected to become waste [14]. The restoration of these components is less complicated and so not addressed here.

The relationship between flood depth and costs from a synthesis of the literature is shown in Fig. 20.2. Here, the water height is considered to extend positively upward

Table 20.1 Literature sources for a single residential building component's possible damaged conditions due to variable flood heights

Institution	Citation	Location	Year	Building type
US Army Corps of Engineers	[15]	United States of America (US)	1992	Residential buildings with basement
US Army Corps of Engineers New Orleans District	[16]	US	1997	Residential buildings with basement
US Army Corps of Engineers New Orleans District	[17]	New Orleans, US	2006	Residential buildings on slab structure
US Department of Homeland Security Federal Emergency Management Agency Mitigation Division	[4]	New Orleans, US	2013	Residential buildings with basement
European Commission, Joint Research Centre (JRC)	[18]	European Union (EU)	2017	One and two stories with basement
Natural Resources Canada (NRCAN)	[14]	Alberta, Canada	2017	Residential buildings with basement
US Department of Homeland Security Federal Emergency Management Agency	[1]	Boulder, Colorado, US	2017	One story with basement

from the main floor level in a range of 0.1–3.6 m and negatively downward in a range from −0.3 to −2.7 m (which is a typical basement depth in North America). Of course, water flows to the lowest possible depth so a depth of −2.7 m would result in no impact assuming sufficiently small flow rates. As the water rises from that depth, building components start to be impacted and damages are cumulative. Damage information was collected for each 0.3 m increment of water depth which is also a standard as observed in most of the sources listed in Table 20.1. Figure 20.2 illustrates the cumulative component failures of a building with respect to depth of water inside a building. At any depth of flood, components presented within boxes at and below that height need to be restored.

In total, fourteen flood depth intervals were identified for which significant information on component failure is gathered from the literature. These fourteen intervals are termed as damage states (DS) and presented as DS-A up to DS-N. Seven DSs affect only the basement level, four DSs include the main floor level and three DSs include the second floor. DSs that mark 0.1 m water depth above a floor are considered critical DSs as damage increases significantly as soon as water passes above another floor. Critical DSs (of a two-story building with basement) are: − 2.6 m (above basement floor), + 0.1 m (above main floor) and + 2.8 m (above second

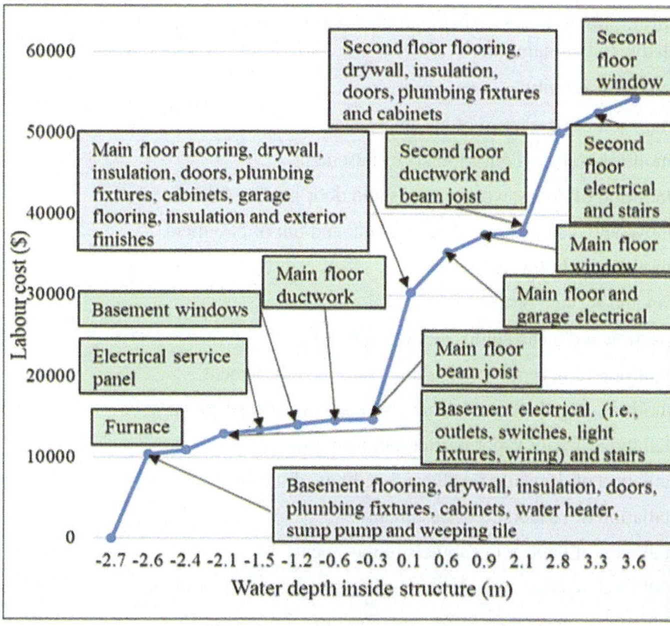

Damage States (DS)	Flood Depth (m) inside the structure
DS-A	-2.6
DS-B	-2.4
DS-C	-2.1
DS-D	-1.5
DS-E	-1.2
DS-F	-0.6
DS-G	-0.3
DS-H	0.1
DS-I	0.6
DS-J	0.9
DS-K	2.1
DS-L	2.8
DS-M	3.3
DS-N	3.6

Fig. 20.2 Damages of a two-story single detached residential building with basement for a range of possible flood heights relative to the main floor. Cumulative labor costs for restoration are plotted against water depth. Depths of water are associated with named damage states (DS)

floor). The building finishes and materials such as interior finishes, interior wall and ceiling, doors and windows are found to be more prone to water damage before most other components [14]. Water damage prone components also include the electrical system components (i.e., service panels, meters, switches and outlets). Electrical system components require complete replacement in most cases even if they are inundated for short periods of time [19]. Some other building system components such as, mechanical (plumbing, heating, ventilating and cooling), and security and fire systems require replacement after inundation reaches certain heights [17].

Another significant aspect of flood water damage is mold contamination which is considered as secondary damage. This situation is severe and occurs due to several reasons, such as lack of immediate response, wicking of moisture upward by semi-permeable building materials and lack of proper dehumidification. This situation may occur regardless of building component exposure to direct flooding [14]. As a result, quick response and complete recovery become crucial.

Detailed restoration schemes for each DS are specified based on the above damage conditions extracted from the literature sources listed in Table 20.1 coupled with construction contractor job tasks from RSMeans data [20]. The schedule of restoration activities for each DS is called a *restoration scheme.* Table 20.2 outlines the restoration scheme developed for damage state A (DS-A) or −2.6 m flood water depth inside the structure. Restoration works (i.e., demolition, installation, clean and

Table 20.2 Physical restoration scheme for damage state A DS-A or −2.6 m depth of water inside structure measured relative to the main floor

Abbreviation	Restoration activity
D wood work	Demolition of baseboard of basement
	Demolition of bathroom cabinets of basement
	Demolition of doors, wood casings and door jambs of basement
RCD plum	Remove and clean bathroom toilet, sink and tub of basement
	Demolition of hot water heater
IS furnace	Clean and service furnace
IS sump_weep	Inspection and clean sumps and weeping tile
D Drywall	Demolition of drywall to walls and ceilings of basement
D Vap_ins	Demolition of poly vapor barrier and insulation of basement
I Vap_ins	Installation of poly vapor barrier and insulation of basement
I Drywall	Installation of drywall to walls and ceilings of basement
I Wood work	Installation of baseboard of basement
	Installation of bathroom cabinets of basement
	Installation of doors, wood casings and door jambs of basement
I Plum	Installation of bathroom toilet, sink and tub of basement
	Installation of hot water heater
D Floor	Demolition of flooring of basement
I Floor	Installation of flooring and Carpeting of basement
CP Int_comp	Clean and paint interior structural and non-structural components of basement

**D* demolition, *RCD* = remove, clean and demolish, *IS* Inspection and Service, *I* Installation, *CP* clean and paint

paint) of the building components are delineated according to the expected damage related to the flood depth.

Similar component restoration tasks are lumped into the general restoration activity groups listed in Table 20.2 for simplicity of the model. For example, plumbing work includes restoration of water heater and plumbing fixtures. Similarly, woodwork includes restoration of baseboard, stairs, doors, windows, and cabinets of kitchen and washroom. The electrical works are divided into two groups, namely wiring and other electrical works; the latter consists of restoration of the service panel, furnace, electrical outlets, switches and light fixtures.

Technically, the overall recovery time of a residential building includes time for damage investigation, clean-up, approval application and the physical restorations of building components [21]. Steps related to application approval and permits have various contextual dependencies that can vary greatly from location to location. As such for this research, analysis is restricted to building specific inspections (i.e., power, furnace and beam joist), clean-up and the physical restorations.

20.2.2 Restoration Time Calculation

Building specifications are used to interlink the damage extent with the restoration time estimation. It has been already established that building characteristics play an important role in determining flood damages [4, 14, 17]. This is because building specifications such as material, age, height and presence of basement define the probability of damage to components. Furthermore, the quantified damage information helps to calculate a damaged component's restoration cost and time. As a result, a representative building selection that is specific to the study area of interest is a significant first step for producing a useful model. For example, Table 3 provides the values of component units (CU) for various building components as per the level within the residential building. These values are specific to the location of the case study area presented in Sect. 20.3.

Determining the restoration time of a building component is the next step. The initial challenge encountered for the flood recovery modeling of a residential building is to obtain reliable data sources for component restoration times (i.e., doors, floor and drywall restoration times). From the literature search, one of the significant gaps that was recognized is the lack of historical recovery data specific to any given location of interest. As a result, an alternative reliable method of restoration time estimation is proposed as an outcome of this research. Namely, time to complete a restoration activity is inferred from construction cost estimation using the RSMeans data [20]. RSMeans has been an evolving database since 1942 for construction and repair-related pricing data. Contractors, facility owners, architects, engineers and researchers can use this information for localized construction and repair-related job cost estimations. RSMeans does, however, restrict the assessment to deterministic estimates.

Building recovery is ultimately a function of building component restorations. A schedule of building restoration activities for each of the DSs is devised using representative building data and pricing information on restoration jobs as shown in Sect. 20.2.1. These schedules are restoration schemes, which are then used for determining a restoration time estimation baseline. Once the restoration schemes are delineated using representative building data (i.e., material type and dimension of floor), the restoration labor cost for a specific task (i.e., demolition or installation of total floor area) can be estimated. Next, the cost of labor can be divided by a specific labor rate for a specific trade (i.e., contractor, plumber). Note that labor rate includes overhead and profit. Hence, the formula for calculating the restoration time where time estimates are essentially back-calculated from RSMeans data is as follows:

$$t = L * CU * \text{unit labor cost/labor rate per day} \tag{20.1}$$

where t is restoration time in days, L is the location factor, CU is the component unit (from Table 20.3). Location factor, unit labor cost and labor rate per day are obtained from RSMean data [20]. Detailed clean-up times and physical restoration times of individual building components are calculated in the same way. Note that to enable

Table 20.3 Building component specification for representative building as per building classification "B" (floor area 166 m^2) [14]

Building components	Basement	Main floor	Second floor	Garage
	(Units)			
Drywall (m^2)	219	336	336	–
Insulation (m2)	87	87	87	349
Doors (unit)	6	8	8	1
Windows (unit)	5	14	14	2
Flooring (m^2)	30	9	9	–
Carpeting (m^2)	53	74	74	–
Baseboard (linear m)	66	106	106	–
Bathroom (unit)	1	2.5	2.5	–

the use of the RSMeans data, more refined specification of building components may be required such as the specification of furnace type, water heater and ductwork dimensions. Where actual specifications are unavailable, values are determined based on average quality components.

20.2.3 Restoration Dependencies

Restoration dependencies are the basis of GMOR's sequencing of recovery activities. The dependencies are determined based on information about residential building component restoration activities given the damaged condition and the restoration techniques involved. This information is based on our knowledge and can also be gathered from flood damage repair databases, construction reports related to residential building renovations and interviews. These dependencies are specified in GMOR as a restoration schemes for each of the damage states (DS-A to DS-N). A graph model is useful to illustrate the mapping, where the nodes of the graph are the restoration activities and the directed arcs are the precedence relationships, such that the activities in a sink node can only be initiated after activities in a source node have completed.

Moreover, the dependence of certain activities on the availability of specialized labor is also included in the same graph. The local labor workforce is of great significance as many specialized trades are involved in the recovery of a single building; recall that the cost of labor is a crucial parameter for restoration time calculation, and later we will see that labor supply affects the prioritizing of restoration works due to the assumed limited supply of workforce during a post-disaster situation. Eight types of specific workforce are identified to be involved throughout the restoration process: (1) contractor, (2) insulation worker, (3) electrician, (4) plumber, (5) pipe fitter, (6) structural contractor, (7) painter and (8) skilled labor force. As an

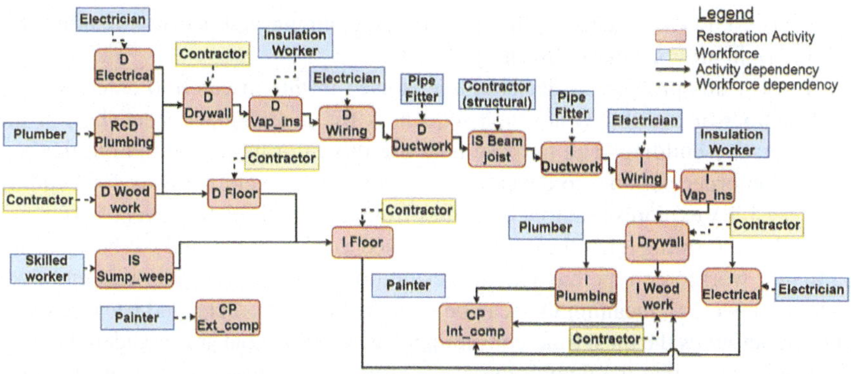

Fig. 20.3 Schematic diagram of dependencies for executing the physical restoration activities for a single building recovery post-encountering DS-H at 0.1 m depth of water inside structure. Here, two types of model entities are presented, these are: restoration activity (rounded rectangle) and workforce (box shape). Connections of entities by arrows defines their dependency relationships. *D = demolition, RCD = remove, clean and demolition, IS = Inspection and Service, I = Installation and CP = clean and paint

example, Fig. 20.3 shows the precedence relationships among restoration activities and workforce dependencies for DS-H (0.1 m depth) of flood inside the structure.

As per GMOR's sequencing of functionality, a restoration process only starts to proceed from "initiation" to "end" once the required workforce (i.e., contractor, plumber etc.,) is made available. Once a given restoration requiring a specific type of workforce is complete, then the next restoration activity requiring that workforce can commence. Critical labor can be easily identified from the workforce and restoration activity dependency map for the physical restoration of a building.

Referring to Fig. 20.3 again as an example, it is found that the contractor is a critical resource of the recovery model. The contractor is needed by six restoration activities. Moreover, it may occur that the contractor is needed at the same time due to its presence in the parallel restoration pathways A and B. Whereas the rest of the workforce are generally required in a series manner, the contractor has jobs that can be performed in parallel to other tasks. For buildings experiencing DS-H, the contractor can be required for drywall and flooring works at the same time as those tasks do not depend on each other. As a result, the decisions affecting the utilization of contractors may significantly impact the recovery process time.

20.2.4 Restoration Prioritizing

The prioritization of restoration works is an important factor as during a post disaster situation, supply is always limited compared to the demand, and prioritization of tasks has a great impact on the recovery process. However, the optimal order is not necessarily known, especially during recovery planning and decision making that follows right after a natural disaster [22, 23].

To solve this problem, a decision tree is used in this research to determine the total number of possible restoration sequences for each damage state (DS) of the representative building. The purpose is to use this information to find out the best restoration order possible to develop the optimized recovery pathway for building recovery. ETA explores possible outcomes from a single initiating event [24].

While performing ETA, the contractor is used as a key to the analysis because this particular workforce type has parallel restoration activities (i.e., drywall and floor restoration works), meaning they can be done in different orders, and because many of these activities the most time-consuming. Other factors that are considered during the analysis are: that the demolition of wood work (i.e., baseboard, cabinets, door, etc.,) need to be ordered first and their replacements installed at the end. Furthermore, the installation of any of the flooring or drywall cannot be performed unless their demolition has already taken place.

The purpose of ETA, as used here, is to determine all the possible restoration sequences in order to optimize the building recovery. The specific resources (i.e., workforces) are ordered according to these permutations. Next, the permutations are taken as input to the recovery model. Finally, the overall recovery time is calculated in GMOR for each of the workforce orderings. The optimal solution for fastest recovery is determined through comparison among all the recovery times.

20.3 Case Study Results and Discussion

20.3.1 Case Study Area and Building Selection

To demonstrate the proposed methodology, flood vulnerable Bridgeview area of Surrey, British Columbia Canada, with a population of 1895 (2005) is selected. This area of 124 ha lies within the Fraser River low lands. Flooding is an important issue in this area since residential homes and businesses are present. Incidents of frequent flooding within this area have been observed by the City Engineering Operations and Maintenance staff. In particular, 112A Avenue west of 124th Street has been marked as the worst area and during almost every rain event water levels reach up to residential front yards, and a significant amount of complaints from residents regarding inaccessibility and nuisance due to the floods occur [25].

A sample set of 30 buildings located within the single family residential zone (RF) in the Bridgeview area of Surrey, BC, is selected. For this research, publicly available sources [26, 27] and several real estate websites were used to investigate the typical building characteristics in the study area. The selected set of buildings were assessed based on the available information related to building height, material quality, age, property value, number of stories and presence of basements.

It was observed that 16 out of these 30 buildings are built after the 1950's and are built following basic building code requirements of the era with average quality building materials [26]. Most of the buildings have floor area between 1200 and 2400 ft^2 and a basement or second story is common. As per assessment from July 2018, these buildings' value varies from $600,000 to $730,000 (CAD). Property related detailed information such as dimension and material of building components tends to be confidential, and in this case, simply unavailable. So, building classification schemes in the Canadian context [14] were compared with the assessment data to determine an equivalent characteristic building class that resembles the characteristics of the representative building of the study area. The same building classification scheme was recently used in the Alberta flood damage assessment [13]. The Bridgeview building characteristics of a representative building were most similar with the characteristics of building classification "B" of the Alberta study. "B" type buildings are average quality units with conventional design, and medium-quality materials and finishes (Table 20.3). As a result, a two-story residential building with basement with B type building characteristics is selected as the representative building for the study area and used as data input for the recovery modeling.

The results of the recovery modeling for a representative building unfold into three output segments. The first segment is a comparison of the workforce order permutations (i.e., the order of the sequence of tasks assigned to the different types of workforces) that output the fastest recovery pathway. Next, following the fastest recovery pathway, the second segment of results outputs the cumulative recovery time of a single building for each DS. Finally, the third segment of results provides a detailed building component restoration timeline which shows by when one can expect certain components to be restored. This can be useful for determining the time until a house is safely habitable.

20.3.2 Optimized Recovery Pathway

The permutations for the assignment of the contractor are shown as a decision tree in Fig. 20.4. The decision tree provides six permutations in total. These permutations are termed: P-1, P-2, P-3, P-4, P-5 and P-6. Complete building recovery times for the highest level of damage (DS-N, 3.6 m depth of water inside the structure) calculated using GMOR for these permutations are 182.1, 179.4, 167.3, 182.1, 169.1 and 182.1 days, respectively. For most of the damage states (DS-C to DS-N), recovery model permutation P-3 provides the shortest recovery time. For DS-A and DS-B, recovery in the least possible time can be achieved by adopting permutation P-2. However, recovery time differences in adopting P-2 or P-3 are less than a single day in these cases. Furthermore, for DS-N, the improvement of recovery time with P-3 over half of the permutations (P-1, P-4 and P-6) is as high as nearly 8.2% (or 15 days). As a result, the ordering of P-3 is nominally the best restoration pathway if restricted to choosing a single strategy in order to simplify the planning process.

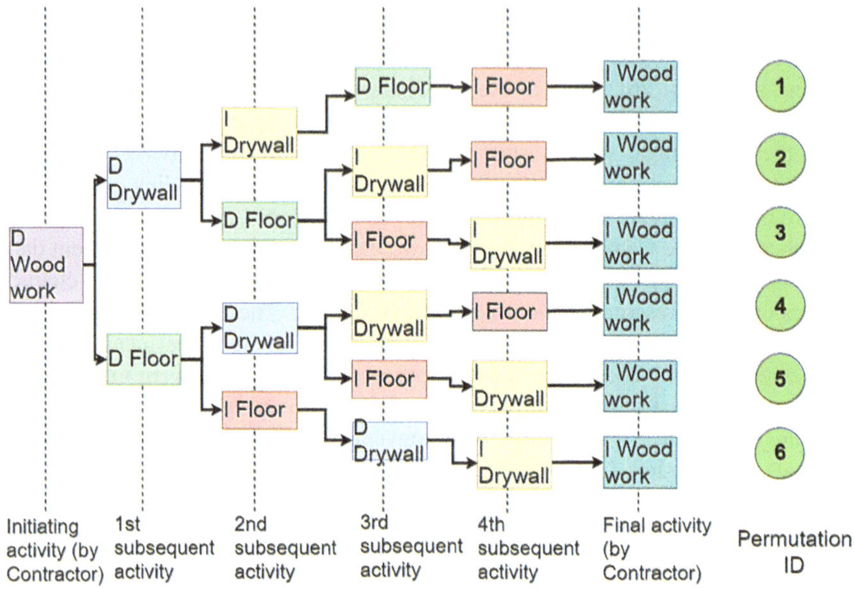

Fig. 20.4 Potential pathways for restoration activity prioritization by the Contractor (decision tree). Here, D = demolition and I = Installation

20.3.3 Complete Building Recovery

In Fig. 20.5, compete building recovery is presented for all possible damages for the recovery model permutation P-3. The graph shows the nonlinear impacts of flood height increase on the recovery time.

Here, pronounced increases in recovery time occur once water reaches the main floor level (0.1 m depth of water) and again when water reaches the second floor (2.8 m depth of water). The jumps would have an impact on the viability of successfully recovering a neighborhood and on the economic prosperity of the region.

The overall results of the recovery time comply with the results of building repair and clean up provided by [5]. This report provides building recovery timelines for three types of repair requirements. These are, 30 days for moderate repair, 90 days for extensive repair and 180 days for complete repair. These are similar to recovery timelines for DS-A (−2.6 m depth), DS-H (0.1 m depth) and DS-N (3.6 m depth), respectively.

20.3.4 Building Function and Component Recovery

The component level (e.g., flooring, drywall) restoration times is another key finding of this research. This finding can be considered as baseline data as no similar data

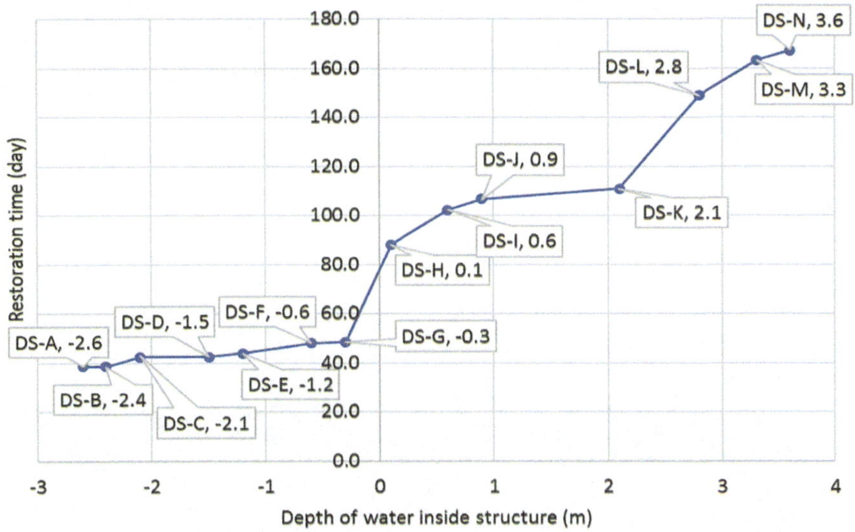

Fig. 20.5 Recovery time of a single detached two story with basement building for variable flood damage states (DS-A, −2.6 m depth of water to DS-N, 3.6 m depth of water). These recovery times were determined using sequence model (permutation 3)

could be found in the literature search. The component level recovery timeline is presented in Fig. 20.6 for the P-3 permutation. DS-A = −2.6 m, DS-H = 0.1 m and DS-N = + 3.6 m are selected for presentation among the fourteen DSs. The timeline presented against each restoration activity provides the expected completion time of that restoration work. For DS-A, some of the restoration activities (e.g., D wiring, I wiring, etc.) the completion times are presented as zero as they are expected not to be damaged for the given damage.

Here, it is determined that the whole building clean-up phase for the main floor level flooding takes at least a few weeks to complete. This complies with Hazus-MH 2.1 assumption for the clean-up phase [4]. However, Hazus does not provide other component level restoration time estimations.

The differences in values of the data points in Fig. 20.6 are large between various phases of recovery (i.e., installation of flooring and installation of drywall). These differences may depict valuable information about the recovery process. The flooring and drywall is seen to be restored quite a long time before the complete recovery of the building (for DS-H about 7 days and for DS-N around 20 days). At this phase, the house may be considered habitable as a structure as most of the crucial building functions, such as structural integrity, heating and ventilation, most of the electric works are restored. The rest of the work that needs to be completed are kitchen and washroom cabinets, baseboard, door and window fixtures; faucets, commode, bathtub installation; electrical outlets, light fixtures and overall cleaning and painting. Although these are crucial works, they can be addressed one room at a time.

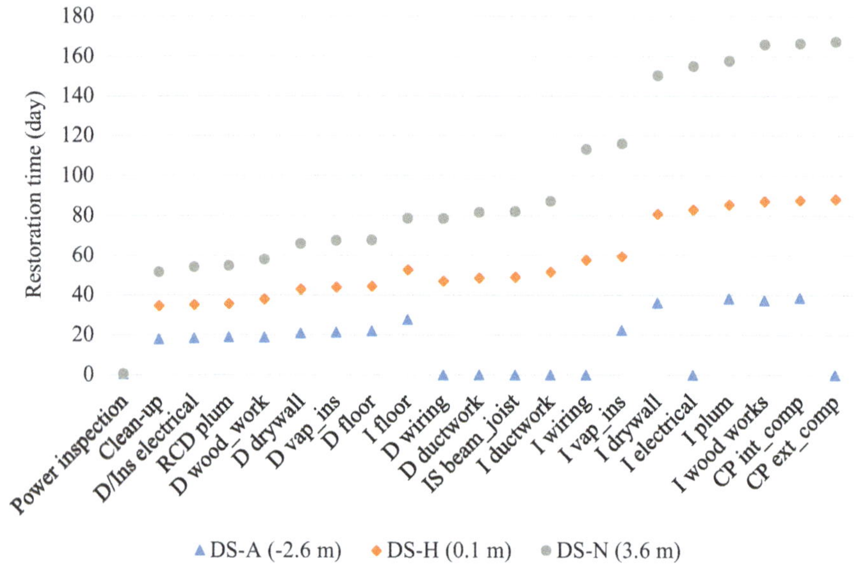

Fig. 20.6 Building component restoration times for the damage states A, H and N (depth of flood for these damage states, DS-A = −2.6 m, DS-H = 0.1 m and DS-N = 3.6 m). D = demolition, RCD = remove, clean and demolition, IS = Inspection and Service, I = Installation and CP = clean and paint

Further, for the estimates of recovery after most of the damage states, the plumbing and architectural functions are restored within a similar amount of time. Although electrical restoration times vary depending on the damage state, it is found that it is constant for 0.6–2.1 m of flooding and constant for 2.8–3.6 m flooding. This is because 0.6 m signifies that effectively all the electrical works existing in the main floor need to be restored. A similar case arises for the 2.8 m flood for the second floor. The structural restoration time is not required until −0.3 m depth of water inside the structure. The restoration time for the heating and insulation function also varies with the damage state (e.g., increasing at −1.5 m and decreasing at −1.2 m). This is a reflection of how the model prioritizes specific restoration activities. These results illustrate some of the patterns and trade-offs that restoration planning decisions have on recovery of the functions.

20.3.5 Future Work

This work is informative for refining regional flood risk assessments to buildings stocks. For instance, this work highlights the need for additional data affecting damage and recovery of buildings due to flooding. For instance, the depth damage curves have sensitivities at the interface between building stories. In particular, DS-H and DS-L should be given careful attention because of the large increase in recovery time that occurs at these flood depths. The main floor elevation from ground level (set at a minimum of 0.3 m height from adjacent road mid-point by local regulations) would also be an important parameter to include in such an assessment. The main floor elevation may vary for building to building as per [28, 29] zoning law. As a result, while considering depth of water inside a structure from an actual flood height, this factor (main floor elevation) may affect the recovery design considerations. Moreover, the accuracy of the recovery time relies on the underlying data accuracy (i.e., building material, dimensions and local contractors per hour unit costs). In this way, any comprehensive probabilistic risk assessments of flood risks to the buildings of a city should account for variation and uncertainty in these key pieces of information and in other parameters associated with the particular flood risk, such as flow velocity.

Furthermore, this work demonstrates how coordination and planning in workforce prioritization could impact recovery timelines of a building. Future work could explore what this type of action might mean to the recovery of an entire building stock. For instance, consider the following two distinct approaches. One, begin the recovery process with the buildings that experienced basement-level flooding (−2.7 m up to −0.3 m depth of water inside structure) only. This may help to return a portion of the population within the shortest possible time and help them to get back to their normal life. Two, prioritize the workforce to first address those buildings whose inhabitants share some characteristics (e.g., low-income, presence of children). There are likely to be trade-offs in any prioritization and scaling this approach to the assessment of a building-stock could help clarify the trade-offs related to overall recovery times of a community as well as the distribution of recovery times to different parts of a community.

20.4 Conclusion

This paper contributes a modeling approach for estimating component restoration times for single-detached houses based on flood depth. The case study recovery analysis of a representative building (located in Bridgeview area, Surrey, BC) following a flooding is performed considering flood depths ranging from the basement to the second floor. The recovery model includes damage inspection, flood clean-up and physical restoration works. A key finding is the pronounced increase in the recovery time for small increase in flood depths (at 0.1 and 2.8 m flood depth). There is evidence that prioritizing workforce resources plays a role in determining the time of restoration. Six permutations for assigning the contractors tasks were assessed for each of the flood heights (damage states A to N). The contractor task ordering permutation illustrates a means to reduce the recovery time.

In the future, this work can be expanded to aid community-wide planning by allowing for a similar assessment at the scale of a building-stock. For such efforts, location-specific recovery cost estimation can be investigated through collaboration with local contractors, insurance providers, local government, home owners and commercial and industrial experts in order to aid in joined-up action on disaster risk reduction.

References

1. Deniz D, Arneson EE, Liel AB et al (2017) Flood loss models for residential buildings, based on the 2013 Colorado floods. Nat Hazards 85:977–1003. https://doi.org/10.1007/s11069-016-2615-3
2. Nastev M, Todorov N (2013) Hazus: a standardized methodology for flood risk assessment in Canada. Can Water Resour J Rev Can Ressour Hydr 38:223–231. https://doi.org/10.1080/07011784.2013.801599
3. Kammouh O, Cimellaro GP (2017) Restoration time of infrastructures following earthquakes. In: 12th international conference on structural safety & reliability (ICOSSAR 2017). Vienna, Austria
4. FEMA (2013) Flood Model Hazus®-MH Technical Manual. US Department of Homeland Security Federal Emergency Management Agency Mitigation Division, Washington, D.C.
5. NiyamIT, Inc (2017) Hazus tsunami model technical guidance. US Department of Homeland Security Federal Emergency Management Agency, Herndon, VA
6. FEMA (2013) Earthquake Model Hazus®–MH 2.1 Technical Manual. US Department of Homeland Security Federal Emergency Management Agency Mitigation Division, Washington, D.C.
7. He X, Cha EJ (2018) Modeling the damage and recovery of interdependent critical infrastructure systems from natural hazards. Reliab Eng Syst Saf 177:162–175. https://doi.org/10.1016/j.ress.2018.04.029
8. Afrin T, Yodo N (2019) Resilience-based recovery assessments of networked infrastructure systems under localized attacks. Infrastructures 4:11. https://doi.org/10.3390/infrastructures4010011
9. Bristow DN, Hay AH (2017) Graph model for probabilistic resilience and recovery planning of multi-infrastructure systems. J Infrastruct Syst 23:04016039. https://doi.org/10.1061/(ASCE)IS.1943-555X.0000338
10. Bristow DN (2019) How spatial and functional dependencies between operations and infrastructure leads to resilient recovery. J Infrastruct Syst 25:04019011. https://doi.org/10.1061/(ASCE)IS.1943-555X.0000490
11. Romali NS, Sulaiman MAK, Yusop Z, Ismail Z (2015) Flood damage assessment: a review of flood stage-damage function curve. In: Abu S, Tahir W, Wahid M et al (eds) International symposium on flood research and management (ISFRAM 2014). Springer, Singapore, pp 147–159
12. Pistrika A, Tsakiris G, Nalbantis I (2014) Flood depth-damage functions for built environment. Environ Process 1:553–572. https://doi.org/10.1007/s40710-014-0038-2
13. IBI Group, Golder Associates Ltd (2015) Provincial flood damage assessment study. Government of Alberta, Environment and Sustainable Resource Development, Calgary, AB
14. Shawcross SW, Sol D, Stante A (2017) Canadian guidelines and database of flood vulnerability functions. Natural Resources Canada, Ottawa, ON, Canada
15. Davis SA, Skaggs LL (1992) Catalog of residential depth-damage functions used by the army corps of engineers in flood damage estimation. US Army Corps of Engineers, Washington, D.C.
16. Gulf Engineers & Consultants (GEC) (1997) Depth-damage relationships for structures, contents and vehicles and content-to-structure value ratios (CSVR) in support of the lower Atchafalaya re-evaluation and Morganza to the gulf, Louisiana feasibility studies. US Army Corps of Engineers New Orleans District, Baton Rough, Louisiana

17. Gulf Engineers & Consultants (GEC) (2006) Depth-damage relationships for structures, contents, and vehicles and content-to-structure value ratios (CSVR) in support of the Donaldsonville to the Gulf, Louisiana, Feasibility Study. US Army Corps of Engineers New Orleans District, Baton Rough, Louisiana
18. Huizinga J, De Moel H, Szewczyk W (2017) Global flood depth-damage functions: methodology and the database with guidelines. Publications Office of the European Union, Luxembourg
19. US Department of Homeland Security (2005) Protecting your home or small business from disasters. US Department of Homeland Security Federal Emergency Management Agency, Washington, D.C.
20. Lane T (2018) 2019 contractor's pricing guide: residential repair & remodeling CD. R S Means Co, Rockland, MA
21. US Department of Housing and Urban Development (2015) Rebuild healthy homes. Department of Housing and Urban Development
22. Lubashevskiy V, Kanno T, Furuta K (2013) Algorithm for resource redistribution required for recovery of society after large scale disasters. ArXiv13100648 Nlin Physicsphysics
23. Lubashevskiy V, Kanno T, Furuta K (2016) Resource redistribution under lack of information: short-term recovery after large scale disasters. J Adv Simul Sci Eng 3:1–16. https://doi.org/10.15748/jasse.3.1
24. Rausand M, Høyland A (2009) System reliability theory: models, statistical methods, and applications, 2nd edn. Wiley-Interscience, Hoboken, NJ
25. Brotherson J, Phang M, Neuman S et al (2015) Bridgeview—North slope integrated storm water management plan. AECOM Canada Ltd., Surrey, BC
26. BC Assessment (2019) Independent, uniform and efficient property assessment. In: BC Assess. https://www.bcassessment.ca/. Accessed 15 Jan 2020
27. City of Surrey (2020) City of surrey mapping online system (COSMOS)
28. City of Surrey (2010) Zoning 5942 (pre 12000). https://www.surrey.ca/city-government/5173.aspx. Accessed 15 Jan 2020
29. City of Surrey (2019) Zoning. https://www.surrey.ca/city-government/5172.aspx. Accessed 15 Jan 2020

Chapter 21
Vibration Isolation of Foundation Systems Using Geosynthetics Barriers

Amarnath Hegde and Hasthi Venkateswarlu

21.1 Introduction

One of the major sources that have the potential to exhibit severe destruction on the structures and the environment is ground born vibration. Nowadays, the activities responsible to induce ground vibration are increasing rapidly by the fact that the increase in growth of industrialization. Few practical considerations include transit systems, blasting activities, construction operations, and machine foundations. Thus, restricting the adverse effects associated with the unfavorable ground-borne vibrations has become a great concern in the modern era. The extent of the destruction prominently depends upon the type of source, the distance between the source and construction, the presence of sensitive equipment, and number of workers and inhabitants in the surrounding area.

Among the various sources, machine foundations are the ones, which support heavy pieces of machinery such as rotary machines, reciprocating engines, turbines, compressors, and forge hammers. Such foundations are continually subjected to dynamic forces by the vibratory motion of the machine parts. Prominently, emanated vibration energy by the machine sources is carried out through body and surface waves [1, 2]. The body waves (compression and shear wave) carry 33% of the overall induced vibration energy through the spherical wavefront in all directions. On the other side, the remaining 67% vibration energy solely transmits through the surface wave (Rayleigh wave). Based on this fact, the attenuation rate of surface waves is very less in comparison with the body waves [3, 4]. Therefore, the disaster caused by the surface waves is more predominant and needs to be mitigated.

A. Hegde (✉) · H. Venkateswarlu
Indian Institute of Technology Patna, Patna 801106, India
e-mail: ahegde@iitp.ac.in

S. Kolathayar et al. (eds.), *Civil Engineering for Disaster Risk Reduction*, Springer Tracts in Civil Engineering, https://doi.org/10.1007/978-981-16-5312-4_21

In practice, isolation and strengthening are the two approaches used by the practitioners to safeguard the structures and equipment from the vibration stipulations [5]. As observed from the literature, the vibration isolation countermeasures have been classified into two categories, namely infield and far field isolations [1, 6]. These techniques are intended to isolate the vibrations originated by the transmission of Rayleigh waves [7, 8]. In the infield isolation, vibration is potentially mitigated at the source of vibration, while the harmful effects of the vibration could be controlled at the location of the structure in the far field isolation. The study of woods [1] described that the open trenches are a cost-effective, simple, and quick solution to isolate the ground vibrations in both the approaches. Further, the required trench depth for obtaining the 75% isolation efficacy in active and passive isolation modes was suggested [8]. Later, the performance of infill trenches has been studied by considering the instability and maintenance problems of open trenches [9, 10].

Although the vibration isolation efficacy of open and infill trenches is considerable, its execution becomes a difficult task in densely populated areas. Under such circumstances, the approach of strengthening is an alternative for reducing the vibration problems. The principle strategy of this method is to enhance the dynamic properties and bearing capacity of the foundation bed. Stiffness of the foundation bed, shear modulus, and the damping ratio represent the dynamic attributes. It is achieved by altering the profile of the bed, soil stabilization, and reinforcing the foundation bed. Many researchers [11, 12] have studied the effect of change in the foundation bed profile on the dynamic behavior of a machine foundation system. The results described that the existence of the stiff layer close to the footing can considerably increase the stiffness of the foundation bed. Various researchers have stated the ability of different stabilization techniques in enhancing the dynamic attributes of the foundation bed [13–15].

In the current scenario, the use of engineered materials, known as geosynthetics, has received the attention of practitioners to enhance the performance of geostructures. Among the different geosynthetics products, geogrid and geocell were majorly designed to serve the reinforcement function. The major applications of such materials include retaining walls, embankments, slope protection, protection of buried pipelines, pavements, foundations, and railway systems. Various studies have described the useful aspects of geosynthetics in the above-mentioned applications [16, 17]. The past studies [18, 19] have recommended that reinforcing the soil bed using iron strips can improve the behavior of the machine foundation bed. Also, limited studies reported the benefits of geosynthetics in enhancing the foundation bed behavior supporting the vibration sources [2, 20, 21].

It is evidenced by the literature that the vibration isolation behavior of geosynthetics is not completely investigated. Thus, a major aim of this study is to highlight the potential of geosynthetics as barriers for isolating the foundation systems from the ground-borne vibration. In this regard, field vibration tests have been carried out over the different reinforced barrier systems. The ground vibration emanated from the high-speed rotary machines has been simulated in the field tests through the oscillator. The effect of dynamic force and location of geosynthetics on the vibration isolation potential of reinforced barrier systems have been studied.

Table 21.1 Characteristics of geomaterials

Material	Fines content (%)	C_u	C_c	Maximum unit weight (kN/m^3)	OMC (%)	Cohesion (kPa)	Friction angle (°)
CDW	5	47.33	4.54	19.3	9	12	43
Slag	4	6.58	1.08	18.9	NA	0	39
Sand	16	22.22	10.1	17.9	12.6	2	32

Note—C_u is the uniformity coefficient; C_c is the curvature coefficient; OMC is optimum moisture content; NA is not applicable

21.2 Material Characterization

In this study, three different geomaterials and two types of geosynthetics were used. Sand, steel slag, and construction and demolition waste (CDW) were the employed geomaterials. Similarly, geosynthetics products utilized in this study were the geogrid and geocells. Geomaterials were used as infill materials in geocells. Following the guidelines of the Unified Soil Classification System (USCS), CDW, slag, and sand were classified as GW, GP, and SM, respectively. The numerous geotechnical properties of these materials are listed in Table 21.1. Standard proctor test based on ASTM D698 [22] guidelines was conducted to evaluate the compaction properties of CDW and sand materials, whereas the maximum dry unit weight of slag was estimated using the standard of ASTM D4253 [23]. Further, to evaluate the shear strength parameters of each geomaterial, the guidelines of ASTM D3080 [24] were followed.

The geogrid used in this study was manufactured from polyester following the square-shaped aperture of 35 mm × 35 mm. The geocell used in the study was manufactured from the novel polymeric alloy (NPA) polymer. It is known for exhibiting unparalleled behavior in terms of service life, stiffness, and creep resistance over HDPE geocell. Each pocket of the geocell reinforcement consists of 1.53 mm strip thickness, 330 mm seam-to-seam distance, 150 mm width, and 120 mm height. To examine the ultimate load-carrying capacity of both the geosynthetics products, tensile test was conducted. To do so, the recommendations of ASTM-6637 [25] for geogrid and ASTM-4885 [26] for geocell were followed. Tensile strength versus axial strain variation of both the geosynthetics is shown in Fig. 21.1. The maximum tensile strength of geogrid and geocell was estimated as 13.6 MPa and 13.4 MPa, respectively.

21.3 Particulars of Field Vibration Test

To envisage the vibration isolation potential of geosynthetics, a series of vibration tests were conducted over the different reinforced barrier conditions. The arrangement adopted to conduct the vibration test is shown schematically in Fig. 21.2. During

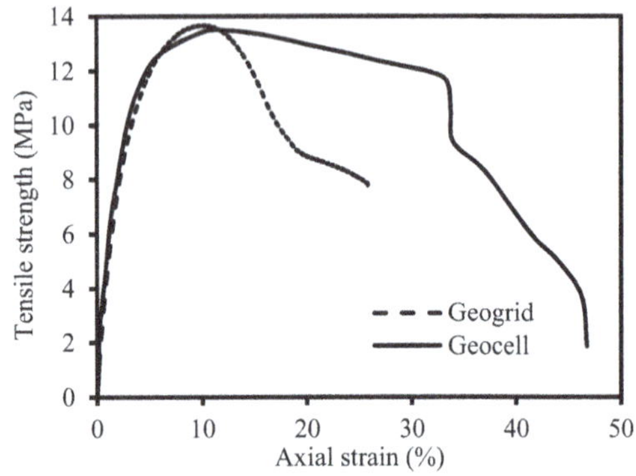

Fig. 21.1 Tensile strength variation of geosynthetics products

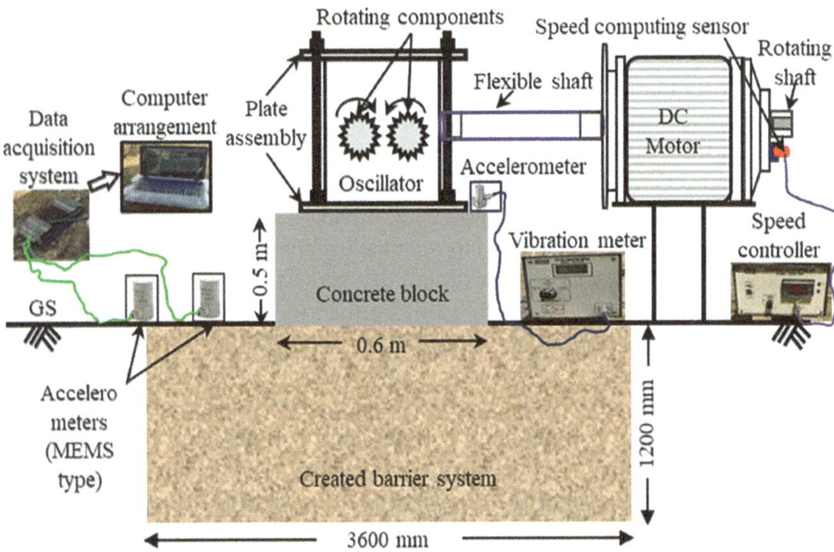

Fig. 21.2 Assembly of field vibration test

the test, the oscillator was mounted over the concrete block to generate the vertical mode of rotating mass-type dynamic excitation. This excitation is the true resemblance of vibration caused by the rotary machines in the practical perspective. To evaluate the dynamic force generated from the oscillator, the approach of a vector combination of centrifugal forces can be followed. With regards to the rotating-type oscillator, induced dynamic force depends on the eccentric setting (θ) and angular velocity (ω). The following expression can be used to determine the total vibrating force emanated by the oscillator.

$$V_d = V_0 \sin\left(\frac{\theta}{2}\right) \tag{21.1}$$

$$V_0 = m_e e\omega^2 \left(m_e = \frac{W_e}{g} \right) \quad (21.2)$$

$$V_d = \frac{W_e}{g} e\omega^2 \sin\left(\frac{\theta}{2} \right) \quad (21.3)$$

where V_d is the total vertical dynamic force, V_0 is the total unbalanced dynamic force generated above the footing, W_e indicates the weight of rotating components in the oscillator, e is the eccentricity (radius) of the rotating mass, and g is the acceleration due to gravity. The magnitude of the dynamic force can be increased by changing the frequency and the eccentric setting as shown in Fig. 21.3.

During the test, one accelerometer was kept over the concrete block for studying the displacement amplitude response at the vibration source. To record the measurements of the accelerometer, it was connected to the vibration meter. Further, two more accelerometers were placed over the ground surface to understand the attenuation response of displacement amplitude. These accelerometers were connected with the data acquisition system (Fig. 21.2). In order to acquire the response recorded by the accelerometers, the acquisition system is attached to the computer having LabVIEW software. The sampling rate of 2 kHz was employed to record the accelerometer measurements. In total, three barrier systems, namely unreinforced, geogrid, and geocell reinforced barriers, were developed in a test pit of 3.6 m × 3.6 m × 1.2 m. The sand (SM) was compacted in the individual layers of 0.12 m up to a depth of 1.2 m to prepare the unreinforced barrier system. Compaction was performed at optimum

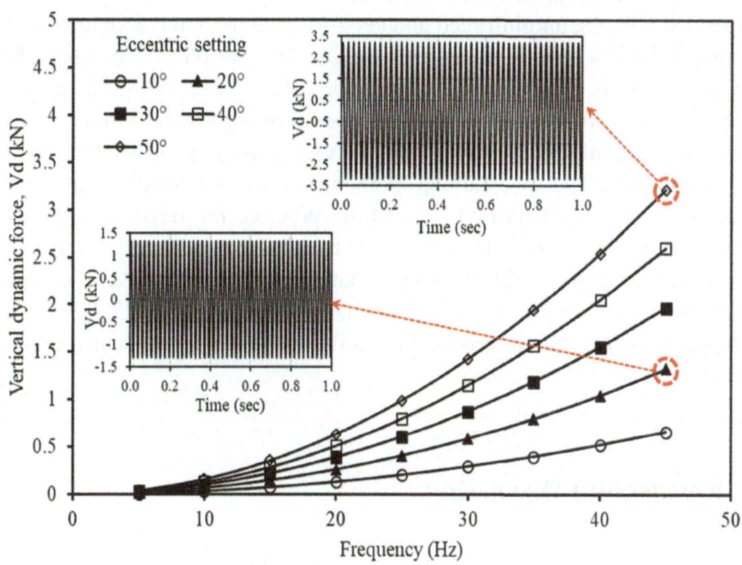

Fig. 21.3 Dynamic force variation with the frequency and eccentric setting

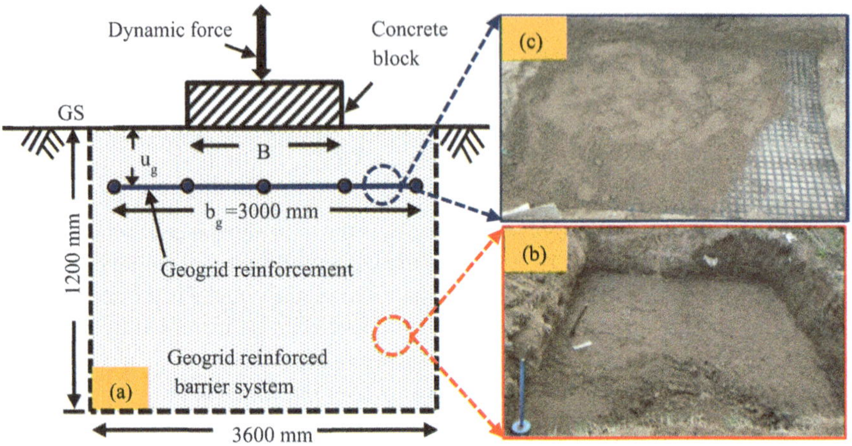

Fig. 21.4 Preparation of geogrid barrier system: **a** Schematic outlook; **b** Compacted bed with silty sand; and **c** Partially filled geogrid reinforcement

moisture content (OMC) to obtain the maximum dry density. After compaction, the achieved average OMC and dry unit weight of the unreinforced barrier were found as 12.1%, and 17.25 kN/m^3, respectively. A similar approach was adopted to prepare the geogrid barrier system. However, the geogrid was buried at a suitable depth of placement from the bottom of the footing. Schematic and field photographs corresponding to the geogrid reinforced barrier system are shown in Fig. 21.4.

The geometrical arrangement of the geocell barrier system is shown in Fig. 21.5a. In comparison with the unreinforced and geogrid systems, a major change in geocell barrier preparation is filling the geocell pockets. In addition to silty sand, slag and CDW materials were employed as infill materials as shown in Fig. 21.5b–d. Each geocell pocket was compacted in three layers employing standard Proctor to fill the silty sand. The obtained average density was considered as a yardstick for filling the other two materials. The sand raining method was adopted to fill the slag material in the pockets of geocell. Prior to the filling process, the required falling height was estimated using trail tests to achieve the baseline density, whereas the tamping process was employed to infill the CDW material in geocell pockets. Overall, the average density variation of each infill material was noticed as 17.25 ± (7%) kN/m^3. The series of field tests conducted over the different barrier systems are summarized in Table 21.2.

21.4 Results and Discussion

Different vibration isolation descriptors were introduced to evaluate the vibration isolation potential of geosynthetics barriers. It includes amplitude attenuation factor

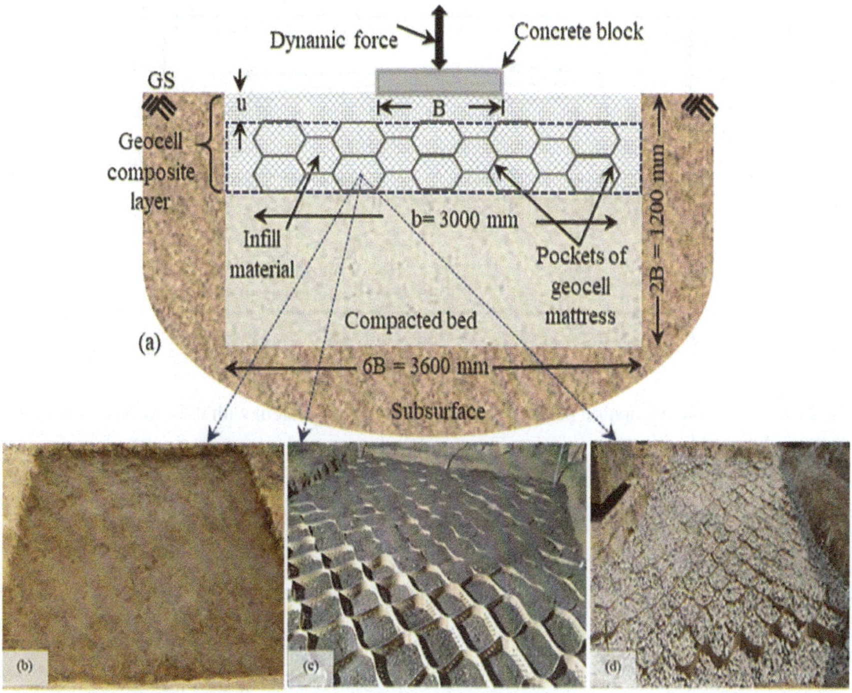

Fig. 21.5 Preparation of geocell barrier system: **a** Schematic outlook; **b** Geocell infilled with silty sand; **c** Geocell infilled with slag; and **d** Geocell infilled with CDW

Table 21.2 Summary of experimental investigation

Barrier condition	Details of a variable parameter		Details of non-variable parameter
	Parameter	Value	
Unreinforced	–	–	–
Geogrid	U_g	$0.15B - 0.45B$	$b_g = 5B$
Geocell			
GSM (geocell + silty sand)	u	$0.1B - 0.5B$	$b = 5B$
GSS (geocell + steel slag)	–	–	$u = 0.1B; b = 5B$
GCDW (geocell + CDW)	–	–	$u = 0.1B; b = 5B$

Note—U_g is the depth of placement of geogrid beneath the footing; u is the depth of placement of geocell beneath the footing; b_g is the width of the geogrid; b is the width of the geocell; B is the width of the concrete block

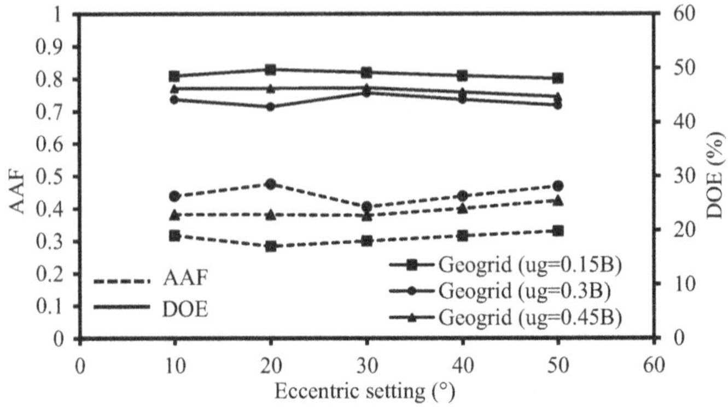

Fig. 21.6 Variation in isolation parameters of geogrid barrier system with the eccentric setting and location of geogrid

(AAF) and degree of efficiency (DOE). AAF and DOE can be calculated by,

$$\mathrm{AAF} = \frac{A_r}{A_u} \tag{21.4}$$

$$\mathrm{DOE} = (1 - \mathrm{AAF}) \times 100 \tag{21.5}$$

where A_r and A_u represent the peak displacement amplitude values corresponding to the reinforced and unreinforced barrier systems. To evaluate the peak displacement amplitude, the frequency of vibration was varied from 5 to 45 Hz during the experiment, irrespective of the eccentric setting value. Generally, maximum displacement was observed at the resonance condition. In practice, the smallest AAF and maximum DOE could be endorsed for achieving better screening performance. Value of AAF corresponding to the unreinforced condition is typically considered as one. Figure 21.6 shows the variation of isolation parameters of the geogrid barrier system with the change in geogrid location and eccentric setting. The attenuation in AAF and increase in DOE were noticed in the presence of a geogrid barrier, irrespective of the depth of placement. Moreover, at the 0.3*B* location of a geogrid barrier, the minimum value of AAF was observed. Further, an increase in the geogrid location from 0.3*B* to 0.45*B* resulted in the amplification of AAF and a decrease in DOE. Therefore, 0.3*B* is recommended as the optimum location of a geogrid for exhibiting better vibration isolation performance of a geogrid barrier system. The possible reason for this behavior is the maximum increase in shear strength behavior of the foundation bed owing to the lateral resistance and open confinement offered by the geogrid at this location. Further, the effect of eccentric setting was found insignificant on the vibration isolation parameters.

The effect of eccentric setting and placement depth of geocell on the isolation parameters of geocell barrier condition are shown in Fig. 21.7. From the figure,

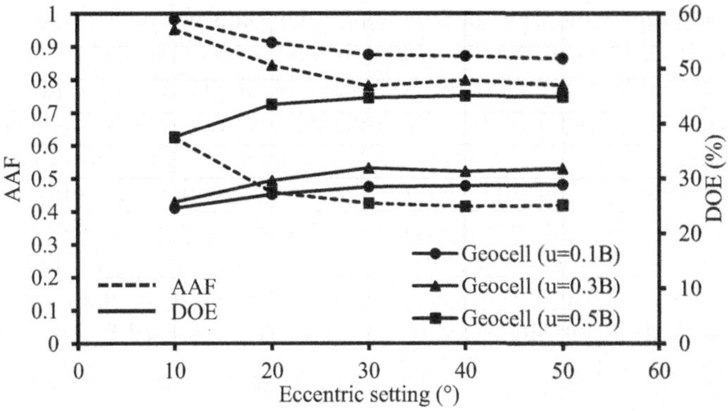

Fig. 21.7 Variation in isolation parameters of geocell barrier system with the eccentric setting and location of the geocell barrier

minimum AAF and the maximum DOE were observed when the geocell located at the placement depth of 0.1*B*. Thus, this placement depth is suggested to obtain the higher isolation performance of the geocell barrier system. At this location, more than 52% of DOE was observed regardless of the eccentric setting. It is worth mentioning that the minimum AAF was noticed in the geocell barrier condition in comparison with the geogrid barrier. It highlights that use of geocell as a barrier provides superior vibration isolation efficacy than the geogrid material. Further, the influence of infill material on the variation in isolation parameters of geocell barrier condition is shown in Fig. 21.8.

From the figure, slag and CDW materials have exhibited a lower AAF value and maximum DOE as compared to the sand material. It signifies that solid waste products like slag and CDW are the viable alternatives for sand to enhance the vibration mitigation behavior. The virtue of higher friction angle of slag and CDW

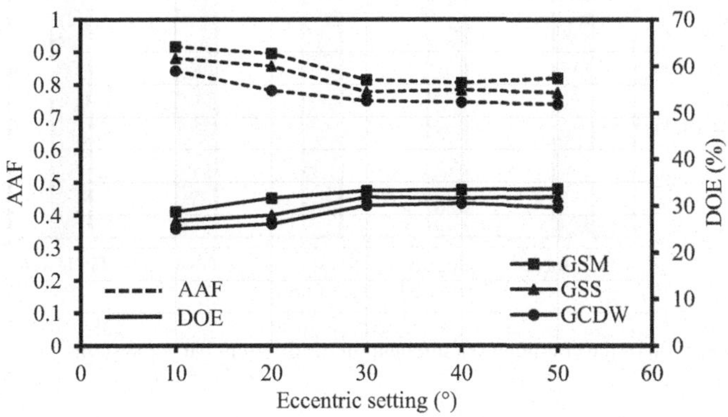

Fig. 21.8 Variation in isolation parameters of geocell barrier system with the eccentric setting and infill material

materials in comparison with sand increased damping behavior of geocell barrier condition [27]. As a result, better isolation behavior was observed.

On the other side, variation in the coefficient of attenuation (α) value of different barrier conditions was studied. It is a measure of the attenuation of induced vibration caused by the absorption of energy as they transmitted through different barrier conditions. Higher the value of α signifies the quicker attenuation ability of a system. In order to measure the α, two accelerometers (MEMS-type) were placed at distances of 1.5 m and 3 m from the center of the vibration source (as shown in Fig. 21.2). The accelerometers were fixed in the same line along the longitudinal direction. Finally, the following expression suggested by IS 5249 [28] was used for determining the coefficient of attenuation.

$$A_2 = A_1\sqrt{\frac{d_1}{d_2}} \times e^{\alpha(d_2-d_1)} \tag{21.6}$$

where A_1 and A_2 are the amplitudes of vibration at the distances of d_1 and d_2, respectively. Figure 21.9 shows the coefficient of attenuation variation for different conditions. The reported results are corresponding to the vertical dynamic force of 1.5 kN. To induce this force, eccentric setting and operating frequency were maintained as 50° and 30 Hz, respectively. Among the considered conditions, a maximum value of α was perceived in the case of GCDW. Hence, the minimum AAF and maximum DOE were observed in this condition. Similarly, a lower α value was found in the presence of unreinforced barrier system.

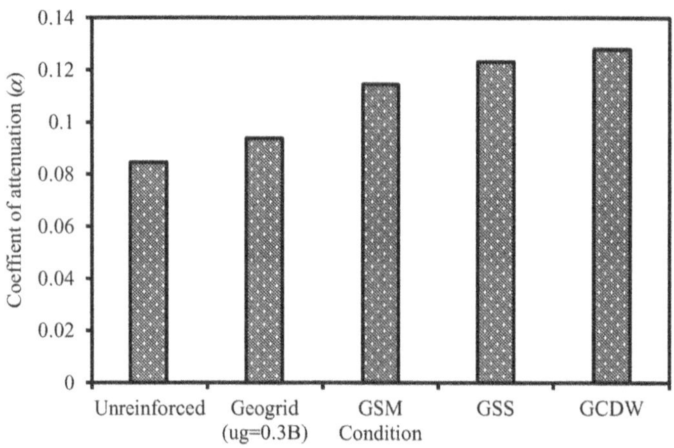

Fig. 21.9 Variation of the coefficient of attenuation for different barrier systems

21.5 Conclusions

Vibration isolation potential of geosynthetics barrier systems, namely geogrid and geocell, were investigated in the present study using field vibration tests. Isolation effectiveness was examined in terms of amplitude attenuation factor (AAF), degree of efficiency (DOE), and attenuation coefficient. Based on the vibration test results, the optimum placement depth of geogrid and geocell for obtaining the maximum isolation efficacy of geogrid and geocell barriers was found as $0.3B$ and $0.1B$, respectively. However, the geocell barrier system has exhibited the maximum vibration isolation potential in comparison with the geogrid barrier system. At the optimum location of geogrid and geocell, the DOE of the reinforced barrier systems was improved by more than 25% and 52%, respectively, as compared to the unreinforced barrier. The increase in the eccentric setting has not shown a significant effect on AAF and DOE response of reinforced barrier systems. Further, the DOE of the geocell barrier was increased with the change in infill material having high friction angle. The maximum was observed for CDW infill material as compared to the slag and sand materials. Further, the maximum attenuation coefficient was noticed in the GCDW condition as compared to the geogrid and unreinforced barrier systems. Finally, it is worth mentioning that the reported observations may vary noticeably with regards to the change in soil conditions and stiffness of the reinforcement products.

References

1. Woods RD (1968) Screening of surface waves in soils. J Soil Mech Found Div 94:951–979
2. Venkateswarlu H, Ujjawal KN, Hegde A (2018) Laboratory and numerical investigation of machine foundations reinforced with geogrids and geocells. Geotext Geomembr 46(6):882–896
3. Miller GF, Pursey H (1955) On the partition of energy between elastic waves in a semi-infinite solid. Proc R Soc Lond A 233:55–69
4. Hegde A, Venkateswarlu H (2019) Mitigation of Traffic Induced Vibration using Geocell Inclusions. Front Built Environ 5:136
5. Sivakumar Babu GL, Srivastava A, Nanjunda Rao KS, Venkatesha S (2010) Analysis and design of vibration isolation system using open trenches. Int J Geomech 11(5):364–369
6. Ahmad S, Al-Hussaini TM, Fishman KL (1996) Investigation on active isolation of machine foundations by open trenches. J Geotech Eng 122(6):454–461
7. Choudhury D, Katdare AD (2013) New approach to determine seismic passive resistance on retaining walls considering seismic waves. Int J Geomech 13(6):852–860
8. Woods RD, Barnett NE, Sagesser R (1974) Holography—a new tool for soil dynamics. J Geotech Eng 100(11):31–47
9. Dasgupta B, Beskos DE, Vardoulakis IG (1990) Vibration isolation using open or filled trenches Part 2: 3-D homogeneous soil. Comput Mech 6(2):129–142
10. Çelebi E, Fırat S, Beyhan G, Çankaya İ, Vural İ, Kırtel O (2009) Field experiments on wave propagation and vibration isolation by using wave barriers. Soil Dyn Earthq Eng 29(5):824–833
11. Baidya DK, Mandal A (2006) Dynamic response of footing resting on a layered soil system. West Indian J Eng 28(2):65–79
12. Mandal A, Baidya DK, Roy D (2012) Dynamic response of the foundations resting on a two-layered soil underlain by a rigid layer. Geotech Geol Eng 30(4):775–786

13. Baker NH (1982) Planning and performing structural chemical grouting. In: ASCE specialty conference on grouting in geotechnical engineering, New Or leans, Louisiana, pp 515–539
14. Mitchell JK (1981) Soil improvement state-of-the art report. In: Proceedings of the tenth international conference on soil mechanics and foundation engineering, Stockholm, pp 509–565
15. Welsh JP (1986) Construction considerations for ground modification projects. In: Proceedings of international conference on deep foundations, Beijing, China
16. Ziegler M (2017) Application of geogrid reinforced constructions: history, recent and future developments. Procedia Eng 172:42–51
17. Hegde A (2017) Geocell reinforced foundation beds-past findings, present trends and future prospects: a state-of-the-art review. Constr Build Mater 154:658–674
18. Haldar S, Sivakumar Babu GL (2009) Improvement of machine foundations using reinforcement. Proc Inst Civ Eng-Ground Improv 162(4):199–204
19. Boominathan S, Senathipathi K, Jayaprakasam V (1991) Field studies on dynamic properties of reinforced earth. Soil Dyn Earthq Eng 10(8):402–406
20. Venkateswarlu H, Hegde A (2019) Block resonance test on geosynthetic reinforced foundation beds. Geotechnical Spaecial Publication 306, ASCE (Geo-Congress 2019), pp 266–276
21. Sreedhar MVS, Abhishek J (2016) Effect of geosynthetic reinforcement on dynamic characteristics through model block resonance tests. In: Indian geotechnical conference, IGC 2016, pp 1–4
22. ASTM D-698 (2007) Standard test methods for laboratory compaction characteristics of soil using standard effort [12 400 ft-lbf/ft^3 (600 kN-m/m^3)]. ASTM International, Philadelphia, USA
23. ASTM D4253 (2006) Standard test methods for maximum index density and unit weight of soils using a vibratory table. ASTM International, West Conshohocken, PA, USA
24. ASTM D-3080 (1998) Standard test method for direct shear test of soils under consolidated drained conditions. ASTM International, West Conshohocken, PA, USA
25. ASTM D-6637 (2011) Standard test method for determining the tensile properties of geogrid by the single or multi-rib tensile method; ASTM International, West Conshohocken, PA, USA
26. ASTM D-4885 (2011) Standard test method for determining performance strength of geomembranes by wide strip tensile method; ASTM International, West Conshohocken, PA, USA
27. Venkateswarlu H, Hegde A (2020) Effect of infill materials on vibration isolation efficacy of geocell-reinforced soil beds. Can Geotech J 57(9):1304–1319
28. Indian Standard Code 5249 (1992) Determination of dynamic properties of soil-method of test. Prabhat Offset Press, Delhi, India

Chapter 22
Response Reductions in Base-Isolated Liquid Storage Tank Under Far and Near Field Seismic Excitations

Sourabh Vern, Sunita Tolani, Shiv Dayal Bharti, and Mahendra Kumar Shrimali

22.1 Introduction

Concrete ground supporting LST is one of the important and widespread civil facility infrastructures that has been used in the production and transportation of numerous materials and products in the oil refining, chemical industry, nuclear power station, wastewater plant, sewage treatment plant, and railway industries. The collapse of these systems leads to serious hazards in the regional environment and also have a long-term impact. Control mechanisms are needed to protect against the failure of LSTs even for the extreme earthquake levels. Damages in the liquid storage tanks during seismic events were observed.

For long-term and effective control of the responses in LSTs, its behavior needs a profound understanding. The structural responses of the LSTs under the earthquake excitations are different from normal load-bearing structures, as the interaction between two different media is involved, i.e., tank and fluid. Thus, to simplify the understanding of the LSTs, it is divided into two major components, i.e., impulsive and convective. The impulsive component is due to the rigid behavior of the LSTs wall and the stationary fluid mass which behaves like solid mass during the earthquake event.

The most common lumped mass approach was presented by Housner [1, 2]; by extending Housner's lumped mass approach, several authors studied the behavior of the LSTs under harmonic and irregular excitations [3, 4].

As the experimental approach gains attention of the researchers, the analytical results and numerical results for the harmonic motions were verified by many authors [5, 6]. Due to the infrastructural limitations of the experimental approach and the inability of the numerical and analytical approaches in solving and capturing the nonlinear behavior of the fluid, finite element method (FEM) quickly gained attention as an effective and computationally less expensive method [7–10]. In the recent

S. Vern · S. Tolani · S. D. Bharti · M. K. Shrimali (✉)
National Centre for Disaster Mitigation and Management, MNIT Jaipur, Jaipur, Rajasthan, India

S. Kolathayar et al. (eds.), *Civil Engineering for Disaster Risk Reduction*, Springer Tracts in Civil Engineering, https://doi.org/10.1007/978-981-16-5312-4_22

past, Rawat et al. [11] studied the earthquake-induced sloshing and hydrodynamic pressure developed in the rigid cylindrical storage tanks with the help of commercial FEM software ABAQUS. The study showed the difference between the two different FEM approaches, namely coupled acoustic-structural (CAS) and coupled Euler–Lagrangian method (CEL). It was concluded that CAS provided the faster results with lower accuracy, whereas CEL proved to be a better option in studying the low-amplitude sloshing effect. Kirtas et al. [12] investigated the modal response of liquid storage steel tanks, and the associated prevailing frequencies in the horizontal impulsive mode of vibration are explored using earthquake recordings.

The use of base isolation in LSTs has drawn the interest of research due to the successful application of base isolation systems in the seismic regulation of many structures. Many researchers have explored thoroughly the seismic study of the base-isolated cylindrical steel LSTs [13–18].

The study of isolated LSTs for bidirectional motion under seismic excitations is comparatively smaller. Many scholars presented some of the early finite element studies [19–21]. Recently, Cheng et al. [22] studied the effect of free-surface sloshing with the help of the velocity potential function that was obtained by the superposition principle and the Navier-Stokes equation obtained by the principle of the mass conservation of the fluid movement. It was concluded that free-surface sloshing plays an important part in estimating the structural responses.

While both steel and concrete LSTs have been studied in the past for a broad range of controlling variables in the seismic response behavior study, the same is comparatively fewer for the base-isolated LSTs. In particular, the efficacy of base isolation in improving the seismic behavior under bidirectional earthquake excitation of 3D concrete LSTs has not been studied in great detail. In this article, to figure out the efficacy of base isolation in LSTs under a range of significant parametric variations, a detailed analysis of a 3D concrete base-isolated LST is performed. These include the use of base isolation in terms of (i) type of earthquake; (ii) the reduction of the various amounts of concern in response; (iii) the spillage of fluid; (iv) the seismic isolation characteristics; and (v) the direction of incidence of seismic events.

22.2 Theory

For the combination of dead and seismic loads, the concrete LST is designed. The LST base is protected by base isolators resting on the hardened surface. The laminated rubber bearing lacks the high dissipation properties of energy; thus, in the New Zealand lead core rubber bearing, a more efficient isolator device is chosen (LRB). Additional energy dissipation properties are pumped into the isolation system due to the central lead center, which results in a greater volume of the isolator system's hysteresis loop. The nonlinear behavior of the LRB isolation system (LRB) has been idealized by various models [23, 24]. Wen's bilinear hysteretic model is shown in Fig. 22.1.

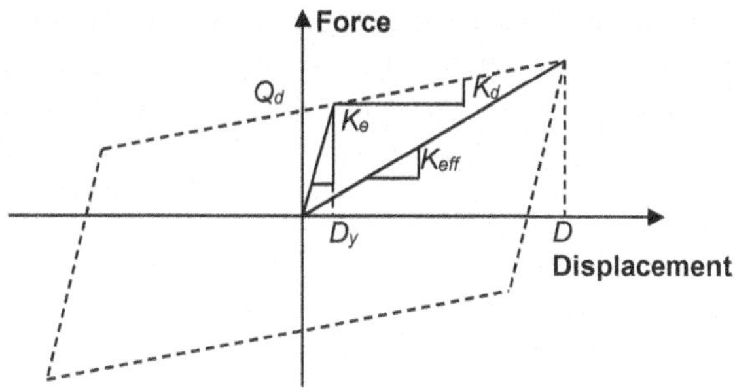

Fig. 22.1 Bilinear behavior of the LRB isolator

There are three salient parameters in the bilinear curve, namely (i) yield strength F_y; (ii) characteristic strength Q_d; and (iii) post-yield rigidity ratio (K_d/K_e or K_2/K_1). At the displacement value of zero, the characteristic strength is the intercept of force. The characteristic strength of the LRB isolator is determined by the yield strength of lead in shear, f_{py}, as shown in Eq. 22.1 for a given area of A_p.

$$Q_d = f_{py} \times A_p \tag{22.1}$$

Equation 22.2 gives the relation between the post-yield stiffness K_d and effective stiffness K_{eff} at the stated design displacement D and Q_d

$$K_d = K_{\text{eff}} - \frac{Q_d}{D} \tag{22.2}$$

The effective damping β_{eff} of the isolator is calculated for the displacement D_y by Eq. (22.3)

$$\beta_{\text{eff}} = \frac{2Q_d}{\pi D^2 K_{\text{eff}}}(D - D_y) \tag{22.3}$$

With the LRB isolator, the time history analysis of the concrete LST is very costly and poses many challenges. An eight-node linear brick element (C3D8R) is defined for the LST, with hourglass control and reduced integration scheme available in ABAQUS. The fluid is also modeled by the same brick element as the tank to grasp and mimic the sloshing in the LST. It provides a combined hourglass regulation with a factor of 0.85 to provide the brick components with fluid behavior. Because the fluid simulation takes the distortion of the mesh to a large factor, an effective mesh control technique is needed. The analysis of finite elements provides a few methods to analysis, generally known as Lagrangian and Eulerian approaches. The material will move through the defined mesh boundaries in the Eulerian analysis while preventing any distortion of the element, whereas in the Lagrangian analysis, the material stays in the closed boundaries of the elements that do not cause the elements to be strongly distorted. The arbitrary Lagrangian–Eulerian (ALE) solution is used in the current simulation of the FSI to prevent high distortion in the finite elements. The material flow and movement caused in the mesh were controlled by the use of ALE to control the distortion in the study, resulting in a reduced disruption in the mesh. In the analysis, this lower distortion of the mesh ensures continuity. For the current case study, the ALE formulation used by the ABAQUS uses second-order advection and element center projection momentum advection that requires fewer iterations.

22.3 Numerical Study

The LST for the present study is taken up for the study of the dimension 6 m × 6 m × 4.8 m with a wall thickness of 0.3 m. The various constitutive material properties of the LST and fluid are shown in Table 22.1. For the present study of the LST, a total of five isolators are used that are attached to the rigid base. For an extensive review of the behavior of the LST under the seismic event, four different types of earthquakes are used for the nonlinear response history analysis in ABAQUS. The four different earthquakes are Kern County (1952), Parkfield (2004), Imperial Valley (1940), and Kocaeli (1999). The earthquakes are scaled to the three different PGA levels, namely 0.2 g, 0.4 g, and 0.6 g. The ratio between the horizontal components of the earthquakes is taken as 1:0.67.

For the analysis, the isolators are modeled by the connector elements which are available in the ABAQUS element library. For an effective isolator model, elasticity, damping, and bilinear behavior are selected. The connector element used in the analysis is built by Cartesian and align boundary conditions. The bottom node of the connector elements is joined to the fixed base, whereas the top node is connected to LST.

The effect of the directionality of the earthquake is an important and influential parameter that can alter the critical response in the LSTs. For the present study, five different angles of incidence are taken which are zero degrees, fifteen degrees, thirty degrees, forty-five degrees, sixty degrees, and seventy-five degrees. For base isolation of the LSTs, as there are no such recommendations for an optimum period, therefore, seven different isolation time periods are used to find out the optimum time period for response control. Different isolation properties are shown in Table 22.2.

Table 22.1 Properties of 3D square tank and fluid

Concrete	Fluid (water)
Modulus of elasticity, E_s = 24.86 GPa	Density, $\rho_w = 983.204$ kg/m^3
Density, $\rho_s = 2450$ kg/m^3	Equation of state: $c_0 = 1450$, $s = 0$, $\gamma_0 = 0$
Poisson's ratio, $\upsilon = 0.17$	Dynamic viscosity = 0.001 N s/m^2

Table 22.2 Properties of isolators designed for different time periods

Isolation period, T_{iso} (s)	Effective stiffness, K_{eff} (kN/m)	Elastic stiffness, K_e (kN/m)	Post-yield stiffness ratio, $\gamma = K_d/K_e$	Characteristic strength, Q_d (kN)
0.85	2250	17,690	0.09	30
1.0	1600	12,780	0.09	25
1.5	725	5700	0.09	18
2.0	400	3200	0.09	16
2.25	320	2525	0.09	14
2.5	250	1980	0.09	12
3.0	180	1420	0.09	9.5

22.4 Results and Discussions

The effectiveness of the LRB isolator for the LSTs under seismic ground motion is evaluated against the effective isolation period, type of ground motion, peak ground acceleration (PGA), and angle of incidence. The response quantities under investigation are shear force, overturning moment, hydrodynamic pressure, and sloshing height.

22.4.1 *Shear Force*

Figure 22.2 shows the effect of the variation of the effective isolation period on the shear force. It is seen from the estimates that with the increase in the T_b, the decrease in the shear force usually increases. However, after the value of $T_b = 2$ s, the rate of increase in the percentage reduction of the shear force is not very important. The curve becomes almost horizontal after a value of $T_b = 2.5$ s. The overall decrease in effective stress is 75% of the order.

For various types of earthquakes, the pattern variation of the percentage reduction of shear force with T_b is different before $T_b = 1.5$ s; after that, for all types of earthquakes, the pattern of variation is observed to be almost the same.

The variation between the percentage reduction of the shear force and the angle of earthquake incidence is seen in Fig. 22.3. The figure reveals that the percentage of reduction in shear force stays the same for all angles of incidence for most earthquakes, beginning from 15° to 75°, except for the Imperial Valley earthquake, which shows further reduction at 15° and 60°, but the percentage reduction rise is not very important.

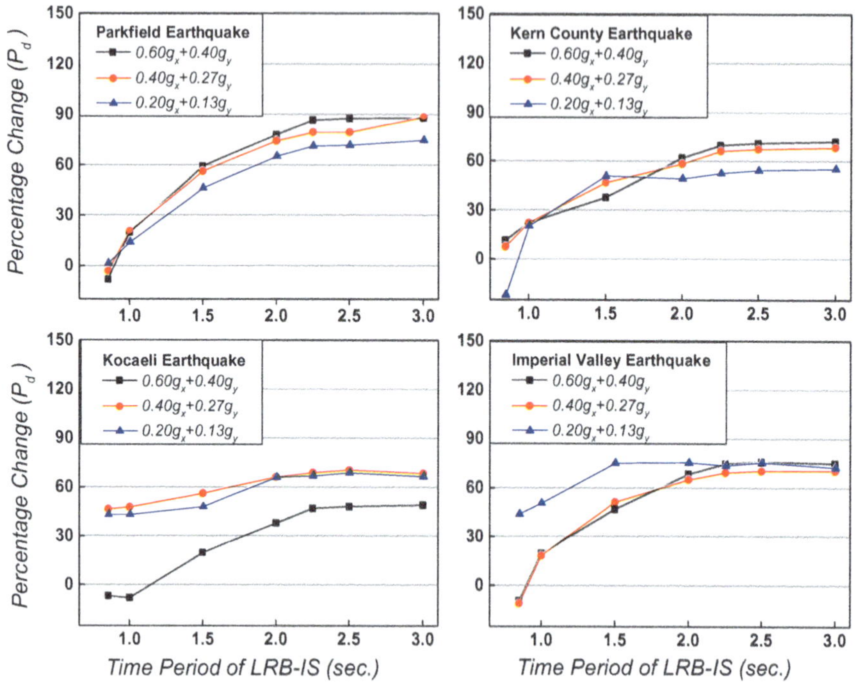

Fig. 22.2 Variation of percentage reduction in shear force with the time period of LRB isolator

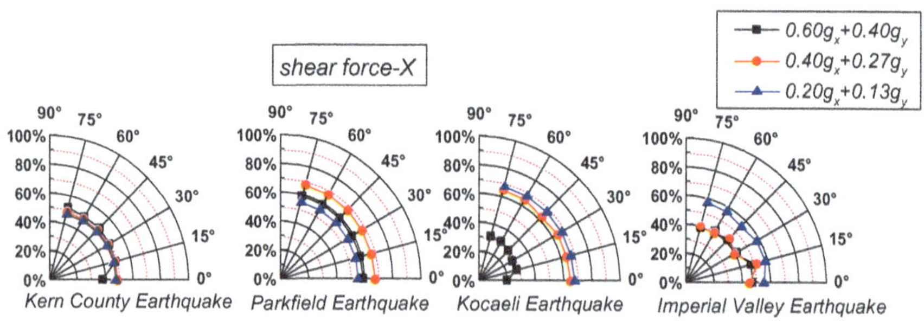

Fig. 22.3 Variation of percentage reduction in shear force with the angle of incidence

22.4.2 Overturning Moment

Figure 22.4 shows the effect of the variation of the effective isolation period on the overturning moment with the effective time period of the isolator. From Fig. 22.4, it can be seen that after the particular value of the isolator time period, the responses are no longer get affected. Like the pattern in the shear force reduction, the overturning moment shows similar behavior. Except for the Parkfield earthquake (near field with

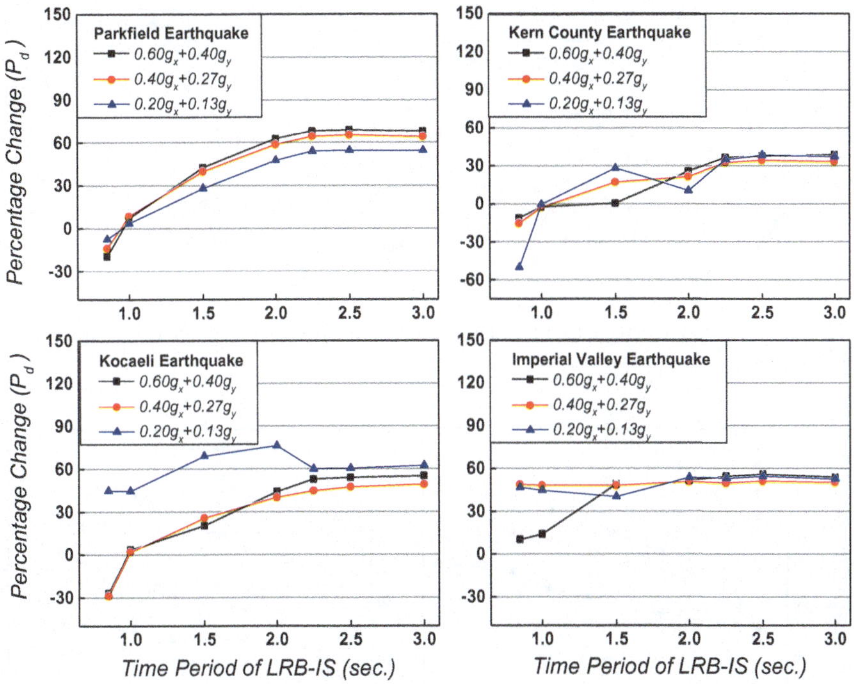

Fig. 22.4 Variation of percentage reduction in the overturning moment with the time period of LRB isolator

directivity effect), the rate of change in the reduction in response is very little in the overturning moments, and the maximum percentage reduction is on the order of 70–75%. The pattern of variation of the percentage reduction of the overturning moments with the T_b is distinct for different types of earthquakes up to $T_b = 1.5$ s; after that, the pattern of variation for all types of earthquakes does not change substantially.

The variations of the percentage decrease in the overturning moments with different PGAs do not show any definite pattern, like for the instance of the shear forces. As regards the difference of the reduction of the overturning moment with the angle of incidence, the percentage decrease, as shown in Fig. 22.5, is not sensitive to the variation of the angle of incidence of earthquake for all the cases under examination.

22.4.3 Hydrodynamic Pressure

The effect of the variation of the isolator time period on the reduction of the hydrodynamic pressure can be seen from Fig. 22.6. It can be seen from the figure that reduction in hydrodynamic pressure reaches a constant reduction value after a $T_b = 2.5$ s. In

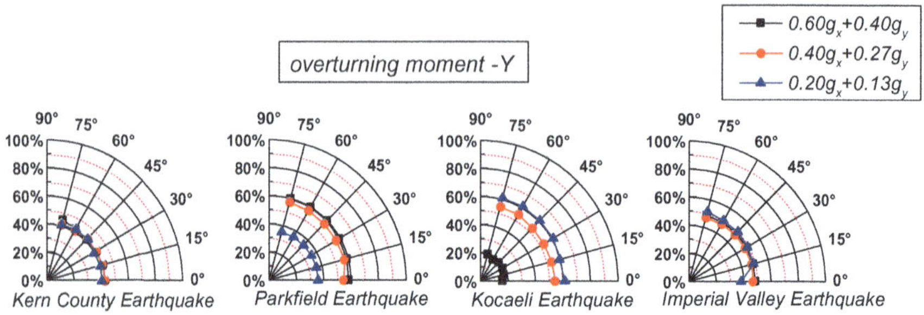

Fig. 22.5 Variation of percentage reduction in the maximum overturning moment with the angle of incidence

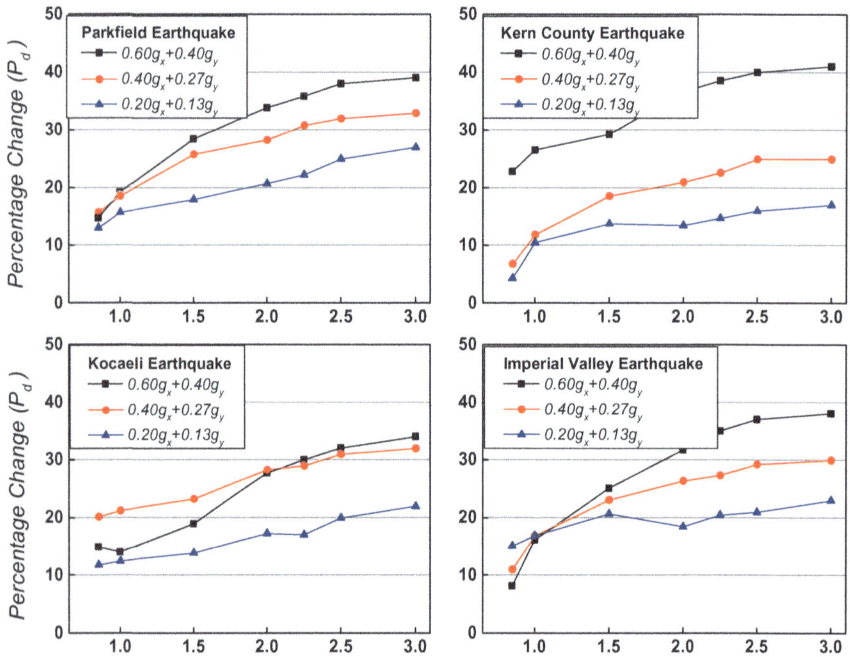

Fig. 22.6 Variation of percentage amplification in sloshing height with the time period of LRB isolator

comparison, it is shown that the magnitude of the difference depends to a greater degree on the type of earthquake relative to other response quantities. Compared to other response quantities, the influence of PGA on hydrodynamic pressure reduction is also prominent.

The effect of the angle of incidence of the earthquake on the reduction of the hydrodynamic pressure can be from Fig. 22.7. It can be seen from the figure that at an angle of incidence of fifteen degrees, the reduction is slightly more, whereas at an

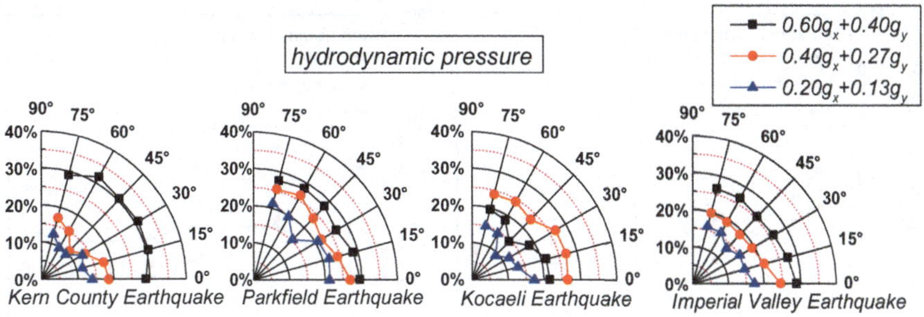

Fig. 22.7 Variation of percentage reduction in maximum hydrodynamic pressure with the angle of incidence

angle of incidence of forty-five degrees in all four cases of earthquakes, percentage reduction becomes reduced. Again, the effect of the PGA variation can be seen in the results of the angle of incidence variations.

22.4.4 Sloshing Height

In the base-isolated LSTs, the sloshing height is amplified in comparison to the other response quantities. The variation of the sloshing height percentage amplification with the effective time period (T_b) for various earthquake types is seen in Fig. 22.8. It can be seen from the figure that as the isolator time period is increased, the amplification in the sloshing height becomes constant, and even after the value of $T_b = 2$ s, there is a very small amplification in the response. The maximum amplification in the sloshing height is for the Kern County earthquake which is a far field earthquake. It can also be seen from the figure that the effect of the PGA on the amplification is also significant.

The effect of the angle of incidence of the earthquake on the amplification of the sloshing height can be seen in Fig. 22.9. It can be seen from the figure that the angle of incidence of the earthquake has little effect on the variation of the amplification in the sloshing height. The amplification in the sloshing height gets lesser as the angle of incidence of earthquake reaches the range of thirty to sixty degrees. The maximum variation of amplification can be seen for the Kocaeli earthquake which is a near field earthquake with fling step effect, whereas the minimum variation is for the Imperial Valley earthquake.

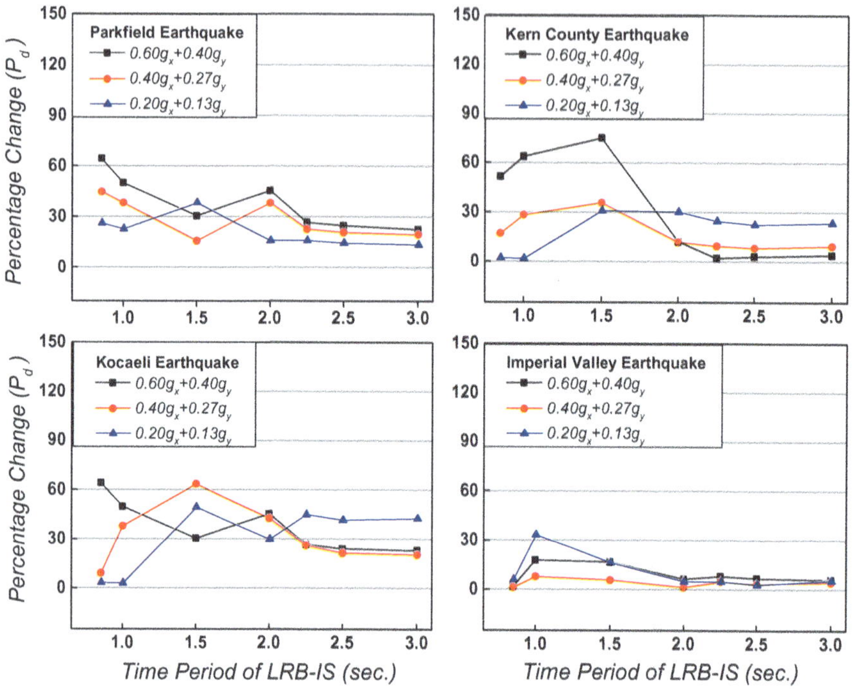

Fig. 22.8 Variation of percentage amplification in sloshing height with the time period of LRB isolator

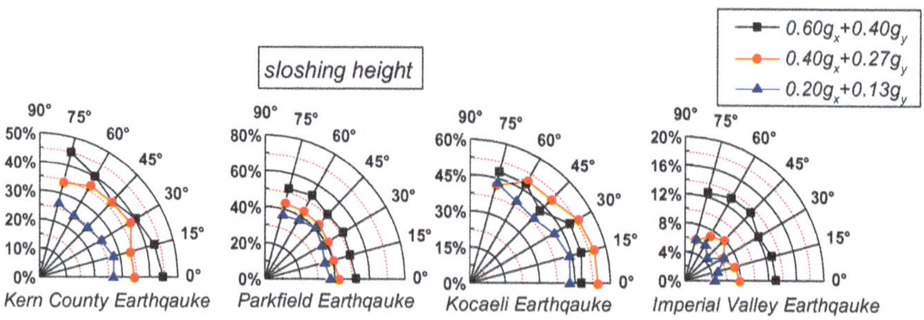

Fig. 22.9 Variation of percentage amplification in sloshing height with the time period of LRB isolator

22.5 Conclusions

The effectiveness of lead rubber bearing (LRB) in mitigation of the structural response and amplification of the sloshing height is studied by varying several important parameters. They include (i) effective time period of LRB isolator, (ii) the type of the ground motion, (iii) angle of incidence of the earthquake, and (iv) PGA. The reductions in the response quantities are shear force, overturning moment, and hydrodynamic pressure. Also, it is known fact that seismic isolation tends to amplify the sloshing height which is also taken in the consideration for the present study. For the numerical study, a concrete LST with dimensions of 6 m ⊗ 6 m ⊗ 4.8 m is studied with a fluid height of 3.6 m and having five LRB base isolators attached to the base of the LST. The study reveals the following notable conclusions:

1. With the increase in the effective time period up to 2.5 s, decreases in the response quantities of interest and the enhancement in sloshing height usually increase, after which they appear to become stationary.
2. The maximum reduction of the order of 60–70% can be seen for the shear force and overturning moment at the optimum time period of the isolator system.
3. There is an ideal effective time period in which it is possible to obtain relatively high reductions in response quantities with much less amplification of the sloshing height; this effective time period is observed to be 2 s for the given study.
4. The effect of the type and PGA of ground motion are found to be an important factor for governing the responses.
5. Except for the drop in maximum hydrodynamic pressure and sloshing height amplification, the decrease in other response quantities is indifferent to the difference in the angle of incidence of the earthquake.

References

1. Housner GW (1963) The dynamic behavior of water tanks. Bull Seismol Soc Am 53:381–387
2. Housner GW (1957) Dynamic pressures on accelerated fluid containers. Bull Seismol Soc Am 47:15–35
3. Gupta RK, Hutchinson GL (1988) Free vibration analysis of liquid storage tanks. J Sound Vib 122:491–506
4. Sakal F, Nishimura M, Ogawa H (1984) Sloshing behavior of floating-roof oil storage tanks. Comput Struct 19:183–192
5. Gavrilyuk IP, Lukovsky IA, Timokha AN (2005) Linear and nonlinear sloshing in a circular conical tank. Fluid Dyn Res 37:399–429. https://doi.org/10.1016/j.fluiddyn.2005.08.004
6. Faltinsen OM, Rognebakke OF, Lukovsky IA, Timokha AN (2000) Multidimensional modal analysis of nonlinear sloshing in a rectangular tank with finite water depth. J Fluid Mech 407:201–234. https://doi.org/10.1017/S0022112099007569
7. Vern S, Shrimali MK, Bharti SD, Datta TK (2021) Seismic behavior of baffled liquid storage tank under far-field and near-field earthquake. In: Recent advances in computational mechanics and simulations. Springer, Singapore, pp 445–456. https://doi.org/10.1007/978-981-15-8138-0_34
8. Vern S, Shrimali MK, Bharti SD, Datta TK (2020) Impact of angle of incidence in rectangular liquid storage tanks. In: Technologies for sustainable development. CRC Press, pp 68–72. https://doi.org/10.1201/9780429321573-13
9. Vern S, Shrimali MK, Bharti SD, Datta TK (2021) Behavior of liquid storage tank under multidirectional excitation. In: Lecture Notes in Civil Engineering. Springer Science and Business Media Deutschland GmbH, pp 203–217. https://doi.org/10.1007/978-981-15-5235-9_16

10. Vern S, Shrimali MK, Bharti SD, Datta TK (accepted) Optimum passive control in the liquid storage tank by using multiple vertical baffles. Pract Period Struct Des Constr ASCE
11. Rawat A, Mittal V, Tanusree C, Matsagar V (2019) Earthquake induced sloshing and hydrodynamic in rigid liquid storage tanks analyed by coupled acoustic-structural and Euler-Lagrange methods. Thin-Walled Struct 333–346
12. Kirtas E, Rovithis E, Makra K (2020) On the modal response of an instrumented steel water-storage tank including soil-structure interaction. Soil Dyn Earthq Eng 135:106198. https://doi.org/10.1016/j.soildyn.2020.106198
13. Harry SW, Hampton FP (1999) Seismic response of isolated elevated water tanks. J Struct Eng 125:965–976
14. Shrimali MK, Jangid RS (2002) Non-linear seismic response of base-isolated liquid storage tanks to bi-directional excitation. Nucl Eng Des. https://doi.org/10.1016/S0029-5493(02)00134-6
15. Shrimali MK, Jangid RS (2003) Seismic response of isolated liquid storage tanks with elastomeric bearings. J Vib Control 2:1201–1218
16. Shrimali MK, Jangid RS (2003) Earthquake response of isolated elevated liquid storage steel tanks. J Constr Steel Res 59:1267–1288. https://doi.org/10.1016/S0143-974X(03)00066-X
17. Panchal VR, Jangid RS (2012) Behaviour of liquid storage tanks with VCFPS under near-fault ground motions. Struct Infrastruct Eng 8:71–88. https://doi.org/10.1080/15732470903300919
18. Shrimali MK, Jangid RS (2002) Seismic response of liquid storage tanks isolated by sliding bearings. Eng Struct 24:909–921. https://doi.org/10.1016/S0141-0296(02)00009-3
19. Liang B, Tang J (1994) xiang: Vibration studies of base-isolated liquid storage tanks. Comput Struct 52:1051–1059. https://doi.org/10.1016/0045-7949(94)90089-2
20. Son I-M, Kim J-M, Lee C (2019) Seismic soil-structure interaction analyses of LNG storage tanks depending on foundation type. J Comput Struct Eng Inst Korea 32:155–164. https://doi.org/10.7734/coseik.2019.32.3.155
21. Villegas-jim O, Tena-colunga A (1999) Dynamic design procedure for the design of base isolated, pp 1–8
22. Cheng X, Jing W, Gong L (2018) Dynamic responses of a sliding base-isolated RLSS considering free surface liquid sloshing. KSCE J Civ Eng 00:1–13. https://doi.org/10.1007/s12205-018-0154-z
23. Wen Y (1976) Method for random vibration of hysteretic systems. J Eng Mech Div 102:246–263
24. Bonet JL, Miguel PF, Fernandez MA, Romero ML (2006) Analytical approach to failure surfaces in reinforced and biaxial bending. J Struct Eng Eng 130:1133–1144. https://doi.org/10.1061/(ASCE)0733-9445(2004)130

Chapter 23
Estimating Service-Life Deterioration of RC Bridges Due to Multi-hazards in Barak Valley Region, Assam, India

Joydeep Das and Arjun Sil

23.1 Introduction

The bridge structure is a necessary resource for the city or country to maintain a proper and efficient transport system. Deterioration of the structure and its system can be the result of unsatisfactory performance of the structure under operating conditions and can lead to accidents and hazards that can be caused by earthquakes, fires, vehicle collisions, and other causes of natural and human-made disasters. The transport infrastructure is aging rapidly. The increase in traffic in both urban and rural areas puts more pressure on the bridge network than the original objectives. Bridge failure can be devastating to both human life and economic loss. The reliability and sustainability of structural infrastructure facilities are critical to the development and growth of any modern society.

Reinforced concrete (RC) bridges deteriorate with time due to aging, material and construction flaws, interaction with extreme loads, lack of flexibility, and aggressive environments. The aging or maturity factor is one of the most noticeable hazards to which RC bridges are exposed during their service life. Furthermore, if the bridges are in an earthquake-prone location, the structures can deteriorate prematurely and ultimately collapse. Bridge caretaker or inspectors need a consistent way of evaluating the structural stability of the bridge in order to keep the traffic system operating properly but simultaneously maintaining public safety.

Many previous studies analyzed structural safety and reliability based on different approaches and concepts. For the design code of bridges [21], Nowak and Lind [24] used probability theory and proposed the standardized safety bridge design for steel and pre-stressed concrete bridges for the Canadian province named Ontario. Thereafter, as per AASHTO (1983) design code, Ghosn and Moses [14] used reliability and analyzed bridge safety based on loading done by an actual truck on the span of bridge. Enright and Frangopol [9–10] and Dey et al. [8] presented the concepts

J. Das (✉) · A. Sil
Department of Civil Engineering, National Institute of Technology Silchar, Silchar, Assam, India

S. Kolathayar et al. (eds.), *Civil Engineering for Disaster Risk Reduction*, Springer Tracts in Civil Engineering, https://doi.org/10.1007/978-981-16-5312-4_23

of service-life improvement of the RC bridges attacked by environmental factors, failure time prediction of structure in damaged states, and time-variant bridge reliability analysis. Moreover, the probabilistic approach is also employed significantly to illustrate deterioration and reliability of the bridge structure. Mauch and Madanat [20] discussed the probabilistic models and types in their study. The deterioration predicting models of infrastructure mainly consist of a discrete-time model, state-based model, and time-based model. Mauch and Madanat [20] presented the suitability of using each type of model based on the empirical nature of the condition data present for modeling and estimating. Sasmal and Ramanjaneyulu [25] scientists from India have also presented a case study that consists of the idea of evaluating and prioritizing condition of existing bridges and their elements with the help of computer programs (fuzzy logic approach and analytic hierarchy process). Nasrollahi and Washer [23] revealed the time-in-CR of the bridges (Oregon, US) for different levels of reliability from the 20 years along with the inspection interval specified in the NBIS and found Weibull statistics could be used for estimating structural adequacy based on CR of the superstructure.

Subsequently, Ayyub et al. [1] presented critical assets and collection of risk analysis frameworks for assessing risk to various disasters or hazards. Li and Ellingwood [19] performed several risk assessments for wood structures endangered mainly to hurricane and earthquake hazard areas. Beaver [3] studied the overall risk of bridges due to earthquakes, hurricanes, and ships. Kameshwar and Padget [17] have proposed a multi-risk approach to earthquake and storm-affected bridges.

Assam is one of the states in northeast India, having 1450 bridges (steel, reinforced concrete, pre-stressed concrete) all around the state including highway and rural regions. The study area is located between longitude 92° 15′–93° 15′ east and latitude 24° 8′–25° 8′ north (Fig. 23.1). The two concerning deteriorations that are observed in the bridges situated in this region are maturity and earthquake impact in the infrastructures. Between 2012 and 2019, Barak valley has felt tremors of 10–17 events of magnitude 3–5.5 (Indian Meteorological Department). The Barak valley mainly comprises three districts namely Cachar, Karimganj, and Hailakandi. As a result, it can be inferred that the region's infrastructures are vulnerable to such unexpected shocking events, which would increase the risk and degradation.

The objective of this paper is to probabilistically compute the bridge survivability, by using condition assessment method and gain insights into the impression of the deterioration in the service life of the bridges (Fig. 23.2). Firstly, the bridge inspection data is needed to get the degree of bridge damage and then categorizing the bridges as per the NBIS condition rating method. Secondly, evaluating the structural survivability of the bridges impacted due to maturity factor by considering probabilistic approach and finally combining the deterioration of maturity and earthquake factor, which will result in getting the impact on service life of the bridges.

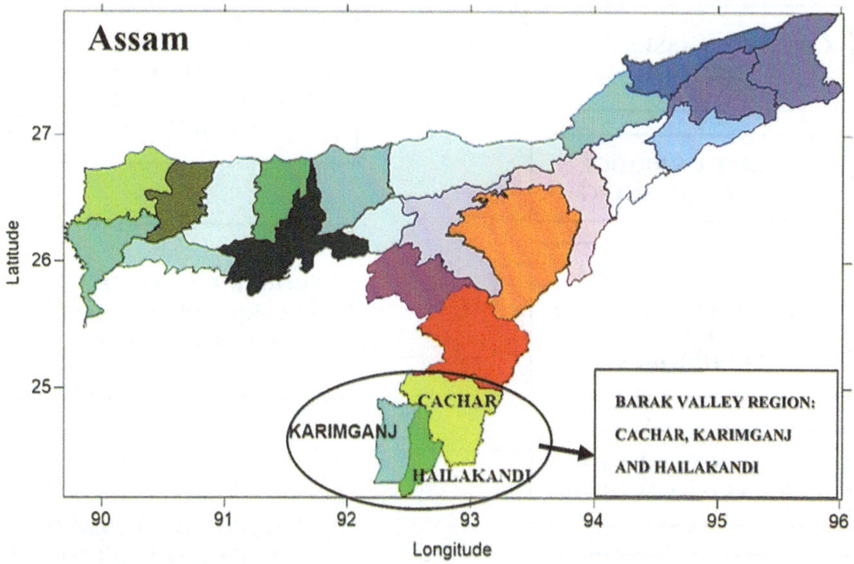

Fig. 23.1 Map of Assam showing the selected study region

23.2 Condition Assessment Method and Condition Rating Technique

Condition assessment method (CAM) has become an important tool to forecast future performance and improve bridge maintenance, restoration, and additional needs. Visual Inspection (VI) is the default and preliminary method used to determine the bridge damage factors, yet its outcomes deeply rest on the expertise and judgment of bridge inspectors, yielding primarily qualitative and subjective results. Although, there is a much-advanced procedure that comes under CAM. VI method is found to be the initial process, producing subjective output, and cost involved is very less. The statutory document providing guidance in bridge inspection procedures in the United States is NBIS (National Bridge Inspection Standard) published by the FHWA [12]. This article serves as an asset of guidelines and standards that must be met in bridge inspections for both state-owned and federal infrastructures.

Even though there is no clear description of visual inspection (VI) as a recommended examination task, from the FHWA article, it can be concluded that VI is the actual initial method of a routine examination. VI is the default method; however, some limitations can affect efficient decision-making and resource utilization. Some of these concerns are summarized in the FHWA report [22].

The BMS adopted in developed countries such as the USA and UK performed satisfactorily well, and the work was done as well as progress reported in many kinds of literature. Other countries such as China, Australia, Serbia, Canada, and Brazil

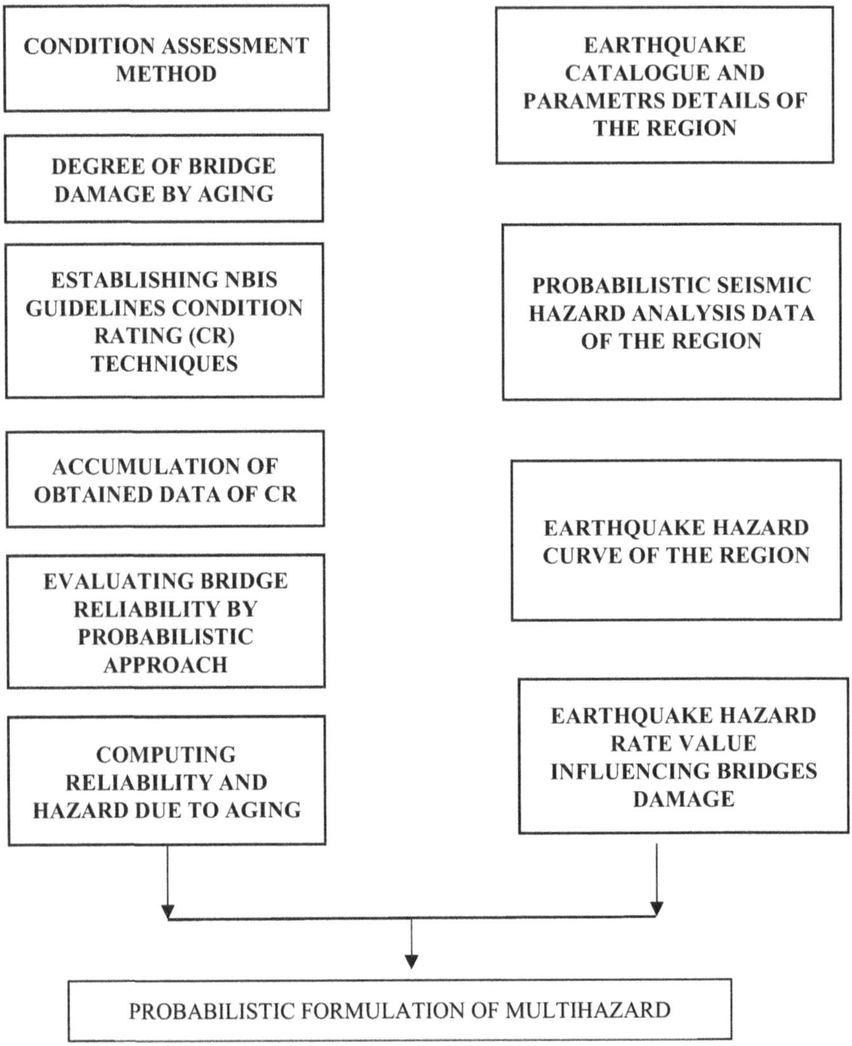

Fig. 23.2 Schematic representation of probabilistic multi-hazard approach

also initiated to work on the bridge deterioration forecasting, reliability prediction, and inspection interval data by referring code from USDOT, NBI, NBIS, and FHWA.

India being the country of emerging economic growth and sustainable development in every aspect consisting of 29 states and 7 union territories. The importance of connectivity or transportation facility is the backbone for movement and growth of urban development.

The Indian BMS remains outdated and the design code such as Indian road congress [IRC: SP 040 (1993), SP 018 (1996) [15], SP 047 (1998), SP 052 (1999) [16], SP 013 (2004), and SP 74 (2007)] suggests the general procedure of inspection

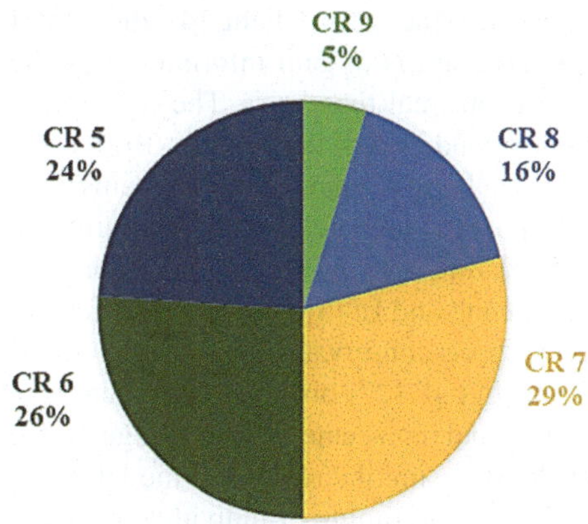

Fig. 23.3 The number of bridges (%) as per the NBIS guidelines condition rating techniques

and maintenance guiding principle and norms; however, it becomes silent about the CR of the existing bridges. Thus, there is a need for such BMS method that will produce insights into the existing bridges facing deterioration.

National bridge inventory includes NBI numerical database assessment of bridge component or whole system provided by inspectors during most semi-annual inspections. Ratings range from 0 to 9. A rating of 0 for the bridge indicates a failed state. Other classifications indicate: (1) represents imminent failure; (2) critical condition; (3) very poor condition; (4) poor condition; (5) fair condition; (6) satisfactory condition; (7) good condition; (8) in very good condition, and (9) shows excellent condition. The bridge is structurally classified as inefficient or unusable if the surface attains a position rating of less than (5).

Bridge inspection is a crucial element of any bridge management system (BMS) and system health monitoring (SHM) principally for aged and deteriorated bridges and a trail to condition rating (CR). NBI coding guide [13] mentioned that CR of the bridge is employed to describe the existing, in-place bridge as compared to the as-built condition. The bridges in the study area have been examined and found the bridge condition ratings are shown in Fig. 23.3. To solve the deterioration problem, the procedures necessary for the introduction of new concepts with proper bridge life cycle assessment, maintenance planning, proper inspection procedures, monitoring, modeling, and reliability became important.

23.3 Combined Survivability of Maturity and Earthquake Impact

A structural reliability assessment is usually the result of identifying a key structural or purposeful deficit during detailed scrutiny. Expected results from this inspection are the depiction of fundamental defects, evaluation of the damage by using a different model that estimates the remaining service life and its current load-carrying capacity.

As per researcher Chang [4], the aim of structural monitoring or bridge inspection program is to gain information on the reliability of in-service structures on a continuous real-time basis. The condition assessment method estimates the existing fitness condition of the bridges. Bridges are a multifarious mixture of systems, but almost all bridge management systems use the evaluation of members or components as inputs to estimate overall structural dependability [28]. The damage prediction process forecasts the future state of bridge elements.

Cheung and Li [5] provided a review on deck recognition dynamics to complete deck space conservation and meet long-term and short-term growth assessments. Stewart et al. [27] observed that there is a process of cracking and splicing of the service and replacement of the bridge deck due to the integration and hobby of the life cycle. By using the reliability model, it is possible to obtain information regarding the needs and timing of individual bridge deck maintenance and rehabilitation. The static evaluation process of bridge deck recognition is minimized and the deck is conservative and helps to increase its repair efficiency.

The term "survival" in general means the state or fact of continuing to live or exist from an earlier period, in engineering terms, called "reliability." In bridge engineering, reliability or survival analysis is significantly important to evaluate the likelihood or probability of a structure or component (structural element) ceasing to fail under various situations. Also, the hazard function is the function of the failure rate that changes the conditional predictions of failure.

Reliability of a bridge, S_b, is:

$$S_b = 1 - P_f \tag{23.1}$$

where P_f is the probability of failure.

The reliability and hazard evaluation can be achieved by developing many deterioration process models; nevertheless, most dependable approaches are of two types: deterministic model and probabilistic model. Both models are well-known and well accepted by many researchers for predicting bridge deterioration.

Probabilistic modeling is used in the deterioration process as it deals with randomness, uncertain factor, and unknown variables to predict service-life and hazard function. The probability also deals with variability (data are spread in a certain way) and uncertainty (percentage of the appropriateness of the data) of the data. Probabilistic analysis is used to evaluate both quantitative and qualitative attributes for the uncertainty model and input.

Here in this study, the data or the information collected to model deterioration is dependent on time and CR. Therefore, the probabilistic approach of time-dependent model is suitable in finding survival or reliability parameters and hazard attributes. There are three approaches regarding probability of survival or reliability analysis: (1) parametric method, (2) semi-parametric method, and (3) non-parametric method.

The parametric method is better in descriptive and brief calculation than the other two methods. The commonly used probability distribution functions are exponential, Weibull, lognormal, normal, extreme values, and gamma. Many researchers have provided the importance of Weibull distribution; this is a special kind of distribution used in finding reliability and hazard prediction for the bridge infrastructures [7, 6].

Weibull distribution is a strictly positive-valued distribution and consists of two characteristics parameters: scale parameter (η) and shape parameter (β). Weibull distribution parameters are calculated by method of maximum likelihood (using MATLAB and Minitab probabilistic-statistical software) and include mean, standard deviation, and both parameters (shape and scale) are evaluated with 95% CI (confidence intervals). The probability density function is given below, where Y is the minimum life.

$$f(t) = \left(\frac{\beta}{\eta}\right)\left(\frac{t-Y}{\eta}\right)\mathrm{e} - \left(\frac{t-Y}{\eta}\right)^{\beta} \tag{23.2}$$

Practically, in most survival function situations, Y is taken as zero (as failure time expected to start at $t = 0$) and Eq. (23.11) becomes,

$$f(t) = \left(\frac{\beta}{\eta}\right)\left(\frac{t}{\eta}\right)\mathrm{e} - \left(\frac{t}{\eta}\right)^{\beta} \tag{23.3}$$

The survival functions $S(t)$ and hazard function $h(t)$ equation become,

$$S(t) = \mathrm{e} - \left(\frac{t}{\eta}\right)^{\beta} \tag{23.4}$$

$$h(t) = \left(\frac{\beta}{\eta}\right)\left(\frac{t}{\eta}\right)^{\beta-1} \tag{23.5}$$

Figures 23.4 and 23.5 illustrating the survivor function and hazard of various CR bridges of NH route. Also, Figs. 23.6 and 23.7 illustrating the survivor function and hazard of various CR bridges of rural route. Tables 23.1 and 23.2 summarized depicting the surviving function and hazard with respect to age for all CR bridges.

Every year, earthquakes inflict major economic damage as well as human casualties. Recently, the rapid population growth and rapid demographic growth scattering are observed as well as the escalation of infrastructure growth led to enormous effects on the susceptibility to earthquake hazards and risk associated with it. The Northeast region is identified as zone V with the highest hazard. The probabilistic models associated with time-dependent earthquake hazards have been employed in recent years. Sitharam and Sil [26] developed seismic hazard maps (data catalog has been gathered from 1731 to 2011) for Tripura and Mizoram states using DSHA and PSHA methodologies. Bahuguna and Sil [2] have carried out DSHA and PSHA of Assam state (data catalog used are gathered from 1761 to 2015) and calculated PGA values for the 2%, 5%, and 10% probability in 50 years and 2%, 5%, 10% probability in 100 years. Bahuguna and Sil [2] reported using PSHA to find PGA values and mentioned that Oldham is the controlling source of Silchar, Hailakandi, and Karimganj with a maximum magnitude (M_w) of 8.6.

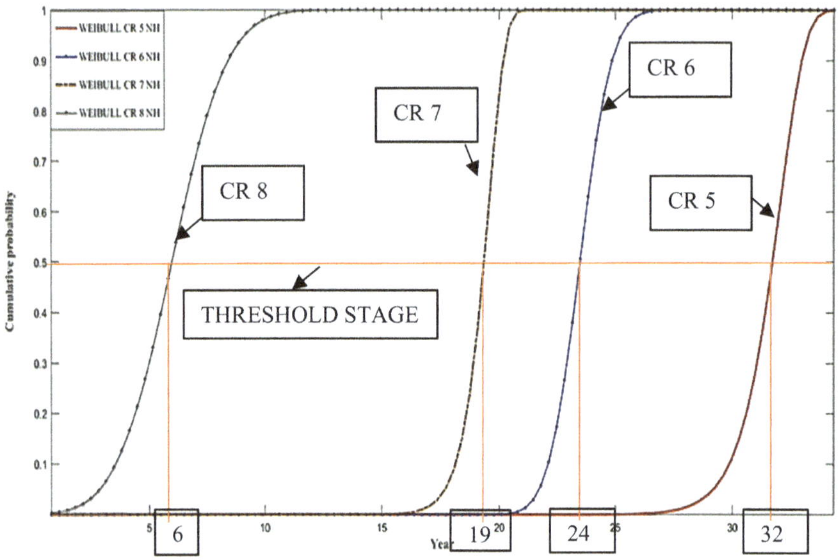

Fig. 23.4 Survivor function plot of Weibull distribution for all CR of NH route bridges

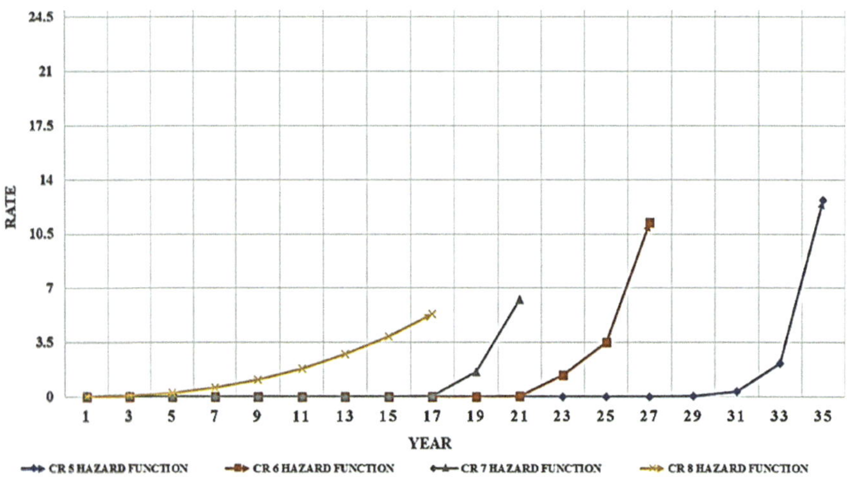

Fig. 23.5 Hazard plot for all CR by Weibull distribution of the NH bridges

In the current study, as mentioned earlier, the active faults parameters, such as maximum magnitude, PGA values (DSHA and PSHA), and knowledge about the faults and lineaments needed for finding the mean probability rate of exceedance obtained from Bahuguna and Sil [2]. The present study is related to the methodology and work are done by Bahuguna and Sil [2] because of the Barak valley region situated

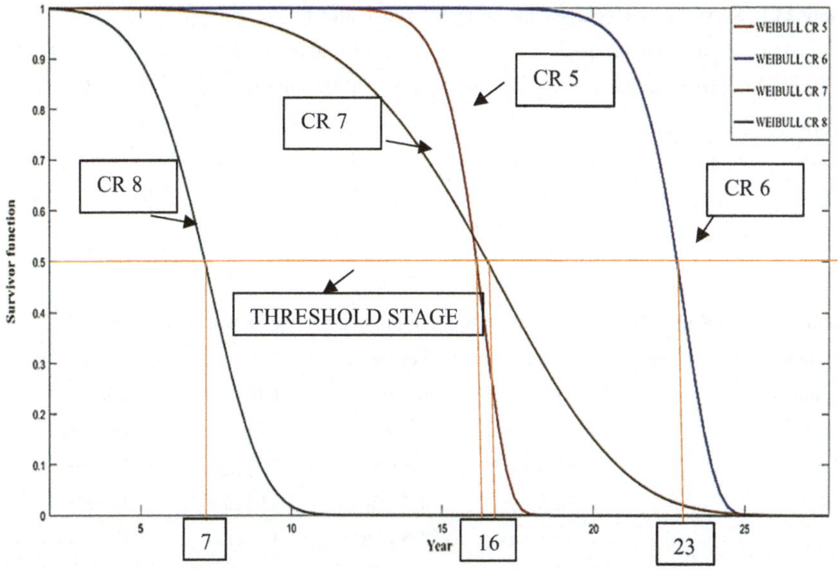

Fig. 23.6 Survivor function plot of Weibull distribution for all CR of rural route bridges

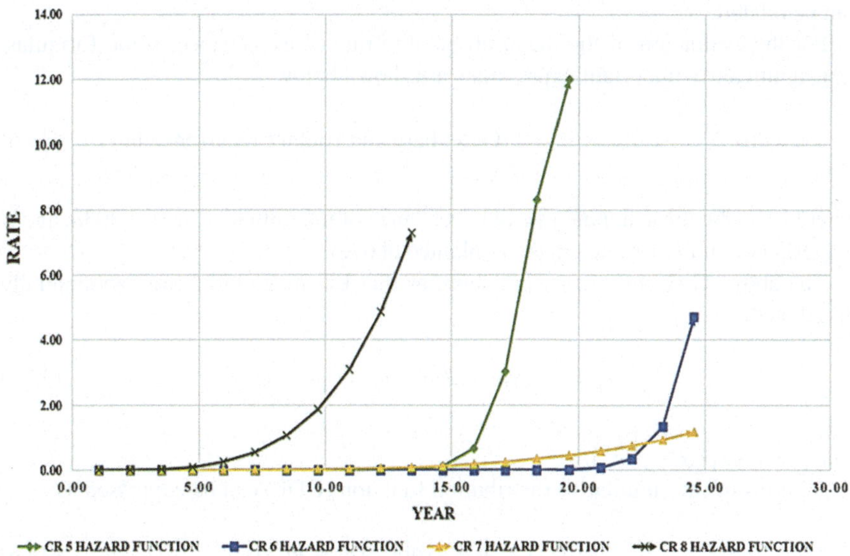

Fig. 23.7 Hazard plot for all CR by Weibull distribution of the rural bridges

Table 23.1 Surviving percentage concerning the age of all CR

Surviving percentage	NH route bridges		Rural route bridges	
	100% till the age	50% till the age	100% till the age	50% till the age
CR 5	21	32	10	16
CR 6	17	24	17	23
CR 7	13	19	6	16
CR 8	1	6	2	7

Table 23.2 Showing cumulative hazard rate for 1 and rate for 5 with respect to years

Cumulative hazard rate	For rate 1	For rate 1	For rate 5	For rate 5
Route type	NH	Rural	NH	Rural
CR 5	2.2 years	2 years	33 years	18 years
CR 6	4.2 years	4.6 years	26 years	24 years
CR 7	5.2 years	4.8 years	21 years	23 years
CR 8	6.6 years	3.6 years	11 years	10 years

in the NE region. Thereby, the seismicity parameters and features are similar and can be related.

For the evaluation of the mean probability rate of exceedance, some formulas, and significantly important expressions are shown below.

$$\lambda_m = 10^{a-bm}, \text{ or, } \lambda_m = e^{\alpha-\beta m} \text{(Gutenberg and Richter recurrence law)} \quad (23.6)$$

where λ_m is the mean annual rate of exceedance of magnitude m, $\alpha = 2.303a$, and $\beta = 2.303b$ (a and b parameters are explained above).

The above G-R law expression signifies that EQ magnitudes are exponentially distributed:

$$\lambda_m = \nu \exp[-\beta(m - m_0)],\ m > m_0 \quad (23.7)$$

where $\nu = \exp(\alpha - \beta m_0)$.

In terms of the cumulative distribution function (CDF) can be expressed as:

$$F_M(m) = 1 - \exp[-\beta(m - m_0)] \quad (23.8)$$

The probability distribution function (PDF) can be stated as:

$$f_M(m) = \beta \exp[-\beta(m - m_0)] \quad (23.9)$$

The standard Gutenberg-Richter law predicts non-zero mean rate of exceedance for magnitude that can be employed in determining the maximum magnitude associated with the source zone and can be articulated as:

$$\lambda_m = \frac{\exp[-\beta(m - m_0)] - \exp[-\beta(m_{\max} - m_0)]}{1 - \exp[-\beta(m_{\max} - m_0)]}, m_0 \leq m \leq m_{\max} \quad (23.10)$$

For magnitude uncertainty, the fault magnitude may vary from m_o to $M_{\max}$ and magnitude is a random variable M; therefore, probability distribution function is:

$$p(m) = \frac{\beta e^{-\beta(m-m_0)}}{1 - e^{-\beta(M_{\max}-m_0)}}, m_0 \leq m \leq M_{\max} \quad (23.11)$$

The probabilities of earthquakes of several sizes happening in finite periods of time are typically calculated by presuming that earthquakes occur as Poisson processes [18]. As per the Poisson distribution, probability of a random variable given as:

$$P(Z > x) = 1 - \exp(-\lambda t) \quad (23.12)$$

where x is the exceeding level, T is time period and λ is annual rate of exceedance. The annual rate of exceedance (λ) expressed as:

$$\lambda = \sum_{x=1}^{M_x} M_x(m_0) \quad (23.13)$$

Let us assume that earthquakes that would affect bridges (sudden shocking events that occurred in the region) are modelled as a system of Poisson events with an average annual rate, which is considered as λ (Table 23.3). Assuming that failure defined for aging and failure due to earthquake events are in a similar manner affecting the bridge deterioration.

Primarily, this is to be noted that the exponential distribution is implied to model failure times for the earthquake hazard, that is,

$$h_{\text{eq}}(t) = \lambda \quad (23.14)$$

Moreover, for the bridge deterioration that is due to the aging effect, the hazard may be described by the Weibull distribution on Eq. (23.5) which is listed earlier, that is,

$$h_{\text{age}}(t) = \left(\frac{\beta}{\eta}\right)\left(\frac{t}{\eta}\right)^{\beta-1} \quad (23.15)$$

Combining both the above equation of hazard rates,

Table 23.3 Calculated λ_m for different magnitude associated with Oldham and Lohiti fault

Calculated mean annual rate of exceedance for different magnitude								Average value			
$\lambda_m > 5$		$\lambda_m > 6$		$\lambda_m > 7$		$\lambda_m > 8$		$\lambda_m > 5$	$\lambda_m > 6$	$\lambda_m > 7$	$\lambda_m > 8$
Oldham	Lohiti	Oldham	Lohiti	Oldham	Lohiti	Oldham	Lohiti				
0.438	0.390	0.225	0.151	0.091	0.056	0.023	0.018	0.414	0.188	0.073	0.020

Table 23.4 Difference between aging effect and combined effect [Aging + EQ ($\lambda > 5$)]

Difference in mode of $f(t)$	Aging effect	Combined effect [Aging + EQ ($\lambda > 5$)]
For CR 5 (rural bridges)	17 years	9 years
For CR 5 (NH bridges)	26 years	11 year
For CR 8 (rural bridges)	7 years	4 year
For CR 8 (NH bridges)	6 years	5 year

$$h_{\text{combined}}(t) = h_{\text{eq}}(t) + h_{\text{age}}(t) = \lambda + \left(\frac{\beta}{\eta}\right)\left(\frac{t}{\eta}\right)^{\beta-1} \tag{23.16}$$

The significant observations that came forward after evaluating the combined hazard function [$h_{\text{combined}}(t)$] and combined probability distribution function [$f(t)_{\text{combined}}$] by using the above equations are

1. The hazard rate increases when both hazards have interacted which are obvious for all the CR of the bridges (as shown in Fig. 23.9).
2. The probability distribution functions (PDFs) mode fluctuates and variations are observed when the EQ magnitude increases. Some cases are tabulated below (Tables 23.4, 23.5, 23.6 and 23.7) which signifies that if occurrences of a particular magnitude are more, then it results in reduction of failure times or service life (that means the failure rate increases). Here, the occurrence of EQ ($\lambda > 5$)

Table 23.5 Difference between aging effect and combined effect [Aging + EQ ($\lambda > 6$)]

Difference in mode of $f(t)$	Aging effect	Combined effect [Aging + EQ ($\lambda > 6$)]
For CR 6 (rural bridges)	23 years	13 years
For CR 6 (NH bridges)	23 years	14 years
For CR 8 (rural bridges)	7 years	6 years
For CR 8 (NH bridges)	6 years	3 years

Table 23.6 Difference between aging effect and combined effect [Aging + EQ ($\lambda > 7$)]

Difference in mode of $f(t)$	Aging effect	Combined effect [Aging + EQ ($\lambda > 7$)]
For CR 5 (rural bridges)	16 years	15 years
For CR 5 (NH bridges)	26 years	24 years
For CR 8 (rural bridges)	7 years	7 years
For CR 8 (NH bridges)	6 years	5 years

Table 23.7 Difference between aging effect and combined effect [Aging + EQ ($\lambda > 8$)]

Difference in mode of $f(t)$	Aging effect	Combined effect [Aging + EQ ($\lambda > 8$)]
For CR 5 (rural bridges)	16 years	15 years
For CR 5 (NH bridges)	26 years	24 years
For CR 7 (rural bridges)	17 years	17 years
For CR 7 (NH bridges)	17 years	16 years

and EQ ($\lambda > 6$) are more frequent than EQ ($\lambda > 7$) and EQ ($\lambda > 8$), thereby the effect of EQ ($\lambda > 5$) and EQ ($\lambda > 6$) are more in the bridges which increase hazard rate and lowering the value of failure time or service life. These observations can be easily identified from the figures presented below (Fig. 23.8, 23.10, and 23.11) typically showing the probability distribution function graph $f(t)$ for both the damaging effect in terms of different magnitudes (Fig. 23.9).

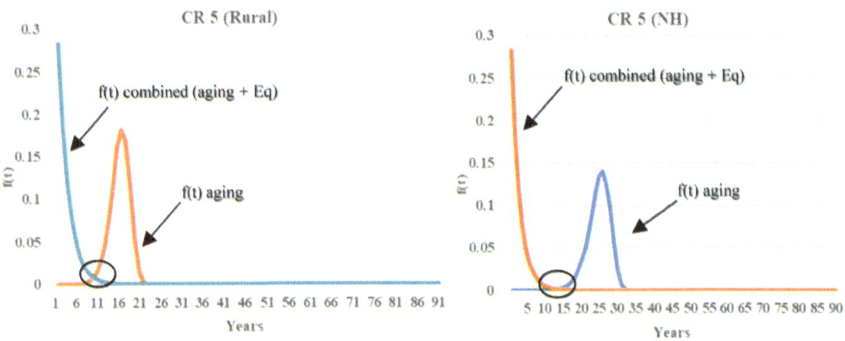

Fig. 23.8 PDF for combined bridge hazards for CR 5 RC bridges in rural and NH routes (Earthquake magnitude > 5)

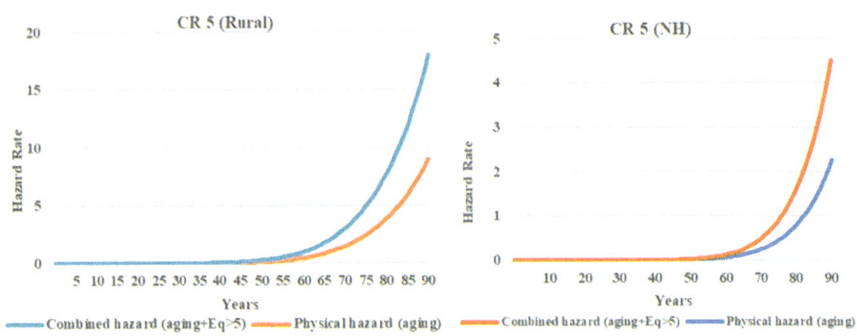

Fig. 23.9 Hazard rate function for combined bridge hazards for CR 5 RC bridges in rural and NH routes

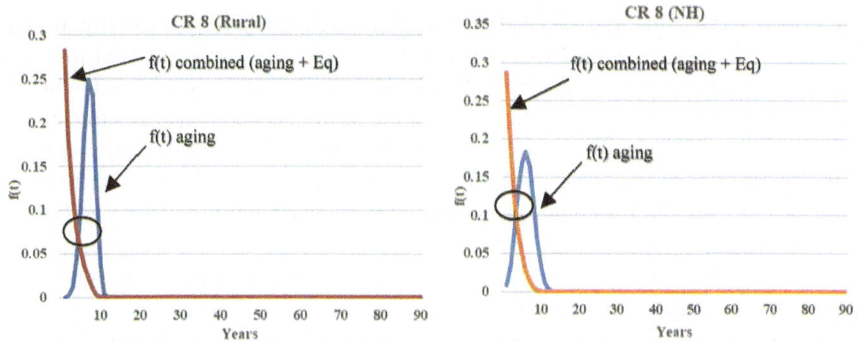

Fig. 23.10 PDF for combined bridge hazards for CR 8 RC bridges in NH routes (Earthquake magnitude > 5)

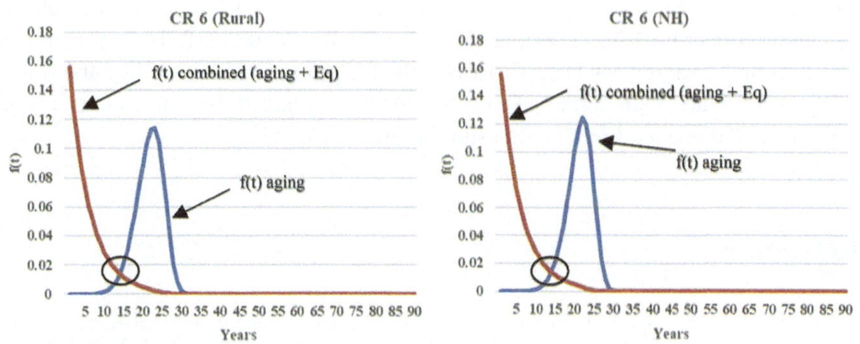

Fig. 23.11 PDF for combined bridge hazards for CR 5 RC bridges in rural and NH routes (Earthquake magnitude > 6)

23.4 Results and Conclusions

In this paper, a probabilistic model for predicting the deterioration of the national highway bridges and rural bridges (RC bridges in the Barak valley region) is proposed. The prediction comprises two main damaging effects physical deterioration (aging effect) and sudden shocking effects (earthquake) which are inter-related by using the probabilistic method. From the above figures and tables, it can clearly state that the aging of the bridge has influenced the different CR of bridges in the area. Also, the combined effect (aging and EQ) is more impactful and bringing drastic deterioration in the service life of bridges. Figures 23.8, 23.9, and 23.10, the PDF model depicts two shapes; one is due to aging and the other is due to combined effect. The line of combined effect which crosses the aging hazard mode shape is the point where service life is affected and evaluated decrease in service life is tabulated in given Tables 23.4, 23.5, 23.6 and 23.7. This study has proposed successfully the multi-hazard assessment of the bridges and how more than one hazard affects the bridge service-life and survival characteristics. As a result, it allows the bridge management department and the structural health-monitoring department to know more about the bridge situation in the Barak valley region and improve maintenance and rehabilitation decisions. Further developments are needed to address the hazard issue in the earthquake-prone area and adopting life-cycle improvement concepts, strategic priority to maintain and improve the safety of the bridges. As infrastructure, systems are the backbone of modern society and help in boosting economic growth and development of the region. These estimates would aid bridge engineers and structural health monitoring departments in taking rational decisions for bridge design and maintenance.

The limitations, which are worth mentioning here are that the bridge inspection is done by visual inspection, and a more advanced inspection method, can be adopted to improve the findings, up-gradation of IBMS, and IRC is needed for a better condition rating system. The future scope is that this preliminary approach of multi-hazard assessment can be further employed for determining the hazard for the whole state of Assam by collecting different bridge data. Overall, it was successfully demonstrated how the effect of aging can bring distress to the RC bridges of Barak valley and when aging is combined with earthquake events, the rate of hazard becomes more severe and affects the service-life by increasing failure times.

References

1. Ayyub BM, McGill WL, Kaminskiy M (2007) Critical assest and portfolio risk analysis: an all-hazards framework. Risk Anal 27(4):789–801. https://doi.org/10.1111/j.1539-6924.2007.00911
2. Bahuguna A, Sil A (2018) Comprehensive seismicity, seismic sources and seismic hazard assessment of Assam, North East India. J Earthquake Eng 1–44. https://doi.org/10.1080/13632469.2018.1453405
3. Beavers JE (2009) Multihazard issues in the central United States: understanding the hazards and reducing the losses. ASCE Council on Disaster Risk Management Monograph No. 3, Reston, VA
4. Chang FK (1999) Structural health monitoring. In: Proceedings of the international workshop on SKIM, Stanford, 1999. Tecnomic Publishing Co.
5. Cheung MS, Li WC (2001) Serviceability reliability of corroded steel bridges. Can J Civ Eng 283:419–424
6. Das J, Sil A (2020) Condition assessment of superstructure component of reinforced concrete bridges through visual inspection in the Assam, India. Bridge Struct 16(1):39–57
7. Das J, Sil A (2020) Condition assessment and failure probability of existing bridges in the Cachar District, Assam. Advances in computer methods and geomechanics. Lecture Notes in Civil Engineering, vol 55. Springer, Singapore
8. Dey A, Miyani G, Sil A (2019) Reliability assessment of reinforced concrete (RC) bridges due to service loading. Innov Infrastruct Solut 4(9). https://doi.org/10.1007/s41062-018-0194-8
9. Enright MP, Frangopol DM (1998) Probabilistic analysis of resistance degradation of reinforced concrete bridge beams under corrosion. Eng Struct 20(11):960–971. https://doi.org/10.1016/S0141-0296(97)00190-9
10. Enright MP, Frangopol DM (1998) Failure time prediction of deteriorating fail-safe structures. J Struct Eng 124(12):1448–1457. https://doi.org/10.1061/(ASCE)0733-9445(1998)124:12(1448)

11. Enright MP, Frangopol DM (1998) Service-life prediction of deteriorating concrete bridges. J Struct Eng 124(3):309–317. https://doi.org/10.1061/(ASCE)0733-9445(1998)124:3(309)
12. Federal Highway Administration (2004) National bridge inspection standards (NBIS). Federal regulations 69 FR74436, USA. 14 Dec 2004
13. Federal Highway Administration (FHWA) (1995) Recording and coding guide for the structure inventory and appraisal of the nation's bridges. Rep. No. FHWA-PD-96-001, Washington, DC
14. Ghosn M, Moses F (1986) Reliability calibration of bridge design code. J Struct Eng 112(4):745–763. https://doi.org/10.1061/(ASCE)0733-9445(1986)112:4(745)
15. IRC: SP:18 (1996) Manual for highway bridge maintenance inspection. The Indian Roads Congress, New Delhi
16. IRC: SP:52 (1999) Bridge inspectors reference manual. The Indian Roads Congress, New Delhi
17. Kameshwar S, Padgett JE (2014) Multi-hazard risk assessment of highway bridges subjected to earthquake and hurricane hazards. Eng Struct 78:154–166
18. Kramer SL (1996) Geotechnical earthquake engineering. Prentice Hall India Private Limited
19. Li Y, Ellingwood BR (2009) Framework for multi-hazard risk assessment and mitigation for wood-frame residential construction. J Struct Eng 135(2):159–168
20. Mauch M, Madanat S (2001) Semiparametric hazard rate models of reinforced concrete bridge deck deterioration. J Infrastruct Syst 7(2):49–57. https://doi.org/10.1061/(ASCE)1076-0342(2001)7:2(49)
21. Ministry of Transportation and Communications (MTO) (1979) Ontario highway bridge design code, Downsview, Canada
22. Moore M, Phares B, Graybeal B, Rolander D, Washer G (2001) Reliability of visual inspection for highway bridges. Final Rep. No. FHWA-RD-01-020, vol 1, Federal Highway Administration (FHWA), Washington, DC
23. Nasrollahi M, Washer G (2014) Estimating inspection intervals for bridges based on statistical analysis of national bridge inventory data. J Bridg Eng 20(9):04014104. https://doi.org/10.1061/(ASCE)be.1943-5592.0000710
24. Nowak AS, Lind NC (1979) Practical code calibration procedures. Can J Civ Eng 6(1):112–119. https://doi.org/10.1139/l79-012
25. Sasmal S, Ramanjaneyulu K (2008) Condition evaluation of existing reinforced concrete bridges using fuzzy based analytic hierarchy approach. Expert Syst Appl 35(3):1430–1443. https://doi.org/10.1016/j.eswa.2007.08.017
26. Sitharam TG, Sil A (2014) Comprehensive seismic hazard assessment of Tripura and Mizoram state. J Earth Syst Sci 123(4):837–857. https://doi.org/10.1007/s12040-014-0438-8
27. Stewart MG, Estes AC, Frangopol DM (2004) Bridge deck replacement for minimum expected cost under multiple reliability constraints. J Struct Eng 1309:1414–1419
28. Yanev B (2007) Bridge management. Wiley, Hoboken

Chapter 24
Seismic Strengthening Solutions for Existing Buildings

T. M. Al-Hussaini, M. M. Hoque, and A. S. Moghadam

24.1 Introduction

The region comprising Northeast India, Nepal, Myanmar, Bhutan and Bangladesh is one of the most seismo-tectonically active regions of the world. Also, it is home to many vulnerable buildings that may crumble should a major earthquake event occur. The updated Bangladesh National Building code (BNBC-2020) divides Bangladesh into four seismic zones, with most of the country subjected to moderate to severe seismic hazard. The major cities of Dhaka, Chittagong and Sylhet have a seismic zone coefficient (maximum considered earthquake) of 0.2, 0.28 and 0.36, respectively [1]. However, non-occurrence of strong earthquakes affecting Bangladesh for more than 90 years has left the current generation unaware of the possibility of a strong earthquake. As a natural consequence, a majority of buildings in the urban areas of Bangladesh are lacking earthquake-resistant design. The effect is further compounded by poor quality of materials and construction in many cases. Should a major earthquake occur, the densely populated cities of Bangladesh with vulnerable infrastructure may be a showcase for a catastrophic disaster. This has also been revealed from earthquake vulnerability assessment surveys for Dhaka, Chittagong and Sylhet cities conducted by ADPC [2] under the Comprehensive Disaster Mitigation Program (CDMP). According to that report, for the capital city of Dhaka with Scenario case 2 motion (PGA = 0.1–0.3 g), the number of collapsed buildings is 14%, while that of moderately damaged buildings is 29%. This presents a very grim picture of probable consequences for a likely earthquake event.

T. M. Al-Hussaini (✉)
Bangladesh University of Engineering & Technology (BUET), Dhaka, Bangladesh

M. M. Hoque
Dhaka University of Engineering & Technology (DUET), Gazipur, Bangladesh

A. S. Moghadam
International Institute of Earthquake Engineering and Seismology (IIEES), Tehran, Iran

S. Kolathayar et al. (eds.), *Civil Engineering for Disaster Risk Reduction*, Springer Tracts in Civil Engineering, https://doi.org/10.1007/978-981-16-5312-4_24

As a testament to inherent structural vulnerabilities, in the last two decades, a number of building collapse incidents have taken place in Dhaka city without the occurrence of earthquakes. Unfortunately, this has been also true for industrial factory buildings which are supposed to provide life safety for workers working inside it. Such structural failures under gravity loading point out the possibility of inherent weaknesses in many buildings of the city, which may easily collapse in the event of additional loading from an earthquake. Building collapse incidents in Bangladesh came into sharp focus when factory building collapsed killing workers working inside the building. The first major factory building collapse occurred on April 11, 2005 when an unauthorized nine-story RCF sweater factory building (each floor area 15,000 sq. ft.) located in Savar in the outskirts of Dhaka totally collapsed within a few minutes [3]. The final casualties were 61 dead and 86 injured. Unfortunately, this incident did not lead to taking tough measures regarding factory building safety. Subsequently, eight years later in April 24, 2013, another factory building collapse occurred only 10 km from the previous site, again in Savar, but this time with much more severe consequences. Rana Plaza, another unauthorized 9 + 1 story RCF building on pile foundation (each floor area 20,000 sq. ft.) in Savar collapsed suddenly in the morning hours killing more than 1100, mostly garment workers and injuring. Figure 24.1a displays a photo [4] of the Rana Plaza building which housed garments factories in 2nd to 7th floors. 8th floor was not completed. Figure 24.1b presents a photo [5] of the collapsed building.

Structural assessment [6] of the collapsed building reveals non-engineered unauthorized construction and severe structural deficiencies including overstressed columns (up to 3rd floor), inadequate column reinforcements, poor concrete (8.76–14.7 MPa), lack of proper detailing, lack of quality control, etc. What is more striking, according to news reports, large cracks developed in structural elements of the building one day earlier creating panic among the garments workers. The building was then evacuated. However, the owner thought the building to be safe and the next day the garments workers were compelled to come in for work at around

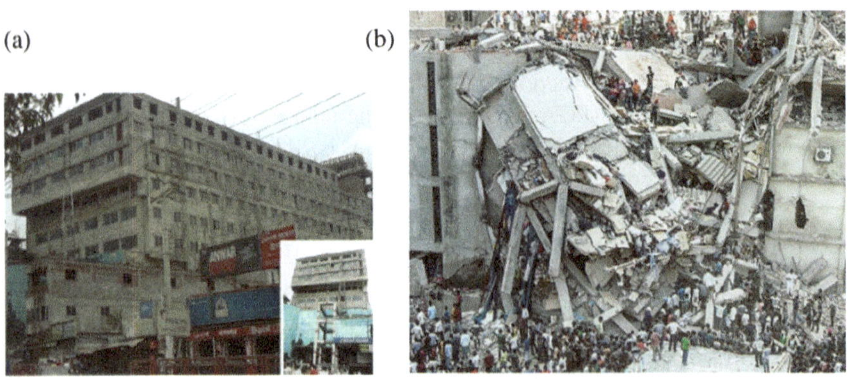

Fig. 24.1 9 + 1storied Rana Plaza Building, **a** before collapse [4], **b** after collapse in April 2013 [5]

8 am. Due to power outage, generators located in upper floors of the building were started, and suddenly the building came down just before 9 am. The vibrations of the generators may have caused widening of the cracks and might have triggered the collapse.

This incident of Rana Plaza collapse, being one of the worst industrial accidents in history, raised a hue and cry, both nationally and internationally. Foreign garment buyers raised serious objections to worker safety issues in Bangladesh. Subsequently, through series of negotiations, a practical plan to identify and upgrade safety of existing vulnerable ready-made garment (RMG) factory buildings was taken up for implementation in 2013. Both structural safety and fire safety has been considered. Several factory buildings that did not meet standards required by building and construction legislation have been identified and have gone through structural strengthening process.

This paper presents a very brief account of the three-party agreement for assessing and ensuring safety of existing RMG buildings in Bangladesh. Methods of structural strengthening adopted for existing RCF factory buildings have been presented. Seismic strengthening of existing URM buildings have been discussed, sharing experience from Iran. Finally, advanced methods for seismic protection of existing buildings are presented.

24.2 Framework for Structural Vulnerability Assessment and Seismic Strengthening of Existing RMG Buildings

Following Rana Plaza incident, a guideline [7] for structural and fire safety assessments for existing RMG factory buildings in Bangladesh has been prepared by the Technical Committee of the National Tripartite Working Group on the National Tripartite Plan of Action on Fire Safety and Structural Integrity for RMG Sector in Bangladesh in November 2013. This Guideline is intended only for the assessment of existing RMG factory buildings in operation. The construction of any new RMG factory building, as well as modification of existing RMG buildings, shall have to be done as per Bangladesh National Building Code (BNBC). A Review Panel consisting of expert engineers shall have the power to review the factory closure decision of any Engineering Team and may overturn the closure. The Guideline gives the minimum criteria for assessing the structural integrity of existing RMG factory buildings against danger to life from building collapse. It also emphasizes the importance of professional experience and expert engineering judgment.

Some key features of the structural integrity assessment and strengthening procedure of existing RMG buildings as outlined in the guideline is summarized below:

1. Structural assessment and certification of safety of an existing RMG factory building shall be carried out in the following three steps:

(a) Preliminary visual assessment
(b) Detailed engineering assessment (if necessary)
(c) Design of retrofit (strengthening) scheme (if necessary)

The first step in the structural assessment procedure is the preliminary visual assessment. This assessment shall include a visual inspection of the structure, collecting data and performing simple calculations to assess the basic capacity of some of the critical member/columns or foundations. After the preliminary assessment, the engineering team shall recommend if a detailed engineering assessment (DEA) is necessary or not. The DEA shall determine if the structure needs to be strengthened (retrofitted) or not.

2. An existing building has already passed the construction phase and significant part of its usable life. This allows the use of a lower live load or lower overload factor for safety assessment of an existing structure provided there is no distress or minor distress in its structural members.
3. Regarding material properties, higher value of strength reduction factor may be used if fieldwork reduces uncertainties regarding material strength of steel and concrete. Such values should not exceed the limits specified by ACI 318 code.
4. For minor cracks in structural members, a strict monitoring plan regarding the future status of the crack (its extent of propagation and width change) may be recommended.
5. Columns with high slenderness, flat plates and footings with inadequate thickness for shear, etc., shall be identified as vulnerable critical members. The capacity of such member shall be critically examined.
6. Guidelines for assessing structures that lack proper documentation have also been provided.
7. Detailed engineering assessment (DEA) should identify structural inadequacies if any. Lateral loadings due to earthquake or wind shall be as per BNBC. When a structural member is identified to have gross design inadequacies in terms of either not meeting the requirements of strength limit states or serviceability limit states, an appropriate retrofit scheme shall be selected based on standard practice.
8. Standard retrofit techniques such as concrete jacketing, micro-concrete encasement, FRP wrapping, etc., for strengthening of columns may be applied. Bracing systems may be considered to reduce effects of column slenderness. Beam and slab cracks may be repaired applying epoxy injection. Ferro cement, micro-concrete and FRP may be used to increase beam and slab strengths. Any distressed member should be properly repaired so as to ensure durability of the structure.

24.3 Seismic Strengthening of RCF Buildings in Bangladesh

Following the guidelines described in the previous section, structural safety of existing RMG buildings in Bangladesh has been assessed by engineering teams. Several buildings have been found to be structurally deficient and structural strengthening measures have been implemented. This paper presents some applications to existing RCF buildings.

Figure 24.2 represents column strengthening works adopting concrete jacketing at an RMG factory building in Savar. This procedure results in increase of both strength and stiffness of the column. This building possessed many strange structural irregularities including having some brick columns. The brick columns were encased with the RCC jacket; Fig. 24.2 (right) shows installation of reinforcement around the brick column. Figure 24.2 (left) represents the same for an RCC column. RCC columns were subjected to removal of plaster, chipping and groove cutting to enable better bonding (grip) between old concrete and new concrete. Shear connectors to hold vertical and horizontal reinforcements were embedded into the existing column with epoxy adhesive. Vertical reinforcements were embedded with epoxy into the RCC footing.

Deteriorated beams due to corrosion of bottom bars have been rehabilitated by replacement of bottom concrete with fresh concrete and ferro-cement. Figure 24.3 (left) shows corrosion preventing process by applying corrosion inhibitor from MAPPIE on bottom beam reinforcements. Figure 24.3 (right) shows application

Fig. 24.2 Reinforcement placement for concrete jacketing of RCC column (left) and brick column (right) of an RMG factory building

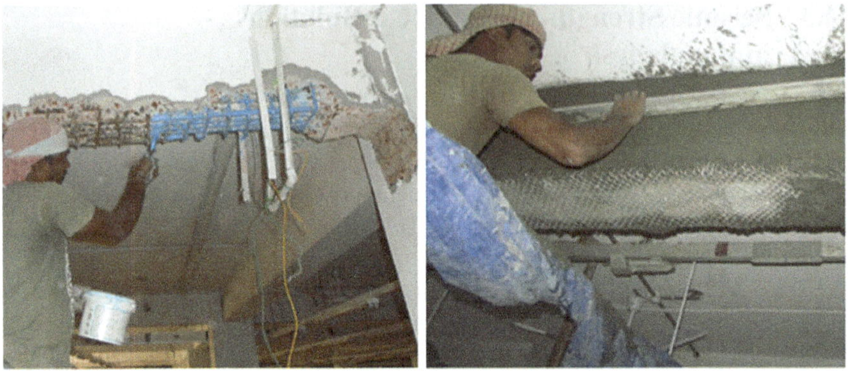

Fig. 24.3 Applying corrosion inhibitor on bottom beam rebars (left), ferro-cementing with wire mesh on beam bottom (right)

of ferro-cement technique on beam bottom using 14 SWG wire mesh with 20 mm opening.

Grade beams have also been found to be structurally deficient and section enlargement of grade beams have been done with additional horizontal reinforcements being epoxy grouted into existing RCC column. High brick masonry walls in stair case have been strengthened using 14 SWG wire mesh nailed to the wall followed by plastering. This has been done to prevent out of plane failure of wall during earthquake shaking.

Retrofitting works for footing and slab have been presented in Fig. 24.4, which represents case of another RMG factory building in Savar. Figure 24.4 (left) shows reinforcements works for strengthening an existing RCC footing, both size and

Fig. 24.4 Reinforcement placement for strengthening of RCC footing (left) and RCC slab/beam (right)

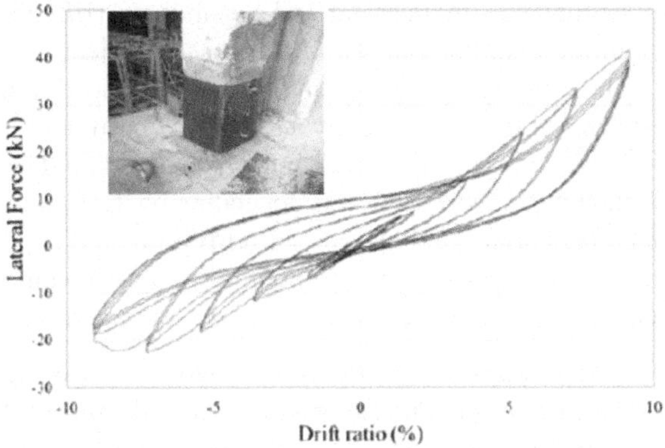

Fig. 24.5 Lateral load-displacement curves for FRP wrapped damaged RCC column, photo of retrofitted column shown in inset

thickness being increased. Figure 24.4 (right) shows reinforcements for strengthening RCC slab/beam and shear connectors embedded with epoxy into the RCC slab supporting bottom reinforcement bars.

Although the application of FRP has been included as a retrofit solution in the Tripartite Plan of Action, only very few cases of FRP applications have taken place. FRP application has later been discouraged due to the risk of fire hazard in RMG factory buildings. The use of steel jacketing for strengthening columns has also been discouraged for the same reason.

However, it may be noted that where fire hazard is not a likely event, FRP applications can be a very effective solution. Figure 24.5 presents encouraging laboratory test results [8] from cyclic lateral loading test of FRP wrapped damaged RCC column conducted at BUET. An RCC column was loaded to failure at its bottom, following which repair of damaged portion of column was done using micro-concrete and then FRP wrapping applied. As shown in Fig. 24.5, FRP wrapped damaged column yielded very stable hysteresis loops, more-over the lateral load exceeded the maximum lateral load obtained for the original RCC column.

24.4 Seismic Strengthening of URM Buildings in Iran

Iran has experienced considerable earthquakes in its long history. Large earthquakes have caused this country many casualties with widespread economic consequences. According to the Iranian code for seismic design, schools are among high importance facilities that are expected to perform much better than ordinary common buildings. Therefore, initially 4 billion US$ was approved by the Iranian Parliament in 2007 to both demolish and reconstruct seismically critical schools and retrofit vulnerable ones. The actual budget of the project increased to more than 10 billion US$ in later stages. In this project, more than 132,000 classrooms were demolished and reconstructed and more than 126,000 classrooms retrofitted. The International Institute of Earthquake Engineering and Seismology (IIEES) oversaw development of the technical guidelines and strategies for the project.

The primary step in starting the project was development of a decision-making mechanism to systematically answer the question of "which school building should be demolished-reconstructed and which one should be retrofitted?" In the developed methodology, the answer was depending on the following three main parameters [9]: price of school building, cost of retrofitting and expected life of the building after retrofit. To implement the methodology, rapid screening forms were derived, and the school buildings were categorized into seven categories.

1. Approved schools: The schools which were designed and constructed based on the latest version of seismic design code of Iran. These schools at the time were mainly those that constructed after 2006.
2. Semi-approved schools: The buildings which were designed and constructed based on the previous versions of seismic design code from 2001 to 2006.
3. Schools that need partial-rehabilitation: Buildings with adequate seismic resistance but without slab integrity, mainly buildings with jack arch slab.
4. Schools suitable for application of the typical retrofitting pattern (TRP) approaches: Single-story masonry buildings having a set of common deficiencies that justify application of a pre-defined details and methods of retrofitting.
5. Schools that are suitable for demolition and reconstruction process: Those with a retrofitting to reconstruction cost of more than 50%, or buildings that architecturally are not suitable, or those schools located in developing areas that soon will need larger school buildings.
6. Schools not worth spending money: Schools in those remote rural areas that are facing a steady rate of population decrease.
7. Schools for complete retrofitting study: These are schools that will go through a complete seismic retrofitting study. Many of such buildings are frame structural buildings, without proper lateral load-resisting system or those with serious deficiencies. It usually takes a long time to complete all steps of complete retrofitting study such as the preliminary screening, detail analysis and preparing school-specific retrofit plans.

Among different retrofitting techniques applied to the Iranian schools, this paper presents three retrofitting methods adopted by IEES mainly as TRP approaches:

1. Shear walls: This method was designed to facilitate application of shear walls as an efficient retrofitting technique for schools. Schools with shear walls are stiffer and their stiffness usually reduces the number of cracks in building during earthquakes. Observing less damages in schools provides a better sense of safety for the students and their parents. But proper design and detailing of shear walls is a relatively complicated task that is beyond capability of many unexperienced engineers. As such, in this methodology, a set of design tables for estimation of capacity of the shear walls, their piles, also a set of pre-defined details for

Fig. 24.6 Retrofitting URM School buildings with shear walls [10]

detailing of reinforcements, based on the concrete and soil condition are developed. This approach eliminates almost any possibility of errors. A typical engineer can simply calculate the base shear of the building and the number and length of shear walls needed. For buildings with jack-arch diaphragms, a procedure is provided to transform the diaphragm to composite concrete diaphragms. The in-plane and out-of-plane masonry wall performances are also improved by special details. This method was started to be implemented in many school buildings since summer 2010 (Fig. 24.6) [10].

2. Shotcrete: This method initially was chosen based on its application in other countries. As there were some ambiguities in its efficiency in special cases, numerous experiments on its application to masonry walls was carried out in Iranian universities and research centers, leading to development and verification of some detailing to improve its application. In the adopted method, usually only the walls in the perimeter area of the URM building is shotcreted. The roof of the most of masonry school buildings is jack-arch which should be transformed to a composite concrete element. The widespread use of this method started in many schools all over the country in 2010 (Fig. 24.7) [11].
3. Center core: This method was first implemented in U.S. in 1987 as a technique for seismic retrofitting of masonry buildings, without affecting their appearances. The most complex step in application of the method is properly drilling vertical circular holes (typically 4″-6″ dia) throughout the height of the wall at specific intervals. Then, rebars (typically #6–#8) are placed in the drilled holes. The holes are filled with grout through grout tubes (Fig. 24.8 left) which are withdrawn as grout rises. The resulting tie-columns will improve greatly the integrity and the seismic performance of walls. Figure 24.8 (right) displays a photo of drilling through a masonry wall [12]. In this method, the outer surface of masonry wall is not affected. Since the architecture of the building is unchanged, it is one of the best options for seismic retrofitting of historical buildings. Experts in the Iranian School Organization have increased the accuracy and speed in

Fig. 24.7 Retrofitting single story URM school buildings with peripheral shotcrete [11]

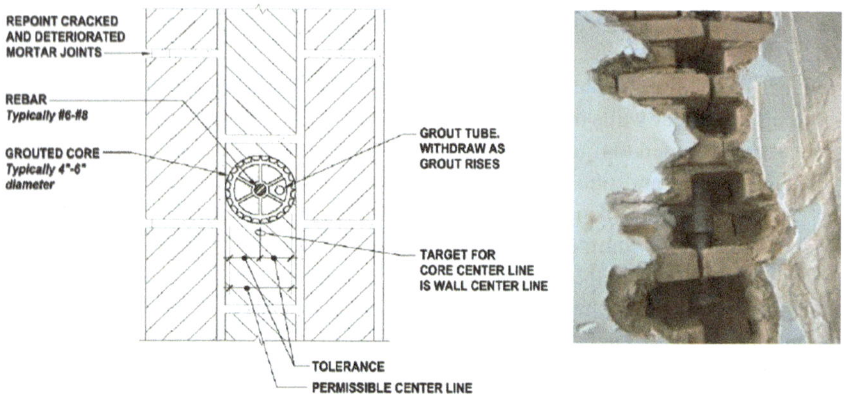

Fig. 24.8 Left: Plan detail of center core method in masonry wall, Right: Applying vertical perforations in masonry walls [12]

execution of the center core technique and applied it in some historic masonry school buildings.

24.5 Advanced Methods of Seismic Strengthening

Advanced methods of seismic retrofitting adopting base isolation and/or energy dissipation devices offer an attractive solution for keeping a critical building facility operational even after a major earthquake. Base isolation bearing isolates the building superstructure from its foundation from the damaging effects of an earthquake, resulting in much smaller shaking of the building floors. This is evident from comparison of shake table test results [13] on base isolated (using friction pendulum system isolators) and non-isolated seven-story moment frame building models, subjected

Table 24.1 Shake table test results on non-isolated and base-isolated 7 story frame structure [13]

Story	Non-isolated (MFF)		Isolated (MFUIS)	
	El Centro SOOE 35%		El Centro SOOE 200%	
	Shear	Drift	Shear	Drift
	Weight	Height (%)	Weight	Height (%)
7	0.077	0.231	0.083	0.318
6	0.138	0.339	0.143	0.42
5	0.181	0.387	0.159	0.453
4	0.21	0.458	0.168	0.501
3	0.218	0.46	0.206	0.569
2	0.22	0.361	0.226	0.435
1	0.235	0.281 (Ex)	0.24	1.396 (Ex)
		0.284 (In)		0.875 (In)

to different strengths of the same earthquake. As shown in Table 24.1 [13], the isolated structure has similar shear forces at different floor levels for an earthquake which is 5.7 times (=200/35) stronger when compared to a non-isolated (fixed base) conventional building.

A 9-storey base-isolated governmental building in Sendai (Fig. 24.9 left), retrofitted with high damping rubber (HDR) isolators, performed well during the 2011 Tohoku Earthquake, reducing the acceleration from 0.25 g and 0.29 g at the base to 0.14 g and 0.18 g at the top of the building. Figure 24.9 right represents seismic retrofit of 5-story historic US Court of Appeals Building in San Francisco using FPS isolators.

Energy dissipation (ED) devices are placed at different story levels of a building to resist and dissipate energy while a story deforms during an earthquake. ED devices may be classified as (i) friction damper, (ii) metallic damper, (iii) viscoelastic solid

Fig. 24.9 Left: 9-story building in Sendai retrofitted with HDRB isolators. Right: 5-story US Court of Appeals building in San Francisco retrofitted with FPS isolators

Fig. 24.10 Seismic retrofit of school buildings in Italy using ED devices shown on the outside [14]

damper and (iv) viscous fluid damper. Figure 24.10 shows an example [14] of seismic retrofit of school buildings in Italy using ED devices visible on the outside.

There are hundreds of base-isolated buildings in some countries with high seismic risk such as Japan, China, Russia, Italy and USA [14]. Base-isolated buildings have proven to respond as expected with little or no damage during major earthquake events. This technology is likely to be adopted in developing countries as well for critical building infrastructures where such buildings need to remain operational during a major earthquake event. ED devices have sometimes been used in combination with base isolation to obtain benefits of both. ED devices alone have also been used in seismic retrofit of buildings. It may be noted that implementation of ED devices in an existing structure is much easier compared to base-isolation.

Figure 24.11 shows comparison of story deformation of a conventional building with that of a building with base isolation and that of a building with ED devices. It may be seen that the story drift (Di) for a base-isolated building is much smaller than that (Dc) of a conventional non-isolated building. This is the principal reason for minimal or no damage of base isolated buildings. Floor accelerations are represented by Ai, Ac while floor inertia forces are represented by Fc, Fi. Also, it is obvious from the figure that floor accelerations rapidly increase with height in conventional buildings, i.e., $Ac2 \gg Ac1$ and $Fc2 \gg Fc1$. The story drift (De) for a building with ED devices is quite larger than that (Di) of a base isolated building but significantly smaller than that (Dc) of a conventional building. Consequently, buildings with ED devices will provide seismic safety but is likely to be subjected to some damage.

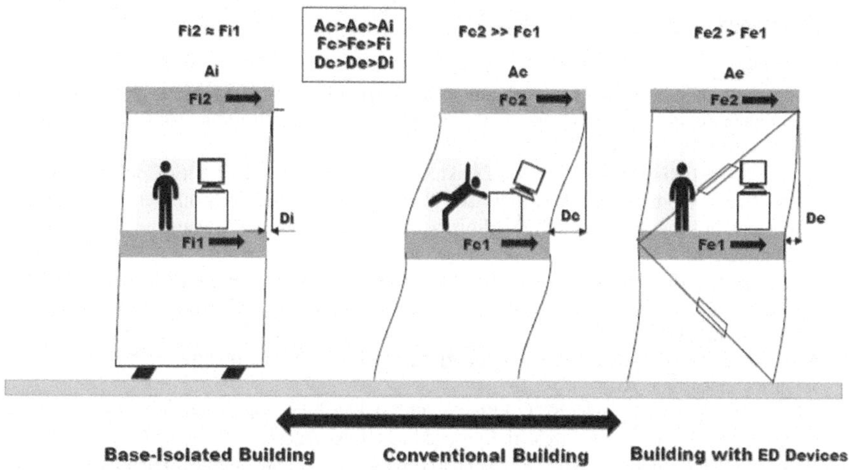

Fig. 24.11 Comparison of story drifts between Conventional building and Base-isolated building and Building using ED device, concept from [14]

24.6 Conclusions

This paper reveals the circumstances under which systematic structural and seismic vulnerability assessment followed by strengthening of existing ready-made garment factory buildings have been taking place in Bangladesh following collapse of Rana Plaza. The building vulnerability situation in the country has been briefly stated, along with methodical steps needed for assessment and strengthening. Factory buildings in Bangladesh are mainly RCF buildings, strengthening techniques adopted include concrete jacketing of columns/beams/shear walls, strengthening of footings, use of wire-mesh/ferro-cement on masonry walls, beams or slabs. Although use of FRP and steel jacketing have also been considered as retrofit solutions, their use has been discouraged considering possibility of fire hazard. Seismic strengthening techniques for URM buildings have also been presented sharing experiences from Iran which include shotcreting, shear wall addition and center core methods. These strengthening techniques have been widely accepted as practical, economical and satisfying building requirements. Finally, advanced methods of seismic retrofitting using base-isolation or energy dissipation devices have been addressed. In developed countries, these devices have found many applications in buildings. These devices have already been applied in some important bridges of Bangladesh and are likely to find its application for critical building infrastructures in future.

References

1. Al-Hussaini TM, Hossain TR, Al Noman MN (2012) Proposed changes to the geotechnical earthquake engineering provisions of the Bangladesh National Building Code. Geotech Eng J SEAGS & AGSSEA 43(2):1–7
2. ADPC (2009) Risk Assessment of Dhaka, Chittagong and Sylhet City Corporation Area. Main Report submitted by Asian Disaster Preparedness Centre for Comprehensive Disaster Management Programme (CDMP), Ministry of Food and Disaster Management, Government of Bangladesh
3. Al-Hussaini TM, Hossain MNA (2008) Post-disaster management issues related to building collapse in Bangladesh. In: Proceedings (CDROM), international disaster and risk conference, Davos, Switzerland
4. Odhikar (2013) Broken Dreams: A report on the Rana Plaza Collapse. Fact finding report released on 19 June 2013
5. Dhaka Tribune homepage (2021) https://www.dhakatribune.com/bangladesh/court/2018/04/24/no-progress-rana-plaza-cases-trial-proceedings. Last accessed 2021/04/15
6. Rahman MM, Halder AK (2013) Rana plaza collapse: causes and our role to ensure construction of safer buildings. Seminar presented at Public Works Department, Dhaka, 8 July 2013
7. DIFE (2013) National tripartite plan of action on fire safety and structural integrity—guidelines for building assessments (structural & fire) for existing RMG factory buildings in Bangladesh. Report published by Department of Inspection for Factories and Establishments, Ministry of Labour and Employment, Government of Bangladesh, 24 Nov 2013
8. Al-Hussaini TM, Mohaimen M, Islam K, Ahmed KS, Ahsan R (2018) New dynamic testing facility at BUET-JIDPUS: cyclic loading of 1/3rd scale damaged RC column retrofitted with externally wrapped CFRP sheets. In Proceedings, 15th international conference on structural and geotechnical engineering, Cairo, Egypt
9. Mahdizadeh A (2010) Report on Retrofit of School Buildings in Islamic Republic of Iran. Technical Report, State Organization of School Renovation, Development and Mobilization of Iran, Tehran, Iran
10. International Institute of Earthquake Engineering and Seismology (2013) Retrofitting techniques for Iranian Schools. Technical Report, IIEES, Tehran, Iran
11. Raissi M, Yekrangnia M, Eghbali M, Mahdizadeh A (2015) Report on retrofit procedure of Iranian School. Technical Report, State Organization of School Renovation, Development and Mobilization of Iran, Tehran, Iran
12. Mahdizadeh A, Borzouie J (2012) Perforating masonry walls for rehabilitation of masonry buildings using center-core method. In: Proceedings of the 6th international conference on seismology and earthquake engineering (SEE6), Tehran, Iran
13. Al-Hussaini TM, Zayas VA, Constantinou MC (1994) Seismic isolation of multi-story frame structures using spherical sliding isolation systems. Technical Report No. NCEER-94-0007, National Center for Earthquake Engineering Research, State University of New York at Buffalo, NY, USA
14. Martelli A (2010) On the need for a reliable seismic input assessment for optimized design and retrofit of seismically isolated civil and industrial structures, equipment and cultural heritage. In: Presentation at the advanced conference on seismic risk mitigation and sustainable development, ICTP, Trieste, Italy

Chapter 25
On Structural Rehabilitation and Retrofitting for Risk Reduction

G. R. Reddy

25.1 Introduction

The disasters/risks are generally occurred due to manmade hazards and natural hazards. The hazards are not only related to the air quality but also with related to the structures it is on load demands. For simple example, the road bridges are not properly maintained resulting in pot holes may induce higher loads causing failures. In industry, handling poison gases and liquids, if the equipment not properly maintained resulting reduced load bearing capacity and eventually may lead to failures as shown in Fig. 25.1. Similarly, increase in machine-induced vibrations and man-induced blasts may cause such failures. One of the failures of bridges across the railway lines is shown in Fig. 25.2 which has failed due to poor maintenance, resulting in low capacity to withstand normal loads. All these come under man-induced extra demands or hazards on the structures resulting failures leading to risk. The risk is high in the case of tank failures whereas in the case of bridge, it is low. If the bridge is a part of lifeline structures for managing the disaster after natural hazards such as cyclonic winds, floods and earthquakes, and its non-availability results in high risk. Some of the failures of structures under cyclonic winds are shown in Figs. 25.3 and 25.4.

Referring to the failures of airport structures, if it is part of disaster management, the risk may increase after the natural hazard. If such failures are seen in structures related to communication systems or industrial structures handling hazardous gases and liquids, it will lead to high risk. Similar to the cyclonic winds, earthquakes will cause damages in the structures such as bridges [1] as shown in Figs. 25.5 and 25.6.

Bridges are very important lifelines and have to be usable during and after the natural hazards. In addition to the earthquake-induced loads, these structures have to be designed for pavement unevenness/roughness, vehicle brake loads, vehicle collision loads and explosion loads. If codes are not clear about these loads, the utility has to make measurements in situ and see the adequacy. There may be a need

G. R. Reddy (✉)
National Institute of Technology Surathkal, Surathkal, Karnataka, India

S. Kolathayar et al. (eds.), *Civil Engineering for Disaster Risk Reduction*, Springer Tracts in Civil Engineering, https://doi.org/10.1007/978-981-16-5312-4_25

Fig. 25.1 Failure of tanks releasing poisonous gases

Fig. 25.2 Failure of bridge crossing the railway lines

Fig. 25.3 Failure of airport structures

to perform laboratory tests to understand the behavior under vehicle explosions, vehicle collisions, etc. Other important lifelines are piping systems [1], and their performance is shown in Fig. 25.7a, b. Though, they are generally ductile, due to poor performance of supports and concentrated loads at joints result in failures.

Regarding flood hazards are shown in Fig. 25.8 at different basins of the country. Presently, due to climatic changes, heavy precipitation at local regions is becoming

Fig. 25.4 Failure of material handling cranes in industry

Fig. 25.5 Sliding of bridge deck and failure of pier cap during Bhuj earthquake

Fig. 25.6 Sliding of bridge deck during Kobe Earthquake

very common. If the structures as shown in Fig. 25.8 are not performing due to the flood levels, the risk will be very high, and risk management is also becomes difficult if lifelines as mentioned are not available.

It is important to mention that such disasters may be avoided if the rivers are inter linked through natural channels. Interlinking involves building large number of structures such as dams, check dams as shown in Fig. 25.9 across the links and channel

(a) Pipe support failures due to ground failure

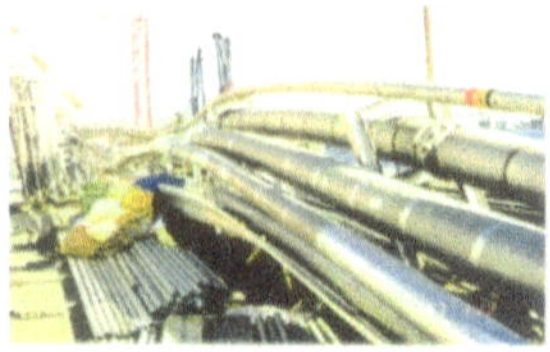

(b) Pipe support failures due to large loads

Fig. 25.7 Failures of piping supports during Kobe Earthquake

(a) Brahmaputra basin

(b) Mahanadi basin

(c) Godavari basin

(d) Krishna basin

Fig. 25.8 Flood situations in different parts of the country

(b) Dams

(a) Check Dams

Fig. 25.9 Water structures

embankments and need to be maintained, re-evaluated, rehabilitated/retrofitted to see its performance and reduce the disasters/risk.

As an off shoot, it is necessary to mention that most of the existing dams have become old and may require special attention. In addition to floods, future disaster/risk related to water is inevitable if not planned, conserved and distributed properly. Interlinking rivers will surely [2] help to address this issue. Other areas which needs attention are large precipitation areas where a clear plan will help to conserve the water and distribute as envisaged in Fig. 25.10. Large precipitation areas generally are located in coastal areas. It is leading to flood disasters also. A crude estimation of rainfall in heavy precipitation areas says that it will suffice to provide one foot of water throughout the country such as India. For this purpose, as shown in the Fig. 25.10, large number of structures have to be built with proper design and have to be maintained and updated with time. Some of the structures to mention are tunnels, reservoirs, bunds, etc. As another offshoot, it is important to make a mention about utilization of solar power for lifting the water from low level to higher level. For power generation, also require structures which need special attention.

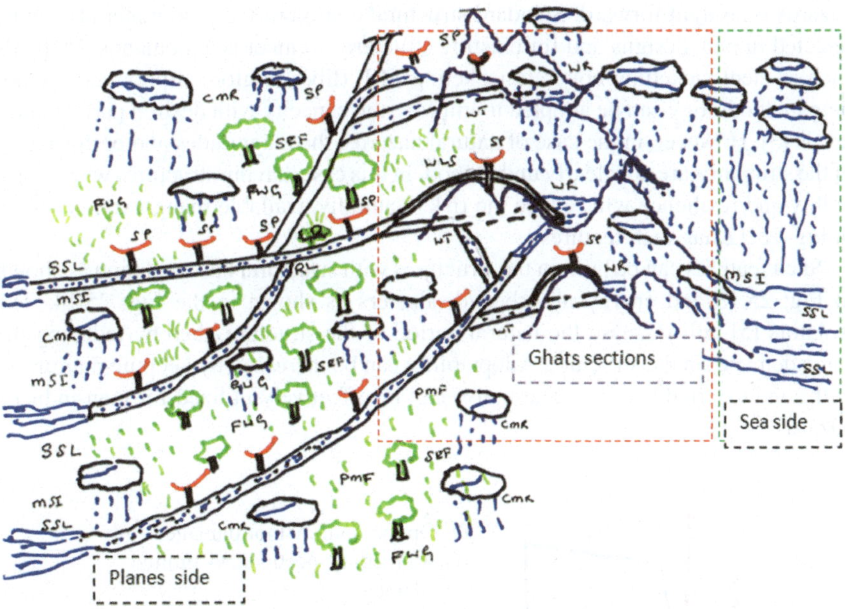

Fig. 25.10 Artistic picture of visionary blue revolution for conservation of rain water

25.2 Design, Repair, Rehabilitation and Retrofitting of Structures for Reduce Risk/Disaster

As mentioned above, the risk/disaster hinges on structures and its performance. The risk is related with structural parameters as follows.

$$\text{Risk} = \text{Hazard} + \text{Vulnerability}(\text{structures}) + \text{exposure}(\text{structures})$$

Man-made hazards can be minimized and reduced with proper designs, maintenance and usage of structures. Whereas, natural hazards are unpredictable and based on the past history, the design basis hazards have to be considered. While designing the structures and with new learning, design basis hazards have to be updated and considered for design of new structures and evaluating the existing structures. In view of this, the risk can be controlled mainly designing resilient structures with high durability. Depending on the risk affordable, the designs or rehabilitation/retrofitting can be performed as per the ASCE and ASTR guidelines [3, 4] on performance as shown in Fig. 25.11. To design or retrofit the structure based on new man-induced hazards is straight forward. Standard structural configurations and materials can be selected in new designs and for existing structures to meet new demands. Methods such as steel jacketing, concrete jacketing FRP (fiber reinforced polymers) wrapping methodology can be adopted to improve the capacity with desired performance levels [3]. However, in the case of natural hazards, the magnitudes and performance of the structures are not clearly understood. In this case, having structures with adoptability and resilience will reduce the risk. Naturally available mangroves as shown in Fig. 25.12 has such features.

Such features can be built in the structures with structural base isolation as shown in Fig. 25.13 and energy absorbers/dissipaters as shown in the Fig. 25.14. The isolators [5] will increase the natural period of the structure, thereby reducing the force demand on the structure. Adoptability can be increased by keeping margins to withstand external hazards. These also can be effectively utilized for man-induced hazards.

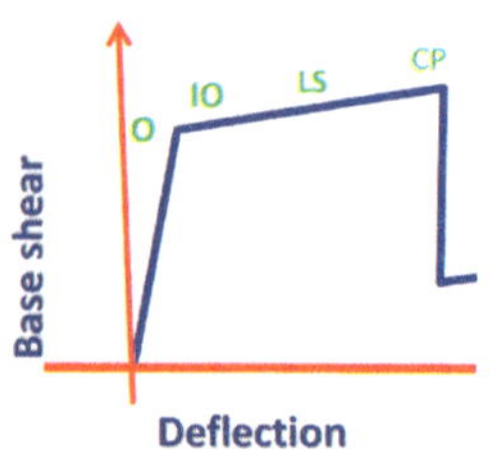

O- Operational- Minor damage
IO- Immediate occupancy- limited damage
LS- Life safety- Significant damage - major repairs
CP- collapse prevention- No margin against collapse

Fig. 25.11 Capacity of the structure along with performance levels

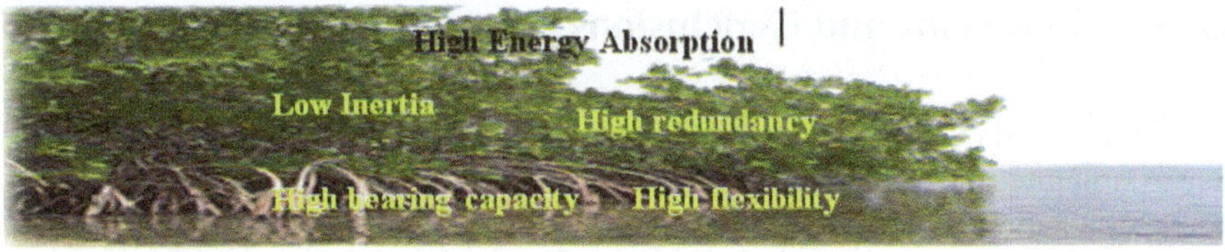

Fig. 25.12 Mangroves with adoptability and resilience

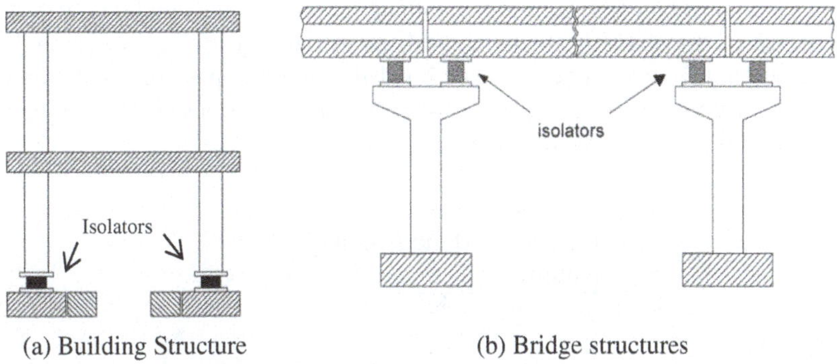

Fig. 25.13 Bases isolation of structures to withstand external hazards

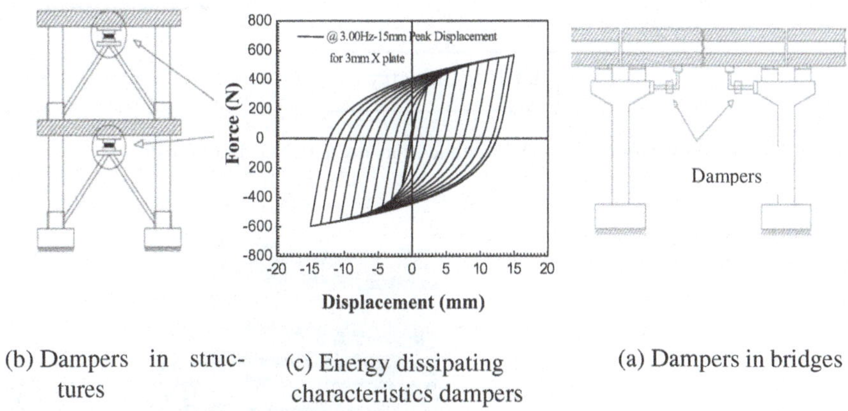

Fig. 25.14 Dampers in structures to with stand external hazards

The dampers [5] will absorb/dissipate the external hazard energy, thereby reducing the demand force on the structure and improves its performance.

25.3 Discussions and Conclusions

Various teams shown in Fig. 25.15 have to work together to reduce the risk/disasters due to structures. For example, in new construction of buildings, to manage fire accidents, evacuation centers are built in higher floors. This introduces weaker sections in the structure. This type of discontinuities may result in failure of structures [5] due to higher energy of high frequency excitations (e.g., earthquakes, winds, etc.) as shown in Fig. 25.16. Such lessons have to be discussed among designers, researchers, utility and improve the designs and rehabilitate/retrofit existing buildings. To reduce the intense of disaster/risk all the normal, industrial and lifeline structures have to be continuously maintained, assessed and if required repair, rehabilitation and retrofitting need to be performed on continuous mode. Various steps in doing this activity are briefly explained hereunder [4].

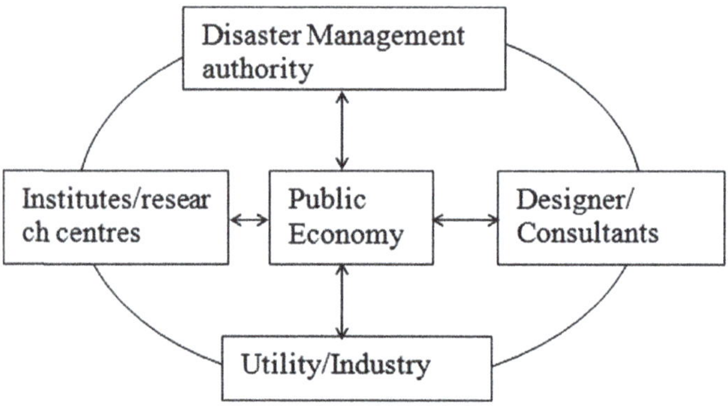

Fig. 25.15 Responsible teams around the public/economy to reduce disaster/risk

Fig. 25.16 Higher level weak floor failures due to higher energy modes

1. Collecting all the structural records including drawings, design reports, maintenance reports. If not, generate the reports by in situ actual data.
2. Estimate/postulate the structural design basis hazards including natural and manmade events.
3. Assess the structure for its capacity to bear the loads meeting the desired performance.
4. If not, perform rehabilitation and retrofitting to improve its performance and eventually improve the safety and reduce the risk/disaster.
5. Manage the structures to ensure the structure performs the intended performance.

The details are discussed in the Association of Structural Rehabilitation (ASTR) Guidelines for structural rehabilitation and retrofitting, 2019 [4].

To perform above activities, various responsible agencies as explained in Fig. 25.15 have to work hand in hand to reduce the risk/disaster. Public has to understand the structures, where they live, they use, they work, etc., and if found disorders, inform the local agencies. All the agencies around the public have to keep well informed on reciprocating basis about the hazards, new codal provisions, new technologies and ensure required performance of the structures as desired.

Acknowledgements I am thankful to all the members of the organizing committee involved in the virtual conference on disaster risk reduction, especially Dr. Sreevalsa K, NITK, Surathkal for inviting me to give key note lecture. Also thankful to all the contributed authors of (ASTR) Guidelines for structural rehabilitation and retrofitting.

References

1. Reddy GR (1994) Advanced approaches for the seismic analysis of nuclear power plant structure, equipment and piping systems. Ph.D. thesis, Tokyo Metropolitan University
2. Economic Impact of Interlinking of Rivers Program, National Council of Applied Economic Research, Revised final report, Apr 2008
3. ASCE Standard, Seismic Evaluation and Retrofit t of Existing Buildings, ASCE/SEI 41–13, 2014
4. Association of Structural Rehabilitation (ASTR) Guidelines for structural rehabilitation and retrofitting, 2019
5. Reddy GR, Prasad H, Verma AK (2019) Textbook of seismic design (structures, piping system and components). Springer, Berlin

Part IV
Disaster Risk and Resilience

Chapter 26
Integrated Cost and Risk Management Enhancing Supply Chain Resilience

Jasmine Siu Lee Lam and Yuwei Yin

26.1 Introduction

Given the increasingly intensive competition between global markets, competition between companies has shifted from a focus on the performance of companies to that of supply chain management (SCM) performance. To gain a competitive edge, companies are required to implement best practice techniques across the supply chain to maximize the overall benefit [1, 15]. Strategic SCM is of great importance, as organizations strive towards providing their products and services in an efficient and cost-effective manner. Moreover, globalization has increased logistics risks and vulnerability in supply chains. As a business becomes more international, there is a growing need for managing supply chain disruptions from a cross-national perspective [2, 27]. Generally, a supply chain is faced with increasing possibilities and severity of disruption. Furthermore, disruption management is complex as disruptions can occur for a wide variety of reasons due to various risks. As the world economy becomes more interconnected, the relationships between supply chain partners become more complex than before [2, 17]. Nowadays, supply chains are exposed to progressively higher risks of disruption, which may be propagated and amplified and thus manifested as an observed "snowball effect". Such kinds of impact can even span several continents [33]. For instance, the COVID-19 pandemic resulted in disruptions of an entirely unprecedented magnitude of global supply chains [9].

Trade facilitators, such as air, road and maritime transportation, will disrupt the smooth flow of supply chains when a breakdown occurs [7]. These are categorized

J. S. L. Lam (✉) · Y. Yin
School of Civil and Environmental Engineering, Nanyang Technological University, 50 Nanyang Avenue, Singapore 639798, Singapore
e-mail: SLLAM@ntu.edu.sg

Y. Yin
e-mail: YINY0010@e.ntu.edu.sg

S. Kolathayar et al. (eds.), *Civil Engineering for Disaster Risk Reduction*, Springer Tracts in Civil Engineering, https://doi.org/10.1007/978-981-16-5312-4_26

as transportation disruptions. As defined by Chopra and Sodhi [4], there is a distinction between transportation disruption and other types of disruptions. Transportation disruption affects the overall flow of multiple supply chains, while other disruptions are likely to affect specific nodes of specific supply chains, such as supply disruption. Considering an example of a manufacturer-related disruption, the Tohoku earthquake in March 2011 rendered a Japanese manufacturing base for General Motors non-functional but did not stop the entire production of goods for this company [18]. In contrast, considering a transportation disruption, in November 2020, due to a COVID-19 case reported, all cargo terminals at Shanghai Pudong airport were shut down temporarily and over 500 flights were cancelled [35]. As Pudong airport is a major hub for air transportation in China, this transportation disruption paused all types of cargo flows to and from China via Pudong, which brought about material and information delays across various supply chains internationally. In this context, transportation disruptions affect the flow of supply chains as a whole, rather than performance at a single node. Transportation disruption distinctively represents that goods and information in transit have been stopped, which paralyses the whole supply chain, although all other operations of the supply chain may be intact [38]. The literature mainly focuses on the general disruption of supply chains, but having a very little investigation into the uniqueness of transportation disruption [9]. Notably, in recent studies, Khemiri et al. [12] chose not to model the transportation network in their risk analysis due to complexity. Namdar et al. [24] also assumed a completely reliable transportation network in their model.

This research gap results in an impetus for the authors to investigate this under-researched area. This paper focuses on increasing the resilience of supply chain performance, in terms of the strategies to help a supply chain "bounce back" in the aftermath of an event that interrupts transportation, in turn facilitating the supply chain. The paper aims to examine the existing strategies which are in place to enhance resilience to transportation disruptions in supply chains. Based on the major limitations of these strategies identified, the study also aims to develop an original approach to address the limitations.

The paper will be organized as follows: after the introduction, the classification of supply chain risks will be presented. Section 26.3 examines the existing resilience strategies. Section 26.4 will introduce an integrated cost and risk management approach. Finally, Sect. 26.5 concludes and suggests future research directions.

26.2 Classification of Supply Chain Risks

Identifying and assessing risks are prerequisites to risk management. It is essential to determine which techniques and instruments are best suited for risk assessment in complex and multi-organizational environments [21]. To have wider coverage in the literature review, this paper not only examines various risks that cause transportation disruption but also supply chain disruption. The study firstly investigates how to classify these risks.

Looking at broad coverage of risks, Chopra and Sodhi [4] classified supply chain risks into 9 categories, namely, disruptions, delays, systems, forecast inaccuracies, intellectual property breaches, procurement failures, system breakdown, inventory problems and capacity issues. Being one of the risks, disruption, in this case, is the disruption of material flows. Our paper shall extend the definition of disruption to include information flows besides material flows. For example, events such as deviation of shipping routes belong to the category of disruption, as it satisfies the definition of interrupting a smooth material flow. Furthermore, the information flow may also be negatively affected, because such a disruption may add difficulties to communication between relevant supply chain parties. For example, due to unexpected disruptive events such as accidents and political tensions, an original shipping route through the Suez Canal has to be re-routed to sail via the Cape instead. Thus, the transport vessel fails to collect cargoes from nearby Middle East countries. At the same time, the bills of lading and the associated information of these cargoes have to be changed. This extension of definition is also supported by the method of risk classification proposed by Spekman and Davis [31], who grouped supply chain risks into 6 areas: physical movement of goods, information flow, money flow, security of a firm's internal information system, relationships between supply chain partners and corporate social responsibility. This classification reveals that supply chain risks do not only interfere with the physical flow of goods but also influence intangible factors, such as information as discussed.

Risk is occasionally also classified based on the reason of event occurrence. Mitroff and Alpaslan [23] divided the causes of disruptions into natural disasters, normal accidents and abnormal accidents. Natural disasters, as its name suggests, are metrological or geographical calamities, such as earthquakes and floods. In terms of normal accidents and abnormal accidents, the classification is based on the intention of the "trouble makers". Normal accidents happen due to error or omission, such as system overload problems, while abnormal accidents refer to evil actions that intentionally lead to damages. Similar classifications can be seen in Sheffi and Rice [30], which categorized disruptions into random events, accidents and intentional disruptions; and in Stecke and Kumar [32], which sorted disruptions into man-made catastrophes (i.e. non-terrorist intentional acts, terrorist acts and accidents) and natural calamities. However, as the intention of the "troublemaker" may usually be hard to identify, it is difficult to determine whether an unnatural event is caused accidentally or intentionally. Therefore, this paper only discusses whether these risks are natural or otherwise, avoiding the ambiguity generated from the analysis of intention.

Another approach of risk classification is based on the stage of supply chain where risk takes place. Christopher and Peck [5] divided risks into process, control, demand, supply and environmental risks. In terms of supply chain disruption, this categorization is useful for an organization to look into the risk to which it is exposed and identify which stage of the supply chain will consequently be affected. In this way, the disruption can be quickly triggered and specific actions can be taken to make the supply chain bounce back efficiently. This approach is simplified by Kleindorfer and Saad [14], which categorized supply chain risks into three areas: supply, demand and disruption risks. Disruption risks consist of operational risks, involving equipment malfunctions, unforeseen discontinuities in supply, human-centric issues and the risks arising from natural hazards, terrorism and political instability.

Therefore, in literature, there exists little to no uniformity observable in the classification of supply chain risks. However, there is a consensus achieved on the definition of disruption, in that it is a risk leading to disturbances in the smooth flow of normal daily activities. Wagner and Bode [36] illustrated disruption as an exceptional and irregular situation compared to daily business. The study also explained disruption as a sudden unexpected occurrence that deviates from the business norms. In our paper, the definition of disruption is an event that causes a sudden interruption in material flow, as well as information flow, which leads to a halt in the movement of cargoes [38] or transit of information in a supply chain. This in turn jeopardizes the firm, by negatively affecting its financial and operational performance. Besides, this paper not only emphasizes the direct disruption impact on a particular supply chain stage but also focuses on measures to control the indirect impact on the supply chain network as a whole. The merging of existing classification methods explains the basis of disruption, as well as the type of risks that firms are exposed to, thereby facilitating one of the objectives of this research—finding strategies to manage disruption risks from various sources and nodes of the supply chain.

To provide a detailed analysis of supply chain risks and disruptions, further classification of such events is required. As mentioned above, disruption risks in the supply chain context can be divided into disruption in material flow and information flow. For each flow risk of disruption, it can be subdivided into external and internal risks, where external factors refer to those driven by events upstream and/or downstream in the supply chain, while internal factors are driven by events within the organization's control [13].

As suggested by Gaudenzi and Borghesi [8], risk evaluation is inherently subjective, because each analyst has his or her own concept of risk constitution. While the literature tried to develop a typology, there is no uniform method of classifying supply chain risks [8, 9]. The most important goal would not be having a consensus of the various methods but to choose the most suitable method according to the situation and point of interest. It is also important to adopt one method consistently throughout an analysis. Based on the research objectives, the method for classification adopted by this paper is presented in Fig. 26.1. The grey shaded boxes will represent the categories of risk that this paper has chosen to focus on.

26.3 Examination of Existing Strategies to Enhance Resilience

After discussing risk classification, this section examines the existing strategies which are in place to enhance resilience to transportation disruptions in supply chains. The analysis based on the literature and industry information is summarized in Tables 26.1, 26.2, 26.3, 26.4 and 26.5.

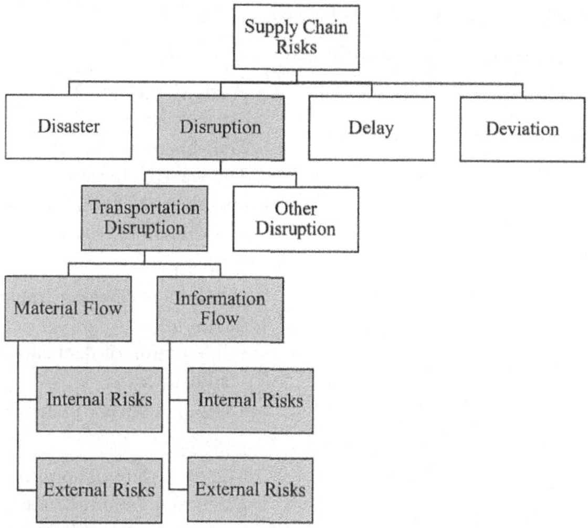

Fig. 26.1 Supply chain risk classification. *Source* Drawn by Authors

From these tables, it can be observed that the five different strategies identified have some common properties. For example, the Cooperation/Alliance strategy and the Investing in redundant capacity strategy share a similar main idea, which is to integrate the capacity and enable flexible adjustments during disruptions. In this way, when the normal capacity breaks down, the backup capacity (obtained from its alliance or its own buffer capacity) can be utilized to maintain daily operations and minimize the disruption impact. From the aspect of cost efficiency, forming an alliance is preferable compared with investing in redundant capacity, because establishing redundant capacity requires a relatively high capital input. Therefore, by forming alliances, the member companies share their capacities, which means each company can reduce their capital investments by building up extra capacity. However, adding capacity by joining an alliance means that the company does not have full control of the redundant capacity. As Lu et al. [19] revealed the key success factor of shipping alliances lies in the potential synergy in partnerships if certain caveats are addressed. For example, there should exist shared consensus, be it in the form of similar organization structure and culture, past collaboration and reputation, teamwork ability, reasonable risk/revenue sharing and/or flexibility of market change. If such factors cannot be satisfied, the alliance will not be effective. For instance, liner shipping alliances operate in such a way that each member of an alliance can share space on the services provided by the other members. However, the uncertainties may increase, since the availability of such spare space is unpredictable.

The strategies of adding control and responsive pricing both require a high standard of control power over the supply chain. At the same time, the responsive pricing strategy also requires the company to have a high degree of control over the whole market, so that they can have a strong bargaining power [5]. Such strategies are hard for small companies to implement. At present, the largest container shipping lines have formulated alliances (i.e. The Alliance, Ocean Alliance and 2M). Their

Table 26.1 Cooperation/alliance as a strategy

Strategy 1: Cooperation/alliance	
Examples	Forming formal alliance or informal cooperation to share capacity (fleet or slot) *Real-life example* The wave of mergers and acquisitions in 2014–2017 of container shipping lines bring about strong market consolidation, leading to dynamic capacity management
Objectives and advantages	*Objective* Quick recovery after transportation disruption *Advantages* • Integration of fleet capacities enabling flexible and timely adjustments • Quick recovery to achieve smooth material flow
Difficulties and disadvantages	*Difficulties* The procedure of applying for alliances is complicated, with much time required, many authorities to approach and many documents to be filed requirements on maintaining the relationship with alliance members *Disadvantages* • Companies need to lower certain control power to form an alliance
Examples of related risk events	Risks interfering with a smooth material flow *Internal risks* • Mechanical breakdown of a company owned fleet (Measures: sharing fleet/slot with co-operative partners) *External risks* • Transport congestion • Natural disasters • Lockdown of specific countries due to a pandemic • Terrorist attacks (Measures: contacting its alliance to take unaffected alternative routes)
References	1. Christopher and Peck [5] 2. Zhang and Lam [40] 3. Lam and Bai [16]

Source Compiled by Authors

combined market share is significant and economies of scale can be achieved. In comparison, the smaller shipping lines have less power of control on the shipping market and more restricted capability to bargain with other supply chain parties. In this instance, they may face great difficulties in deploying these two strategies in the foreseeable future.

The strategy of a resilient information system also faces a dilemma between enhancing the degree of resilience and the amount of capital invested. Majorities of logistics service providers now have backup information systems. However, there exist great differences among these companies, in terms of how complete and efficient the backup system is [3]. With a higher degree of comprehensiveness in the backup

Table 26.2 Investing in redundant capacity as a strategy

Strategy 2: Investing in redundant capacity	
Examples	Investment of redundant capacity so that it can be used as a backup when the normal operation is disrupted *Real-life example* Majority of transport facilities and terminals build extra cargo handling capacity to minimize congestion and service breakdown
Objectives and advantages	*Objective* Timely recovery of service capability after disruption with the backup capacity *Advantages* • Compared with cooperation/alliance, control power is completely within the organization • Reducing potential loss of revenues due to disruptions of normal operation
Difficulties and disadvantages	*Difficulties* The redundant capacity may fail to perform if destroyed together with the normal capacity (e.g. earthquake that impacts the whole transport infrastructure) *Disadvantages* • Relatively high capital investments • Redundant capacity does not generate revenue and requires expense for maintenance
Examples of related risk events	Risks interfering with a smooth material flow *Internal risks* • Mechanical breakdown of normal operating capacity (Measures: using redundant capacities during the repairing period) *External risks* • Transport congestion • Natural disasters • Terrorist attacks (Measures: using backup capacities to replace the disrupted capacity)
References	1. Rice and Caniato [28] 2. Mansouri et al. [20] 3. Ganin et al.[7]

Source Compiled by Authors

system, fewer data will be lost when a disruption occurs. The same goes for the enhanced degree of backup system efficiency, resulting in a shorter data-recovery period following a disruption occurrence. However, more resilience in information systems means more time, energy, personnel and capital inputs. There is considerable cost required to maintain and update the backup information system. Moreover, the backup information system does not generate any revenue for the company most of the time, during normal operations. Therefore, there is a trade-off between capital investment and degree of resilience, which requires the decision-makers to analyze

Table 26.3 Responsive pricing strategy

Strategy 3: Responsive pricing strategy	
Examples	Adjusting the freight rate corresponding to the current market condition *Real-life example* Common practice promoted by the US Energy Information and Administration (EIA), that shipping companies cap oil surcharge in the transportation contract to stabilize price and thus stabilize demand during recession
Objectives and advantages	*Objective* After disruption occurs, using available controlling power to adjust the market rate responding to the supply/demand shocks *Advantages* • Immediate impact on the market • Maintained profitability to facilitate the recovery from disruption
Difficulties and disadvantages	*Difficulties* Execution of this approach is difficult because demand is hard to control and the company only has limited power to influence the price *Disadvantages* • Price fluctuations might amplify the supply and demand shocks and bring about unexpected consequences
Examples of related risk events	Risks interfering with a smooth material flow *Internal risks* • Inability of major carriers to provide adequate supplies • Demand shock or inability of major customers to receive cargoes • Both risks above may impact the supply-demand equilibrium of the market and cause freight rate fluctuations *External risks* • Policies leading to fluctuations of currency, or increase in port charge, etc. • Poor economic conditions that greatly shrink supply and demand (Measures: controlling the supply/demand volumes to avoid shocks in freight rates)
References	1. Tang and Tomlin [34] 2. Gong and Janssen [10]

Source Compiled by Authors

the optimal point of investment, which may add complexity to the implementation process of this strategy.

Based on the analysis of the existing strategies, we find that there are several unsolved problems. Firstly, some of the strategies require high-level capital investments. However, the existing studies merely focus on the rationale of the strategy and the suggested procedures for companies to put these strategies into practice. Nevertheless, for a company considering which strategy to adopt, it is of great importance

Table 26.4 Adding control as a strategy

Strategy 4: Adding control	
Examples	One player performs as multiple supply chain parties or adopts contractual terms to restrict other parties *Real-life example* Singapore Airlines has detailed contracts with all parties on its supply chain, from suppliers of air freight containers (ULD) for cargo transport, as well as crew training centre for passenger transport
Objectives and advantages	*Objective* To increase the bargaining power over the whole supply chain, so to have strong control of the recovery process after disruption *Advantages* • Enabling a quick trace of disruption problems with high transparency and control of the supply chain • Exerting greater impacts on various supply chain parties (e.g. suppliers and customers)
Difficulties and disadvantages	*Difficulties* To achieve great control and bargaining power, the company needs to have a high market share and a strong financial performance *Disadvantages* • Requiring relatively large amounts of time, personnel and capital investments to be devoted
Examples of related risk events	Risks interfering with a smooth material flow *Internal risks* • Mechanical breakdown of aircraft, vehicle, infrastructure, etc. *External risks* • Shutdown of warehouses • Delivery of unqualified containers • Poor economic conditions that greatly shrink supply and demand (Measures: seeking for compensation in disruptions according to contractual terms)
References	1. Wirtz and Johnston [39] 2. Revilla and Sáenz [27]

Source Compiled by Authors

to calculate the cost involved for each strategy. Currently, there is a limited discussion about the cost involved in each strategy to cope with transportation disruptions. There is also a lack of effective cost management suggestions that may help the company to save costs and facilitate capital investment while enhancing resilience. Secondly, each strategy corresponds to different scenarios, such as high/low power of control, high/low bargaining power, etc. The existing studies do not provide tools for such scenario analysis and thus the ability of companies to choose the most appropriate strategies based on their current operating conditions is impeded.

Table 26.5 Resilient information system strategy

Strategy 5: Resilient information system strategy	
Examples	Backup data and enhance the security of the information system so that it can be recovered despite system outages *Real-life example* More than half of the business with headquarters located in World Trade Centre started its normal operation within one month after the 911 attack, thanks to their backup information system
Objectives and advantages	*Objective* To quickly recover from the information disruption *Advantages* • Largely reducing the risk of data loss • The disrupted information can be recovered efficiently
Difficulties and disadvantages	*Difficulties* Without much difficulty to have a backup system of core information, but greater difficulties are met when seeking detailed and complete information backup *Disadvantages* • Needs to hire professionals for operation and maintenance • Require extra capital investment for system establishment and maintenance
Examples of related risk events	Risks interfering with a smooth information flow *Internal risks* • Malfunction or breakdown of computer systems *External risks* • Computer virus • Natural disasters • Terrorist attacks Measures: data backups utilizing cloud storage services; advanced technologies (e.g. blockchains) to enhance information security
References	1. Wakolbinger and Cruz [37] 2. Carvalhoa et al. [3] 3. Pu and Lam [26]

Source Compiled by Authors

This paper contributes by seeking solutions for these two unsolved problems. Firstly, an effective approach to cost analysis for supply chain resilience strategies will be proposed to help companies evaluate and manage costs. Secondly, useful techniques will be introduced for managers and researchers to conduct scenario-based analysis for the choice of a resilience strategy.

26.4 Integrated Cost and Risk Management Approach

Based on the literature review, there is a common disadvantage, in that some of the strategies require extra inputs of capital investment. While those strategies may help a company recover from disruptions in a timely manner, it increases the implementation costs significantly at the same time, which may result in decreased profitability. For example, the current economic downturn caused by pandemic led to a significant contraction in the global demand for transportation services. In spite of that, the operating costs did not mirror the traffic trends and kept increasing during the same period. In this situation, the need for increased capital investments in resilience strategies to manage transportation disruptions will become an additional burden that companies may not be able to bear. With this background, this paper aims at developing a cost-effective resilient approach. The cost management strategy does not function separately in a company but can be used in conjunction with existing strategies. The resultant integrated cost and risk management approach will help to analyze the cost of each resilient strategy, as well as provide effective concepts to control costs to support the capital investment required.

26.4.1 Effective Way of Cost Analysis for Supply Chain Resilience Strategies

First and foremost, collecting information about costs is of great importance. To maintain profitability, as well as guarantee sufficient sales and market share, companies are required to precisely determine the achievable selling prices and margins for each product by collecting detailed information about related costs. Cost is not limited to the cost of producing the product or service, but also includes the cost of overall management, like the cost spent to increase resilience, in the context of this paper. The major considerations of cost management are listed below: market influences on certain business activity, the extent to which costs are competitive compared with industrial average, the company's profit margin according to its production structure and developments costs [11]. Development costs are related to developments in improving the overall function of the business, such as supply chain improvement. For instance, although still unproven on a broader scale, blockchain is an advanced technology to significantly improve supply chain resilience in terms of secure and undisrupted information flows [26]. However, the development cost is difficult to budget for, unlike the other conventional costs that can be easily captured. When implementing a new strategy, which has never been undertaken by a company, it is challenging to pre-empt the expected costs, perhaps largely due to limited available historical data.

Taking a retrospective look, the paper identifies a few shortcomings of the existing resilience management system, applied by the vast majority of companies in the current logistics field [9, 29]:

1. Organizational development and cost management are allocated to separate departments, with divergent management objectives;
2. Accurate records of direct cost (e.g. cost of goods sold), while estimation and allocation of indirect costs are challenging (e.g. investment in enhancing overall supply chain performance);
3. The conventional cost management techniques have difficulties to be applied for the development costs, especially for newly emerging technologies.

Observing the traditional cost management strategies, we emphasize that such traditional approaches no longer meet the needs of resilience management in a newly competitive environment [6, 7]. Therefore, it is necessary to improve the management accounting mentality and achieve a more realistic view of cost management strategies to be implemented [7]. In this respect, the use of strategic management accounting techniques related to aspects of external information, such as industrial and economic environments, supports the decisions of a strategic nature. This can be done by estimating a short-run stochastic cost frontier over a sample database of transport operators, which adopts the listed flexibility strategies. The total change in expenditure on such strategies found on their financial statements can be used as a proxy for the development cost (greater sample size reduces the deviation of inaccurate records of such indirect cost). The impact of variables such as ownership, outsourcing, market dominance and revenue diversification is tested in a second-stage regression, so the relevance of such factors can be found [22]. Given the operating conditions of certain businesses, with a given form of ownership, degree of outsourcing, percentage of market dominance and calculated revenue diversification, the cost of implementing a specific strategy can be approximately calculated.

26.4.2 Strategic Cost Management—Useful Techniques

Strategic cost management is a consequence of mutations that occur in the current environment and requires a permanent adaptation of tools and techniques to improve the strategic position of organizations [6]. Cost management systems should be designed according to two aspects: (1) the mission statement and business objectives of the entire organization; (2) objectives of the particular resilience strategy to be implemented, such as reducing transport operation downtime or information recovery time to a maximum tolerable timeframe using a specified cost budget.

To select the optimal strategy, the collective efforts of an organization will be required. This involves choosing relevant information, discussion among organizational executives, material and financial means necessary to analyze the overall conditions of each strategy, highlighting the differentiated structures and mechanisms for the following sub-systems. The cost is easier to manage if the costs involved are divided into four stages of each strategy implementation, as follows:

1. Formulation of strategies;
2. Communication of these strategies throughout the organizations;

3. Development and implementation of tactical solutions to achieve these strategies;
4. Monitory process during implementation and feedback collection after implementation.

Strategic cost management, as a tool for effective management, is the combination of three main components:

1. Value chain analysis of the supply chain;
2. Analysis of strategic positioning;
3. Sources of cost analysis.

Strategic cost management focuses on the integration and coordination of activities of all organizations or companies along the supply chain. For example, to reduce storage costs and increases transport efficiency, the move towards lean supply chains reduces storage costs by increasing inter-dependency among supply chain parties [25].

The concept of strategic positioning is about finding the answer to the question: what role does cost management play under a particular scenario? This concept guides decision-makers to specify the aim of cost management. Under this condition, the decision-makers need to analyze the approximate cost of each strategy, so that they can form a basic understanding of how much cost is involved to support the capital investment.

The third component of strategic cost management is an analysis of the sources of costs because the sources of cost help to explain the behaviour of costs [29]. For example, to reduce the cost of computer system maintenance, the company should make adjustments to the operation of IT department, as it is the major source of cost generation. By targeting the source of each cost, the efficiency of cost management can be enhanced.

In sum, based on the approach of cost analysis and cost management, a company can improve the power of control over the cost of the resilience strategies so that each strategy can be well budgeted and financed.

26.5 Conclusion

This paper contributes to the risk management literature and related industries by examining the existing risk classification approaches and combines several approaches to systematically classify different types of risks. It specifically focuses on transportation disruption risks that interfere with material and information flows, from external or internal sources. This paper also investigates existing strategies to enhance resilience, with feasible solutions and examples of the current practice of transport providers. The benefits, drawbacks and related risks of each strategy have been analyzed. We point out that a common shortcoming of existing strategies in the literature is that it does not pay sufficient attention to the cost involved by adopting certain strategies and miss out on the scenario analysis about which strategy is appropriate for which type of business. Therefore, with a specific focus on disruption risks, this paper proposes an integrated cost and risk management approach for supply chain resilience. The approach is able to help decision-makers to choose the optimal strategies under a given scenario.

As for future research, more work is being undertaken to advance the proposed approach, by introducing the costs and expected values involved in each strategy, based on empirical research for data collection. Future research may also consider finding out the probabilities of each scenario. All in all, researchers are recommended to consider effective cost and risk management to enhance resilience in transportation, supply chain and logistics systems.

References

1. Agrell PJ, Hatami-Marbini A (2013) Frontier-based performance analysis models for supply chain management: state of the art and research directions. Comput Ind Eng 66(3):567–583
2. Bode C, Macdonald JR (2017) Stages of supply chain disruption response: direct, constraining, and mediating factors for impact mitigation. Decis Sci 48(5):836–874
3. Carvalho H, Barroso AP, Machado VH, Azevedo S, Cruz-Machado V (2012) Supply chain redesign for resilience using simulation. Comput Ind Eng 62(1):329–341
4. Chopra S, Sodhi MS (2004) Managing risk to avoid supply-chain breakdown. MIT Sloan Manag Rev 46(1):53–61
5. Christopher M, Peck H (2004) Building the resilient supply chain. Int J Logist Manag 15(2):1–14
6. Ellram LM (2012) Strategic cost management in the supply chain: a purchasing and supply management perspective. CAPS Research
7. Ganin AA, Kitsak M, Marchese D, Keisler JM, Seager T, Linkov I (2017) Resilience and efficiency in transportation networks. Sci Adv 3(12):e1701079
8. Gaudenzi B, Borghesi A (2006) Managing risks in the supply chain using the AHP method. Int J Logist Manag 17(1):114–136
9. Golan MS, Jernegan LH, Linkov I (2020) Trends and applications of resilience analytics in supply chain modeling: systematic literature review in the context of the COVID-19 pandemic. Environ Syst Decis 40:222–243
10. Gong Y, Janssen M (2012) From policy implementation to business process management: principles for creating flexibility and agility. Gov Inf Q 29(1):S61–S71
11. Horngren CT, Foster G, Datar SM, Rajan M, Ittner C (2009) Cost accounting: a managerial emphasis, 13th edn. Pearson Prentice Hall, Hoboken, NJ
12. Khemiri R, Elbedoui-Maktouf K, Grabot B, Zouari B (2017) A fuzzy multi-criteria decision-making approach for managing performance and risk in integrated procurement–production planning. Int J Prod Res 55(18):5305–5329
13. Kiser J, Cantrell G (2006) 6 steps to managing risk. Supply Chain Manag Rev 10(3):12–17
14. Kleindorfer PR, Saad GH (2005) Managing disruption risks in supply chains. Prod Oper Manag 14(1):53–68
15. Kumar V, Chibuzo EN, Garza-Reyes JA, Kumari A, Rocha-Lona L, Lopez-Torres GC (2017) The impact of supply chain integration on performance: evidence from the UK food sector. Procedia Manuf 11:814–821
16. Lam JSL, Bai X (2016) A quality function deployment approach to improve maritime supply chain resilience. Transp Res Part E 92:16–27

17. Lam JSL, Van de Voorde E (2011) Scenario analysis for supply chain integration in container shipping. Marit Policy Manag 38(7):705–725
18. Lohr S (2011) Stress test for the global supply chain. The New York Times, 20 Mar. Retrieved from https://www.nytimes.com/2011/03/20/business/20supply.html
19. Lu HA, Cheng J, Lee TS (2006) An evaluation of strategic alliances in liner shipping—an empirical study of CKYH. J Mar Sci Technol 14(4):202–212
20. Mansouri M, Nilchiani R, Mostashari A (2010) A policy making framework for resilient port infrastructure systems. Mar Policy 34(6):1125–1134
21. Markmann Inga C, Darkow L, von der Gracht H (2013) A Delphi-based risk analysis—identifying and assessing future challenges for supply chain security in a multi-stakeholder environment. Technol Forecast Soc Change 80(9):1815–1833
22. Martín JC, Rodríguez-Déniz H, Voltes-Dorta A (2013) Determinants of airport cost flexibility in a context of economic recession. Transp Res Part E Logist Transp Rev 57(1):70–84
23. Mitroff II, Alpaslan MC (2003) Preparing for evil. Harvard Business School Pub., Boston, MA
24. Namdar J, Li X, Sawhney R, Pradhan N (2018) Supply chain resilience for single and multiple sourcing in the presence of disruption risks. Int J Prod Res 56(6):2339–2360
25. Otto C, Willner SN, Wenz L, Frieler K, Levermann A (2017) Modeling loss-propagation in the global supply network: the dynamic agent-based model acclimate. J Econ Dyn Control 83:232–269
26. Pu S, Lam JSL (2020) Blockchain adoptions in the maritime industry: a conceptual framework. Marit Policy Manag 1–18
27. Revilla E, Sáenz MJ (2014) Supply chain disruption management: global convergence vs national specificity. J Bus Res 67(6):1123–1135
28. Rice JB, Caniato F (2003) Building a secure and resilient supply network. Supply Chain Manag Rev 7(5):22–30
29. Seuring S, Goldbach M (eds) (2002) Cost management in supply chains. Physica-Verlag, Heidelberg
30. Sheffi Y, Rice JB (2005) A supply chain view of the resilient enterprise. MIT Sloan Manag Rev 47(1):41–48
31. Spekman RE, Davis EW (2004) Risky business: expanding the discussion on risk and the extended enterprise. Int J Phys Distrib Logist Manag 34(5):414–433
32. Stecke KE, Kumar S (2009) Sources of supply chain disruptions, factors that breed vulnerability, and mitigating strategies. J Mark Channels 16(3):193–226
33. Świerczek A (2014) The impact of supply chain integration on the "snowball effect" in the transmission of disruptions: an empirical evaluation of the model. Int J Prod Econ 157:89–104
34. Tang C, Tomlin B (2008) The power of flexibility for mitigating supply chain risks. Int J Prod Econ 116(1):12–27
35. The Straits Times (2020) Hundreds of flights cancelled as Shanghai tackles coronavirus outbreak, 24 Nov. Retrieved from https://www.straitstimes.com/asia/east-asia/hundreds-of-flights-cancelled-as-shanghai-tackles-coronavirus-outbreak
36. Wagner SM, Bode C (2006) An empirical investigation into supply chain vulnerability. J Purch Supply Manag 12(6):301–312
37. Wakolbinger T, Cruz JM (2011) Supply chain disruption risk management through strategic information acquisition and sharing and risk-sharing contracts. Int J Prod Res 49(13):4063–4084
38. Wilson MC (2007) The impact of transportation disruptions on supply chain performance. Transp Res Part E Logist Transp Rev 43(4):295–320
39. Wirtz J, Johnston R (2003) Singapore Airlines: what it takes to sustain service excellence—a senior management perspective. Manag Serv Qual 13(1):10–19
40. Zhang A, Lam JSL (2014) Daily Maersk's impacts on shipper's supply chain inventories and implications for the liner shipping industry. Marit Policy Manag 42(3):246–262

Chapter 27
Modelling Climate Change and Glacier Melt for Sustainable Development of a Himalayan Region

Sakkeri Ramya, Virupaxi Bagodi, and V. Devadas

27.1 Introduction

27.1.1 Climate Change, Glacier Melt, and Sustainable Development Planning

Change in state of climate can be referred as climate change. It is often identified (using statistical tests) by considering and using the indicators such as the change in mean and the variability of climate properties typically that persists for a decade or longer [20]. Generally, climate change is observed due to internal processes or external forces in the composition if the atmosphere, involves, volcanic eruptions, anthropogenic changes, solar cycles, etc. The major characteristics of the climate include the rise in average global temperature, an increase in sea level due to the melting of ice caps and glaciers, changes in precipitation, and the ocean temperature. The system of water resources for sustainable development is a large, complex, and nonlinear system, and the development of the social economy can be severely restricted by a potential shortage of water resources [7, 37]. It is observed that more than one-sixth of the world population depends on snowmelt or glacier-fed rivers, and the glaciers are expected to continue to melt as the temperature rise. Despite the understanding of climate change associated problems has grown in recent years, the researcher's/scientist's ability to improve decision-making still

---

S. Ramya (✉)
Department of Urban Planning, School of Planning and Architecture, New Delhi 110002, India
e-mail: sakkeri.ramya@spa.ac.in

V. Bagodi
Government Engineering College, Talakal, Koppal, Karnataka 583231, India

V. Devadas
Department of Architecture and Planning, Indian Institute of Technology Roorkee, Roorkee, Uttarakhand 247667, India

S. Kolathayar et al. (eds.), *Civil Engineering for Disaster Risk Reduction*, Springer Tracts in Civil Engineering, https://doi.org/10.1007/978-981-16-5312-4_27

seems limited. However, it is observed that the difficulties still exist in understanding the major processes of climate change despite the considerable knowledge of the broad characteristics of the climate.

27.1.2 State of the Art (Informetric Analysis)

This section majorly aims to consolidate the state of the art in climate change, glacier melt, and sustainable development planning. This study identifies the critical knowledge gaps for effective decision-making in the development of a district/region. Current literature on climate change, glacier melt, and applications of system dynamics is brought together in this section to distinguish characteristics that allow decision-maker in the sustainable development planning of a region. In this study, a methodology is developed for assessing the climate change and its impact on a region to address the complex sustainable development challenges that transcend individual sectors and administrative borders. Further, an attempt is made to present the review of research done on the use of an integrated approach by using system dynamic techniques in the context of evolving development plans. A methodical analysis of literature has been performed on climate change, glacier melt, and development planning (Fig. 27.1) and is presented as follows: (i) scientific publications search on the Engineering Village platform (COMPENDEX and INSPEC) using set of key words, such as system dynamics; climate change, and controlled terms, such as water resource; sustainable development; planning; (ii) initial search resulted 204,377 records (excluding duplicates) between the years 1884 and 2020, including all document types, viz. journal articles, conference proceedings and dissertations; (iii) later, to avail the most updated literature on this subject, the search was limited to past decade, i.e., between 2009 and 2018, includes, 143,387 records; (iv) further, the author conducted a search limiting to journal articles only (66,417 records), as they signify a more mature research on this subject; (v) to refine the results, the search was limited based on controlled vocabulary, such as socio-economic effects, environment factors, climate mitigation, and glacier melt; (vi) content analysis for 614 article abstracts was conducted to validate whether it focused on the role of system dynamics in the sustainable development process. The full text was further read by authors to understand its focus when in doubt; (vii) excluded articles that do not focus on the subject (208 records were excluded); (viii) categorization of literature (406 records) was done based on focus sector-based linkages involved in climate change resilient/sustainable development planning process; (ix) to observe the trends in research, informetric analysis of literature for 406 records was conducted. In addition, it identified top journals, determined classification code based on word frequency, and analyzed the dispersion of articles in previous years and also with respect to publishers using the same; (x) critical analysis carried out based on defined categories.

In this study, informetric analysis is carried for the 406 records that are obtained after excluding articles that are out of focus on the subject. Content analysis was used

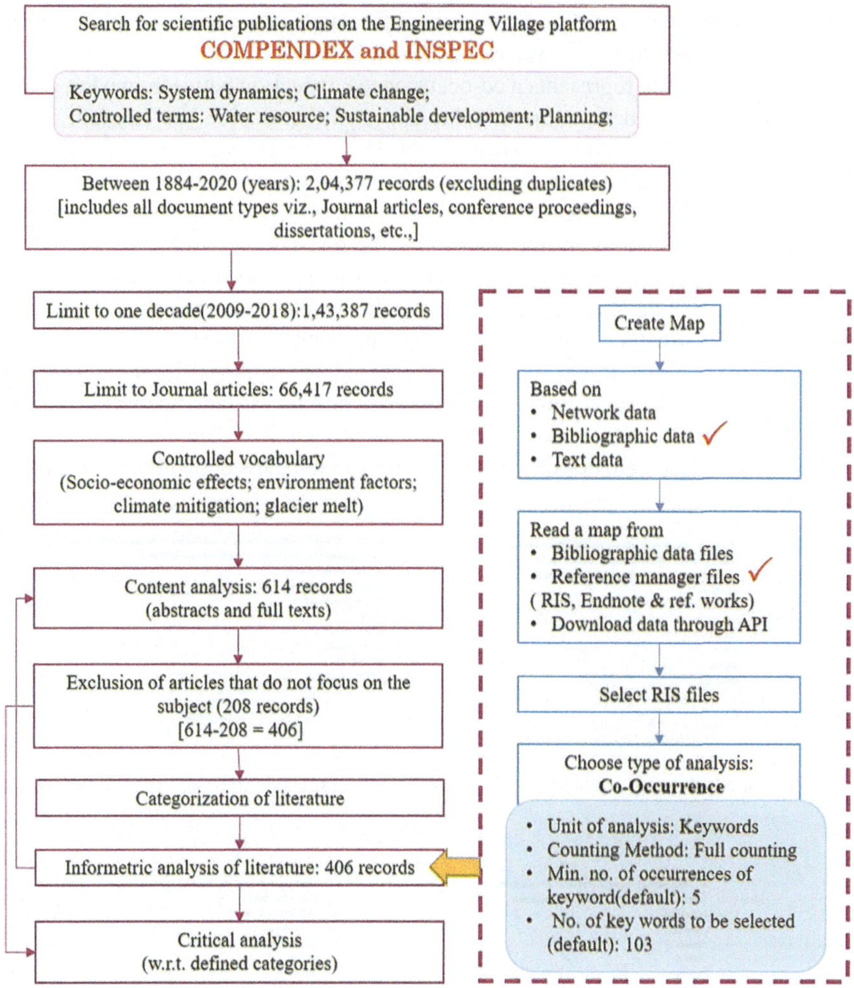

Fig. 27.1 Methodology adopted for a systematic analysis of literature on sustainable development planning and system dynamics

to further validate the information obtained from informetric analysis and identified the gaps and trends. It is observed that the articles are primarily concentrated on the three major pillars of sustainable development, such as environment, economic, and social dimensions. However, it is important to note that there is a possibility that an article can fall under more than one category too. Further, it is observed that 77.5% of the articles are being published in the last five years with an exponential growth rate. A total of 1308 authors were identified as contributing to 406 published articles, as high as 60 articles are from the Journal of Cleaner Production followed by sustainability with 52 articles and energy policy with 12 articles. Also, Elsevier publications accommodate for 34% of the total articles reviewed and are followed by Emerald, MDPI, Springer, Wiley, etc. (Figs. 27.2 and 27.3).

In Fig. 27.4, a link represents a co-occurrence connection between two keywords for which there would be numeric value assigned that depicts the strength of the link. The higher this value, the stronger the link. The articles largely focused on keywords (top five) are socio-economic effects, sustainable development, environmental factors, climate mitigation, and government policies. Lastly, Fig. 27.5 shows the network analysis and cluster analysis of the top words used by the authors in keywords. The items in the network visualization are represented by their label (by a circle). The size of the same is determined by the weight of the item. The color code for the keywords/items is done based on the clustering, and the lines between them represent links. A maximum of 1000 lines would be displayed by default, representing the 1000 strongest links between items, and the distance between the two journals signifies the relatedness in terms of co-citation links. The cluster visualization is similar to the item density visualization in which the weight given to the color

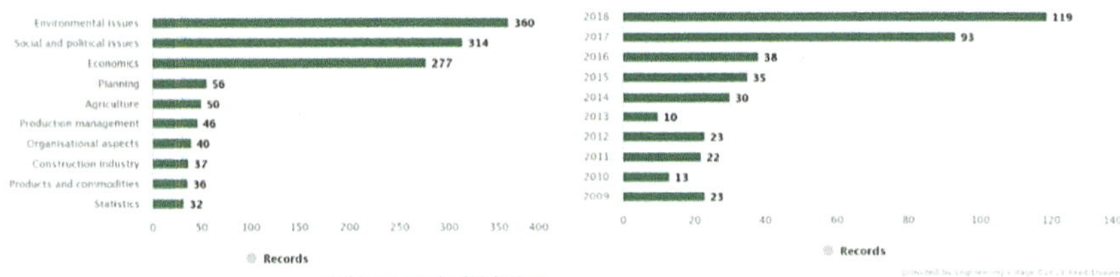

Fig. 27.2 Graph showing articles count based on classification code (left) and year of publication (right)

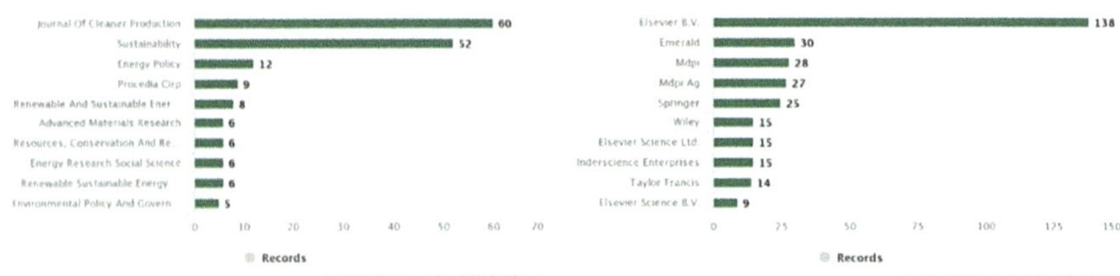

Fig. 27.3 Graph showing articles count based on journal title (left) and publisher (right)

Selected	Keyword	Occurrences	Total link strength
☑	socio-economic effects	406	898
☑	sustainable development	393	884
☑	environmental factors	306	675
☑	climate mitigation	127	313
☑	government policies	72	224

Fig. 27.4 No. of occurrences and total link strength of top five keywords identified

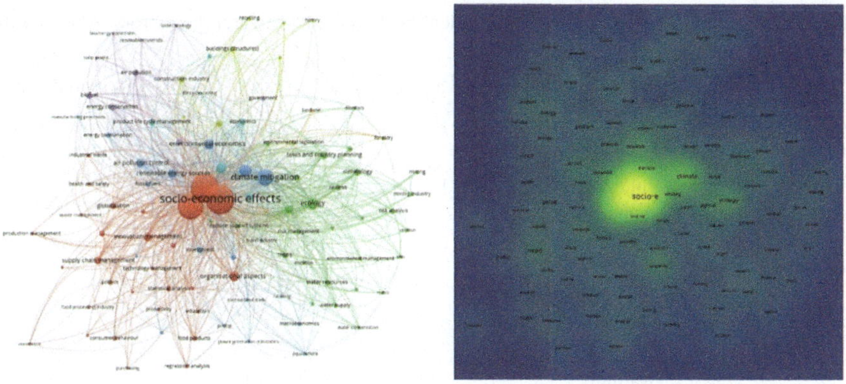

Fig. 27.5 Network analysis (left) and cluster analysis (right) of the top words used by the authors in the keywords

of certain cluster is determined by the number of items belonging to the cluster in the neighborhood of the point.

The major responsibility of the planning community is to adopt a rational planning approach addressing the complexity of the system. Yet, it is appearing that the models used at various stages are not well-developed to keep the same pace. This demands the acknowledgment and a better understanding of the dynamic inter-linkage and interdependence of the complex systems and subsystems (namely physical, social, economic, ecology, environment, infrastructure, and institutional subsystems) using system dynamics technique. System dynamics techniques are seen as an essential tool that can facilitate strategic decision-making in the region/district development system.

The comprehensive study of the literature reveals that researches have been conducted in the field of sustainable regional planning by considering only one or two subsystems and lacks an integrated approach that considers all independent subsystems of the system. Some of the key recent studies done by various researchers/organizations are presented in Table 27.1. Also, the authors identified a gap in the literature pertaining to the integration of the climate change, glacier melt, water resource management, and sustainable development planning by using system dynamics technique; however, the literature on the aforementioned has grown considerably in recent years, which indicates that system dynamics technique is increasingly being used as a reliable method for sustainable development practices. The present research focus on the Himalayan district of Uttarakhand state (consists of almost 90% of hilly terrain). The mountain/hilly people have high dependency on natural ecosystems and ecosystem services that are severely influenced/affected by climate change. In the past few decades, it has been observed that Uttarakhand state landscape which is characterized by scattered, inaccessible, and independent settlements has been transformed. Rapid urbanization has increased connectivity and a connection with markets, but at the same time, it also increased vulnerability by moving away from the self-sufficiency that characterized the communities of this region. Further,

Table 27.1 Summary of major topics and identification of the research gaps from the literature reviewed

Type of linkage	Focus of studies	Sources	Research gap
Impact of climate change on glaciers	Glacier melting and growing crises	[6, 15, 21, 31]	An integrated approach considering these linkages
Impact of climate change/glacier melting on water resources	Integrated regional water management planning; assessing regional vulnerabilities and measuring regional impacts	[4, 30]	
	Rainfall, temperature, and catchment hydrological responses	[14, 21, 26, 29, 36, 38]	
	Evaluation of snowmelt runoff	[1, 11, 12, 16, 18, 24, 25, 27, 33]	
Impact of water resources on sustainable development of the region	Physical (climate, topography, land use, soil type)	[23, 28, 35]	
	Social (demography, health, education, occupation, culture/customs)	[8, 22]	
	Economic (resources, GSDP/NSDP/per capita income, sectors of economy, finance)	[3, 13, 15, 32]	
	Ecology (flora and fauna)	–	
	Environment (pollution, water resources, forestry)	[17]	
	Infrastructure Physical (electricity, water supply, sewerage, transport, energy) Social (education health, recreational, etc.) Economy (banks, co-operations, markets/SEZ's)	Physical: [2] Social: – Economy: –	

(continued)

Table 27.1 (continued)

Type of linkage	Focus of studies	Sources	Research gap
	Institutional (local self-government, DRDA, MoUD, regional planning board, MoEF, CWC, CPCB/SPCB, planning commission)	–	

Source Compiled by the authors

not many researches have been done in identification of growth and development in Uttarakhand since the creation of the state.

Uttarakhand recorded the lowest HDI (0.49) among the glaciated states of India (2007–08). In addition, among the glaciated districts of Uttarakhand, Tehri Garhwal (Fig. 27.6) shows the highest impact of snow cover change on water bodies contribution in the district (−1:0.14), and also, there is a lack of researches conducted in the study area of Tehri Garhwal district at regional level development planning. With due consideration of the aforementioned facts, the study area of the Tehri Garhwal region

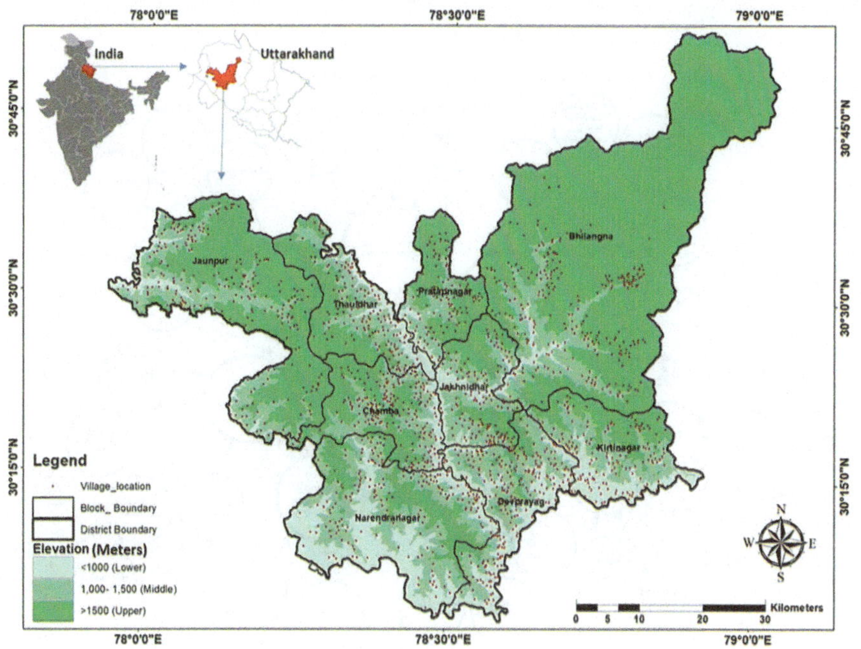

Fig. 27.6 Location map of Tehri Garhwal district

has been selected for planning sustainable development in the present investigation, and a set of objectives have been framed to achieve the same.

27.2 Approach (System Dynamics)

The term "system" means composition or whole body/organization compounded of several components. According to George Chadwick, "A system is a set of objects together with the relationship between the objects and between their attributes." A "system" can be defined as "a complex of elements standing in interactions" [5]. According to Devadas [9, 10], "A system functions as a whole with the interaction of several sub-systems. All the sub-systems of the system are interconnected, and interdependent to each other, and forming system (Fig. 27.7). A system can be a static or dynamic one. In a dynamic system, if one of the sub-systems of the system defunct or functions with a higher degree (taking a lead role during its function), its effect can be visualized in the entire system over a period" [9].

System dynamics is one of the techniques that is effective in studying and managing dynamic complex systems. This method includes information feedback loops to capture the dynamic behavior. Computer simulation modeling helps particularly in understanding the time delays and nonlinearities in the system. System dynamics is a method in which time is a very important factor in analyzing problems

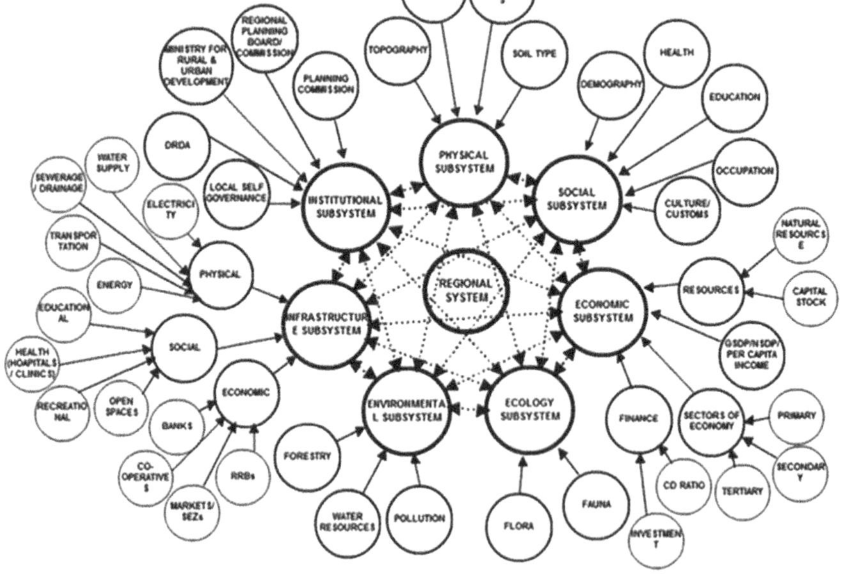

Fig. 27.7 System [10]

and assesses how a system can be defended or mad benefit from the shocks that fall from the outside world. System dynamics has been widely used in business, public policy, and environmental policymaking.

27.2.1 *Methodology*

The methodology developed in this research involves various stages as follows: Initially, the problem is identified through either literature review or theories/concepts or by knowing the current condition/existing scenario of the region through various stakeholders and then a goal is formulated accordingly. Parameters concerning climate change, glacier melt, and sustainable development planning are thereby identified, and the quantitative and qualitative data for the identified parameters are collected by using primary and secondary surveys. Further, data analysis is carried out by applying analytical methods/techniques that lead to the identification of control parameters of the system (region). The identified control parameters are further distributed among seven subsystems, namely physical, social, economic, ecology, environment, infrastructure, and institutional. A qualitative/conceptual model is developed, which is represented in the form of causal-loop diagrams with a defined system boundary. Later, qualitative simulation model described as stock-flow diagram is developed by using the following stages: (1) formulation of base year system dynamic model; (2) model validation (evaluate and test model); (3) evolving projected year system dynamic model; (4) application of the model under different alternative conditions. Results obtained from the analysis are studied in detail, and in case the obtained results are unsatisfactory or contain any error, then go to the data analysis stage to rectify the error or change the analytical method/technique adopted in accordance with the problem statement requirement. Once the results obtained are found satisfactory, then the findings were synthesized to evolve plausible solutions and implement the suggested recommendations.

27.2.2 *Population and Sampling Method/Design*

- Study area- Tehri Garhwal district, divisions (towns, villages): its socio-economic condition
- Multistage sampling method is employed.

Multistage sampling has been done for conducting household survey in the study area. In the first stage, cluster random sampling has been done on the basis of geographical area. For instance, villages within elevation zones, namely less than 1000 m, 1000–1500 m, and greater than 1500 m, are selected randomly for conducting the survey. Secondly, based on the growth potential index is calculated using the facilities index and basic index of all the villages in the district. The growth potential index accounts for the high-pressure areas within the district by analyzing the potential for fast-growing regions. This clustering reduces travel and administrative costs.

27.2.3 Survey Tools

Pre-tested schedules and questionnaires are used.

27.2.4 Analytical Tools and Techniques

Tools: Different types of software (SPSS, EXCEL, STELLA, Arc GIS) are used.

Techniques: Multiple regression and system dynamics techniques are used.

27.3 Application of System Dynamics Model

System dynamics is a method of learning complex processes that relate to the interaction of the elements of the system and dynamic concepts of its main components, i.e., stock-flow and feedback-delay. The system dynamics approach considers a holistic view to analyze the functional aspect through the spread and backwash effects in all subsystems of the system. The present study describes the development of a system dynamic model that is capable of capturing interaction and the feedbacks between the key variables of the Tehri Garhwal district (i.e., the population sub-sector, water resource sub-sector, agricultural production sub-sector, employment sub-section, tourism sub-sector, industries sub-sector, land use sub-sector, climate sub-sector, GDP sub-sector, poverty sub-sector, and infrastructure sub-sector). Out of the aforesaid sub-sector models, the three very important models have been detailed out in following sections.

27.3.1 Models (Stock-Flow)

27.3.1.1 Population Model

The population is considered as the one important parameter that has the potential to influence every part of the system. Also, in turn, it is influenced by every subsystem within the system. A system dynamic model is built to calculate population (stock/level variable) by considering the influential variables, such as births (flow), deaths (flow), birth rate fraction (converter), death rate fraction (converter),

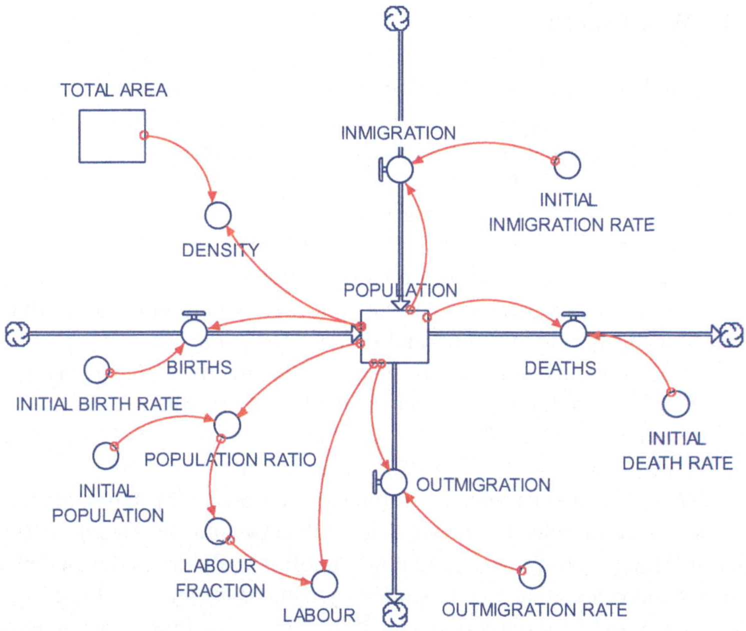

Fig. 27.8 Population model

in-migration (flow), in-migration rate fraction (converter), outmigration (flow), and outmigration rate fraction (converter). Further, in-migration and outmigration are influenced by the overall district attractiveness index. A stock-flow diagram is developed and presented in Fig. 27.8. Population is considered as stock or the level variable and is measured as a function of births, deaths, in-migration, and outmigration, which are taken as rate variables. The base year (2011) population considered is 618,931. The definitions of each variable and mathematical (algebraic) equation are described in the model equations; they are:

Population (t) = population $(t - \mathrm{d}t)$ + (births + in-migration − deaths − outmigration) * dt

Birth = f (population, births)

Birth_rate = f (population, birth_rate_fraction)

In-migration = f (population, in-migration)

In-migration_rate = f (in-migration_fraction, district_attractiveness_index)

Death = f (population, death_rate)

Death_rate = f (population, death_rate_fraction)

Outmigration = f (population, outmigration_rate)

On-migration_rate = f (outmigration_fraction, district_attractivenes_index)

27.3.1.2 Water Model

There are number of factors that contribute to the total surface water availability, such as water inflows and out flows, runoff, ground water discharge, and climate-related factors (such as the amount of precipitation, temperature, and evaporation). Water is considered as the basic need of the people. Water demand in the district is majorly from domestic activities, agriculture, livestock, and industrial purposes. Total water available in the district is calculated using rainfall and the runoff as the inflows, and the water withdrawal for various purposes, evaporation, and excess water to downstream are considered as the outflows. The aforementioned dynamics are modeled using system dynamics and is presented in Fig. 27.9.

Katling glacier is situated within the district where the river Bilangana originates. The district has the runoff contribution from all three rivers, Bhagirathi River, Bilangana River, and Alakananda River that originate from glaciers. The total surface water available in the district is 79 MCM having an inflow of 4.20E + 1 L and out flow of 4.21E + 12 L at Tehri dam. According to Khan et al. [16], approximately 11% of total runoff is contributed by snow/glacier melt at Devprayag. Therefore, runoff is contributed through snow/glacier, and other runoff is evaluated and included as one of the inflows in calculating water availability. Total domestic water used is calculated with the standard assumption of 135 LPCD, agriculture water requirement estimated using the total cropped area ever year, livestock water estimated based on the total livestock in the district, and lastly, industrial water requirement is estimated

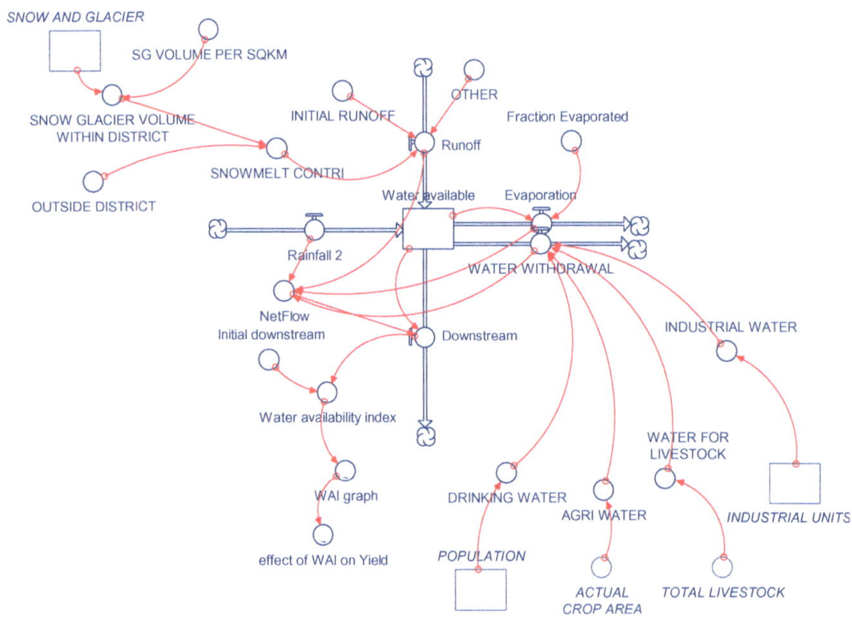

Fig. 27.9 Water model

with respect to total registered industrial units. All the above estimates change as per the change in population dynamics, land use change, industrial units registered, and lastly the change in livestock availability. Subsequently, the water availability index is assessed. The aforementioned dynamics are all well-addressed in the present model, and equations presented below describe the definition of each variable and the associated mathematical (algebraic) equations:

Water_available (t) = Water_available (t − dt) + (rainfall_2 + runoff − downstream − domestic − evaporation) * dt

Runoff = f ((initial + other) runoff, snowmelt contribution)

Snowmelt contribution = f (snow glacier area)

Water withdrawal = f (drinking water, agriculture water, water for livestock, industrial water)

Drinking water = f (population)

Agriculture water = f (actual crop area)

Water for livestock = f (total livestock)

Industrial water = f (industrial units)

Water availability index = f (total water availability)

27.3.1.3 Climate Model

One of the significant parameters to assess the climate change scenario within a district is through the contribution of CO_2 emissions into the atmosphere. To capture the aforesaid dynamics, the CO_2 generated within the district is estimated using three major sources, i.e., LPG consumption, electric energy consumption, and transportation, and the CO_2 sequestration dynamics was also included within the system (Fig. 27.10). LPG consumption is directly dependent on the total housing units in the district, which, in turn, is influenced by the population dynamics. The average LPG gas consumption per housing unit is, therefore, considered in calculating the total CO_2 emissions from LPG consumption within the district. Likewise, CO_2 emissions from electric energy consumption are calculated. This is carried out by using the electrical energy consumption trend of domestic purpose, public lights and miscellaneous, and industrial purpose. The quantities of consumption for various purposes are captured using the population dynamics that are modeled in the system. Further, CO_2 emissions from transportation are calculated as the summation of emissions from personal vehicles and public transport vehicles, which are dependent on total personal vehicles available, total length traveled per year, total gasoline consumption, and trips traveled per day. All the aforementioned variables dynamics change as the population dynamics change within the system and subsequently having a significant effect on the number of housing units too. The CO_2 sequestration through the forest is measured by combing the carbon stock from very dense forest cover area, moderate forest cover area, and other forest cover areas.

The aggregate CO_2 stock in the district is the summation of CO_2 emissions from electrical energy consumption, LPG consumption and transportation, and deletion of the total CO_2 sequestration through forests. The aggregate CO_2 in PPM is later used

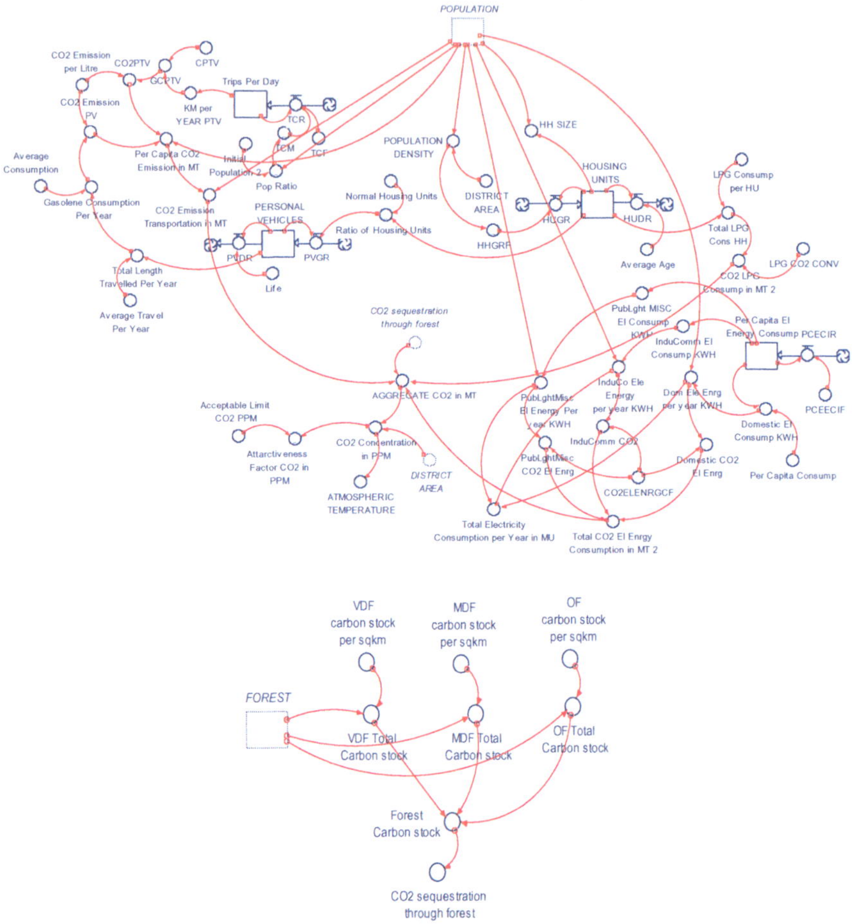

Fig. 27.10 Climate model

to estimate the atmospheric temperature using a log-linear equation, adapted from the work of Kumarr [19], which is developed for the micro-level climatic conditions such as Tehri Garhwal district. Furthermore, the attractiveness factor CO_2 is worked out using the aggregate CO_2 in PPM and the acceptable limits of CO_2 in PPM in rural areas of India. The model equations presented below describe the definition of each variable and the associated mathematical (algebraic) equations:

Aggregate CO_2 in MT $= f$ (CO_2 LPG consumption, CO_2 electric energy consumption, CO_2 transportation, CO_2 sequestration through the forest)

CO_2 LPG consumption $= f$ (housing units, LPG consumption per HU)

Housing units $= f$ (HUGR, HUDR)

HUGR $= f$ (population density)

HUDR $= f$ (average age of HU)

CO_2 electric energy consumption = *f* (public lights, industries, domestic energy consumption)

CO_2 transportation = *f* (emission from personal vehicles, public transport vehicles)

Emission from personal vehicles = *f* (gasoline consumption per year)

Gasoline consumption per year = *f* (total length traveled per year, avg. consumption)

Total length traveled per year = *f* (personal vehicles, avg. travel per year)

Personal vehicles = *f* (PVGR, PVDR)

PVGR = *f* (housing units)

PVDR = *f* (vehicle life)

CO_2 sequestration through forest = *f* (forest carbon stock)

Forest carbon stock = *f* (VDF forest stock, MDF forest stock, OF forest stock)

Atmospheric temperature = *f* (CO_2 concentration in PPM)

CO_2 concentration in PPM = *f* (CO_2 in MT, district area)

Attractiveness factor for CO_2 in PPM = *f* (CO_2 concentration in PPM, acceptable limit CO_2 PPM)

27.3.2 Model Result

The investigator has developed the system dynamic model with the base year 2011, to understand the functions of the regional system by using various control variables. Table 27.2 summarizes the base year model results of the important parameters.

27.3.3 Model Validation

According to Sterman [34], synthesis/testing of the model is considered as an essential tool to identify the flaws in the system dynamic model and leaves a scope for better understanding. In principle, more tests help to gain more confidence that the model cannot be flawed. The consolidated model was validated by performing several tests such as parameter-confirmation tests, behavior pattern tests, and sensitivity analysis using an extreme-condition test. The parameter-confirmation test involves checking the equations contained in the model that are ensured to correspond to the relationships illustrated in the conceptual model/causal-loop diagram.

Extreme values were assigned to the input variables, and subsequently, it is checked with the output variables, whether the behavior is logical in the real system under the same extreme conditions. Validity of the selected model equations is done using sensitivity analysis. On other hand, behavior pattern test (comparing simulated results with the reference modes) is used to measure how well the developed model exhibits the real system. The base data for the year 2011 used in the model have been taken from secondary sources. The projections generated by the model for the

Table 27.2 Model result for the base year 2011

S. No.	Parameter	Unit	2011 (base year)
1	Population	Number	618,931
2	Housing units	Number	132,714
3	Density	No./km^2	169.94
4	Total labor	Number	280,676
5	Arable land	km^2	862.22
6	Barren/waste land	km^2	96.87
7	Built-up land	km^2	16.33
8	Forest land	km^2	2378.62
9	Grass/grazing	km^2	111.13
10	Snow/glacier	km^2	123.03
11	Water bodies	km^2	53.83
12	GDP (in lakh)	Rupees	390,458
13	Primary sector		8,157,999,492
14	Secondary sector		13,546,200,000
15	Tertiary sector		17,151,112,000
16	Subsidies/taxes		233,900,000
17	Primary schools	Number	1753
18	High schools	Number	483
19	Colleges	Number	269
20	Primary health center	Number	30
21	Community health center	Number	220
22	Hospital	Number	142
23	Livestock	Number	457,753
24	Hotels	Number	312
25	Hotels and restaurants revenue	Rupees	4,906,512,000
26	Industrial units	Number	2198
27	Industrial revenue	Rupees	5,275,200,000
28	Industrial area	km^2	0.32
29	Tourist destinations	Number	25
30	Tourists	Number	914,608
31	Water withdrawal	L	53,430,865,881
32	Evaporation	L	7,900,000,000
33	Runoff	L	2,896,488,633,238
34	Crop area	km^2	951.62
35	Cereals area	km^2	691.83
36	Pulses area	km^2	41.87

(continued)

Table 27.2 (continued)

S. No.	Parameter	Unit	2011 (base year)
37	Oil seeds area	km^2	7.61
38	Fruits area	km^2	201.74
39	Potato area	km^2	7.61
40	Cereals yield	Quintal per km^2	1569.34
41	Pulses yield	Quintal per km^2	816.07
42	Oil seeds yield	Quintal per km^2	1489.11
43	Fruits yield	Quintal per km^2	1349.40
44	Potato yield	Quintal per km^2	9425.70
45	Cereals production	Quintal	1,085,715
46	Pulses production	Quintal	34,170
47	Oilseeds production	Quintal	11,337
48	Fruits production	Quintal	272,233
49	Potato production	Quintal	71,757
50	Crop production	Quintal	1,475,212
51	Meat production	Quintal	60,238
52	Milk production	Quintal	4,998,611
53	Agriculture and allied activities revenue	Rupees	5,021,061,684
54	Food available	kg	653,509,370
55	CO_2 MT	Metric ton	130,075
56	CO_2 sequestration	Metric ton	10,569
57	CO_2 LPG MT	Metric ton	42,787
58	CO_2 electricity MT	Metric ton	57,207
59	CO_2 transportation MT	Metric ton	40,650
60	Temperature	°C	16.52

year 2011 have been compared with the data from the secondary source to find the estimated error that is present in the model. In this study, sixteen important parameters (population, births, deaths, gross domestic product, primary sector, secondary sector, tertiary sector, cereals production, primary schools, high schools, colleges, tourist destinations, primary health center, hospital, community health center, and industries) were selected for behavioral pattern testing based on the availability of historical data, and it is observed that all sixteen parameters have R^2 between 0.7 and 0.99 and percentage error less than or equal to five.

27.3.4 Forecasting

The business as usual (BAU) scenario, i.e., with no new policy interventions has been projected from the base year 2011 to 2041 A.D. to understand the behavior of each sub-model and its impact on the attractiveness of the district. It is observed from the model result that there is an increase in population from 618,931 to 811,704 between the years 2011 and 2041, and also, a similar trend is observed in all the flow variables in this sub-model, i.e., births, deaths, in-migration, and outmigration. Consequently, housing units in the district are expected to increase from 132,714 to 164,136; the land use change projections display the change in snow/glacier area from 123.03 to 1.5 km^2 resulting an increase of water bodies area within the district from 53.83 to 74.01 km^2 (Error! Reference source not found.). Also, it is observed that there is a significant decrease in the total arable land from 862.22 to 694.87 km^2, followed by the total grass/gazing area. On the other hand, there is an increase in built-up land, forest land, and barren/wasteland. Especially, the increase of barren land from 96.87 to 148.89 km^2 captures the scenario of change of land use from arable and snow/glacier area and contributing for the same; an increase in district's economy, i.e., GDP (in lakhs) rise from 390,458 to 1,013,402 is observed with a per capita GDP change from 63,085 to 124,848. Further, there is an increase in primary sector from 8.15 to 11.8 crores, secondary sector from 13.54 to 34.09 crores, and in tertiary sector from 17.15 to 51.02 crores; there is an increasing trend observed in the primary and community health centers and a decline in the number of hospitals within the district. The primary schools, secondary schools, and colleges displayed an increasing trend within the system. However, the respective number of institutions is compared against the URDPFI guidelines, and subsequently, health and education attractiveness index has been calculated. The health and education attractiveness index remained constant over the years as the aforementioned institution's areas being up to the URDPFI standards; with a significant increase in CO_2 emissions, i.e., from 130,075 to 457,825, MT resulted in an increase of temperature from 16.52 to 19.24 °C, and consequently, a decrease in snow/glacier area is observed; an increasing trend observed in temperature and soil erosion had a significant effect on the yield. In reference to the water sub-model, water availability index is showing a decreasing trend, i.e., from 0.87 to 0.77, and this variable is highly being influenced by the change in snow/glacier area and its consequent change in effect on the total runoff in the district. In addition, the water availability index is also influenced by the total water withdrawal variable. It constitutes water use for agriculture and livestock purpose and is experiencing a decreasing trend as the total crop area and the livestock population are declining. In contrast, water used for industrial and drinking water is witnessing an increasing trend considering the population and industries requirement dynamics. A decreasing trend is witnessed in the total arable land and, therefore, reducing the total crop area in the district. It is further influencing the total production quantities of cereals, pulses, oilseed, potato, and fruits in the district. The similar scenario is reflected in total livestock reflecting on the decrease of total milk and meat production; the decreasing production quantities and agriculture employment (from

97,523 to 68,040) are effecting the total food available (decrease from 653,509,370 to 542,241,817) and subsequent decrease in revenue generated from the crops in the district. This is resulting in a decrease in income per capita generated from the agriculture; Tehri Garhwal district is known for its tourism, and the same is reflected in the projections of the number of tourists and the increasing number of tourist destinations over the decades; it is observed that attractiveness index of tourism and industries is increasing, and a decreasing trend is witnessed in the attractiveness index of CO_2, employment, and farm income. Lastly, the district attractiveness index is decreasing from 0.68 to 0.56 between the years 2011 and 2041, therefore, presenting the need for interventions or policy implications to improve district attractiveness index.

27.3.5 Recommended Policy and Discussion

Apart from BAU scenario, fifteen additional policy scenarios were designed and simulated. Inputs for creating these scenarios were received during the participatory stakeholder/modelling. In addition, important information has been taken from the existing literature relating to the effective management of the district. The policy scenarios were developed on the basis of the system's prevailing trend and dynamic behavior, using selected control parameters among the indicators of different subsystems of the regional system, determining policy interventions, and making recommendations for sustainable development. Different policy scenarios are designed and simulated for a period of 20 years (i.e., between 2021 and 2041) to arrive at plausible decisions.

All the policy scenarios analyzed in this study are experiencing an increase in the total population, driving the increase in water demand (agricultural, industrial, and domestic). It is imperative in this situation that not just the supply of water without effective demand management is leading to water shortages. For this reason, in the district of Tehri Garhwal, a combination of population growth control, water conservation, and strategic water demand management is required to meet its current and future water needs in a sustainable manner. Detailed analysis of the evolved policies and their perceived outcomes by the investigator signify that the policy (S12) (−30% emissions from LPG consumption − 35% emissions from electric energy consumption − 50% emissions from transportation + 20% working population (labors) + 10% agriculture labor percent + 0.5% high yield seeds + 54% farm machinery + 30% fertilizers + 35% poultry fraction rate + 40% cows − buffalos fraction rate + 10% eggs production + 30% fish production + 20% industrial units fraction + 50% hotels and restaurants growth fraction + 35% employment change rate fraction + 50% tourist destination fraction) would be more suitable for the social, economic, physical, infrastructure, institutional, environmental, and ecological subsystems, and overall development of the regional system is having the highest district attractiveness index (i.e., 0.64) when compared to other 14 scenarios. The results suggest that in the case of Tehri Garhwal district, cropland expansion, reducing the number of private vehicles trips, and reduction in electric energy consumption are likely to lead to an outcome that is more than business as usual (BUA) and will maximize the benefits to people living within the district. The selected scenario has been tested in the projected year (2041 A.D.) model, and it has been observed from the result that there is a decrease of temperature by 1.25 °C; increase in GDP by 4.96%; increase in arable land by 8.77%; increase in revenue of primary, secondary, and tertiary sectors as by 17.84%, 0.61%, and 5.78%, respectively; increase in food available by 39.54%; increase in the effect of tourism index, CO_2 emission index, agriculture income index, employment index, and industries index on district attractiveness index by 11.9%, 33.85%, 56%, 11.43%, and 4.88%, respectively; and lastly, increase in district attractiveness index by 0.08 (i.e., from 0.56 to 0.64) as compared to the projected year model result.

This scenario illustrates the improved crop yields and net-farm income, resulting in maximum benefits to the local people, especially in the areas of education, health (nutrition), and social equity. Also, improved water facility/infrastructure results in enhanced food security. Consequently, they have a positive impact on socio-economic development and poverty reduction. If the most optimistic scenario (S12) is supported by the local governments/authorities through initiatives such as micro-credit, soil fertility management, infrastructure management, awareness/training programs for knowledge improvements of the advanced technology, and innovation in the agriculture sector would enhance the growth and development of Tehri Garhwal district. It is important to consider that water infrastructure comes with a cost as it competes with other priorities. Also, it has to associate with the water allocation management and good governance supported by institutional development in the district.

27.4 Conclusion

The present study describes the development of a system dynamic model that is capable of capturing interaction and the feedbacks between the key variables of the Tehri Garhwal district (i.e., the population sub-sector, employment sub-section, tourism sub-sector, industries sub-sector, land use sub-sector, water resource sub-sector, climate sub-sector, GDP sub-sector, poverty sub-sector, agricultural production sub-sector, and infrastructure sub-sector). System dynamic models are built for various sectors and have wide application across the globe. However, this study is one of the few applications in India and the first application in the Tehri Garhwal district, Uttarakhand. This study has best captured the dynamics of interactions between the district subsystems over time. Also, this model reproduces the response properties of state-of-the-art climate models very well.

Stakeholders through the participatory process have played a major role in the model development phase. Understanding/capturing the long-term dynamics if the district was the major objective of the model. The same has been the basis for developing alternative policy scenarios for the sustainable development of the Tehri Garhwal district. The developed model was further tested. The system dynamic model has been used to simulate the outcomes of fifteen different policy scenarios for over 20 years (i.e., between 2021 and 2041); out of which, scenario 12 is selected as more suitable for the social, economic, physical, infrastructure, institutional, environmental and ecological subsystems, and overall development of the regional system. The scenario 12 has been tested in the projected year (2041 A.D.) model, and it has been observed from the result that there is a decrease in temperature by 1.25 °C; increase in GDP by 4.96%; increase in arable land by 8.77%; increase in revenue of primary, secondary, and tertiary sectors as by 17.84%, 0.61%, and 5.78%, respectively; increase in food available by 39.54%; increase in the effect of tourism index, CO_2 emission index, agriculture income index, employment index, and industries index on district attractiveness index by 11.9%, 33.85%, 56%, 11.43%, and 4.88%, respectively; and lastly, increase in district attractiveness index by 0.08 (i.e., from 0.56 to 0.64) as compared to the projected year model result.

The model results shown in this study play a significant role in the sustainable development of the district, including water resource management, agriculture development, growth in the industrial and tourism sector, and gross GDP growth. It is to be noted that like any other collaboratively constructed model, it is imperfect and incomplete. Hence, the findings obtained from the present research have been understood with due consideration to assumptions and limitations used in the process of modelling. This suggests that there should be some degree of caution to be maintained while using the model results (the assumption and scenarios) to other hill districts of India. In conclusion, it is expected that if the proposed planning model is implemented successfully in the study area, it will ensure sustainable development.

Acknowledgements We would like to thank to Tehri Garhwal District Collector (Ms. Sonika) and Indian Institute of Technology, Roorkee, for extending their support all throughout the research.

References

1. Arora M (2010) Estimation of melt contribution to total streamflow in river Bhagirathi and river DhauliGanga at Loharinag Pala and Tapovan Vishnugad project sites. J Water Resour Prot 02:636–643. https://doi.org/10.4236/jwarp.2010.27073
2. Asian Development Bank (2014) India: support for the implementation of the national water mission by state governments in India: scoping study for a national water use. Asian Development Bank
3. Bajracharya SR, Shrestha M, Shrestha AB (n.d.) Impact of climate change on water resources and livelihood in the HKH region: introduction. https://doi.org/10.13140/RG.2.1.1330.0082
4. CDM (2011) Climate change handbook fo regional water planning. U.S. Environmental Protection Agency, California Department of Water Resources, U.S. Army Corps of Engineers
5. Chadwick GF (1971) A systems view of planning: towards a theory of the urban and regional planning process. Pergamon Press
6. Chandrappa R, Kulshrestha UC, Gupta S (2011) Coping with climate change. Agriculture 213–224. https://doi.org/10.1007/978-3-642-19674-4
7. Chen Z, Wei S (2014) Application of system dynamics to water security research. Water Resour Manage. https://doi.org/10.1007/s11269-013-0496-8

8. Cox WE (1987) The role of water in socio-economic development. United Nations Educational, Scientific and Cultural Organization (UNESCO). https://doi.org/10.1017/CBO9781107415324.004
9. Devadas V (2001a) Planning for rural energy system: part III. Renew Sustain Energy Rev 5:271–297. https://doi.org/10.1016/S1364-0321(00)00016-2
10. Devadas V (2001b) Planning for rural energy system: part I. Renew Sustain Energy Rev 5:203–226. https://doi.org/10.1016/S1364-0321(00)00014-9
11. Duethmann D, Bolch T, Farinotti D, Kriegel D, Vorogushyn S, Merz B, Pieczonka T, Jiang T, Su B, Güntner A (2015) Attribution of streamflow trends in snow and glacier melt-dominated catchments of the Tarim River, Central Asia. Water Resour Res 51:4727–4750. https://doi.org/10.1002/2014WR016716
12. Fujita K, Sakai A (2014) Modelling runoff from a Himalayan debris-covered glacier. Hydrol Earth Syst Sci 18:2679–2694. https://doi.org/10.5194/hess-18-2679-2014
13. Harender Raj Gautam HLS (2012) Environment and rural issues. Kurukshetra J Rural Dev 60:3–5
14. Jansson P, Hock R, Schneider T (2003) The concept of glacier storage: a review. J Hydrol 282:116–129. https://doi.org/10.1016/S0022-1694(03)00258-0
15. Kaltenborn BP, Nellemann C, Vistnes II (2010) High mountain glaciers and climate change—challenges to human livelihoods and adaptation. United Nations Environment Programme
16. Khan AA, Pant NC, Sarkar A, Tandon SK, Thamban M, Mahalinganathan K (2016) The Himalayan cryosphere: a critical assessment and evaluation of glacial melt fraction in the Bhagirathi basin. Geosci Front. https://doi.org/10.1016/j.gsf.2015.12.009
17. Kotir JH (2017) Systems thinking and modelling for sustainable [WWW document]. URL espace.library.uq.edu.au. Accessed 12.13.19
18. Kumar V, Singh P, Singh V (2007) Snow and glacier melt contribution in the Beas River at Pandoh Dam, Himachal Pradesh, India. Hydrol Sci J 52:376–388. https://doi.org/10.1623/hysj.52.2.376
19. Kumarr AS (2017) Planning for energy efficient sustainable development of Chennai. Anna University
20. Masson-Delmotte V, Zhai P, Pörtner H-O, Roberts D, Skea J, Shukla PR, Pirani A, Moufouma-Okia W, Péan C, Pidcock R, Connors S, Matthews JBR, Chen Y, Zhou X, Gomis MI, Lonnoy E, Maycock T, Tignor M, Waterfield T (2019) Special report: global warming of 1.5 °C
21. Miller JD, Immerzeel WW, Rees G (2012) Climate change impacts on glacier hydrology and river discharge in the Hindu Kush—Himalayas: a synthesis of the scientific basis. Mt Res Dev 32:461–467. https://doi.org/10.1659/MRD-JOURNAL-D-12-00027.1
22. Mittal S, Tripathi G, Sethi D (2008) Development strategy for the hill districts of Uttarakhand. Indian Council for Research on International Economic Relations
23. Nepal S, Flügel W-A, Shrestha AB (2014) Upstream-downstream linkages of hydrological processes in the Himalayan region. Ecol Process 3:1–16. https://doi.org/10.1186/s13717-014-0019-4
24. Omani N, Srinivasan R, Smith PK (2016) Glacier mass balance simulation using SWAT distributed snow algorithm. Hydrol Sci J 6667. https://doi.org/10.1080/02626667.2016.1162907
25. Panday PK, Williams CA, Frey KE, Brown ME (2014) Application and evaluation of a snowmelt runoff model in the Tamor River basin, Eastern Himalaya using a Markov Chain Monte Carlo (MCMC) data assimilation approach. Hydrol Process 28:5337–5353. https://doi.org/10.1002/hyp.10005
26. Prasad AK, Yang KHS, El-Askary HM, Kafatos M (2009) Melting of major glaciers in the western Himalayas: evidence of climatic changes from long term MSU derived tropospheric temperature trend (1979–2008). Ann Geophys 27:4505–4519. https://doi.org/10.5194/angeo-27-4505-2009
27. Prasch M, Mauser W, Weber M (2013) Quantifying present and future glacier melt-water contribution to runoff in a central Himalayan river basin. Cryosphere 7:889–904. https://doi.org/10.5194/tc-7-889-2013

28. Rahman K, da Silva AG, Tejeda EM, Gobiet A, Beniston M, Lehmann A (2015) An independent and combined effect analysis of land use and climate change in the upper Rhone River watershed, Switzerland. Appl Geogr 63:264–272. https://doi.org/10.1016/j.apgeog.2015.06.021
29. Rajesh B (2012) Jawaharlal Nehru University research objective. Center for Studies in Science Policy, School of Social Sciences
30. Sharma BR, Sharma D (2008) Impact of climate change on water resources and glacier melt and potential adaptations for Indian agriculture, pp 86–101
31. Singh BR, Singh O (2013) A study about realities of climate change: glacier melting and growing crises. In: Climate change—realities, impacts over ice cap, sea level and risks
32. Singh M, Pallavi P (2015) Reservoir induced impact on agricultural patterns and livelihood practices: a case study of Pratapnagar block in Tehri dam rim area. Int J Sci Res Publ 5:1–8
33. Singh P, Jain SK (2003) Modelling of streamflow and its components for a large Himalayan basin with predominant snowmelt yields. Hydrol Sci J 48:257–276. https://doi.org/10.1623/hysj.48.2.257.44693
34. Sterman J (2000) Business dynamics: systems thinking and modelling for a complex world. McGraw-Hill, Boston
35. Tadele K, Förch G (2007) Impact of land use/cover change on streamflow: the case of Hare River Watershed, Ethiopia. In: Symposium (LARS), Arba Minch, Ethiopia, pp 80–85
36. Watts G, Battarbee RW, Bloomfield JP, Crossman J, Daccache A, Durance I, Elliott JA, Garner G, Hannaford J, Hannah DM, Hess T, Jackson CR, Kay AL, Kernan M, Knox J, Mackay J, Monteith DT, Ormerod SJ, Rance J, Stuart ME, Wade AJ, Wade SD, Weatherhead K, Whitehead PG, Wilby RL (2015) Climate change and water in the UK—past changes and future prospects. Prog Phys Geogr 39:6–28. https://doi.org/10.1177/0309133314542957
37. Winz I, Brierley G, Trowsdale S (2008) The use of system dynamics simulation in water resources management. Water Resour Manag 23:1301–1323. https://doi.org/10.1007/s11269-008-9328-7
38. Zeng R, Cai X (2014) Analyzing streamflow changes: irrigation-enhanced interaction between aquifer and streamflow in the Republican River basin. Hydrol Earth Syst Sci 18:493–502. https://doi.org/10.5194/hess-18-493-2014

Chapter 28
Lessons from a Century-Tradition on Ecosystem-Based Disaster Risk Reduction (Eco-DRR) in Mountains: The Case of the Torrential System *Los Arañones* (Canfranc, Pyrenees)

José Luis García Rodríguez, Álvaro Esteban Muñoz, Fernando Garcia-Robredo, José Carlos Robredo Sánchez, Santiago Fábregas Reigosa, Guillermo Tardío Cerrillo, Miguel Esteban Herrero, Francisco Arriaga Martitegui, Guillermo Íñiguez-González, Rocío Hurtado Roa, and Juan Antonio Ballesteros Cánovas

28.1 Introduction

Mountain regions are characterized by the occurrence of extreme hydrogeomorphic events, favored by the high energy relief, complex geology, and extreme climate conditions [1, 2]. The occurrence of torrential floods, snow avalanches, rockfall, or landslides have the potential to affect exposed infrastructure, cultural and economic assets as well as human life [3]. Such extreme events can disrupt the status quo of communities and may put the future welfare of people living in mountain valleys

J. L. G. Rodríguez (✉) · Á. E. Muñoz · F. Garcia-Robredo · J. C. R. Sánchez · G. T. Cerrillo · M. E. Herrero · F. A. Martitegui · G. Íñiguez-González · R. H. Roa
Department of Forest and Environmental Engineering and Management, School of Forest Engineering and Natural Resources, Universidad Politécnica de Madrid, Calle José Antonio Novais 10, Ciudad Universitaria, ES-28040 Madrid, Spain
e-mail: josel.garcia@upm.es

S. F. Reigosa
AECT Espacio Portalet, Frontera de El Portalet Carretera A-136 PK 27, ES-22640 Sallent de Gállego, Huesca, Spain

J. A. B. Cánovas (✉)
Dendrolab.ch, Department of Earth Sciences, University of Geneva, 13 rue des Maraîchers, CH-1205 Geneva, Switzerland
e-mail: juan.ballesteros@unige.ch

Climate Change Impacts and Risks in the Anthropocene (C-CIA), Institute for Environmental Sciences, University of Geneva, 66 Boulevard Carl Vogt, CH-1205 Geneva, Switzerland

S. Kolathayar et al. (eds.), *Civil Engineering for Disaster Risk Reduction*, Springer Tracts in Civil Engineering, https://doi.org/10.1007/978-981-16-5312-4_28

at risk [4, 5]. Torrential floods are especially common, representing about half of all-weather related disasters that occurred between 1995 and 2015 [5]. The impact of torrential floods, often connected to large floods down the valley, have affected millions of people and caused many tolls worldwide [6].

People living in mountain areas are especially vulnerable to the impact of natural hazards [7]. In these regions, social systems are characterized by isolating settlements which livelihood is linked to agriculture, animal husbandry, and/or gathering, especially in low-income countries [8]. Besides, in medium–high income countries, mountain regions are well connected to major population centers favoring permanent residents, economic migrants, amenity migrants, tourists, and other transients [9], which increases the elements-at-risk. In the next decades, the occupation of prone areas to natural hazards with the expected climate change impact could lead to a potential increase in losses and damages, as identified already in global trends [10]. According to the IPCC, climate change is both increasing the frequency, intensity, spatial extent, duration and changing the seasonality of extreme weather events [10]. The rate of change is dissimilar across the world, but there is a certain confidence that hydrological extreme events will be more common, with mountain regions being especially sensitive to these changes [11]. Besides, mountain regions are highly sensitive to land-use changes and present a higher hazard for coalescence of high energy processes and cascading events. People living in mountain regions are, therefore, especially vulnerable to these extreme events [12] (Fig. 28.1).

Improving our capacity to mitigate future disaster in mountain regions will be essential for sustainable development, representing one of the main goals of different international and national frameworks, such as Disaster Risk Management Sendai Framework and the 2030 Agenda for Sustainable Development [13]. Overall, all these policies underline the need to improve our knowledge on extreme scenarios with preparedness purposes [3, 14], considering the efficiency of land-use planning as a tool to reduce damages and death tolls. Assertive land-use maps, where the hazard and risk for each process are represented, allow to refrain the installation of new settlements in places where exposure is too high, as well as the construction of defense structures following the principle of environmental and economical effectiveness. Hazard planning and long-term protective measurements are especially relevant in mountain regions, where the implementation of warning systems could be challenging due to the rapid hydrological response of catchments [15] and the high energy processes taking place in them [16–18]. Finding the right balance between safety and sustainability is, therefore, a real scientific-technical challenge with large implications in many socio-economic sectors.

In this context, the implementation of Ecosystem-based disaster risk reduction (Eco-DRR) strategies represents a paradigm shift to reduce vulnerability and promote resilience-disaster communities [19, 20]. Eco-DRR refers to the reduction of risk by (i) reducing the exposure of elements-at-risk in natural hazard prone areas; and (ii) improving ecosystem health and functioning by means of the use of environmentally and economically viable measurements, as a buffer for protecting lives and infrastructures. Being promoted in newer DRR-frameworks, the implementation of Eco-DRR

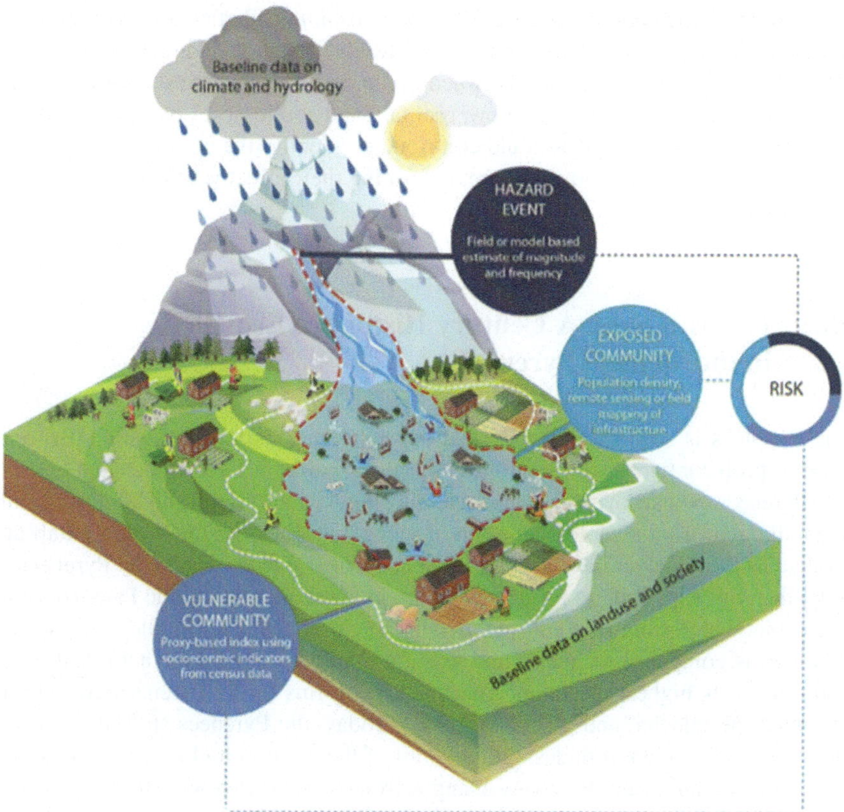

Fig. 28.1 Schematic framework for risk assessment of hydrological-related hazards [12]

can capitalize the long-term forest engineering experience in classical hydrological-forest correction strategies [21] aimed at improving the ecosystem's state of highly degraded mountain catchments. Thus, under such scheme, hydrological-correction dams, levees. or anti-erosion corrective works have been installed to reduce soil losses, promote forest establishment, and, ultimately, reduce impacts from natural disasters [22]. Thus, the implementation of structural and non-structural measures are currently increasing as a more flexible, effective, and efficient solution for increasing resilience, especially in mountain regions [23]. However, even considering this as the most appropriate way to cope with future disasters, little is still known about the reliability of such Eco-DRR solutions for long term. The reasons are (i) the systematic evaluation of such Eco-DRR measurements have not been broadly performed, and (ii) when such analyses have been done, the monitoring period is short, as such analyses are mostly focused on recently implemented Eco-DRR.

Here, we review the long-term Eco-DRR implementation combined with the installation of large measurements to protect critical infrastructure in the Pyrenees mountain range. In particular, we present the century accumulated experience derived from the implementation of Eco-DRR in *Los Arañones*. Particularly, we show the protection strategy case of the International Railway Station and related infrastructures of Canfranc on the border between Spain and France against recurrent snow avalanches, debris and torrential flow, rockfall, and soil losses. Long-term analyses of the reliability of Eco-DRR could contribute to identify the weakness and robustness of such approach, and therefore, determine their effectiveness under ongoing climate change scenarios.

28.2 *Los Arañones*: A Century History of Eco-DRR in the Spanish Pyrenees

The Pyrenees is a mountain range that separates the Iberian Peninsula (IP) from the European continent and constitutes the natural border between Spain and France. The Pyrenees extends over 500 km connecting the Atlantic Ocean and Mediterranean Sea, although, from a geological point of view, it is an extension of the Cantabrian mountains located over the entire northern coast of IP. Geologically, the Pyrenees is older than the Alps and are formed by sediment deposited during the Paleozoic and Mesozoic, which were lifted during the Hercinic orogeny. The middle sector of the Pyrenees is composed of gneiss and granitic material, being the Aneto Peak (i.e., 3404 masl) its highest elevation. The main landforms in the Pyrenees are related to fluvial, periglacial, and glacier activities. Today, the Pyrenees still have several small glaciers at high altitudes, although all of them are in highly recessive stages [24]. The climate in the Pyrenees changes from oceanic (Norwest sectors) toward Mediterranean climate (in South-east sectors), with certain continental climate in central sectors. These conditions define temperate oceanic and sub Mediterranean bioclimes which, together with altitudinal gradient, condition the existing vegetation. Overall, vegetation can be divided into a montane belt (mixed deciduous and conifer trees, Fagus sp, Pinus sp), subalpine belt (mostly conifer trees, *Pinus* sp, *Abies* sp, *Piceas* sp), alpine belt (conifer trees and the presence of *Alnus* sp), and finally a sub and nival belt.

The Pyrenees are intersected by several valleys that connect France and Spain throughout mountain passes. Although there is physical evidence (Roman bridges, Roman roads) that indicates that some valleys were used for trading since the Roman period, the agreement "Treaty of the Pyrenees" signed between France and Spain in 1659 A.D. stipulated the international border and gave the importance of some valleys as routes for international trade between countries. Since then, corridors of communication have been developed. An example was the route to the Somport pass built in 1876 A.D. following the path of the Caesar Augusta Roman road. In particular, the Somport pass became a strategic mountain pass since it was the pass with the lowest altitude in the central Pyrenees. In 1882 A.D., during the reign of Alfonso XII (1874–1885), the construction of a railway line to link France with Spain through the Somport tunnel was planned. The works started in 1882, although the perforation of the tunnel started several decades later (in 1908) after several discussions about the convenience of the project. These works included the construction of an International Railway station in the south exit of the tunnel, in a flat place called *Los Arañones* within the municipality of Canfranc. The works ended in 1928, and the International Railway of Canfranc quickly became an important spot for international trading. In parallel, the village of *Los Arañones* grew significantly, becoming an important population center at that time. After the Spanish Civil War (1936–1939), this infrastructure experienced a golden age, especially during the 1950s and 1960s. However, a train accident in 1970 on the French side affected a bridge seriously and since then the communication has not been re-established. Without a willingness to solve this issue between both countries in the following years, the infrastructures have seen their aging and progressive deterioration. Currently, the existing International Railway of Canfranc (and related infrastructures) forms part of the tourist attraction and only works on the Spanish side, while the international communication has been changed to road transportation.

The planning of the International Railway of Canfranc supposed a real scientific-technical challenge due to the intrinsic complexity of works in mountain terrains and the existing hazards related to the occurrence of natural disasters [25]. The engineers in charge understood that the feasibility of the infrastructures was linked to their protection against frequent floods and snow avalanches, enhanced by the intense deforestation by the end of nineteenth century [25]. Innovatively, Spanish forest engineers (Benito Ayerbe—1872–1917—and followers, see [22]) designed and planned nature-based solutions to improve the hydrologic and forest ecosystem functioning, aimed at reducing the risk of natural hazards. Overall, these measurements consisted in (i) the design and implementation of small dikes to consolidate and contain erosive processes, rockfall activity, and intense flood-like processes, as well as the innovative design of empty dikes (non seen so far in other mountain regions worldwide) to retain mainly dense or non-inertial snow avalanches; and (ii) the intensive hillslope reforestation of more than 600 ha in the triggering zones to control the occurrence of phenomena with natives species such as *Pinus sylvestris* L., *Pinus uncinata* Ramond ex DC., *Abies alba* Mill., *Picea abies* (L.) Karsten, *Larix decidua* Mill. and *Fagus sylvatica* L. While measurements described in (i) were considered measures to protect a highly degraded zone in the short term, reforestation measures described in (ii) were planning to work in the long term. Interestingly, such strategy of combining structural and non-structural measures, which was planned more than one hundred years ago, is seen now as "a flexible, effective, and efficient solution to reducing impacts in the mountains" [i.e., 26]. Recognizing the innovative Eco-DRR approach and contribution realized in this and analog places and learning from their performance during the entire twentieth century would bring light about future and reliable strategies on DRR.

The study site is placed in the public forest (MUP n° 406) with an extension of 760 ha (Fig. 28.2). This area is formed by different torrents and valleys located in a north-south direction mountain chain, which favors north winds (higher humidity, lower temperature). The climate is defined as type D according to Köppen-Geiger [27] characterized by a cold and wet winter season and a mild summer. The annual

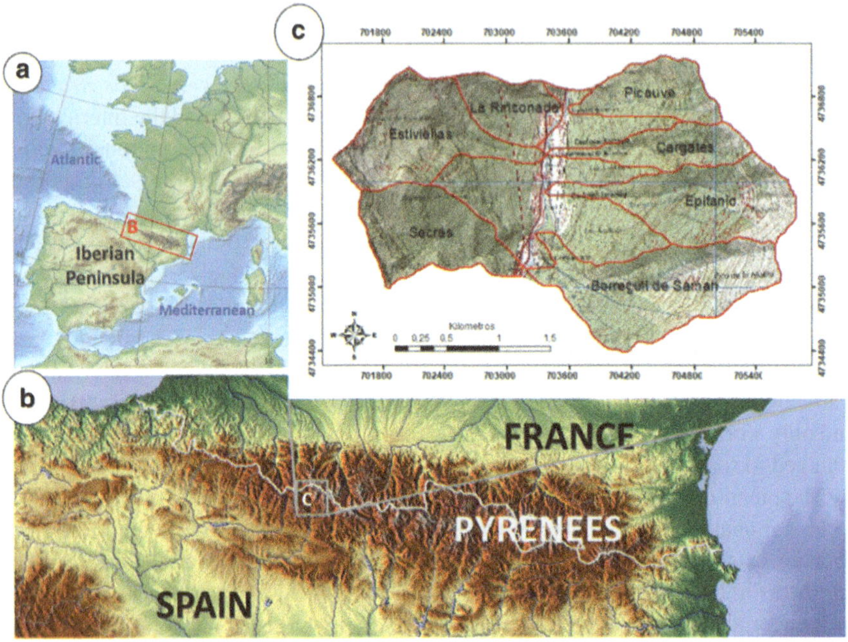

Fig. 28.2 Location of the Pyrenees in the Iberian Peninsula (**a**), the Canfranc valley between the Spanish-France border (**b**) and the torrent system of *Los Arañones* (**c**). *Source* **a** and **b** from available files under public domain (CC0), and **c** from [29]

mean precipitation is 1664 mm while the average annual temperature stands at 7.9 °C. *Los Arañones* is composed by seven torrential systems (Fig. 28.2), from which four represent a threat to the existing infrastructure. Overall, the torrent system called Picauvé and Borreguil de Samán does not affect current infrastructures. The torrent Rinconada is relatively smaller and much more inactive. The torrent Secras may affect the south exit of the current tunnel, while the torrent system called Cargates, San Epifanio, and Estiviellas represent a serious threat to the entire complex infrastructures. Table 28.1 represents the main physical characteristics of these torrential systems, including the Melton index M_{i}, as an indicator of the torrentiality defined as Eq. (28.1), as well as the fraction of covered canopy (fcc).

Table 28.1 Main physical characteristic of the most danger torrents

Torrent	Max. altitude (m)	Relief (m)	Length (m)	Slope (%)	Area (km^2)	Melton index	Fcc
Estiviellas	2358	743	1.771	41.9	~ 1.15	0.69	0.65
Cargates	2328	974	1.862	52.3	~0.53	1.33	0.87
Epifanio	2477	1050	2.25	46.7	~1.26	0.93	0.71

$$M_{\mathrm{i}} = \frac{\mathrm{WR}}{\sqrt{A}} \tag{28.1}$$

where WR is the watershed relief; and A represents the area of the watershed.

Over the last centuries, the landscape evolution of Canfranc valley has followed an analog tendency than the Pyrenees, which clearly is linked to human activities [27]. Thus, this evolution has been characterized by a progressive abandonment of crops and pasture lands from the late nineteenth century that continued until the 1980s. Since this moment, this tendency has been established or even slightly increased [27], including the livestock densities [28]. As a consequence, a progressive increase in vegetation cover can be observed in these regions. This simplified view of recent land-use changes (see [27] for more details about environmental and socio-economic factors related to these changes) was similar in *Los Arañones*, where, moreover, the reforestation practices and soil erosion control in the context of the implemented Eco-DRR have contributed to establishing dense and mature forest stands.

28.3 Natural Hazards and Infrastructures in Los Arañones

Among all processes, snow avalanches and torrential floods are especially common in Los Arañones. At the beginning of the twentieth century, powder (inertial flow) and wet/slab snow avalanches (non-inertial flow) were recurrent in Estivielles, San Epifanio, Secras, and Cargates catchments. Examples of such events took place on January 24th, 1915, when an intense snow avalanche reached the bottom valley affecting housing and other assets. Torrential floods were also recurrent, especially during summertime. The forest service recorded such events, as the torrential flood that took place on August 16th, 1916; when an intense thunderstorm triggered intense floods in all torrents, but especially in Estivielles.

The first Eco-DRR project was designed by Benito Ayerbe. The initial project was only focused on the San Epifanio catchment, but it was extended to surrounding torrents after checking the reliability of the works against snow avalanches during the winter of 1915. The final project was finished in 1919, defining measurements against rockfall, snow avalanches, and torrential floods. These measurements were:

(i) to protect against rockfall, the government bought and reforested the cropland (up to 235 ha) located in the lower parts of the torrents;

(ii) to protect against torrential floods, measurements were carried out in the catchment, at the channel, and on the debris flow cones. At the catchment level, reforestation practices were carried out with the aim of controlling erosion processes (Fig. 28.3). Innovatively, bands of *Salix eleagnos* were planted along contour lines. At channel level, transversal dikes were built to laminate the peak flow, control remontant erosion and retain sediments. At cone level, longitudinal dikes and channelization were built to direct the flood directly to the Aragon river (Fig. 28.4).

Fig. 28.3 **a** View of the *Los Arañones* landscape in early twentieth century. **b** View of the forested torrents in early twenty-first century. Pictures from Arrazola-Herreros et al. [30]

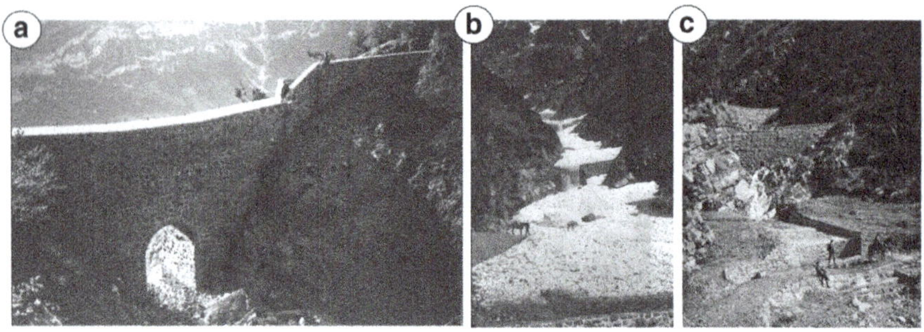

Fig. 28.4 **a** View of an empty dike designed and implemented in *Los Arañones*. **b** Example of snow-avalanche retention by the dikes. **c** Example of channelization at cone level. Pictures from Arrazola-Herreros et al. [30]

(iii) to protect against snow avalanches, measurements were carried out in the starting zone to prevent triggering and in the transport zone to limit the runout of events. In the starting zone, different snowpack stabilizers such as small walls or stakes were implemented, but also the reforestation of 360 ha with native species. In the transport zone different masonry walls, dikes, or reforestation were implemented. Innovatively, large empty dikes (up to 20 m height) were built with a large central culvert to retain the non-inertial snow avalanches but allow the flow of water-like events (Fig. 28.4).

After the implementation of the project, several studies were performed to evaluate the performance of the measures and for maintenance purposes. Yet, in 1933, the first extension of the initial project was designed which mainly consisted of the construction of a new empty dike in the Epifanio catchment. The Spanish Civil War reduced investment in the area, and only a few works technical reviews were done afterward. Over the last decades, the tourism presence has steadily increased, augmenting the elements-at-risk. This new situation has triggered the realization of several studies and projects, such as (1) repair project existing works; (2) built new measures in the Torrente Estiviellas; and (3) the publication of new scientific-technical studies.

Since the implementation of the first Eco-DRR project in Canfranc, the impact of snow avalanches and torrential floods seems to be changed. Thus, while the impact of snow avalanches still has been present during the entire twentieth century and last decades, the impact of torrential floods seems to be reduced. Thus, during the last extreme flood event in 2015, the measures in the torrents were able to reduce damages in Canfranc station, while large damages were reported downvalley in the Aragon river. However, although the impact of minor and moderate snow avalanches events seems to be significantly reduced, yet large snow avalanches have had the potential to reach the bottom valley. For instance, during winter 1954, a massive snow avalanche was triggered in Estiviellas, causing large damages to the vegetation along the transport zone and runout. On 1st February 1963, a refuge in Estiviellas was destroyed, while in 1970 the snow avalanches reached the railway installation. The largest snow avalanches occurred in the winter of 1986 when a snow avalanche caused important damages to the church of Canfranc; and more recently in 2013, when the snow reached the village.

28.4 Current and Future Performance of *Los Arañones*

As previously mentioned, the steady increase in tourism has attracted a larger socio-economic activity in the region. This has resulted in newer urban developments in the bottom valley, leading to an important increase in the elements-at-risk. The situation is worrisome since the records suggest that extreme events still eventually take place. Besides, the aging of the infrastructures could reduce their reliability [23] (Fig. 28.5), and the forest stands could have reached their mature state. On the other hand, climate change may modify the hydrological response of the catchments in *Los Arañones*, as seen in analog zones [31]. Overall, climate change scenarios are expected to have an impact on temperature trends, specifically with an increase in maximum temperatures [32], while the uncertainties related to the projected precipitation avoid to see any clear tendency [33]. Yet, intense precipitation events due to convective processes could increase in the region, probably favored by the provided moisture from the ocean at this latitude [34]. As result, the snowpack in the Pyrenees may be strongly affected, especially in low-mid altitudes [32] as well as the soil moisture, which is expected to be reduced.

All these abovementioned issues have opened the question about the reliability of the implemented Eco-DRR actions to protect the existing and new settlements and infrastructure. Specifically, questions about the changes in the frequency and type of snow avalanches processes due to climate change [5], as well as the impact of soil moisture on soil erosion (i.e., sediment laden) under torrential precipitation, are

Fig. 28.5 Current state reveals the aging of the transversal dikes in San Epifanio torrent

on the table. As a first analytical step, an analysis of the reliability of the Eco-DRR approach under current conditions has been performed. Particularly, a classical risk assessment was implemented to evaluate the feasibility of the current scenario (implemented Eco-DRR) and scenario defined without any implementation (Fig. 28.6). To this end, hydraulic (i.e., IBER model) and snow-avalanche numerical models (RAMMS::AVALANCHE) have been calibrated based on more than a hundred year' records from the Forest Service. Additionally, losses from current exposed elements have been analyzed, by including local and national vulnerability curves. The risk analyses of each scenario have been evaluated based on Avoided Costs Method (ACM) as well as the use of the Human Capital Method (HCM), the calculation of Replacement Costs (RC), and the revision of market prices (RMP). Finally, the economic indicators related to the implementation of a hypothetical analog original project with current costs and an approach to the maintenance costs were also evaluated. A detailed methodological description can be found on Esteban-Muñoz et al. [29].

Overall, results show that the Eco-DRR works and the designed maintenance actions (which included check dams reconstruction) are economically and socially justified. Maintenance costs reach up to € 1.089.848 during works lifetime. Thus, the approach developed to provide evidence that the Net Present Value (NPV), which corresponds to the protective effect of the Eco-DRR is € 606.79 per year and ha; while the estimated risk reduction reaches up to € 1,522,188/year. These indicators demonstrate the economic and social viability of the Eco-DRR. Furthermore, the values of the indicators are high due, in part, to the age of the works and the local development that has allowed the defensive nature of the works. Thus, the long-term cost-benefit ratio of the project was 0.31, with an investment return period of 19 years.

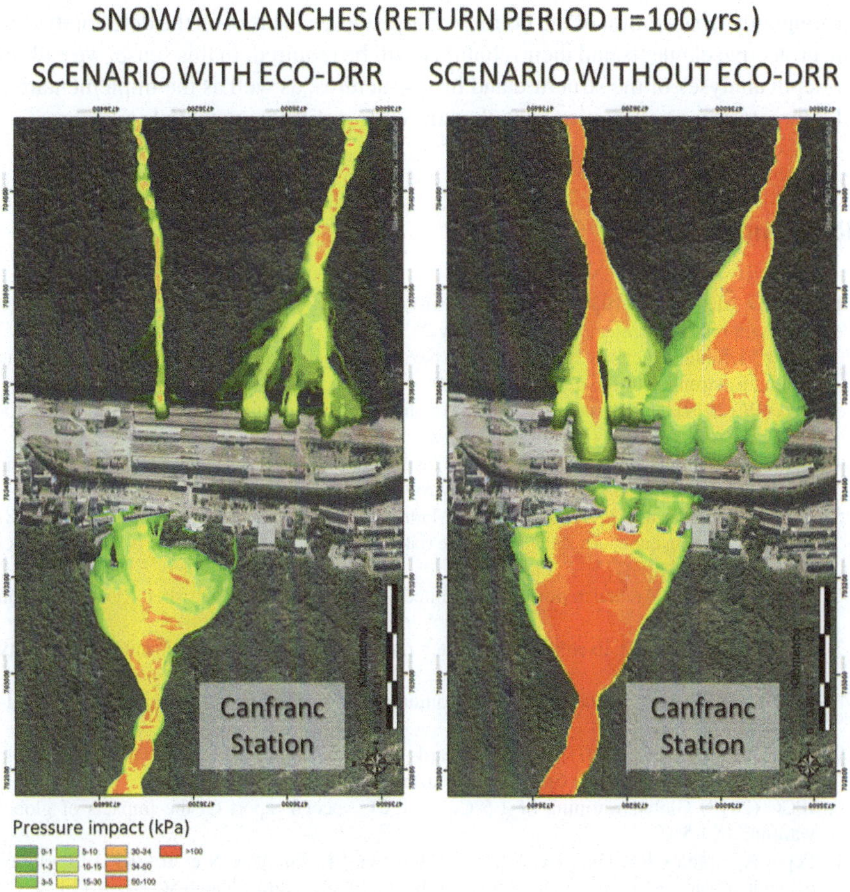

Fig. 28.6 Comparison between the impact of snow avalanches (T = 100 yrs) considering the Eco-DRR (left) and without Eco-DRR (right)

It was also estimated that carrying out the same works with current material costs would be 37% more expensive, thereby reducing economic indicators such as NPV.

28.5 Conclusions

With more than one hundred years old, the Eco-DRR implemented in Los Arañones constitutes physical evidence on how the combination of soft and hard measurements can be implemented to reduce the risk in mountain areas. The updated risk analyses suggest that such an approach still is still functioning, and is willing to be effective at reduce reducing the risk. Yet, the passing of time has resulted in a loss of functionality, and cost of maintenance is increasing. Under climate change conditions, and given the ongoing urbanization of the bottom valley, studies focused on the deterioration of the protective elements and their reliability will be required. In this sense, we call for in-depth analyses of the structure and infrastructures, as well as the implementation of a monitoring an evaluation plan that allows to anticipate potential failures.

References

1. Ives JD, Messerli B (1981) Mountain hazards mapping in Nepal introduction to an applied mountain research project. Mount Res Dev 223–230
2. Holub M, Suda J, Fuchs S (2012) Mountain hazards: reducing vulnerability by adapted building design. Environ Earth Sci 66(7):1853–1870
3. UNDRR (2019) Global assessment report on disaster risk reduction. United Nations Office for Disaster Risk Reduction (UNDRR), Geneva
4. Klein RJ, Nicholls RJ, Thomalla F (2003) Resilience to natural hazards: how useful is this concept? Glob Environ Change Part B Environ Hazards 5(1):35–45
5. Ballesteros-Cánovas JA, Trappmann D, Madrigal-González J, Eckert N, Stoffel M (2018) Climate warming enhances snow avalanche risk in the Western Himalayas. Proc Natl Acad Sci USA 115(13):3410–3415
6. UNISDR (2006) Annual report 2016. United Nations Office for Disaster Risk Reduction (UNISDR)
7. FAO (2013) Resilient livelihoods—disaster risk reduction for food and nutrition security framework programme
8. Lindell MK, Prater CS (2003) Assessing community impacts of natural disasters. Nat Hazards Rev 4(4):176–185
9. Gardner JS, Dekens J (2007) Mountain hazards and the resilience of social–ecological systems: lessons learned in India and Canada. Nat Hazards 41(2):317–336
10. IPCC (2018) Global warming of 1.5°C. An IPCC special report on the impacts of global warming of 1.5°C
11. Pepin N, Bradley RS, Diaz HF, Baraër M, Caceres EB, Forsythe N et al (2015) Elevation-dependent warming in mountain regions of the world. Nat Clim Change 5(5):424–430
12. Allen SK, Ballesteros-Canovas J, Randhawa SS, Singha AK, Huggel C, Stoffel M (2018) Translating the concept of climate risk into an assessment framework to inform adaptation planning: insights from a pilot study of flood risk in Himachal Pradesh, Northern India. Environ Sci Policy 87:1–10
13. Du CR (2016) Sustainable development goals
14. Ballesteros-Cánovas JA, Allen S, Stoffel M (2019) The importance of robust baseline data on past flood events for regional risk assessment: a study case from Indian Himalayas. UNISDR global assessment report
15. Borga M, Gaume E, Creutin JD, Marchi L (2008) Surveying flash floods: gauging the ungauged extremes. Hydrol Process 22(18):3883
16. Marchi L, Borga M, Preciso E, Gaume E (2010) Characterisation of selected extreme flash floods in Europe and implications for flood risk management. J Hydrol 394(1–2):118–133
17. Zoccatelli D, Borga M, Zanon F, Antonescu B, Stancalie G (2010) Which rainfall spatial information for flash flood response modelling? A numerical investigation based on data from the Carpathian range, Romania. J Hydrol 394(1–2):148–161
18. Zaginaev V, Ballesteros-Cánovas JA, Erokhin S, Matov E, Petrakov D, Stoffel M (2016) Reconstruction of glacial lake outburst floods in northern Tien Shan: implications for hazard assessment. Geomorphology 269:75–84
19. Ha KM (2020) Evaluating ecosystem-based natural disaster management. Hum Ecol Risk Assess Int J 26(7):1896–1906

20. Dalimunthe SA (2018) Who manages space? Eco-DRR and the local community. Sustainability 10(6):1705
21. Bathurst JC, Iroumé A, Cisneros F, Fallas J, Iturraspe R, Novillo MG et al (2011) Forest impact on floods due to extreme rainfall and snowmelt in four Latin American environments 1: field data analysis. J Hydrol 400(3–4):281–291
22. Rogger M, Agnoletti M, Alaoui A, Bathurst JC, Bodner G, Borga M et al (2017) Land use change impacts on floods at the catchment scale: challenges and opportunities for future research. Water Resour Res 53(7):5209–5219
23. Ballesteros-Cánovas JA, Stoffel M, Corona C, Schraml K, Gobiet A, Tani S et al (2016) Debris-flow risk analysis in a managed torrent based on a stochastic life-cycle performance. Sci Total Environ 557:142–153
24. López-Moreno JI, Beniston M, García-Ruiz JM (2008) Environmental change and water management in the Pyrenees: facts and future perspectives for Mediterranean mountains. Glob Planet Change 61(3–4):300–312
25. Pemán García J, Pérez-Soba I (2013) Los Ayerbe: tres ingenieros de montes altoaragoneses y su aportación a la hidrología forestal española. Lucas Mallada rev ciencias 15:99–160
26. Peel MC, Finlayson BL, McMahon TA (2007) Updated world map of the Köppen-Geiger climate classification
27. García-Ruiz JM, López-Moreno JI, Lasanta Martínez T, Vicente Serrano SM, González-Sampériz P, Valero-Garcés BL et al (2015) Los efectos geoecológicos del cambio global en el Pirineo Central español: una revisión a distintas escalas espaciales y temporales
28. Kouba Y, Camarero JJ, Alados CL (2012) Roles of land-use and climate change on the establishment and regeneration dynamics of Mediterranean semi-deciduous oak forests. For Ecol Manage 274:143–150
29. Muñoz-Esteban et al (2020) Evaulación del Valor Protector de los Trabajos de Defensa del Núcleo Canfranc-Estación. UPM, Madrid
30. Arrazola-Herreros et al (2006) Revisión de los Trabajos de Corrección del Torrente Epifanio en Canfranc (Huesca). UPM, Madrid
31. López-Moreno JI, Pomeroy JW, Revuelto J, Vicente-Serrano SM (2013) Response of snow processes to climate change: spatial variability in a small basin in the Spanish Pyrenees. Hydrol Process 27(18):2637–2650
32. López-Moreno JI, Goyette S, Beniston M (2009) Impact of climate change on snowpack in the Pyrenees: horizontal spatial variability and vertical gradients. J Hydrol 374(3–4):384–396
33. López-Moreno JI, Goyette S, Beniston M (2008) Climate change prediction over complex areas: spatial variability of uncertainties and predictions over the Pyrenees from a set of regional climate models. Int J Climatol J R Meteorol Soc 28(11):1535–1550
34. Gimeno L, Nieto R, Sorí R (2020) The growing importance of oceanic moisture sources for continental precipitation. NPJ Clim Atmos Sci 3(1):1–9
35. Moos C, Bebi P, Schwarz M, Stoffel M, Sudmeier-Rieux K, Dorren L (2018) Ecosystem-based disaster risk reduction in mountains. Earth Sci Rev 177:497–513

Chapter 29
Filling in the Gaps of the Tsunamigenic Sources in 2018 Palu Bay Tsunami

Pablo Higuera, Ignacio Sepúlveda, and Philip L.-F. Liu

29.1 Introduction

On 30 September 2018, an Mw 7.5 strike-slip earthquake struck towns and villages within Palu Bay, accompanied by landslides and a local tsunami a couple of minutes later. The death toll of the event has been estimated in approximately 4340 people. The infrastructural damage was also significant, as reported by post-tsunami surveys [2, 15, 22, 25]. The map of Palu Bay is included for reference in Fig. 29.1. Names of the coastal villages/cities which may be used in this paper are included in white, and the mapped Palu-Koro fault line, extracted from Valkaniotis et al. [38], is shown in red line.

While the earthquake magnitude was not a surprise in this tectonically active region with similar events in the past [4, 26, 40], the peculiar supershear earthquake rupture process [18], the unexpectedly high local tsunami waves and the possible relationship of both phenomena prompted a long-lasting research effort which remains until today. Preliminary simulations right after the event [12, 30] showed that the tsunami initial condition described by static co-seismic deformation of the bay seafloor cannot explain the size, periods and arrival times observed by eye-witnesses and inferred from post-tsunami surveys. Two possible sources were hypothesised then as the cause of such a devastating tsunami. First, the complex geometric and

P. Higuera (✉)
The University of Auckland, Auckland, New Zealand

P. Higuera · P. L.-F. Liu
National University of Singapore, Singapore, Singapore

I. Sepúlveda
San Diego State University, San Diego, CA, USA

P. L.-F. Liu
Cornell University, Ithaca, NY, USA

National Central University in Taiwan, Taoyuan, Taiwan

S. Kolathayar et al. (eds.), *Civil Engineering for Disaster Risk Reduction*, Springer Tracts in Civil Engineering, https://doi.org/10.1007/978-981-16-5312-4_29

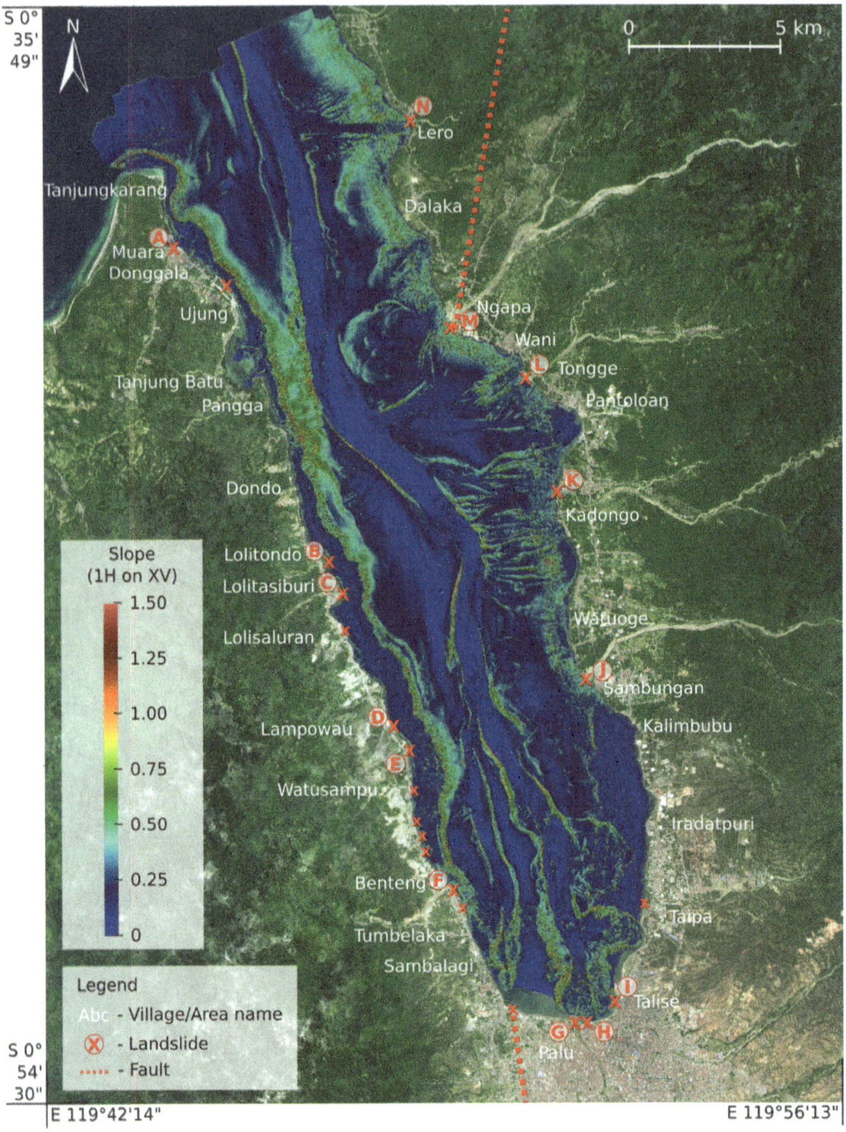

Fig. 29.1 Map of Palu Bay and slopes of the bathymetry. The coastal landslides are marked with a red cross, and the most relevant ones named with letters from A to N. Figure after [15]

dynamic characteristics of the earthquake rupture caused the large tsunami waves (e.g. [33, 34, 37]). Second, several landslides near the bay coastline and triggered by the earthquake contributed to high tsunami waves at different locations. Hypothetical submarine landslides in the deep end of the bay have also been suggested [12, 23], but they are likely a secondary factor, since there is no evidence of large submarine landslides in the most recent bathymetry surveys [15]. It must be noted that some authors use the term submarine landslide to refer to the coastal landslides, even if these are often only partially submerged; in this work, we will only use the term submarine landslides to indicate those which may exist in the deep end of the bay. In short, the validation of the hypotheses in the early days was extremely challenging due to the scarce amount of local earthquake and tsunami data.

The earthquake information collected for this event includes tele-seismic data, InSAR data and photo imagery in the surroundings of the Palu Bay [3, 31, 32]. While the static modelling of the earthquake co-seismic deformation is straightforward by means of classic dislocation models (e.g. [21]), the data is located far from the bay seafloor and a unique fault geometry and earthquake slip distribution cannot be solved [31]. Furthermore, although surface ruptures have been thoroughly studied [13], the contribution of secondary faults beneath the bay or the possibility of a complex dynamic contribution of the seafloor displacement to the tsunami is still unclear, as the deepwater portion of the fault remains unmapped. Information about the tsunami was firstly provided by a tide gauge located at the port of Pantoloan, inside the bay, with a sampling interval of 1 min. As shown by eye-witness videos, the first and most destructive tsunami waves were short and, therefore, the available gauge data could only provide partial information of the tsunami evolution. Carvajal et al. [5] used the video footage from social media to reconstruct tsunami time histories at different locations within Palu Bay. The footage evidenced the issues of the tide gauge capturing the first tsunami waves at Pantoloan and provided further information to constrain earthquake and tsunami models. This study and several others (e.g. [2, 22, 28]) also evidenced signs of localised subsidence at the coast, presumably along with submarine landslides. All these datasets were supported by post-tsunami surveys which provided key information about the inundation and coastal subsidence [8, 22]. The partial earthquake and landslide information from these datasets were fully exploited in new tsunami simulations using a variety of models and approaches [29].

Until today, the studies have arrived to disparate conclusions, the landslides being the primary source [15, 36] or a combination of both landslide and co-seismic deformation (e.g. [1, 29]). Early works, e.g. [37], also hypothesised with the co-seismic deformation being the main source of the tsunami; however, the authors are not aware of any new references using all the new data available presently to support this conclusion. In that sense, the three options seem plausible from a theoretical point of view, since all these mechanisms are known to be capable of producing tsunamis and have been observed before, e.g. in the 1998 Papua New Guinea [11, 35] or 2010 Haiti tsunamis [7]. The key in this analysis should be, in the authors' opinion, defining precisely the meaning of "main", since some degree of contribution of both factors is highly likely. However, we also acknowledge that given the major uncertainties in the existing data, "minor" contributions may be significantly diluted or amplified because of them.

While multiple landslides were identified along the coastline of Palu from post-tsunami surveys and photo imagery, their actual size, tsunamigenic potential and triggering time and evolution are not well constrained. Optimization methods have been presented to characterise the landslide sources from the available tsunami data ranging from trial and error to inversion methods. Sepúlveda et al. [31] employed a linear least square inversion to solve for the tsunami initial elevation at photo-identified landslide locations. The inversion results suggested that the triggering time and evolution of the landslides play a critical role to fit the observations from video footage and post-tsunami survey inferred runups. The delayed triggering time and finite duration of landslides were evident from the video footage analysis [5, 31]. A set of co-seismic deformation models inferred from InSAR data were also combined with the inverted landslides tsunami initial conditions. The simulations showed a minor contribution of the earthquake co-seismic deformation as compared to the landslide sources. A step forward to accurately identifying the source of the Palu tsunami is the inclusion of more geological data and the introduction of more sophisticated landslide tsunami models. Liu et al. [15] conducted a thorough bathymetry survey within Palu Bay, including shallow areas where landslides possibly occurred. The study characterised the possible dimensions of the landslides which constitute key information to further constrain landslide tsunami simulations. The high resolution of the bathymetry also provided complementary information of the fault system beneath Palu Bay [20].

In this study, we aim to use the new landslide data inferred by Liu et al. [15] and solve for the landslide conditions and timing which better fit tsunami observation by solving an optimization problem based on genetic algorithms (GA) [19] and numerical simulations performed with the nonlinear shallow water solver COMCOT [39]. In this chapter, we present preliminary results. Section 29.2 introduces the methodology to estimate the missing characteristics of the landslides and the waves that they produce. The following section discusses the most relevant findings from the new simulations. Finally, Sect. 29.4 draws the final conclusions and describes the future worklines.

29.2 Methodology

This work aims to complement and extend the previous analysis performed in Liu et al. [15], most of which was condensed in Table 2 of that work, and partially reproduced in this work in Table 29.1.

In Liu et al. [15], a detailed bathymetry of Palu Bay, acquired after the event, was compared with existing, less detailed, bathymetries from 2014 to 2017. Pre- and post-earthquake satellite images were also compared. The analysis allowed to detect all coastal landslides and to characterise their relevant magnitudes, such as the emerged and submerged dimensions or the volume. Based on that data, the landslide-generated

Table 29.1 Landslide characteristics, partially reproduced from Liu et al. [15]

	A	B	C	D	E	F	G	H	I	J	K	L	M	N
L_{ls}	255	295	430	1100	475	705	725	505	250	440	430	460	700	615
L_{cs}	72	125	40	135	75	115	90	85	90	140	70	160	125	70
A	1.17	1.82	0.89	6.40	1.39	3.23	2.56	2.83	1.59	3.21	1.30	6.30	4.94	1.62
E_{ls}	260	380	440	1220	335*	755	**670**	**435**	310	**370**	**360**	515	350	390*
E_{cs}	350	340	405	410	175*	340	**280**	**470**	**390**	**620**	**275**	830	800	295*
V	0.41	1.44	2.26	3.07	0.37*	2.22	**1.79**	**2.03**	2.54*	0.67*	1.11*	6.66	3.44	0.36*
H	3.9*	5.8	5.2	10.2	2.5	13.6	**6.2**	**6.8**	**8.1**	**1.8**	**2.9**	11.1	4.2	**1.4**
λ	118	658	1618	644	252	511	**788**	**808**	**834**	**1054**	**1068**	1382	1997	**736**

Data in bold has been estimated in the present work. Data with an asterisk indicates a significant level of uncertainty

wave (LGW) characteristics were calculated using the semi-empirical formulation by Lo and Liu [17] and were finally simulated using the nonlinear shallow water model COMCOT [39].

The data presented in Liu et al. [15] had significant sources of uncertainty (see SM Text 1 in [15] for the full list), and some data items were missing because they could not be inferred from the data available. Table 29.1 depicts this situation. The data in black has been extracted from Liu et al. [15]. L_{ls} and L_{cs}, and E_{ls} and E_{cs} are the emerged and total (emerged + submerged) dimensions of the landslide, respectively, in the longshore (ls) and crosshore (cs) directions, in metres. A is the surface area lost, in hectares (10,000 m^2), and V is the total volume of the landslide, in millions of m^3. H is the wave height produced by the landslide and λ is the wavelength, both in metres. Finally, t_s is the starting time of the LGW after the earthquake start, in seconds. This time does not indicate when the coastline collapses, but instead, when the LGW has fully developed and become independent of the landslide, as defined in Lo [16].

The objective of this work is to fill the gaps and complete the missing data Table 29.1, shown in bold font. In order to do so, we analyse the existing landslide data and develop ad hoc scaling laws for the landslides in Palu Bay to estimate the missing values (Sect. 29.2.1). With the existing and new LGW characteristics, we simulate the propagation of each tsunami wave individually and analyse the effects that they cause all over Palu Bay and the influence at several locations in which free surface elevation data has been estimated from different sources in Carvajal et al. [5] (Sect. 29.2.2). Finally, empirical adjustments to the wave height, wave length and the initiation time of the landslide-generated tsunamis are sought using an optimization technique based on GA to identify if it is possible to explain better the free surface elevation data measured/estimated at the key locations (Sect. 29.2.3).

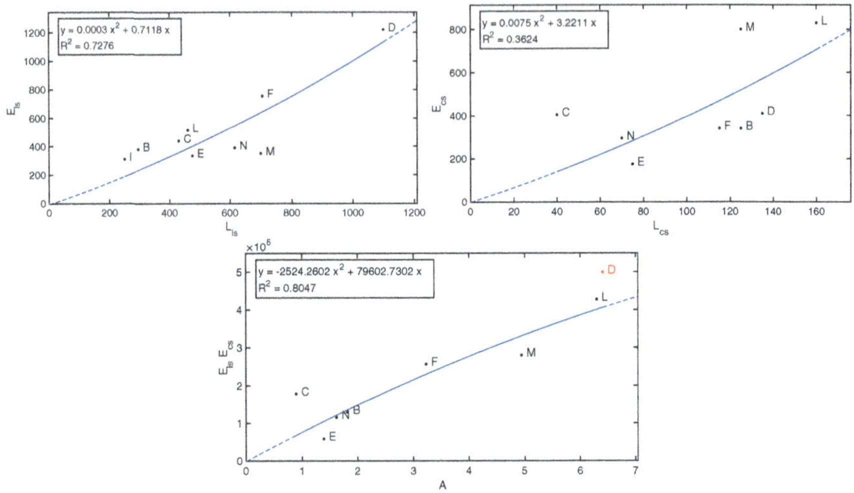

Fig. 29.2 Correlation between the emerged and total landslide dimensions (top panels). Correlation between the emerged area lost and the product of total landslide dimensions (bottom panel). Outliers in red

29.2.1 Analysis of the Existing Landslide Data

The first step of the methodology is to explore the existing data (in black font in Table 29.1) and seek simple nonlinear regression models to develop ad hoc scaling laws for the coastal landslides. We only include those landslides for which the data that we use at that stage is complete to forecast the missing magnitudes of the rest. Moreover, only the translational landslides are accounted for, which is why landslide A is not included in the present analysis.

The procedure that we follow in this work is simple and only requires simple techniques that are completely established in data science. Another option would have been to perform numerical simulations of the landslides, to simulate their interaction with water and measure the wave conditions directly. Numerical modelling of LGW has been developed quite extensively, see Romano [27] for a comprehensive review. However, one of the latest examples in the literature, Ghaïtanellis et al. [10], highlights that the computational cost is usually very high, and simulations require knowledge of the soil properties to accurately model the landslide mechanics. Therefore, we believe that the simple approach applied in this work is aligned with the level of uncertainty of the present data and is adequate to provide meaningful insights.

The analysis of the data begins seeking correlations between the emerged dimensions of the landslides (L_{ls} and L_{cs}), which are known for all the landslides, and the total dimensions (E_{ls} and E_{cs}), some of which are known only partially. As can be seen in the top left panel in Fig. 29.2, there is a fair correlation between E_{ls} and L_{ls} for a parabolic model in terms of the coefficient of determination ($R^2 = 0.7276$). The model has been chosen to be parabolic because it allows fitting slightly more

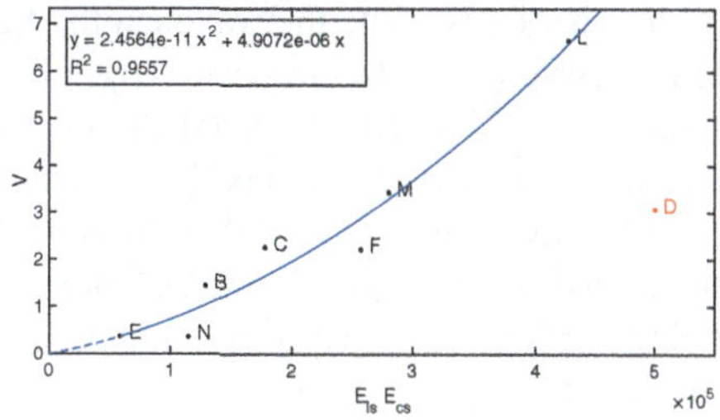

Fig. 29.3 Correlation between the product of total landslide dimensions and the volume of the landslide. Outliers in red

complex data than a linear model, although in this case, the resulting curvature is very small. Moreover, the parabolic model has been constrained to pass through (0,0), since when L_{ls} approaches to zero (meaning that the emerged landslide vanishes), so will E_{ls}. This constraint is also applied when linking the rest of the variables in this section. The equation in the top left corner of the panel has been applied to estimate E_{ls} in Table 29.1 (in bold).

The correlation between E_{cs} and L_{cs} (top right panel in Fig. 29.2) is quite low ($R^2 = 0.3624$), probably due to the diversity in slopes and local geological aspects; therefore, there is a need to find an alternative to estimate the missing values for E_{cs}, without leaving the physics of problem out of sight. The solution was to try to correlate the land area lost (A) with the product of E_{ls} and E_{cs}, in the bottom panel. In this case, landslide D is considered an outlier in terms of the product of the total landslide dimensions, as later it will be shown and discussed in Fig. 29.3 and as so is represented in red. Disregarding D, the correlation is very good ($R^2 = 0.8047$) and the equation of the parabola at the top left box of the panel has been used to estimate E_{cs} in Table 29.1.

Once the total dimensions of the landslides have been obtained, the next step was to estimate the missing landslide volumes (V). Several combinations of variables were tested, but in the end, the best correlation ($R^2 = 0.9557$) was found with the product $E_{ls} \cdot E_{cs}$, as shown in Fig. 29.3. Observing this figure, it is obvious why landslide D has been deemed an outlier in terms of $E_{ls} \cdot E_{cs}$, as it is too far away from its closest volumetric neighbours (C, F, M). The missing landslide volumes (V) are obtained using the parabolic expression and included in bold font in Table 29.1.

Finally and most importantly, the missing LGW characteristics, wave height (H) and wavelength (L), need to be estimated too. Again, a number of variable combinations were tested looking for reasonable correlations based on physical insights. In that sense, Panizzo et al. [24] indicate that both in their experiments and in Kamphuis and Bowering [14], "greater wave heights are expected at smaller slope inclination angles". This means that the wave height should be proportional to the landslide volume and inversely proportional to the local slope (obtained from the bathymetry data shown in Fig. 29.1), as we find in the left panel in Fig. 29.4. Based on observation (left panel), the largest wave height (landslide F) has been deemed as an outlier and left out of the fitting, which ends up having a high correlation with the data ($R^2 = 0.8548$).

In order to estimate the wavelength, the product between wave height and wavelength correlates extremely well (right panel, $R^2 = 0.9352$) with the landslide volume. Since H can be calculated already thanks to the expression in the left panel, L can also be calculated, and both variables are included in bold font in Table 29.1.

It is important to remark that the new estimated values for all variables lie within the bounds of existing landslides. Therefore, the simple models developed are not being applied out of their range and used to interpolate rather than to extrapolate the magnitudes.

29.2.2 Landslide-Generated Tsunami Wave Propagation

The tsunami wave propagation has been simulated using the NLSW model COMCOT [39]. The numerical setup is the same used in Liu et al. [15], namely with a bathymetry derived from the BIG14 dataset, with a resolution of 20 m (0.0012 arc-minutes) and a constant bottom friction Manning coefficient of 0.013 [9]. The simulation time is 20 min, enough to observe the development and evolution of the largest tsunami waves, and requires approximately 5 h to compute.

Each of the 14 waves has been simulated independently to establish their relative contribution to the waves observed in Palu, Wani and Pantoloan (see Fig. 29.1). As a starting point, the wave characteristics (H and λ) tested are those included in Table 29.1, and the waves have been considered to start propagating perpendicularly to the shoreline where they were generated.

The most relevant results from the simulations are included in Figs. 29.5 and 29.6. Figure 29.5 shows the highest free surface elevation (FSE) induced across Palu Bay and the inland areas inundated over the 20 min of the simulation. Figure 29.6 shows the time series of FSE produced by each of the waves at Palu, Wani and Pantoloan, where data from the tide gauge and Carvajal et al. [5] exists.

LGWs from locations A and N barely make any contributions to any of the relevant locations, as expected due to location and orientation and previously reported in Liu

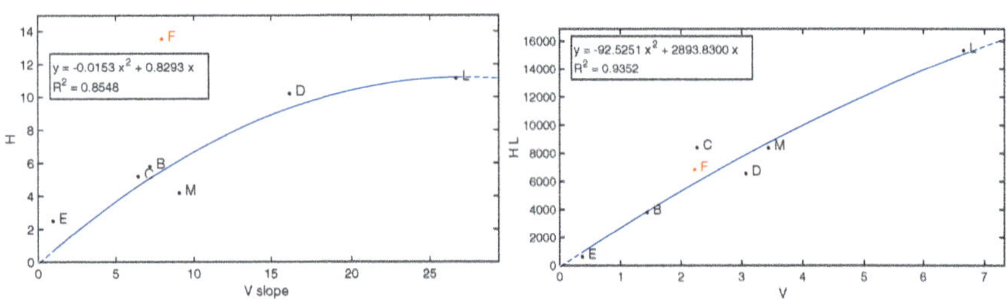

Fig. 29.4 Correlation between the landslide volume by the slope and the wave height induced (left panel). Correlation between the landslide volume and the product of the wave height and the wavelength induced (right panel). Outliers in red

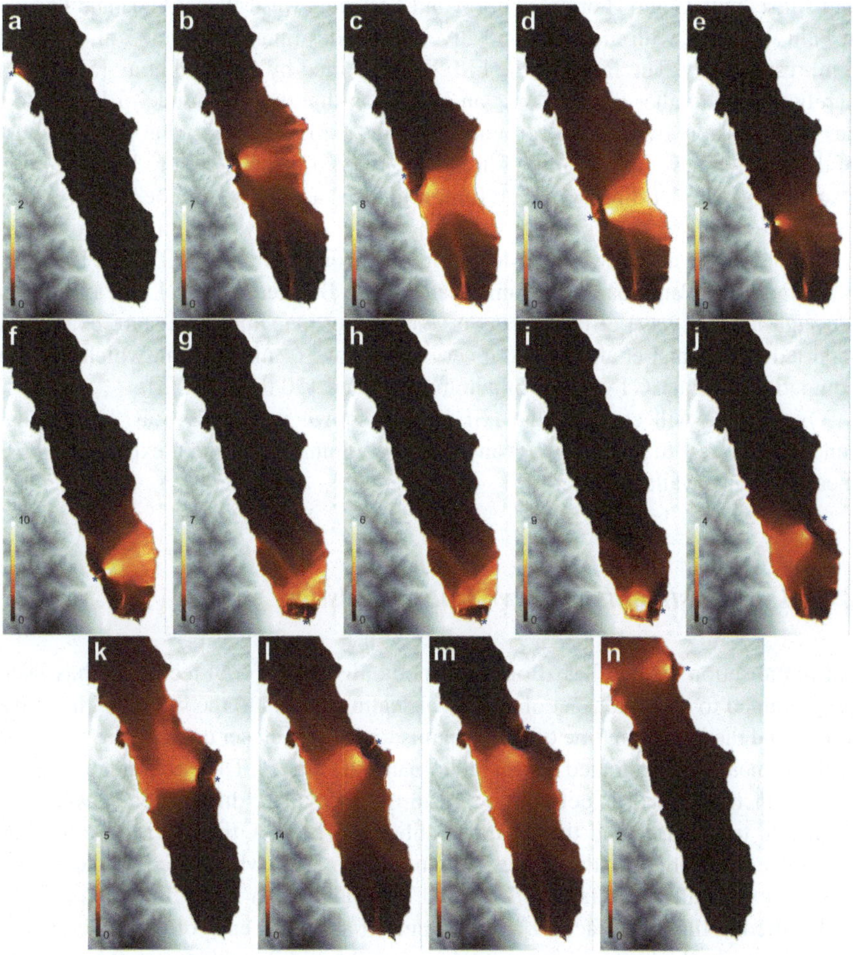

Fig. 29.5 Highest free surface elevation induced independently by each of the landslides in Palu Bay. Colorbar scale in metres. Landslide location denoted by a blue star

et al. [15]. In fact, the LGW A seems to behave like a trapped wave, refracting back to the shoreline instead of propagating towards the deep end of the bay due to the very steep slopes in the area. LGWs B–F are generated along the west coast of Palu Bay, therefore, impact the east coast directly. The highest wave contribution of B–D at Pantoloan and Wani is between 1 and 2 m, whereas it is much smaller for E and F, which are located further south. Interestingly, it is very noticeable LGWs B–F are deeply affected by refraction due to the bathymetry, causing the energy of the tsunami to focus at Palu Mall, despite the main wave initially pointing in another direction. Waves from these landslides arrive at Palu at different times and with maximum amplitudes around 2 m. LGWs G, H and I occur between Palu's bridge and Talise, with G and H oriented north and I oriented west. Waves G, H and I arrive at Palu very fast, with maximum positive amplitudes ranging from 0.5 m, 2 m and 6 m, respectively. The fact that the FSE data reported by Carvajal et al. [5] at Palu barely reaches higher than +2 m is an indication that the magnitude of landslide I has been possibly overestimated. Finally, LGWs J–M are produced at the east coast of the bay and have diverse initial directions. LGW J is the closest to Palu, and it is oriented in a similar direction; therefore, it induces a wave amplitude of 1 m at that location (energy focussing due to the bathymetry also plays a significant role) and negligible waves at the two other locations. The contribution of LGWs K, L and M is the largest at Pantoloan and Wani together with B. Interestingly, LGW L induces a trough-leading wave at Pantoloan very soon after the simulation start, exactly as reported in Carvajal et al. [5], and a very large wave (4 m) at Wani, which may be compatible with a CCTV video of a home recorded 150 m inland [5].

Overall, we will consider the FSE results shown in Fig. 29.6 as "orthogonal functions" to test to what degree can a linear combination of them explain the FSE measurements available.

29.2.3 Adjustments in Wave Characteristics

An optimisation routine based on the genetic algorithm (GA) technique has been programmed to estimate a set of empirical scaling factors to the wave height, wave length and the initiation time of the tsunamis to obtain a closer degree of accordance with the measured/estimated data at Pantoloan and Palu [5]. This program minimises a cost function, which is defined as the sum of the square differences between the estimated time series and the data available [5]. The new time series are built as a linear combination of the curves shown in Fig. 29.6, where each of the curves is scaled vertically by a factor $f(H)$ and horizontally (in time) by a factor $f(\lambda)$, and shifted in time by a factor t_s. The parameter space for each of these factors has been constrained between 0.25 and 2, between 0.5 and 2 and between 0 and 180 s, respectively. Following this approach, two major assumptions have been made: that wave evolution is linear (e.g. not affected by wave–wave interactions) and that wave period and wavelength are linked linearly, although both assumptions will be relaxed by performing a final simulation with the new estimated parameters.

The cost function has been designed to add together the contributions of Palu and Pantoloan time series, giving a higher weight to Pantoloan's, since the characteristics of the video analysed in Carvajal et al. [5] produce smaller uncertainties in the free surface elevation than the videos in Palu. Moreover, larger weight has also been applied to the first part of the signals, in which the level is 0 because the waves have not arrived yet, to penalise an early arrival of the waves.

The runs have 100 individuals, each with 3 chromosomes and last for a total of 2000 time steps. In each step, the individuals are sorted from low to high in terms of the cost function. Then, 25% of the top individuals are retained, along with 10% randomly selected from the rest. Next, 35% of the total individuals are created by

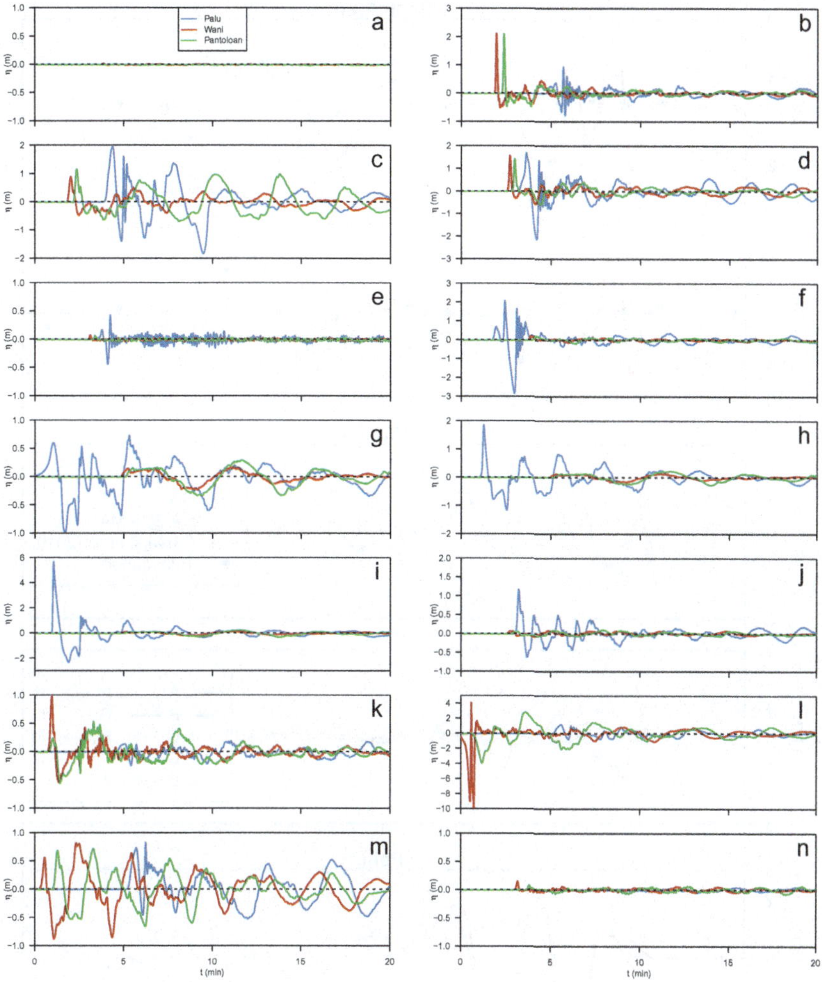

Fig. 29.6 Free surface elevation induced independently by each of the landslides at the three locations of interest

crossing the selected ones (i.e. exchanging chromosomes), with each chromosome having a 20% chance of mutating randomly. Finally, the rest of individuals needed to complete the population (30%) are created randomly. The whole process takes approximately 4 h in parallel using 12 cores of a workstation.

The parameters ($f(H)$, $f(\lambda)$ and t_s) which yielded the best solution are summarised in Table 29.2, along with the new wave heights (H_1) and wavelengths (λ_1) to be tested next in new COMCOT simulations. Overall, the GA optimization technique may indicate that the largest wave heights estimated in Table 29.1 (D, F, I, L) had been probably overestimated, as was also noted in Liu et al. [15].

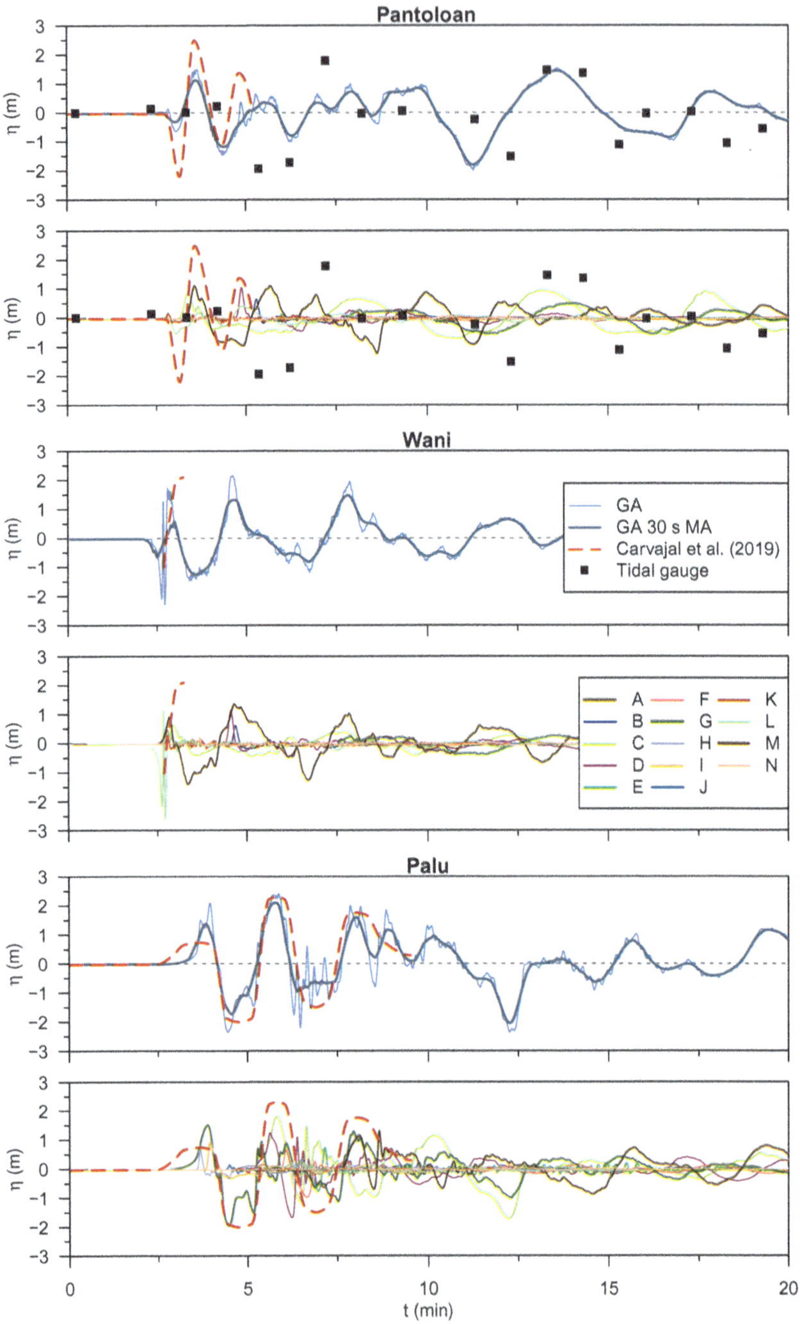

Fig. 29.7 Free surface elevation at Pantoloan, Wani and Palu with the best fit calculated by the GA method. Total contribution (odd number panels) and individual landslide contribution (even number panels)

The results of the GA technique can also be represented graphically, as displayed in Fig. 29.7, where the odd number panels show the global result from adding all the contributions from landslides A–N at Pantoloan, Wani and Palu compared with the data presented in Carvajal et al. [5]. The even number panels show the contributions of the individual landslides before aggregating them. The results in Pantoloan capture well the arrival time of the first wave, which is lead by trough, although the magnitude is much smaller, due to scaling the magnitude of the wave height of landslide L down. The first wave crest and the following trough are also captured well, although the second crest occurs slightly later. The time series in Wani is also led by a trough and has a significantly large peak, with magnitude and occurring time very close to those estimated in Carvajal et al. [5] from the CCTV of a house. The greatest contributors to the initial waves are landslides L and M, which are the closest, and C, which is right opposite to Wani. We note that the estimated time series in Wani is so short that it has not been used in the optimization process; therefore, the matching is reassuring. Finally, the agreement in Palu is the best among all. The first wave has a leading crest, originated by the combination of landslides G, H and I, the closest to Palu, with the wave created by G largely contributing to the first wave tough too. The second crest is mainly produced by the waves from landslides C, D and G (which is now reflected), while the third crest has a primary contribution from landslide M and a subsequent re-reflection of wave the wave from G.

Overall, the optimization technique has shown a satisfactory performance, with promising results to be further evaluated after new COMCOT simulations. The new wave characteristics presented in Table 29.2 have been simulated individually again to obtain the new set of "orthogonal functions", since after changing both wave heights and wavelengths, the FSE at all locations will likely be very different. With the new data from the 14 simulations, the GA optimization program has been run again, but this time only being able to adjust the lag, in order to preserve the wave height and wavelength calculated before. This is because the new wave characteristics impact the shape and location of the LGWs, which are calculated using the method of Lo [16]; therefore, even if COMCOT does not account for amplitude dispersion, the arrival times of the waves will be slightly different than before. The new optimal lags (rounded to the closest 5 s for practical reasons) are denoted as t_{s_1} in Table 29.2. Finally, all the waves have been reproduced in a single COMCOT simulation, whose

Table 29.2 Factors of variation for the landslide characteristics to obtain the best fit with the data available

	A	B	C	D	E	F	G	H	I	J	K	L	M	N
$f(H)$	0.33	0.40	0.81	0.82	1.19	0.47	1.39	0.61	0.32	0.48	0.71	0.51	1.28	1.37
$f(\lambda)$	0.71	1.36	1.25	1.13	1.34	1.29	0.96	0.65	1.06	0.58	0.50	0.67	1.02	0.72
t_s	7	126	33	90	91	177	169	162	165	148	133	140	137	80
H_1	1.3	2.3	4.2	8.4	3.0	6.4	8.6	4.1	2.6	0.9	2.1	5.7	5.4	1.9
λ_1	85	900	2025	730	335	660	760	500	885	610	535	925	2035	530
t_{s_1}	50	75	150	115	10	75	160	155	170	100	150	115	135	50

goal is to capture the nonlinear effects due to wave–wave interactions that have been disregarded so far because of the initial assumptions. In this simulation, each of the waves is introduced at the time t_{S_1}. Doing so requires stopping and restarting the simulation when adding the new waves, and finally gathering and combining all the data, which is presented in the next section.

29.3 Results

The main results from the final simulation are shown in Figs. 29.8 and 29.9.

Figure 29.8 shows the evolution of the LGWs on the bay at different stages during the simulation. The panel on the top left is for reference only and includes the initial stage of all the LGWs, which are then introduced progressively at the time t_{S_1} in Table 29.2. Because of this, the snapshot at time equals 2:00 min does not include all the waves yet, since they have just finished being introduced at time 2:50 (top right panel). By that time the wave N (which is short and small) is propagating out of the bay, waves F and D are about to reach the east coast and the largest waves (C, D, L and M) are still propagating across the bay. By times 3:20 and 3:50, most of the waves have started propagating in the north to south direction towards Palu, forming long wave fronts and experiencing diffraction due to the bathymetry. The cluster of waves generated at landslides G, H and I radiates from the south end of the bay, propagating north and impacting the south-most eastern and western shorelines. However, the most interesting effect is observing those clustered waves bend due to heavy refraction in front of Palu Mall and start propagating south (time 4:20), finally reaching the coast. The snapshots at times 4:50 and 5:50 show the final waves arriving at Palu, and how they become undular bores as they propagate over the shoal in front of Palu Mall. Moreover, an oscillatory pattern in the FSE can be observed adjacent to the eastern and western coasts. These oscillations resemble trapped waves due to the shallow coastal shelf and very short transition to the deep end of the bay (see the steepness of the profile in Fig. 29.1). These trapped waves seem to have between 2.5 and 5 min of wave period and at times appear to propagate along the coastal shelf from north to south as edge waves would do. Such oscillations take a long time to dissipate, as they are still noticeable at time 7:50 and at the end of the 20-min simulation.

Figure 29.9 shows the time series of FSE simulated at Pantoloan, Wani and Palu along with the data from Carvajal et al. [5]. The figure also includes the old data from Liu et al. [15] in grey dotted line, to evaluate the differences that the changes in the magnitude and timing of the LGWs make. Moreover, comparisons can be also made with Fig. 29.7, which is the idealised target applying the linearity assumption.

The tsunami waves in Pantoloan start with a trough, which as also shown in Fig. 29.7 is not as shallow as the wave measured in place. The first crest takes place at the correct timing and has a comparable magnitude, although it is slightly shorter compared with the measurements. The second trough is reasonably well timed, and although the magnitude of the third crest is well captured (improving the estimates in

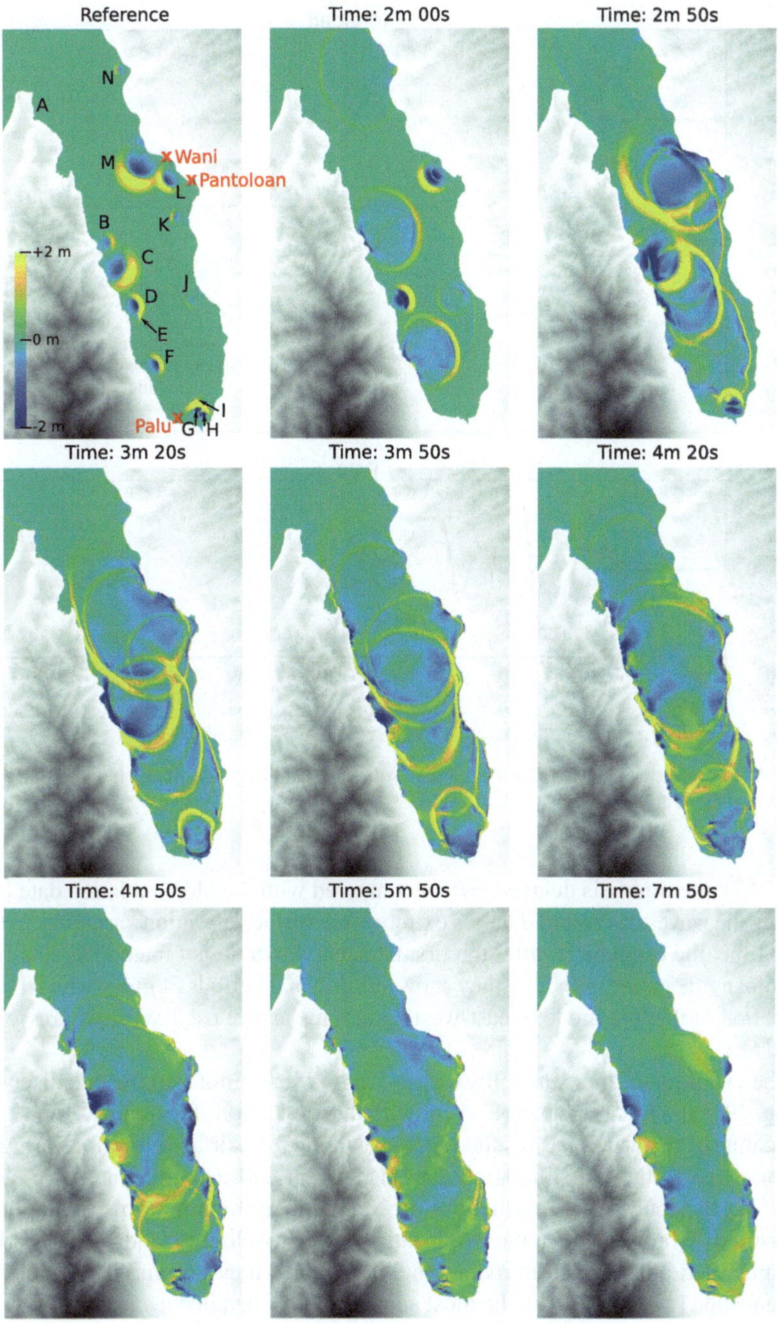

Fig. 29.8 Free surface elevation throughout Palu Bay at different stages of the simulation

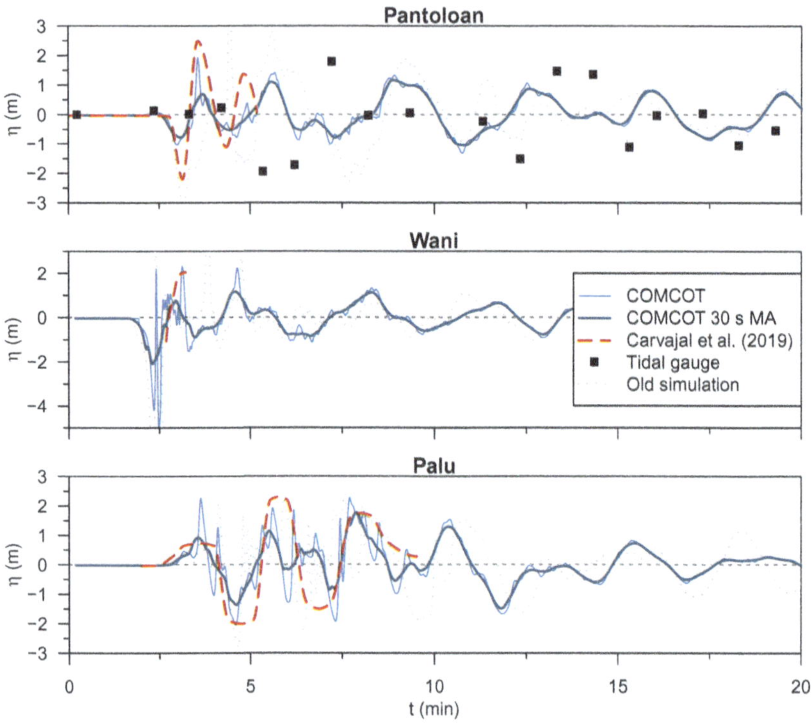

Fig. 29.9 Free surface elevation at Pantoloan, Wani and Palu from the final simulation in which all the landslides are included

Fig. 29.7), the timing is delayed. When compared with the old simulation data [15], scaling the waves has resulted in less extreme free surface elevation variations. At the same time, the oscillations after the first tsunami waves do not match the data from Pantoloan tide gauge, although they show similar long periods (4 min), which match the period of the first mode of east-west natural resonance oscillation, as previously noted [15].

The evolution of the tsunami waves in Wani is very similar to the one discussed in Fig. 29.7 and has a compatible event with the inundation shown in CCTV. Again, the scaling of the waves has resulted in less extreme free surface elevation variations as compared to the old simulation data [15]. Wani and Pantoloan are very close; therefore, the long-period oscillations are very similar, both in terms of amplitude and period. Moreover, since the two locations are very close to landslides L and M, the simplified approach to introduce the waves once they become independent of the landslide [17] may not be the most suitable, as the detailed evolution during the first stages of wave development may be of importance too to reproduce the initial wave arrival. However, obtaining such details will require a significant increase in computational cost, as previously discussed.

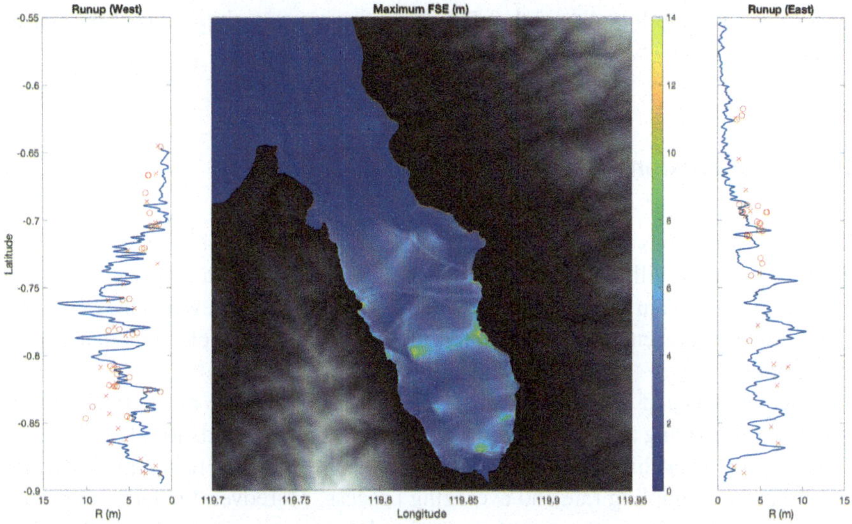

Fig. 29.10 Maximum free surface elevation in Palu Bay produced by the final simulation (central panel). Analysis of runup at the east and west coasts in the lateral panels. Data from surveys in red markers (circles from Omira et al. [22], crosses from Fritz et al. [8])

The initial tsunami wave in Palu, caused by landslides G–I, induces an increase in the water level which is well captured in Fig. 29.9 and very close to the measurements [5]. However, the evolution in the final simulation is less ideal than the that estimated in Fig. 29.7, indicating that nonlinear effects may be most important in Palu. This is reasonable, since the local bathymetry is very complex and LGWs become undular bores in the shallow area in front of the mall. Overall, the second wave crest and both troughs at either side lack the long wave shape and height of Carvajal et al. [5] measurements, but feature oscillations that are not present on the red dashed line but can be seen in the videos recorded at Palu Mall.

Finally, Fig. 29.10 shows the maximum free surface elevation throughout Palu Bay during the final simulation, along with the runup analysis at the east and west coasts of the bay. In the central panel, the marks of the initiation stages of the tsunamis are clearly visible, because the waves would start travelling towards deeper waters, thus decreasing in height due to reverse shoaling. Certain points in which waves crossed one another (e.g. near landslides L and M) are also evident. However, it is often close to the coast where the amplitudes are the largest, where the shoaling effects are maximal and runup occurs. The analysis of the simulation yields maximum runups of 13 m in the west coast and 10 m in the east coast, which are only 30% larger than the maximum values measured at both locations, which do not necessarily occur at the same locations, since runup shows a very localised nature. This point is supported by the large scatter that the measurements from surveys present. Overall, the numerically simulated runup shows an adequate agreement with the field measurements, which may strengthen the hypothesis that Palu tsunamis were mainly caused by coastal landslides.

29.4 Conclusions and Future Work

Due to the limitations in the data available, especially bathymetric survey data before the 2018 event, it will be difficult to get a full understanding and a perfect fit to the measurements of Palu Bay tsunami. In view of the literature review and the numerical modelling data presented in this chapter, the most relevant conclusions that we can draw are as follows.

Video evidence [5] indicates that the Pantoloan tide gauge did not capture the initial waves arriving at the port, which are likely caused by landslides L and M. It is often discussed that this may be because the waves were too short to be captured by the low-frequency sampling rate and averaging process. Moreover, the large amplitude and long-period waves measured later at that location are compatible with LGWs exciting the east-west resonant mode of the bay.The waves measured in Palu can also be explained to a great extent by the landslides, being those the closest to the mall (G, H and I) the most probable contributors to the initial waves that reached the area just 3 min after the earthquake struck. Additionally, the runup data produced by the simulation is coherent with field measurements. Despite some disagreements in the shape of the runup profile, we believe that agreement may be improved with more detailed simulations of the landslide progression or with a finer tuning of landslide timing [31]. In view of the results, the final conclusion would be that coastal landslides can explain most of the measurements and phenomena observed, and thus, they should be considered the main source of Palu tsunami. These results can also be considered a proof against the arguments invalidating the landslide source hypothesis in some references (e.g. [33, 37]) which often ignore the important role played by the timing of the landslides.

Regarding any other secondary sources, the Pantoloan tide gauge record shows no significant co-seismic deformation, whereas in references sustaining the co-seismic hypothesis modelling usually shows a co-seismic deformation of half a metre at that location. In the authors' opinion, even if the contribution of the co-seismic deformation cannot be completely ruled out, we can also conclude that this had most probably been secondary and that further developments in this hypothesis may require a completely new, more complex, earthquake model and validation in order to be able to assess its relative importance. In that sense, although ideally the fault portion located below the sea would need to be surveyed and mapped accurately, the recent work by Natawidjaja et al. [20] may be the most promising starting point to gain more insights about the faulting beneath the bay as reflected by the seafloor geometry. Another hypothesis, submarine landslides deep within the bay [12] can probably at this point be discarded, since there is no evidence of large enough events in the detailed post-event bathymetry surveys [6, 15].

Finally, we would like to conclude this chapter by outlining future work that will help improve the understanding of Palu tsunami event.The most relevant item, which has already been discussed, is performing a detailed mapping of the underwater portion of the fault, which will help to quantify the contribution of the co-seismic deformation hypothesis. Additionally, tsunami models including sophisticated governing equations have demonstrated to be relevant to accurately model the Palu tsunami [29], as they can introduce frequency and amplitude dispersion. In that sense, employing a Boussinesq wave-based model in the future instead of the nonlinear shallow water wave solver COMCOT will be advantageous and a step forward to fully explore the potential of the methodology presented.

Acknowledgements P. L.-F. Liu would like to acknowledge the funding from the National University of Singapore to support this research.

References

1. Aránguiz R, Esteban M, Takagi H, Mikami T, Takabatake T, Gómez M, González J, Shibayama T, Okuwaki R, Yagi Y, Shimizu K, Achiari H, Stolle J, Robertson I, Ohira K, Nakamura R, Nishida Y, Krautwald C, Goseberg N, Nistor I (2020) The 2018 Sulawesi tsunami in Palu city as a result of several landslides and coseismic tsunamis. Coast Eng J 62(4):445–459. https://doi.org/10.1080/21664250.2020.1780719
2. Arikawa T, Muhari A, Okumura Y, Dohi Y, Afriyanto B, Sujatmiko KA, Imamura F (2018) Coastal subsidence induced several tsunamis during the 2018 Sulawesi earthquake. J Disaster Res 13:sc20181204. https://doi.org/10.20965/jdr.2018
3. Bao H, Ampuero J-P, Meng L, Fielding EJ, Liang C, Milliner CWD, Feng TR, Huang H (2019) Early and persistent supershear rupture of the 2018 magnitude 7.5 Palu earthquake. Nat Geosci 12:200–205. https://doi.org/10.1038/s41561-018-0297-z
4. Bellier O, Sébrier M, Beaudouin T, Villeneuve M, Braucher R, Bourles D, Siame L, Putranto E, Pratomo I (2001) High slip rate for a low seismicity along the Palu-Koro active fault in central Sulawesi (Indonesia). Terra Nova 13:463–470. https://doi.org/10.1046/j.1365-3121.2001.00382.x
5. Carvajal M, Araya-Cornejo C, Sepúlveda I, Melnick D, Haase JS (2019) Nearly-instantaneous tsunamis following the Mw 7.5 2018 Palu earthquake. Geophys Res Lett 46:5117–5126. https://doi.org/10.1029/2019GL082578
6. Frederik MCG, Udrekh, Adhitama R, Hananto ND, Asrafil, Sahabuddin S, Irfan M, Moefti O, Putra DB, Riyalda BF (2019) First results of a bathymetric survey of Palu Bay, Central Sulawesi, Indonesia following the tsunamigenic earthquake of 28 September 2018. Pure Appl Geophys 176:3277–3290. https://doi.org/10.1007/s00024-019-02280-7
7. Fritz HM, Hillaire JV, MoliÈre E, Wei Y, Mohammed F (2013) Twin tsunamis triggered by the 12 January 2010 Haiti earthquake. Pure Appl Geophys 170:1463–1474. https://doi.org/10.1007/s00024-012-0479-3
8. Fritz HM, Synolakis C, Kalligeris N, Skanavis V, Mohammad Rizal FS, Prasetya GS, Liu PL-F (2018) Field survey of the 28 September 2018 Sulawesi tsunami. In: American geophysical union fall meeting 2018. Abstract NH22B-04
9. Garzon J, Ferreira C (2016) Storm surge modeling in large estuaries: sensitivity analyses to parameters and physical processes in the Chesapeake Bay. J Mar Sci Eng 4. https://doi.org/10.3390/jmse4030045

10. Ghaïtanellis A, Violeau D, Liu PL-F, Viard T (2021) SPH simulation of the 2007 Chehalis lake landslide and subsequent tsunami. J Hydraul Res (in press). https://doi.org/10.1080/00221686.2020.1844814
11. Harbitz CB, Løovholt F, Pedersen G, Masson DG (2006) Mechanisms of tsunami generation by submarine landslides: a short review. Nor J Geol 86:255–264
12. Heidarzadeh M, Muhari A, Wijanarto AB (2019) Insights on the source of the 28 September 2018 Sulawesi tsunami, Indonesia, based on spectral analyses and numerical simulations. Pure Appl Geophys 176:25–42. https://doi.org/10.1007/s00024-018-2065-9
13. Jaya A, Nishikawa O, Jumadil S (2019) Distribution and morphology of the surface ruptures of the 2018 Donggala-Palu earthquake, Central Sulawesi, Indonesia. Earth Planets Space 71. https://doi.org/10.1186/s40623-019-1126-3
14. Kamphuis JW, Bowering RJ (1970) Impulse waves generated by landslides. In: Proceedings of the 12th international conference on coastal engineering, Washington DC, pp 575–588
15. Liu PL-F, Higuera P, Husrin S, Prasetya GS, Prihantono J, Diastomo H, Pryambodo DG, Susmoro H (2020) Coastal landslides in Palu Bay during 2018 Sulawesi earthquake and tsunami. Landslides 17:2085–2098. https://doi.org/10.1007/s10346-020-01417-3
16. Lo HY (2018) Modeling landslide-generated tsunamis with long-wave equations. PhD thesis, Cornell University
17. Lo HY, Liu PL-F (2017) On the analytical solutions for water waves generated by a prescribed landslide. J Fluid Mech 821:85–116. https://doi.org/10.1017/jfm.2017.251
18. Mai PM (2019) Supershear tsunami disaster. Nat Geosci 12:150–151. https://doi.org/10.1038/s41561-019-0308-8
19. Mitchell M (1998) An introduction to genetic algorithms. MIT Press
20. Natawidjaja DH, Daryono MR, Prasetya G, Udrekh, Liu PL-F, Hananto ND, Kongko W, Triyoso W, Puji AR, Meilano I, Gunawan E, Supendi P, Pamumpuni A, Irsyam M, Faizal L, Hidayati S, Sapiie B, Kusuma MA, Tawil S (2020) The 2018 Mw7.5 Palu 'supershear' earthquake ruptures geological fault's multisegment separated by large bends: results from integrating field measurements, LiDAR, swath bathymetry and seismic-reflection data. Geophys J Int 224(2):985–1002. https://doi.org/10.1093/gji/ggaa498
21. Okada Y (1985) Surface deformation due to shear and tensile faults in a half-space. Bull Seismol Soc Am 75:1135–1154
22. Omira R, Dogan GG, Hidayat R, Husrin S, Prasetya GS, Annunziato A, Proietti C, Probst P, Paparo MA, Wronna M, Zaytsev A, Pronin P, Giniyatullin A, Putra PS, Hartanto D, Ginanjar G, Kongko W, Pelinovsky E, Yalciner AC (2019) The September 28th, 2018, tsunami in Palu-Sulawesi, Indonesia: a post-event field survey. Pure Appl Geophys 176:1379–1395. https://doi.org/10.1007/s00024-019-02145-z
23. Pakoksung K, Suppasri A, Imamura F, Athanasius C, Omang A, Muhari A (2019) Simulation of the submarine landslide tsunami on 28 September 2018 in Palu Bay, Sulawesi Island, Indonesia, using a two-layer model. Pure Appl Geophys 176:3323–3350. https://doi.org/10.1007/s00024-019-02235-y
24. Panizzo A, De Girolamo P, Petaccia A (2005) Forecasting impulse waves generated by subaerial landslides. J Geophys Res Oceans 110(C12). https://doi.org/10.1029/2004JC002778
25. Paulik R, Gusman A, Williams JH, Pratama GM, Lin S-L, Prawirabhakti A, Sulendraand K, Zachari MY, Fortuna ZED, Layuk NBP, Suwarni NWI (2019) Tsunami hazard and built environment damage observations from Palu City after the September 28 2018 Sulawesi earthquake and tsunami. Pure Appl Geophy 176:3305–3321. https://doi.org/10.1007/s00024-019-02254-9
26. Prasetya G, De Lange W, Healy T (2001) The Makassar strait tsunamigenic region, Indonesia. Nat Hazards 24:295–307. https://doi.org/10.1023/A:1012297413280
27. Romano A (2020) Physical and numerical modeling of landslide-generated tsunamis: a review. https://doi.org/10.5772/intechopen.93878
28. Sassa S, Takagawa T (2019) Liquefied gravity flow-induced tsunami: first evidence and comparison from the 2018 Indonesia Sulawesi earthquake and tsunami disasters. Landslides 16:195–200. https://doi.org/10.1007/s10346-018-1114-x

29. Schambach L, Grilli ST, Tappin DR (2021) New high-resolution modeling of the 2018 Palu tsunami, based on supershear earthquake mechanisms and mapped coastal landslides, supports a dual source. Front Earth Sci 8:627. https://doi.org/10.3389/feart.2020.598839
30. Sepúlveda I, Haase JS, Liu PL-F, Xu X, Carvajal M (2018) On the contribution of co-seismic displacements to the 2018 Palu tsunami. In: American geophysical union fall meeting 2018. Abstract NH23F-3552
31. Sepúlveda I, Haase JS, Carvajal M, Xu X, Liu PL-F (2020) Modeling the sources of the 2018 Palu, Indonesia, tsunami using videos from social media. J Geophys Res Solid Earth 125:e2019JB018675. https://doi.org/10.1029/2019JB018675
32. Socquet A, Hollingsworth J, Pathier E, Bouchon M (2019) Evidence of supershear during the 2018 magnitude 7.5 Palu earthquake from space geodesy. Nat Geosci 12:192–199. https://doi.org/10.1038/s41561-018-0296-0
33. Song X, Zhang Y, Shan X, Liu Y, Gong W, Qu C (2019) Geodetic observations of the 2018 Mw 7.5 Sulawesi earthquake and its implications for the kinematics of the Palu fault. Geophys Res Lett 46:4212–4220. https://doi.org/10.1029/2019GL082045
34. Song YT, Chen K, Liu Z, Roback K, Avouac JP (2018) The 28 September 2018 Mw 7.5 Sulawesi (Indonesia) earthquake and its implication for tsunami early warning. In: American geophysical union fall meeting 2018. Abstract NH23F-3545
35. Synolakis CE, Bardet J-P, Borrero JC, Davies HL, Okal EA, Silver EA, Sweet S, Tappin DR (2002) The slump origin of the 1998 Papua new guinea tsunami. Proc Math Phys Eng Sci 458:763–789.https://doi.org/10.1098/rspa.2001.0915
36. Takagi H, Pratama MB, Kurobe S, Esteban M, Aránguiz R, Ke B (2019) Analysis of generation and arrival time of landslide tsunami to Palu City due to the 2018 Sulawesi earthquake. Landslides 16:983–991. https://doi.org/10.1007/s10346-019-01166-y
37. Ulrich T, Vater S, Madden EH, Behrens J, van Dinther Y, van Zelst I, Fielding EJ, Liang C, Gabriel A-A (2019) Coupled, physics-based modeling reveals earthquake displacements are critical to the 2018 Palu, Sulawesi tsunami. Pure Appl Geophys 176:4069–4109. https://doi.org/10.1007/s00024-019-02290-5
38. Valkaniotis S, Ganas A, Tsironi V, Barberopoulou A (2018) A preliminary report on the M7.5 Palu earthquake co-seismic ruptures and landslides using image correlation techniques on optical satellite data. Technical report. https://doi.org/10.5281/zenodo.1467128
39. Wang X (2009) User manual for comcot version 1.7. Technical report, Institute of Geological & Nuclear Science, New Zealand
40. Watkinson IM, Hall R (2017). Fault systems of the eastern Indonesian triple junction: evaluation of quaternary activity and implications for seismic hazards. Geol Soc Lond Spec Publ. https://doi.org/10.1144/SP441.8

Chapter 30
Demystifying Impacts of Cyclone Amphan 2019 Amid COVID-19 Pandemic in West Bengal, India

Anil Kumar, Neshma Tuladhar, and Indrajit Pal

30.1 Introduction

The word cyclone is derived from the Greek word 'cyclos', meaning coiling of a snake. The term was first coined by Henry Piddigton, President of Marine courts, Kolkata, in 1848 [1]. Cyclone or tropical cyclone is used globally to cover tropical weather systems in which wind speed equals or exceeds the gale-force (62 km/h) [2]. As per the National Hurricane Center (NHC) of the National Oceanic and Atmospheric Administration (NOAA), a cyclone is defined as a closed atmospheric circulation rotating counter-clockwise in the northern hemisphere and clockwise in the southern hemisphere. Cyclones are generated when sea surface temperature is above 260 °C with high relative humidity, the spiral inflow of winds at a low level, and divergent winds. Further, they get momentum from differences in the speed of earth rotation [3].

The Bay of Bengal (BOB) basin of north Indian Ocean faces high number of cyclones in the region. The frequency of cyclones in this region is almost five times than the cyclones of the Arabian Sea. Semi-enclosed nature of the basin with its funnel shape steers the cyclone for striking the land. Most deadly cyclones with highest destruction and death tolls have occurred in the basin [4]. All the states on the eastern coast of India are highly vulnerable to cyclones (Table 30.1).

West Bengal's coasts are highly prone to tropical cyclones due to their peculiar nature of coastal features such as bathymetry. The coast of the Bay of Bengal is very shallow and landlocked on three sides; thus, when a cyclone approaches the region, the storm surge generated by the wind submerges the coastal belt. The high storm surge contributes to a higher impact in coastal areas, which is significant because of low flat coastal terrain, less community awareness, high population density, inadequate preparedness and response measures.

A. Kumar (✉) · N. Tuladhar · I. Pal
Disaster Preparedness, Mitigation and Management, Asian Institute of Technology, Khlong Luang, Thailand

S. Kolathayar et al. (eds.), *Civil Engineering for Disaster Risk Reduction*, Springer Tracts in Civil Engineering, https://doi.org/10.1007/978-981-16-5312-4_30

Table 30.1 Cyclonic events in the Bay of Bengal that impacted eastern coast (West Bengal and Odisha)

S. No.	Cyclone	Date year	Category of cyclone	Most impacted states
1	Amphan	16–21 May, 2020	Very severe cyclonic storm	Odisha, West Bengal
2	Fani	3 May, 2019	Extremely severe cyclone	Odisha, West Bengal
3	Bulbul	8–10 November, 2019	Very severe cyclonic storm	Odisha
4	Titli	10–12 October, 2018	Very severe cyclonic storm	Odisha
5	Hudhud	12–14 October, 2014	Very severe cyclonic storm	Andhra Pradesh, Odisha
6	Phailin	12–14 October, 2013	Very severe cyclonic storm	Odisha, Andhra Pradesh
7	Super cyclone (BOB 06)	25–31 October, 1999	Super cyclone	Odisha
8	West Bengal cyclone (BOB 03)	10–12 November, 2002	Severe cyclonic storm	West Bengal

Source Amphan Memorandum, Govt. of Odisha, 2020

In addition to the unprecedented effects of the COVID-19 pandemic, the region was recently affected by cyclone Amphan (20 May, 2020), one of the strongest tropical cyclones in the east coast of India [5]. The response carried out for his mega-disaster event was unforeseen in the past, considering the duality and snow-bowl effect of the disasters occurring at the same time. The impact of hydrometeorological and biological disasters co-occurring caused a cascading impact on the community.

30.2 Objectives

The paper brings together the dimensions of cyclone and COVID-19 impact together. How this dual disaster affected lives of people and stretched the administration to its fullest for the effective response. The key objectives of the paper are mentioned below:

i. To highlight the impact of cyclone Amphan in the state of West Bengal, India
ii. To investigate influence of COVID-19 pandemic on Amphan response measures of the government
iii. To provide insights on the risk governance mechanism adapted for management of COVID-19 crisis and Cyclone Amphan.

30.3 The Super-Cyclone Amphan

Amphan was considered the first of its kind after the Odisha Super cyclone 1999 in severity and scale. On 13 May, 2020, Amphan originated from a low-pressure area persisting around 300 km east of Colombo, Sri Lanka. On 18 May, 2020, Amphan reached its peak intensity with sustained wind speeds of 240 km/h [6].

On 20 May, 2020, Amphan made landfall south of Kolkata, India, and Hathiya Islands in Bangladesh [7]. The United States Joint Typhoon Center (USJTC) measured the peak wind speed of Amphan at 270 km/h, which was the highest ever recorded wind speed of a cyclone in the Bay of Bengal. However, the Indian Meteorological Department (IMD) recorded the maximum wind speed of 240 km/h, 30 lower than the USJTC measurement (Fig. 30.1).

30.3.1 Pre-landfall Preparation and Evacuation by the Administration

As soon as the administration received the first warning of Amphan from the IMD, the central and state governments swung into action for cyclone response planning. Prime Minister took the preparatory meeting with Home Minister and Chief Minister

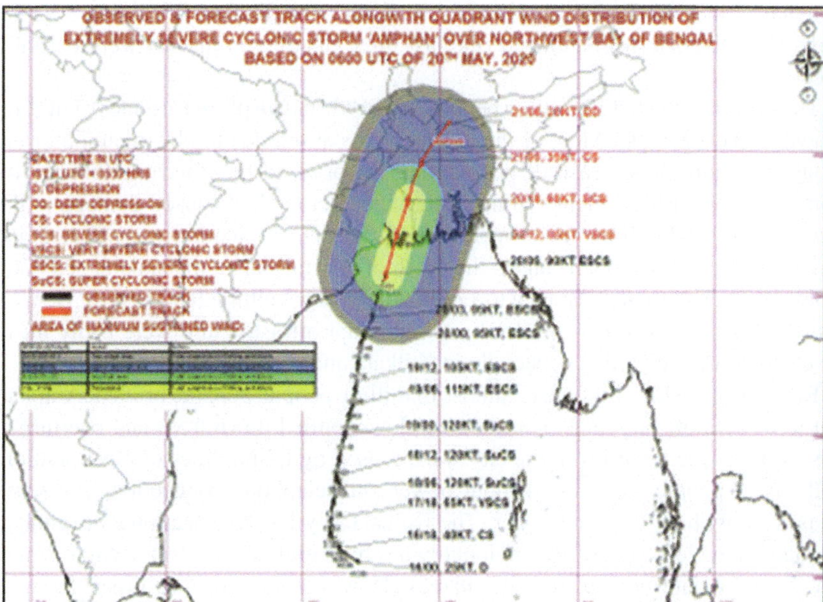

Fig. 30.1 Track of cyclone Amphan. *Source* IMD

of West Bengal and Odisha on 18 May, 2020. National Disaster Response Force (NDRF) was pre-positioned in the cyclone-prone districts, and additional teams of NDRF were also mobilized from other states to West Bengal. A huge evacuation exercise was carried out, and people from vulnerable areas were shifted to cyclone shelters.

The Kolkata International Airport was closed till 21 May, 2020, as a precautionary measure, and the airport terminal roof fortified to minimize damage [8]. The ships, fishing boats were directed to harbour the coast in their respective administrative boundaries of Odisha and West Bengal. All ports were cleared, and their operations were suspended along the Bay of Bengal. To avoid damage in some of the ports, ships were ordered to stay out till the cyclone landfall. Rail and vehicular traffic were halted or rerouted in the coastal districts.

30.3.2 Amphan Impact

Amphan impacted several countries in South Asia; Bangladesh, Bhutan, India and Sri Lanka were affected mainly by the region's cyclone. In India, coastal districts of Odisha and West Bengal were worst affected. Amphan made landfall near Bakkhali in West Bengal at 2:30 p.m. IST on 20 May, 2020, with heavy rainfall and gusty winds. West Bengal being the epicentre of cyclone landfall saw the most widespread damage. Ten districts of West Bengal were affected by the cyclone, including 24 North Paragnas, 24 South Paragnas, Howrah, Kolkata, Hoogli, East Medinipur, Nadia, Purba Barddhaman, Murshidabad (the last three districts in the list were partially affected).

The cyclone affected 13 million people, killed 98, and displaced around 0.7 million people in West Bengal. Most of the deaths were caused due to electrocution or the collapse of homes. At least 19 people were reported to be killed in Kolkata city alone. The state government estimated that the cyclone had caused damage to the tune of USD 13.2 billion, and around 60% of the state population was affected due to the cyclone [9]. Amphan caused the death of 2.1 million animals, damaged 8007 fishing boats, and caused damage to 17,000 km^2 of agricultural land in West Bengal. Industrial warehouses, raw material and industrial infrastructure were badly hit in the cyclone-affected areas, especially in Kolkata and surrounding areas [10].

Kolkata city, which is home to almost 15 million people, bore the brunt of Amphan that tore roofs off buildings, smashed windows, pulled down trees and overturned cars. Most of the fatalities in the city were caused by falling trees or electrocution. Millions of people were left without power and telephone connection, and some areas were without drinking water. The Kolkata city has the locational advantage; the land on the left bank of River Hooghly has high ground and river section, which can accommodate upland high discharges. However, the city has other drainage problems due to the bowl shape's geographical feature, which does not allow the gravitational flow for drainage [11]. Around 83% of city areas require forced/pumped drainage linked with significant pumping stations. These pumping stations dispose of accumulated drainage water into a higher level of the outfall drainage canals; however, due to power outages in several parts of the city, pumping stations remained dysfunctional for several hours, which led to city flooding [12]. In Kolkata city, areas like Behala, College Street, New Alipore and Sukia Street were submerged with high water depth. Similarly, many central, north and south Kolkata areas were also severely waterlogged due to incessant rainfall and clogged drainage [12]. There is serious need to streamline the coordination between state departments responsible for flood alleviation and prevention [13].

30.3.3 Damage to Houses and Infrastructure

Amphan shattered the houses, dwellings and residential buildings of affected districts. More than one million houses were reported to be damaged due to cyclones. Kutcha, thatched, mud-brick and tin-shed houses mainly were affected due to high wind speed and tree fall.

The gusty winds and high storm resulted in uprooting trees over electric poles that led to a power outage and electricity supply disruption to mobile service providers [14]. According to Kolkata Municipal Corporation's reports, cyclone toppled more than 4000 electric poles, leaving much of the city without power for more than 14 hours.

Cyclone also affected mobile tower networks in the city. 240 of 273 damaged substations in Amphan-hit districts and 94% of Kolkata were restored within a week. In several parts of the city, a lack of drinking water and insufficient water supply for household purposes were reported. Through social media, it was also reported that the administration was quick to restore water and electricity supply for the government officials and politicians compared to other affected communities. Citizen groups mentioned that IMD had issued the warning of cyclone well in advance. However, the administration failed to prepare well and mitigate the impact of the cyclone.

30.4 COVID-19 in West Bengal

The COVID-19 outbreak was first reported in Wuhan, Hubei, China, in December 2019 and recognized as a pandemic by the World Health Organization (WHO) on 11 March, 2020. Caused by severe acute respiratory syndrome coronavirus-2, the virus has overshadowed the world in unprecedented ways, forcing change and adaptability never witnessed before [15]. Governments, organizations and people caught unaware are now scrambling for ways and solutions to arrest its impact.

The first case of COVID-19 in the state was reported in Kolkata on 17 March, 2020, after a teenager returned from the United Kingdom tested positive just a day after the

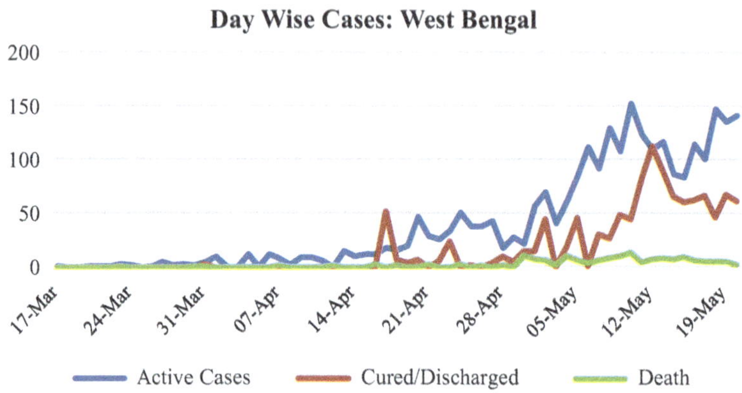

Fig. 30.2 COVID-19 day-wise cases in West Bengal. *Source* Developed by authors with information from Department of Health and Family Welfare, Government of West Bengal [18]

State Government notified the West Bengal Epidemic Disease COVID-19 Regulation [16]. The cases started rapidly increasing from 17 May, 2020. The surge has resulted from various attributing factors like foreign returnees through repatriation flights, migrant labourers' influx and improving COVID testing services [17]. The day-wise cases the total number of cases, cured cases and deaths are depicted in Fig. 30.2.

30.5 Governance and Legal Frameworks for Amphan and COVID-19 Management

The Disaster Management Act, 2005 and the Epidemic Diseases Act, 1897 was enforced to combat COVID-19 by the central and state governments. The legislative framework for epidemic control is 'twofold' in India; first, it is administered through the Indian Penal Code 1860-fundamental criminal law under Section 188, 269, 270 and 271, followed by the legislation for public health like the 'Livestock Importation Act 1898, Indian Ports Act of 1908, Drugs and Cosmetics Act of 1940, Essential Services Maintenance Act, 1968, Indian Aircraft (Public Health) Rules, 2015' [19]. The recently enacted national policy Disaster Management Act 2005 and age-old Epidemic Diseases Act put in place necessary institutional mechanisms for drawing up and monitoring the implementation of disaster management plans, ensuring measures by various department and units of government for prevention and mitigating the effects of anticipated disasters [20].

30.5.1 Disaster Management Act of 2005

The Disaster Management Act 2005 is the national law which provides institutional, legal, financial and coordination mechanisms for disaster management in India at the national, state and local level. The National Act and Policy on Disaster Management define the institutional mechanism in the country [21].

30.5.2 The Epidemic Diseases Act of 1897

The Epidemic Diseases Act 1897 was created to fight the bubonic plague in India and prevent dangerous contagious diseases. The act allows the central and state governments to define the legislation to prevent the outbreak of the disease under Section 2 and provides legal power to punish the ones "for disobeying the order of a public servant and malignant act likely to spread infection of disease dangerous to life" under section 188 and 270 of the Indian Penal Code accordingly [22].

30.5.3 National Disaster Management Guideline for Management of Biological Disaster 2008

Response and management of biological disasters and epidemics are made based on National Disaster Management Guidelines—Management of Biological Disasters 2008. According to the guideline, the Central, State and Local-level bodies and the nodal ministry—Ministry of Health and Family Welfare are responsible for handling epidemics and biological disaster [23]. It also provides an operational framework of how the flow of information and responsibilities are fulfilled.

30.5.4 National Disaster Management Plan 2019

The 2019 National Disaster Management Plan provides a framework for the response and management of biological and public health emergencies by mainstreaming the Sendai Framework for Disaster Risk Reduction [24]. The framework defines the thematic areas under the four priority areas of the Sendai framework specific to the biological and public health emergencies, as shown in Table 30.2. The framework also defines the central and state agencies along with their responsibilities under each thematic area.

Table 30.2 Framework for biological and public health emergencies (BPHE)

Priority areas of action	Thematic areas	Activities
Understanding risk	• Observation networks, information systems, monitoring, research, forecasting, early warning and zoning/mapping • Hazard risk vulnerability and capacity assessment (HRVCA) • Dissemination of warnings, data and information • Disaster data collection and management	• Work on the preventive measures, health system surveillance—adaptation and addition of facilities • Awareness of preventive measures and what ought to be done via print, electronic and social media • Research and study work started by the institutions—health and others • Preparation and stockpiling of materials, equipment, etc. • Systematic data collection, management, and transparency to the public with daily updates
Interagency coordination	• Overall disaster governance—role of national and state bodies • Preparedness and response • Warnings, information, and data	• Preparation of guidelines by the central government—followed by the state and union territories • Rapid health assessment, testing started • Public awareness and community outreach, database, local languages use
Investing in DRR structural measures	• Strengthening response • Upgrading medical facilities	• Hospital preparedness, infrastructure requirements identified • Series of medical units, labs were created for testing and treatment • Preparation of equipment, ICUs, isolation wards, equipping medical responders with all the necessary equipment
Investing in DRR non-structural measures	• Techno-legal regimes • Biosafety and biosecurity measures and environmental management • Risk transfer	• Amendment of laws and regulations • Environmental and social management framework for COVID-19 response and health emergency, adoption of safe disposal and others • Insurance and measures for the health professionals and first responders was done

(continued)

Table 30.2 (continued)

Priority areas of action	Thematic areas	Activities
Capacity development	• Human resource development training • Knowledge management and curriculum development • Awareness generation • Mock drills/exercises, community-based disaster management • Hospital preparedness • Applied research • Empowering women, marginalized communities, SC/ST, and persons with disabilities	• Dos and don'ts to mitigate the effects of medical emergencies caused by biological agents • Awareness about the importance of personal hygiene and preventive measures • Hospital preparedness guidelines and addition of facilities, satellite health facilities • Continuous study on vaccine, research, and scientific study • Measures to support the life of the migrant labourers through Sneh Prahar, Prachesta

30.6 Response Measures for Amphan

The state government mobilized resources for effective response and to minimize the suffering of people. NDRF played a critical role in providing rescue and response measures. 21 teams of NDRF were deployed in West Bengal, and one team of NDRF was placed in Kolkata for emergency response. NDRF teams were deployed with the personnel protective equipment (PPE) to avoid COVID-19 infection during the response. Ten disaster relief teams of the coast guard were sent to West Bengal by central government to aid in the response and recovery efforts.

People staying in the cyclone shelters were provided cooked food, and fodder was provided to livestock by the administration. An ex-gratia payment of Rs. 4 lakhs was made to the kin of the deceased person and Rs. 50,000 to those who got seriously injured during the cyclone was provided by the government.

More than 1000 teams worked to restore infrastructure and services in West Bengal after Amphan. Due to safety and social distancing norms, ground teams were not working in total capacity. Less number of deployment delayed recovery efforts, especially in the electricity sector. Based on the request of the state government, additional teams were provided from Indian Railways for restoration work. Five brigades of the Indian Army were deployed in Kolkata to support recovery efforts. The concerned line departments restored drinking water supply and drainage. Roads were cleared from fallen trees and debris and state departments followed strict adherence to COVID-19 regulations.

Considering the severity and impact of Amphan, the Government of India released an amount of Rs. 1000 crore for West Bengal [25]. West Bengal also received support from the neighbouring state of Odisha for mitigating the effects of the cyclone. Ten Units of Odisha Disaster Rapid Action Force (ODRAF) and 300 Fire Service Personnel and equipment like inflatable light, cutter, power saw and other equipment were provided for search and rescue and road clearance. Fifty thousand pieces of low-density polyethylene sheets were provided for people affected by cyclone.

Social media was used extensively by the affected communities in several important ways, including inquiring about their friends and family members' safety, organizing donation camps and advocating for a government response. Although communities using social media for various purposes on a wide-scale, government departments did not use social media channels in reaching out to the public [26]. It was also noticed that private electricity company (CESC) was more efficient in restoring services than their government counterparts (WBSEDCL) in Kolkata. But, in the restoration of telecommunication services, a new player (Jio) was more efficient than the old players (Airtel, Vodafone-Idea) of the sector [26].

30.6.1 Measures and Guidelines to Combat COVID-19 Outbreak

The Ministry of Health and Family Welfare and the National Disaster Management Authority started putting out the guidelines and measures adopted by the Central Departments and State/Union territories. Based on the Epidemic Diseases Act, 1897, social distancing and voluntary public curfew norms were practised nationwide since March 2020 [27]. Under the Indian Penal Code, 1860 and Criminal Procedure Code, 1973, quarantine facilities were enforced. The Indian Penal Code 1860 (IPC Section 188, 269, 270, 271), Epidemic Diseases Act, 1897, and the Disaster Management Act 2005 were enforced to control COVID-19 [28]. As per the DM Act 2005, NDMA enforced nationwide lockdown and released measures on COVID-19 related advisories, do's and don'ts, to SDMAs to take the measures required [29].

Measures like the travel ban, closure of education, training, public/government institutes at the central and state-level, transportation facilities, private and commercial establishments and others were carried out with enforcement of lockdown. The Ministry of Health and Family Welfare, Government of India, released guidelines since January 2020 for COVID-19 management like sampling, packaging, and transportation of samples, surveillance of the infection, hospital preparedness, cluster containment plans, Indian Council for Medical Research Strategies for COVID testing, maintenance of Indian ports, surveillance at the international airports, dead body management, rational use of personal protective equipment, disinfection of public places and private institutions, quarantine facilities setup, self-isolation and home-isolation, measures to ensure the safety of the first-hand responders and health workers, rapid antibody testing, along with standard operating procedures for offices, public places, entertainment sectors, during festivities, the opening of internal and international flights, post-COVID-19 management protocol and others [30].

Health and Family Welfare Department, Government of West Bengal, and local administration also released various COVID-19-related guidelines. 'Sneher Paras'—scheme to provide financial support to the labourers and workers from West Bengal stranded in other cities due to lockdown measures, 'Prachesta' scheme was introduced to provide financial support to the daily wage earners/labourers—permanent residents of West Bengal, adoption of delivery systems for essentials, defining hospitals specific for COVID-19 treatment [31].

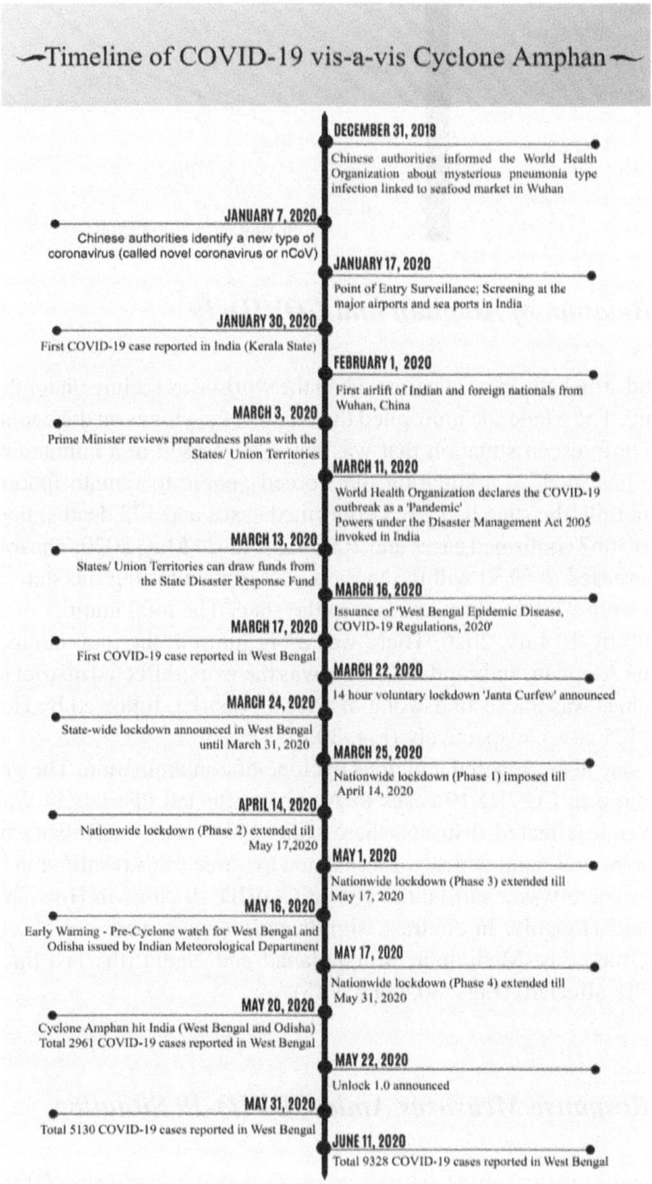

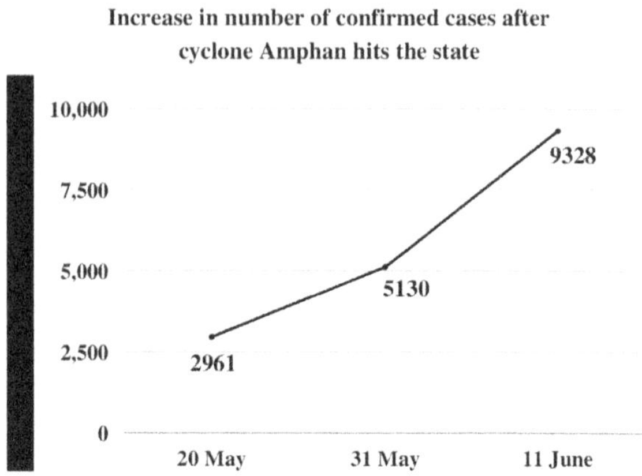

Fig. 30.3 COVID-19 cases projection of West Bengal. *Source* Developed by authors with information from Department of Health and Family Welfare, Government of West Bengal

30.6.2 Relation of Amphan and COVID-19

Amphan had struck the state at a time when the world was reeling under the COVID-19 pandemic. The pandemic multiplied the effects of cyclones on the community and showed an unforeseen situation that was never witnessed in a humanitarian crisis. COVID-19 has induced a situation that forced people to remain indoors. Before Amphan landfall, the state had 2825 confirmed cases and 172 deaths; however, this increased to 3667 confirmed cases and 200 deaths on 24 May, 2020. The total number of cases increased to 5130 within ten days of Amphan hitting the state, and by 11 June, there were 9328 confirmed cases in the state. The total number of cases went up to 27,109 by 10 July, 2020. There was a big jump in the total number of cases after cyclone Amphan, and capital Kolkata was the worst-affected district in the state. North Parganas was the second-worst district-hit district, followed by Howrah with 5273 and 3825 cases, respectively (Fig. 30.3).

The cyclone also triggered a vicious cyclone of contamination. The graph showcases the surge in COVID-19 cases in Amphan affected districts in West Bengal. Out of the nine affected districts, there were 1433 cases in Kolkata on the day Amphan hit West Bengal which was escalated by three folds resulting in 3245 cases by June 11. There was a substantial rise in COVID-19 cases in Howrah, 24 North Parganas, and Hooghly. In contrast, significant changes were not observed in 24 South Paragnas, East Medinipur, Murshidabad and Nadia (the last three districts were partially affected) (Figs. 30.4 and 30.5).

30.6.3 Response Measures Amid COVID-19 Situation

People from the cyclone prone districts were evacuated to cyclone and flood shelters; however, maintaining social distancing in evacuation shelters was difficult for the administration [33]. These evacuation shelters were already occupied by migrant

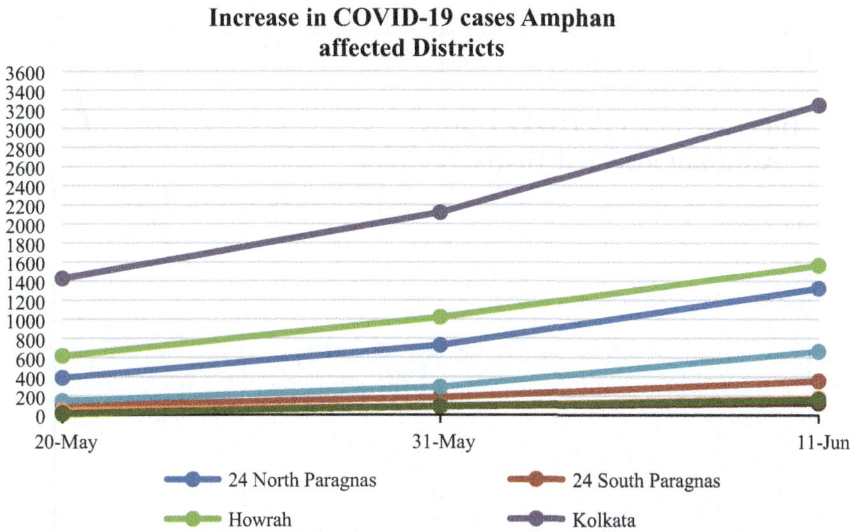

Fig. 30.4 Increase in COVID-19 cases in Amphan affected districts. *Source* COVIDIndia.org [32]

workers returning from other states due to COVID-19 lockdown restrictions. Health and Family Welfare Department of the Government of West Bengal also admitted that the cyclone Amphan coupled with migrant labourers and partial lifting of lockdown led to spike in COVID-19 cases in the state.

Due to social distancing norms, cyclone shelters were occupied one-third of their capacity. These restrictions in West Bengal reduced evacuation capacity in shelters from 500,000 people to 200,000 people. During cyclone Amphan, more than 3 lakh people were evacuated in the 100 evacuation centres in the state's cyclone prone districts; however, during cyclone Bulbul in 2019, 1.8 lakh people were sheltered in these centres [34]. This reflects people's cramming in cyclone shelters to save lives from cyclones during COVID-19 when maintaining social distancing was equally important to minimize coronavirus spread. To augment cyclone shelters' capacity, state administration converted schools and community centres into temporary cyclone shelters. More than 2000 such shelters were used in West Bengal, including schools and community centres.

The administration in various cyclone affected districts was overwhelmed in the first few days of cyclone impact. Restoration of key infrastructure facilities was of prime importance in the first few days after the cyclone landfall. Several households were without electricity and water connections. This led to protests in some places, especially in Kolkata city, to restore these basic services. These protest gatherings violated social distancing norms for COVID-19 [34].

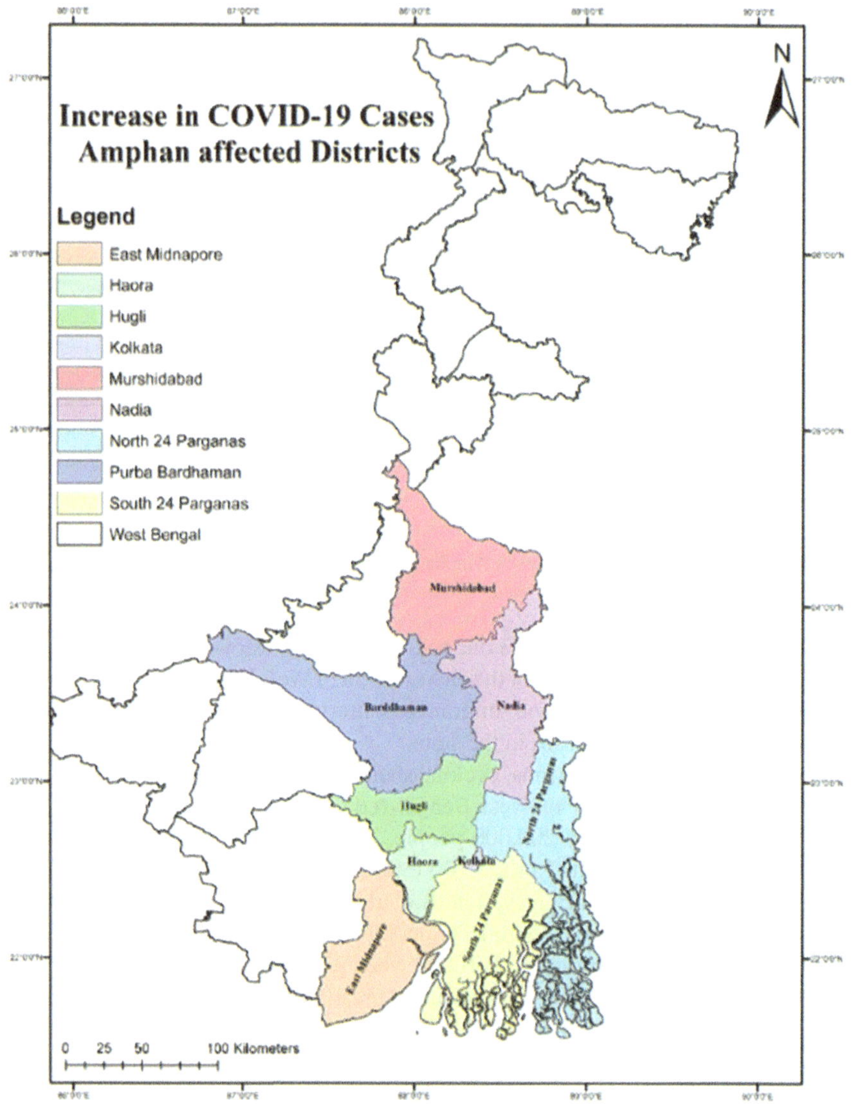

Fig. 30.5 Increase in COVID-19 cases in Amphan affected districts

30.7 Discussion

The time of Amphan landfall coincided with the COVID-19 lockdown. The state was already reeling with the mounting pressure to manage the infectious disease. Due to lockdown and loss of livelihood, migrant workers were also returning to their native places. These migrants were quarantined in existing cyclone or flood shelters in the state. It was the same time when unlock-I was announced from 1 June to 30 June, 2020, and people had started to come out with fewer restrictions. The situation arising out of returning migrants, unlock-I and evacuation of people to shelters due to Amphan magnified the situation of COVID-19 in the state. This led to an increase in the total number of infections in West Bengal.

The COVID-19 pandemic took away many lives, impacted livelihood and brought dramatic changes worldwide in every sector. Based on the Epidemic Diseases Act 1897, Disaster Management Act 2005, and Indian Penal Code series of measures and guidelines were adopted and implemented at the national, state and local level. Despite the implementation and efforts, issues of the political clashes, leniency in lockdown measures, and unequal distribution of ration, slow testing and others were raised.

It was difficult for the state functionaries to ramp up the capacity of first responders for Amphan management with the threat of COVID-19. Imparting training to the state functionaries for COVID-19 aspects such as social-distancing norms, hygiene maintenance and personal protective equipment during the response was also of prime importance. The NDRF and State Disaster Response Force (SDRF) played a crucial role in rescue and relief work. NDRF took the utmost care to minimize transmission of coronavirus during the response phase. To strengthen the capacity of newly constituted SDRF on the lines of NDRF is also paramount for the state.

Implementing the guidelines and measures based on the 123-year-old Epidemics Disease Act and Disaster Management Act, 2005 raised issues for managing the situation. The Epidemics Disease Act was amended in April 2020 to include protecting healthcare personnel combatting epidemic diseases and expanding the central government's powers to prevent the spread of such diseases [35].

Issues and clashes of the State with the Union Government have been raised repeatedly during central guidelines and measures. It was also seen that the West Bengal lockdown measures were made lenient during the time of festivities, and a different system of biweekly and lockdown on destined dates were also done. The insurance policy for the medical staff, ASHA workers, police and their families, social pensions in advance, police stations delivering to the elderly and official cells for smooth distribution of essentials are some of the initiatives taken by the state government.

Clashes between the central and state government have been reported during the implementation of COVID-19 measures and different allegations on the rule of Chief Minister Mamata Banerjee in West Bengal. Despite such situations, the measures adopted and steps taken to contain the contagious disease from the national government to municipal authority efforts are important to manage the outbreak. The coordination between multi-tier stakeholders and continuous efforts being carried at the city from the survey, surveillance, clinical management, economic and material support schemes helped in emergency management.

30.8 Conclusion

The COVID-19 situation brought together the fundamental lacunae of the disaster management systems in the countries. Disaster management has been a multidisciplinary subject involving physical sciences, social sciences, medical and biological sciences, environmental studies and so on. The strengths and weaknesses of these subjects are explored to build the idea of resilience from multiple dimensions in the system. Amphan, along with the COVID-19 pandemic, invoked a situation that needed concerted efforts from disaster management and health departments and all the government departments. Better coordination and communication among the departments of government could have eased out the suffering of people. This complex and dual-disaster also need special attention to revisit the risk management efforts regarding preparedness and response measures. The disaster management plan, standard operating procedures (SOP), existing training modules and awareness generation strategies need to be relooked from future composite multi-hazard management scenarios.

References

1. Asokan (2018) Names of the cyclone and its origin—a review. Int J Comput Eng Res 08(02):42–45
2. Montgomery MT, Farrell BF (1993) Tropical cyclone formation. J Atmos Sci 50(2):285–310
3. Neumann CJ, Pelissier JM (1981) Models for the prediction of tropical cyclone motion over the North Atlantic: an operational evaluation. Mon Weather Rev 109(3):522–538
4. Sahoo B, Bhaskaran PK (2016) Assessment on historical cyclone tracks in the Bay of Bengal, east coast of India. Int J Climatol 36(1):95–109
5. Pramanik M, Szabo S, Pal I, Udmale P, Pongsiri M, Susan C (2021) Population health risks in multi-hazard environments: action needed in the cyclone Amphan and COVID-19—hit Sundarbans region, India. Clim Dev. https://doi.org/10.1080/17565529.2021.1889948
6. Pundir D, Chowdhury S (2020) Cyclone Amphan live updates: Amphan weakens into extremely severe cyclonic storm. NDTV.Com, 19 May 2020. https://www.ndtv.com/india-news/cyclone-amphan-live-updates-super-cyclone-moves-closer-to-odisha-bengal-pm-holds-review-meet-2231174
7. IMD (2020) Indian Meteorological Department. Retrieved from India Meteorological Department, IMD: https://mausam.imd.gov.in/
8. Srinivasan C (2020) Cyclone Amphan: airport, markets shut, people to stay inside as Kolkata prepares for storm. NDTV, 20 May 2020. https://www.ndtv.com/india-news/cyclone-amphan-airport-markets-shut-people-to-stay-inside-as-kolkata-prepares-for-storm-2231968
9. Sud V, Rajaram P (2020) Cyclone Amphan caused an estimated $13.2 billion in damage in India's West Bengal: government source. CNN, 22 May 2020. https://edition.cnn.com/2020/05/22/weather/cyclone-amphan-damage-intl-hnk/index.html
10. Dey S (2020) West Bengal govt shares Amphan damage details of over Rs 1 lakh crore with centre. Times Now, 7 June 2020. https://www.timesnownews.com/india/article/west-bengal-govt-shares-amphan-damage-details-of-over-rs-1-lakh-crore-with-centre/602606
11. Shaw A (2015) Inner-city and outer-city neighbourhoods in Kolkata: their changing dynamics post liberalization. Environ Urban Asia 6(2):139–153

12. Chatterjee A (2020) The cyclone Amphan and its impact in the city Kolkata. Medium, 14 June 2020. https://medium.com/@anjan.chatterjee/the-cyclone-amphan-and-its-impact-in-the-city-kolkata-c63aadc51dfa
13. Thanvisitthpon N, Shrestha S, Pal I (2018) Urban flooding and climate change: a case study of Bangkok, Thailand. Environ Urban Asia 9(1):86–100. https://doi.org/10.1177/0975425317748532
14. Sen S (2020) Kolkata greens, post 'Amphan'. Int J Res Anal Rev 7(2)
15. Dong Z-Q et al (2020) The social psychological impact of the COVID-19 pandemic on medical staff in China: a cross-sectional study. European Psychiatry. Cambridge University Press 63(1). https://doi.org/10.1192/j.eurpsy.2020.59
16. Singh SS (2020) Coronavirus | Kolkata reports first case. The Hindu, 17 Mar 2020. Retrieved from https://www.thehindu.com/news/national/other-states/coronavirus-west-bengal-reports-first-case/article31093265.ece
17. Mathur B (2020) Hit by a pandemic and a super cyclone, here's how West Bengal is tackling the twin challenge of COVID-19 and Amphan. NDTV, 28 July 2020. Retrieved from https://swachhindia.ndtv.com/hit-by-a-pandemic-and-a-super-cyclone-here-is-how-west-bengal-is-tackling-the-twin-challenge-of-covid-and-amphan-47475/
18. Corona Bulletin (2020) https://www.wbhealth.gov.in/pages/corona/bulletin
19. Nomani MZM, Parveen R (2020) Legal dimensions of public health with special reference to COVID-19 pandemic in India. Syst Rev Pharm 11(7):131–134
20. Pal I, Ghosh T, Ghosh C (2017) Institutional framework and administrative systems for effective disaster risk governance—perspectives of 2013 cyclone Phailin in India. Int J Disaster Risk Reduct 21:350–359. https://doi.org/10.1016/j.ijdrr.2017.01.002
21. Disaster Management Act, Government of India (2005)
22. Epidemics Disease Act, Government of India (1897)
23. National Disaster Management Authority (2008) National disaster management guidelines—management of biological disasters. https://nidm.gov.in/pdf/guidelines/new/biological_disasters.pdf
24. National Disaster Management Authority (2019) National disaster management plan 2019. https://ndma.gov.in/sites/default/files/PDF/ndmp-2019.pdf
25. Singh V (2020) Cyclone Amphan | NDRF relief disbursed to West Bengal, Odisha. The Hindu, 25 May 2020. https://www.thehindu.com/news/national/other-states/cyclone-amphan-ndrf-relief-disbursed-to-west-bengal-odisha/article31672776.ece
26. Poddar S, Mondal M, Ghosh S (2020) A survey on disaster: understanding the after-effects of super-cyclone Amphan and helping hand of social media. arXiv preprint arXiv:2007.14910
27. A 123-yr-old act to combat coronavirus in India; experts say nothing wrong. Mint, 14 Mar 2020. https://www.livemint.com/news/india/a-123-yr-old-act-to-combat-coronavirus-in-india-experts-say-nothing-wrong-11584182501707.html
28. Singh KD, Goel V, Kumar H, Gettleman J (2020) India, Day 1: world's largest coronavirus lockdown begins. New York Times
29. National Disaster Management Authority, Government of India (2021) NDMA orders and advisories. https://ndma.gov.in/covid/NDMA-Orders_Advisories
30. MoHFW (2021) Resources. Retrieved from Ministry of Health and Family Welfare, Government of India: https://www.mohfw.gov.in/
31. Health and Family Welfare Department, Government of West Bengal (2021) Corona bulletin. Retrieved from Government of West Bengal—Health and Family Department: https://www.wbhealth.gov.in/pages/corona/bulletin
32. COVIDIndia.org (2021) West Bengal district wise COVID data. Retrieved 01 11, 2021, from COVIDIndia.org: https://covidindia.org/
33. Mishra A (2020) Odisha: cyclone Amphan compounds miseries of those reeling from effect of lockdown, 20 May 2020. Retrieved from The Wire: https://thewire.in/environment/odisha-cyclone-amphan-covid-19
34. Majumdar B, DasGupta S (2020) Let Bengal be heard: dealing with Covid and cyclone Amphan together. South Asian Hist Cult 11(3):317–322

35. MoHFW (2020) The Epidemic Diseases (Amendment) Bill, 2020, 21 Oct 2020. Retrieved from https://www.prsindia.org/billtrack/epidemic-diseases-amendment-bill-2020
36. Sikka DR (2006) Major advances in understanding and prediction of tropical cyclones over the north Indian Ocean: a perspective. Mausam 57(1):165
37. Das G (2020) Amphan—maiden super cyclone of the century. Frontier Weekly
38. Gray WM (1998) The formation of tropical cyclones. Meteorol Atmos Phys 67(1–4):37–69
39. The Epidemic Diseases (Amendment) Bill, 2020 (India)

Chapter 31
Resilient Urbanism from the Perspective of Climate Change in Spain—The Case of Floods

Isabela Beatriz Rufato Machado, Miguel Fernandes Maroto, and Antonio Martinez Graña

31.1 Concepts and Characteristics of Resilient Urbanism

The concept of resilience comes from the field of physics and refers to the capacity of bodies to return to their original form after being deformed [1]. The word is also synonymous with strength, invulnerability and resistance and has several meanings in other fields. When it is applied to urbanism, it is something more than just resistance; it consists of reconstructing the course from a rupture, but, without the pretension of maintaining everything the same or recovering a supposed lost stability, but, to propose a positive adaptation to the new conditions, implying necessarily in a process of learning and adaptability as a consequence of such transformation.

The continued growth in risk and exposure of people and property, combined with lessons from past disasters, highlights the need to strengthen, prepare and adapt cities while also ensuring effective response and recovery at all levels. However, the preservation and regeneration of basic structures and functions also influence urban resilience, as well as the accelerated population growth, the large environmental changes and the growing concentration of population in urban centers that result in greater complexity and uncertainty in the scale of risk; aside from the adaptation strategies implemented, that seem to boost resistance in the short term but lead to serious erosions in the long term, resulting in the collapse of environmental and social systems.

A disaster-resistant city is a place where disasters are minimized, with organized services and infrastructure, respecting safety standards and building rules; there are no irregular occupations built on flood plains or on steep slopes due to lack of other available land [2]. Local government should be competent, inclusive and transparent, focusing on sustainable urbanization and investing in the resources needed

I. B. R. Machado (✉) · M. F. Maroto
Universidad de Valladolid, Valladolid, Spain

I. B. R. Machado · A. M. Graña
Universidad de Salamanca, Salamanca, Spain

S. Kolathayar et al. (eds.), *Civil Engineering for Disaster Risk Reduction*, Springer Tracts in Civil Engineering, https://doi.org/10.1007/978-981-16-5312-4_31

to build municipal management and organizational capacity before, during and after an adverse event or natural threat. People must understand the risks they face and develop information and empowerment processes for citizens, anticipating and trying to mitigate the effects of possible disasters by incorporating surveillance technologies. We mean the compassionate agencies, together with citizens, should be able to respond, implement immediate reconstruction strategies and quickly restore basic services to resume their activities after an adverse event and, above all, implement solutions aimed at resisting environmental change.

Nevertheless, there is the issue that increasing resilience on one scale may interfere with or even reduce resilience on another scale. An advantage would be to use the potential of innovation in cities to develop and apply strategies that promote the resilience of urban ecosystems between the humans and the nature. Of course, these efforts need forms of governance, which are also considered critical for development, as urban resilience to climate change, for example, can consider unintended consequences in space and at different analytical scales, so it is important to ensure that these activities do not reinforce existing inequalities or create new ones. Attention is also needed to the distributional consequences of measures to promote urban resilience, including identifying the social groups, industries and neighborhoods that will benefit from resilience or support the cost of it; furthermore, when the topic is the relationship between resilience and poverty, even more careful note should be taken as questions of power and inequality that arise with the application of resilience approaches themselves [3].

Already on the resilience applied to the phenomenon of floods, it can effectively increase the capacity of resilience of the urban environment, with the intention of minimizing them in projects that value and include the risk factor [4]. However, investment in strengthening community resilience to flooding reduces both the impact and the costs of response and recovery.

When resilience develops from risk-informed decision-making and is driven by the community itself, a holistic approach is adopted and is, therefore, more effective in reducing underlying risks and vulnerabilities and ultimately strengthening resilience.[1] In addition, mitigating such a phenomenon can be done through techniques that rationalize the relationship of water with buildings and urban space, generating structural and non-structural measures.

In relation to flood control, the change of strategies is urgent and cannot be postponed and tactics such as the use of green roofs, rainwater tanks, permeable urban surfaces, surface conduction of pluvial waters, disposition in rainwater situation, as well as emphasize the importance of architects who incorporate in their urban projects the management of pluvial waters [5]. The importance and direct link to green infrastructure, a key resource in climate change risk adaptation planning, is frequently noted [6]; in addition to "article 67" of the New Urban Agenda, we commit ourselves to promote the creation and maintenance of well-connected and distributed networks of quality, open, safe, inclusive, accessible, green and multipurpose public

[1] For more information: http://repo.floodalliance.net/jspui/bitstream/44111/2237/1/05%20IFRC%20-%20Spanish%20-%20FINAL-with%20links-reduced.pdf.

spaces, to increase the resilience of cities to climate change and disasters, such as floods, drought risks and heat waves [7].

In parallel, the promotion of urban resilience to climate change, and in consequence, to floods, will require cities to become resilient to a larger range of interactive impacts and tensions. All characteristics of resilient cities, towns, neighborhoods and systems include diversity, flexibility, adaptive governance and capacity for learning and innovation including both adaptation/mitigation strategies and also accompanied by broader development policies and plans [3].

In a globalized world, where all societies are conscious of the limited resources of the planet and suffer the consequences due to the lack of capacity to protect themselves against the disasters that threaten them, it is time to take steps toward resilience by involving citizens and investing in education and training, so that our cities can be transformed into safe and habitable places with the best quality of life and, ultimately, more resilient.

31.2 Climate Change and the Main Manifestations in Spain

The definition of climate refers to the state of the conditions of the atmosphere influencing a certain area, normally linked to temperature and the recording or not of precipitations. The term "climate change" is associated with the variation of the earth's global climate or of regional climates over time, which affects the balance of long-established systems and ecosystems.

The projections and evaluations carried out to the present time relate that climate change will be quite severe in Spain, which, however, has a socioeconomic and ecological framework that makes it especially vulnerable to these impacts. Moreover, it is very difficult to represent a plan of climatically homogeneous zones in the country because the situations are very diverse, and every region will suffer in its own way.

In summary, the Iberian Peninsula has a predominance of Cfb climate (temperate without dry season with temperate summer) more in the north, with fresher temperatures compared to the south, where there is a predominance of Csa (temperate with dry and hot summer). With this analysis, it should be noted that the increase in temperatures in the south will have a greater impact on the quality of people's urban life, while in the north, its effect will be crucial mainly for the natural environment, collaborating with the melting of the Pyrenees.

In fact, Spain is the scene of very diverse extremes of weather and climate. Its geographical configuration (peninsula and archipelagos) and its location between temperate and subtropical atmospheric influences favor the development of varied meteorological episodes that, on occasion, reach extreme ranges. Because of their socioeconomic and territorial importance, floods, drought sequences, windstorms, episodes of abnormal cold and heat, violent storms with hail fall, tornadoes, etc. can be highlighted and along with them, some dangers of multiple causes, where the meteorological origin has a prominent role, such as snow avalanches or landslides [8].

Climate change generates consequences that are expressed in different ways through varied phenomena that, without a doubt, directly affect the quality of life of citizens, as well as the infrastructure and maintenance of cities. The impacts observed and projected for the Mediterranean area, such as decreases in total annual precipitation, river runoff (and the associated capacity to produce hydropower), crops and tourism; and increases in forest fires, risk of desertification, water demand for agriculture (which contrasts with the reduced availability of water), health problems associated with both extreme events and vector diseases and risk of biodiversity loss. In the Spanish Atlantic area (the Cantabrian Atlantic cornice) and in the rest of the mountain areas, impacts on tourism, biodiversity, coasts or the availability of water resources are also noted, among others [9].

We mean, that is, the use and management of several sectors, resources and services, which guarantee the development and welfare of the country, can be greatly affected, depending directly on the climate, such as tourism, agriculture and health, not to mention the fact that the costs associated with extreme events that aggregate problems, facing the availability and quality of water, coastal stability, soil conservation against desertification, fire risk or the conservation and sustainable use of biodiversity [10].

Among so many possible consequences, there is always a relationship between one phenomenon influencing the other, and enhancing it, negatively. In fact, the discussion of floods, their risks, causes, damages and mitigation measures will be dealt with in detail later on. However, the most certain thing we can say about the subject in general is that everything always translates into losses, whether for the local population, visitors, or in the quality of life, infrastructure, health, government services or the economy.

31.3 The Case of the Floods in Spain and the Principal Mechanisms Applied to Be Resilient

More intense precipitation, more severe storms and rising sea levels are one of the consequences of climate change that increase the risk in the intensity and frequency of flooding, making this issue the most prominent example of a disaster that is occurring increasingly in much of the world, given its high social and economic impact, and this is one of the main reasons that cities are preparing to be resilient.

Flooding should be understood as a process where water temporarily occupies normally dry land, as a result of the unusual contribution of a volume of water greater than that existing in a given area, considered as an exceptional event [11]. The origin of a flood can be derived from natural causes: such as rain, runoff, landslides, thaw, tidal waves, seaquakes; or anthropogenic, due to the failure of man-made infrastructures, such as dam breaks and pipeline leaks, which can also aggravate natural floods with induced actions that interfere or modify them.

The risks associated with water and coastal phenomena in Spain involve both the number of victims produced and very significant property and environmental damage, at a global and national level, where there are more than 50 deaths on average per year related to this type of risk, damaging around 2.5% of GDP on average per year in the long term and a sustained upward trend. In addition, in Spain, we have more than 11,000 km with this type of risk, corresponding to the river areas to about 9555 km (914 locations) and the coastal areas to about 11,027 km (1342 locations) [12]. The result is that, according to this information, around 6% of the Spanish population is located in a flood risk zone for the 500-year return period flood, and this figure is being 4.2% and 1.5% for the 100-year and 10-year floods, respectively [13].

The main flooding phenomena that have occurred in Spain have a common characteristic that of being lightning-floods, torrential, in medium and small hydrographic basins, highlighting the events of 1962 in Besós (Barcelona) with almost 800 deaths; 1973 in Granada-Almería-Murcia with 300; 1982 in the Tous reservoir with 38; 1983 in the Basque country and 1989 in the Mediterranean area, both with 440 deaths. However, there has not been an increase in the frequency of development of flooding episodes, but the risk has increased due to increased vulnerability and exposure to its effects [14].

In general, there are two different positions with respect to the increase in the number of floods. The technical-administrative position defends that there would be more episodes of flooding due to climate causes, the result of the unpredictability and randomness of nature itself, bringing about greater disasters; and the ethical-geographical position based on the analysis of the territorial reality of the geographical space from the field work, the risk of flooding episodes would have increased, greatly, in large areas of the Spanish territory with the implementation of urban uses and leisure, and this would have caused an increase in the vulnerability of populations settled in risk territories [15].

Vulnerability in this area can be distributed into three categories: economic, with a strictly monetary value (according to the relationship between the land uses established in the SIOSE and the magnitude of the flood); social, based on strategic equipment and services (taking into account the vulnerability of users and the impediment to the normal development of services); and the environmental variable, for damage to the natural environment (contamination, deterioration, loss, etc.) [11].

Socioeconomic factors such as the increase in insured properties exposed in Spain, the value of these assets and the penetration of insurance are postulated as the main causes of increased losses for the future, confirming also an influence of anthropogenic climate change on losses in insured properties, emerging the need to improve the knowledge of losses caused by floods, modeling their behavior to anthropogenic factors and estimate benefits (savings) derived from early warning, education in damage prevention and quick mobilization of assets by the population at risk [16].

In reality, it is more logical to think about investing money in urban planning, good infrastructure, sustainable forms of civil construction, reuse of resources and implementation of measures for a better quality of life and safety of the population, than having to spend millions after a disaster to recover and rebuild only the basics. We mean, why not prevent and focus on prevention instead of quickly having to repair a damage that will probably, in the next episode, occur in the same vicious circle?

In this way, it is easy to point out that the city's infrastructure is bad when we throw garbage in the sewers; that it is not good to walk on impermeable ground and also want good drainage when it rains; to discard plastic in the rivers and oceans and criticize the beaches for not being clean or beautiful; to build in irregular areas and accuse the rain when some disaster happens. Blaming nature is very convenient when we are the ones who are responsible for human negligence with the environment, mismanagement of resources and the main drivers of climate alterations.

What should be taken into account, however, when considering this question, are urban plans and current legislation, which prohibit (or should prohibit) construction in areas exposed to risk and in preservation zones. Increasing resilience in the urban environment is related to the review and update of urban plans, regulations of the autonomous communities coordinated with the development of protocols that are essential for adaptation to the challenge [17].

It is worth mentioning that, first of all, Spain lacks a framework law on natural risks, and the sectorial legislation has to be consulted for references. The laws on water, soil and civil protection are the most basic sources on a national scale, in addition to the regional regulations on soil and land use planning for flood risk. There are sufficient legal regulations for the incorporation and effective treatment of natural hazards in land use planning processes[2]; another thing is the real fulfillment of all that, as well as its fiscalization.

There was, therefore, a legal basis for the hydraulic administrations, containing, in general, the zoning and limitation of the uses of flood zones, which should serve for the development of more precise regulations regarding land use planning and land use of flood plains; however, it was never applied. In this context, when in 2007 emerges the new European Directive 2007/60/EC, of the European Parliament and the Council, on the assessment and management of flood risks. As the autonomous communities are sovereign and are expressed in the drafting of regional and sub-regional planning documents between the last decade of the twentieth century and the first decade of the twenty-first century, some of them developed flood hazard maps and regulations for use in flood areas.[3]

In order to mitigate the risks of flooding, we have structural measures (channels, canals, dams), and not structural ones (education, land planning and risk mapping). There are places where the man modifies the river, and others with which he lives together with the river. In the first case, there are engineering works and, hydraulic

[2] Ley del Seguro Agrario Combinado (1978), Ley de Costas (1988), Gestión Medioambiental (Reglamento UE 1836/93), Normativa sismo resistente (1994), Directrices Básicas de Protección Civil (1994 and 1995), Ley de Aguas (2001), Plan Hidrológico Nacional (2001), Legislación de impacto ambiental (2001), Ley del Suelo (2007); etc.

[3] In particular, on the urban developments planned in the cities, in anticipation of the regulation of Law 8/2007 of 28 May, on land, and Royal Legislative Decree 2/2008 of 20 June, approving the revised text of the Land Act, and its successor the Royal Legislative Decree 7/2015 of 30 October, approving the revised text of the Land and Urban Rehabilitation Law.

works with greater economic achievement, such as dams, reservoirs, dikes and channeling. In the second case, it is basically what there is, nature, which only needs a zoning of floodable areas, alert and safe; these are preventive measures, not costing millions and suggesting countless benefits to the quality of life.

One of the keys to effective risk management and enhanced resilience is to promote responsible occupation of space affected by natural processes with hazard potential. In complement to providing more information to the citizens, especially on the risk and the effect of climate change, we have the importance of insurance as a tool to counteract the socioeconomic effects of floods, the need to improve and implement compliance with legislation in a correct manner, as well as to consider floods in spatial planning.

Experience in recent years has shown an inefficient and unsustainable use of water, as well as the prevention of major flooding. Although human action in general aggravates the consequences of climate change and many of these problems are irreversible, requiring complex solutions, especially in densely urbanized areas, it is very important the development and balance between structural and non-structural measures. We must first take into consideration the importance of an adequate risk analysis and then choose and use the best implementation strategy to deal with the problem, based on responsible planning between traditional systems and technological innovations while still being compatible with urban development and nature, to finally generate more intelligent, sustainable and resilient cities.

Additionally, we still emphasize the importance of converting unsafe places into permeable areas every time it is possible, returning to the land its own space, carrying out actions based, mainly and on a first scale, on nature, focusing on reducing the risk in the most vulnerable areas and creating recreation zones that provide a higher quality of life to the population.

31.4 Case Studies

It was chosen to expose three different Spanish cities that face the same problem, the pluvial floods (Alicante) and fluvial floods (Vitoria-Gasteiz and Zaragoza).

Alicante, located on the coast next to the Mediterranean, suffers from the problem of torrential rain, where in a few minutes, it is greatly affected by the large amount of precipitation that often comes unexpectedly. The measurable intervention adopted was the La Marjal Park, designed to collect water from the autumn floods, and the most interesting aspect of this initiative is that it combines the landscape, aesthetic and recreational functions with the solution of a problem in the area, being a park reference in many ways.

Vitoria-Gasteiz is a city in the interior, with a variety of small rivers, located in the north. It has a configuration composed of a "green ring" around it, with an innovative form to organize the peri-urban space and, in general, all the territory, which gradually incorporated good river planning together with its own green infrastructure. Many interventions have been proposed, the hydraulic adaptation and environmental restoration of the Zadorra river, the derivation of the Santo Tomás and Errekaleor rivers to the Zadorra river, the derivation of the Olaizu river to the Errekaleor river, the urban reform of the Av. Gasteiz, regulation of the flows of the Ali, Eskibel, Batán, Zapardiel rivers, which, in general, combine the use of infrastructural measures with solutions based on naturalness in front of floods, which were caused in part by the measures of canalization of the rivers, which not only did not solve the problem, but also did intensified it.

Zaragoza, a large city also in the interior, located in the northeast, with one of the largest rivers of the Iberian Peninsula, the Ebro. Its relationship with the river was replanted in 1990, investing more in green infrastructure, especially, promoted by Expo 2008 and the realization of the project for the Recovery of Las Riberas Del Ebro, which must be taken into account the great transformation achieved where there was a dirty and abandoned place with great potential, which was converted into a site of good environmental and urban quality, highlighting the transitions between natural, rural and urban landscapes (Table 31.1).

No matter how specific each case may be, there is no doubt about the fact that they all approach the challenge with a vision involving mainly nature-based solutions, urban planning measures, and of course, legislation and also structural ones. In general, the three cases include a recovery of the green space within the urban environment and the respect, especially, to the space of the margins of the rivers Zadorra and Ebro. Vitoria implemented a series of actions to return the spaces of the rivers within the city to the rivers themselves; Zaragoza also recovered the streams

Table 31.1 Comparison of the case studies

City	Alicante	Vitória	Zaragoza
Autonomous community	Valencian community	Basque country	Aragon
Area (km^2)	201.27	276.08	973.78
Density (rooms/km^2)	1.677	894.58	678.9
Population (Ine, 2019)	334,887 hab.	251,774 hab.	674,997 hab.
Type of climate	Bsh (hot steppe)	Cfb (temperate without dry season with temperate summer)	Bsk (cold steppe)
Floor type	Fluvisols, regosols and calcissols	Varied	Varied
Type of flooding	Pluvial	Fluvial	Fluvial
Main solution adopted	La Marjal Park	Hydraulic adaptation and environmental restoration of the Zadorra river	Recovery of the banks of the Ebro

Fig. 31.1 Location of the three case studies (Alicante, Vitoria and Zaragoza) on the map of Spain

along the river Ebro and Alicante, which suffers from the large amount of precipitation in a short time, created a huge green area through a park to absorb all this water (Fig. 31.1).

However, the restoration and recovery of the abandoned and unused green spaces near the rivers, as well as the hydraulic adaptation measures, in Alicante, Vitoria or Zaragoza, have one element in common. All these projects envisage the control of floods, of course, but the most brilliant thing is that along with the successful use of actions to combat the problem, was incorporating into these spaces the use of citizens, the valuation of the landscape, and the principal, the certainty that you can take advantage of the strengths of the places and still friendly, makes them for a better quality of life for people and the urban and natural environment itself.

Nevertheless, the research supports, through the analysis of these three case studies, that the most conventional is to always value to the maximum the measures that are in favor of the environment, which include the recovery and expansion of the permeable areas and actions that prioritize mechanisms of green and blue infrastructure. If the problem persists, the ideal is to include the structural tools, such as dikes, dams, subway reservoirs and others already mentioned. And in the last case, when it is not possible in any way to use nature-based solutions or to mix them with infrastructure solutions, we make an extremely detailed study of the area so that such intervention, only structural, does not generate an even greater problem.

31.5 Conclusions

Firstly, it was observed that in priority is necessary to consider the scale of the problem that involves the consequences of climate change, the impact and where it interferes. When examining the history of natural disasters, it is clear that the vulnerability of a region depends on multiple factors and is no doubt connected to the magnitude of the phenomenon, but the proportionally opposite relationship is also noted that the less developed the place where the event occurs, the greater the damage.

In this way, urban resilience comes from the necessity to develop the stability of the natural, built, economic and social components, in front of the extreme natural phenomena, making possible, besides the reconstruction of the space, a significant improvement in the physical-spatial and socio-political dimension of the cities. It is also worth mentioning the dynamic and integrative character that urban resilience must have, with different levels of powers, from public to private, from the individual to the collective and also the variety of different solutions that it must employ for each specific problem.

Flooding in cities is a serious challenge and is increasingly evolving due to climate change. Meanwhile, the results obtained in that research emphasize the certainty that decision-maker must do much more to understand the challenge and improve the solutions adopted in a more integrated and effective way in the face of current and future risks. In fact, elements such as the geographical location of the city, the climate, the soil components, the urban configuration and the constructions, the rivers, the local vegetation and the amount of green areas directly influence the study to choose which are the solutions that better fit in a certain place, besides the frequency and the types of floods that the study zone can suffer.

In conclusion, the most effective method of managing flood risk is to adopt an integrated approach that combines structural measures with the application of non-structural methods, we mean, solutions based on and integrated with nature. But, we underline the importance of the careful study of the factors that influence the behavior of the city over time, related to land occupation, political and administrative structures and socioeconomic dynamics.

The different perspectives on the solutions in the three case studies were essential to generate a greater notion that, no matter how different the situation, the measures that are worthy of appreciation are all related to respect for nature and the promotion of the relationship between river and city. Resulting, as demonstrated in the three examples, in a significant improvement in the quality of life for citizens, urban and natural space, because more than solving, improving or controlling the problem of flooding, is possible to transform these abandoned or disused zones, providing the creation of pleasant places of great environmental and urban value, enhancing the landscape, with recreation spaces, hiking and playing areas, etc.

Finally, it is considered necessary to keep advancing in the analysis of how resilient urbanism can be realized; what impacts can be mitigated; how resilience is incorporated; what are the appropriate efforts to target development, monitoring, policies and people; and how are they done to ensure that increased resilience to climate change continues to promote a positive social attitude while contributing to long-term sustainability.

References

1. Cippec, Progama De Ciudades (2016) Resiliencia Urbana, Diálogos Institucionales, Argentina
2. Naciones Unidas. Cómo desarrollar ciudades más resilientes, un manual para líderes de los gobiernos locales. In: En La Contribución a la Campaña Mundial de Desarrollo de Ciudades Resistentes 2010–2015
3. Leichenko R (2011) Climate change and urban resilience. Curr Opin Environ Sustain 3(3):164–168
4. Jha AK, Miner TW, Geddes ZS (2012) Building urban resilience: principles, tools and practice. The World Bank, Washington, DC, p 209. ISBN 978-0-8213-9826-5
5. Molina-Prieto LF (2016) Resiliencia a inundaciones: nuevo paradigma para el diseño urbano. Rev Arquitect Univ Católica Colom 18(2):1–136. ISSN 1657-0308
6. Sánchez FJG (2019) Planeamiento Urbanístico Y Cambio Climático: La Infraestructura Verde Como Estrategia De Adaptación. Cuaderno de Investigación Urbanística CIUR nº 122, p 102. ISSN 2174-509
7. Nueva Agenda Urbana (2016) Conferencia de las Naciones Unidas sobre la Vivienda y el Desarrollo Urbano Sostenible (Hábitat III). Quito, Ecuador. ISBN 978-92-1-132736-6
8. ANE, Atlas Nacional De España (2009) http://atlasnacional.ign.es/wane/P%C3%A1gina_principal, 20 Sept 2020
9. AEMA, Agencia Europea De Medio Ambiente (2020) https://www.eea.europa.eu/pt, 20 Sept 2020
10. Acuña SC, Gutiérrez A, Picatoste JR (2011) La adaptación al cambio climático en España, cambio climático: aspectos económicos e internacionales. ICE rev econ 862:81–96. ISSN 0019-977x
11. Patricova (2019) Guía de aplicación del plan acción territorial sobre prevención del riesgo de inundación en la comunidad valenciana. http://politicaterritorial.gva.es/documents/20551069/167206402/Gu%C3%ADa+PATRICOVA+%28castellano%29/e1a0b83a-8846-45fa-aff7-0d98b5b4fcc0, 10 Sept 2020
12. Rodríguez AS (2017) España, Un Territorio Diverso Poco Resiliente Ante Los Riesgos Futuros—Resiliencia Territorial Ante Catástrofes. Inundaciones Y Temporales. Revista de Obras Públicas: órgano profesional de los ingenieros de caminos, canales y puertos 3587:23–32. ISSN 1695-4408
13. Magrama (2015) Ministerio de agricultura, alimentación y medio ambiente. https://www.miteco.gob.es/es/agua/temas/gestion-de-los-riesgos-de-inundacion, 21 July 2020
14. Cantos JO (2008) Prevención de riesgos: cambio climático, sequías e inundaciones. En panel científico-técnico de seguimiento de la política del agua. Departamento de análisis geográfico regional y geografía física, Universidad de Alicante, p 240
15. Cantos JO (2004) Riesgo de inundaciones y ordenación del territorio en la escala local, el papel del planeamiento urbano municipal. Boletín de la A.G.E. Instituto Universitario de Geografía, Universidad de Alicante, n 37, pp 49–84
16. Marquínez J, Vuelta AC, Iglesias EF, Fuente LG (2017) Resiiencia Territorial Ante Catástrofes. Inundaciones Y Temporales. Revista de Obras Públicas: órgano profesional de los ingenieros de caminos, canales y puertos 3587:54–63. ISSN 0034-8619
17. Palencia JS (2017) Resiliencia Urbana ante inundaciones y temporales marítimos. Revista de Obras Públicas: órgano profesional de los ingenieros de caminos, canales y puertos. Departamento de Urbanismo, Instituto del Transporte y Territorio, Universidad Politécnica de Valencia, n 3591, pp 30–39. ISSN 0034-8619

Retraction Note to: Vulnerability of Interspersed Railway Tracks Exposed to Flood and Washaway Conditions

Sakdirat Kaewunruen, Yuki Nishinomiya, and Mitsuru Hosoda

Retraction Note to:
Chapter 19 in: S.Kolathayar et al. (eds.), *Civil Engineering for Disaster Risk Reduction*, Springer Tracts in Civil Engineering, https://doi.org/10.1007/978-981-16-5312-4_19

The Volume Editor has retracted this Chapter, as it was published here following a conference presentation, but the contents were also revised, submitted and published as a journal article within a close time frame [1]. This Chapter is therefore redundant.

Yuki Nishinomiya and Mitsuru Hosoda agree to this retraction. Sakdirat Kaewunruen has agreed to this retraction, but not the wording of this retraction notice.

The retracted version of this chapter can be found at
https://doi.org/10.1007/978-981-16-5312-4_19

S. Kolathayar et al. (eds.), *Civil Engineering for Disaster Risk Reduction*, Springer Tracts in Civil Engineering, https://doi.org/10.1007/978-981-16-5312-4_32

The manufacturer's authorised representative in the EU is Springer Nature Customer Service Centre GmbH, Europaplatz 3, 69115 Heidelberg, Germany. If you have any concerns regarding our products, please contact ProductSafety@springernature.com

Printed and bound by CPI Group (UK) Ltd, Croydon, CR0 4YY

30/09/2026

02225288-0015